STRUCTURAL DESIGN
OF STEELWORK
TO BS 5950

L.H. MARTIN
BSc, PhD, CEng FICE

&

J.A. PURKISS
BSc(Eng), PhD, CEng, MIStrucE, MICE, MIFS

Edward Arnold
A division of Hodder & Stoughton
LONDON   MELBOURNE   AUCKLAND

© 1992 L H Martin and J A Purkiss

First published in Great Britain 1992

British Library Cataloguing-in-Publication Data

Martin, L.H.
  Structural design of steelwork to BS 5950
  I. Title   II. Purkiss, J.A.
  624.1

  ISBN 0-340-54443-0

Typeset in 10/12 Times by H Charlesworth & Co. Ltd.,
Huddersfield, Great Britain
Printed in Great Britain for Edward Arnold, a division
of Hodder and Stoughton Limited, Mill Road, Dunton
Green, Sevenoaks, Kent TN13 2YA by
St Edmundsbury Press Ltd, Bury St Edmunds, Suffolk,
and bound by Hartnolls Ltd, Bodmin, Cornwall.

Contents

4.   Laterally Restrained Beams

5.   Laterally Unrestrained Beams

6.   Axially Loaded Members

7.   Structural Connections

28

73

170

194

Preface

The book conforms with the latest recommendations for the limit state design of steel structures as set out in British Standard 5950 (1990) *Structural Use of Steelwork in Building*. Verbatim quotations to BS 5950 are printed in different type and references within the quotations use the numbers in the BS publication. The symbols used are the same as the BS publication.

The book is intended to cover the work required in the second and third years of a typical course leading to a degree in civil engineering but also should be useful to builders, architects, mechanical engineers and electrical engineers. The book should also be useful for third year work where the student is involved in design projects. It is assumed that the student has studied structural analysis or structural mechanics for at least one year prior to studying with the use of this book.

The theoretical content of the book is sufficiently detailed for the book to be used as a stand-alone text. However, the emphasis is on solving problems, which the authors believe is the only way that this subject, or any other mathematically based subject for that matter, can be learnt. Problems are fully worked, and are of two kinds. Those which appear at the end of each section are designed to demonstrate the application of the preceding information and theory. The tutorial problems, which are placed at the end of most chapters, are intended for self assessment by the student, and the solutions are given.

There are research references for each topic and the book should therefore be useful for the postgraduate who requires basic information on a topic in structural steel, or who requires background information. In particular the authors have extensive research experience in fire engineering and structural connections.

The book should be useful to practising structural design engineers who require some explanation of the clauses in BS 5950. Design examples make reference to the BS 5950 clause that justifies the method of design. It should also be of benefit to other practising engineers, builders and architects who are occasionally called upon to design, or understand the design of, steel structures and who may wish to understand the basic principles.

Acknowledgements

Extracts from British standards are reproduced with the permission of the British Standards Institution. Complete copies can be obtained by post from BSI Sales, Linford Wood, Milton Keynes, MK14 6LE; telex 825777 BS1MK: telefax 0908 320856.

Figures 11.10 and 11.11 are reproduced from Figures 12, 13, and 14 of the Institution of Structural Engineers and Concrete Society (1978), *Design and Detailing of Concrete Structures for Fire Resistance*, by permission of the Institution of Structural Engineers.

Figure 5.46 is reproduced from Chart 20 of the *Design of Castellated Beams* by P. R. Knowles by permission of the Director of the Steel Construction Institute.

The design data for Richard Lees Super Holorib decking are reproduced by permission of Richard Lees Ltd.

Symbols

Subscripts b, c, d and w are used to refer to beam (or bolt), column, plate and weld respectively

A	area
A_e	effective area
A_g	gross area
A_g	tensile stress area (bolts)
A_g	shear area (sections)
A_s	shear area (bolts)
a	spacing of transverse stiffeners or effective throat size of weld
B	breadth
b	outstand, or width of panel
b_1	stiff bearing length
C_v	Charpy impact value
D	depth of section, or diameter of section, or diameter of hole
d	depth of web, or nominal diameter of bolt
E	modulus of elasticity of steel
e	end distance
F_c	compressive force due to axial load
F_s	shear force (bolts)
F_t	tensile force
F_v	average shear force (sections)
f_c	compressive stress due to axial load
f_v	shear stress
G	shear modulus of steel
H	warping constant of section
h	storey height

I_x	second moment of area about the major axis
I_y	second moment of area about the minor axis
J	torsion constant
L	length of span
L_E	effective length
M	larger moment
M_{ax}, M_{ay}	maximum buckling moment about the major or minor axis in the presence of axial load
M_{ax}, M_{ay}	buckling resistance (lateral torsional)
M_{cx}, M_{cy}	moment capacity of section about the major or minor axis in the absence of axial load
M_E	elastic critical moment
M_o	midspan moment on a simply supported span equal to the unrestrained length
M_{rx}, M_{ry}	reduced capacity of the section about the major and minor axes in the presence of axial load
M_x, M_y	applied moment about the major and minor axes
$\bar{M}_x, \bar{M}_y$	equivalent uniform moment about the major and minor axes
m	equivalent uniform moment factor
n	slenderness correction factor
P_{bb}	bearing capacity of a bolt
P_{bg}	bearing capacity of parts connected by friction grip fasteners
P_{bs}	bearing capacity of parts connected by ordinary bolts
P_{cx}, P_{cy}	compression resistance about the major and minor axes
P_s	shear capacity of a bolt
P_{sL}	slip resistance provided by a friction grip fastener
P_t	tension capacity of member, or fastener
P_v	shear capacity of a section
p_b	bending strength
p_{bb}	bearing strength of a bolt
p_{bg}	bearing strength of parts connected by friction grip fasteners
p_{bs}	bearing strength of parts connected by ordinary bolts
p_c	compressive strength
p_E	Euler strength
p_s	shear strength of a bolt
p_t	tension strength of a bolt
p_w	design strength of a fillet weld
p_y	design strength of steel
q_b	basic strength of a web panel
q_{cr}	critical shear strength of a web panel
q_e	elastic critical shear strength of web panel
q_f	flange dependent shear strength factor

r_x, r_y	radius of gyration of a member about its major and minor axes
S_x, S_y	plastic modulus about the major and minor axes
s	leg length of a fillet weld
T	thickness of flange or leg
t	thickness of web
U_s	specified minimum ultimate tensile strength of the steel
u	buckling parameter of the section
V_b	shear buckling resistance of stiffened web utilizing tension field action
V_{cr}	shear buckling resistance of stiffened or unstiffened web without utilizing tension field action
v	slenderness factor for beam
x	torsional index of section
Y_s	specified minimum yield strength
Z_x, Z_y	elastic modulus about the major and minor axes
α	coefficient of linear thermal expansion
β	ratio of smaller to larger end moment
γ_f	overall load factor
γ_l	load variation factor, i.e. function of γ_{l1} and γ_{l2}
γ_m	material strength factor
γ_0	ratio M/M_0, i.e. the ratio of the larger end moment to the midspan moment on a simply supported beam
Δ	midspan deflection
δ	deflection
ε	factor $\sqrt{(275/p_y)}$
θ	angle of rotation
λ	slenderness, i.e. the effective length L_E divided by the radius of gyration r
λ_{cr}	elastic critical load factor
λ_{LO}	limiting equivalent slenderness
λ_{LT}	equivalent slenderness
λ_0	limiting slenderness
μ	slip factor
ν	Poisson's ratio
Ω	arching ratio

1

General

1.1 Description of steel structures

1.1.1 Load bearing frames

The introduction of structural steel, circa 1856, provided an additional building material to stone, brick, timber, wrought iron and cast iron. The advantages of steel are high strength, high stiffness and good ductility, combined with relative ease of fabrication and competitive cost. Steel is most often used for structures where loads and spans are large and therefore it is not often used for domestic architecture.

Steel structures include low rise and high rise buildings, bridges, towers, pylons, floors and oil rigs, and are essentially composed of load bearing frames which support the self weight, dead loads and external imposed loads (wind, snow, traffic). For convenience load bearing frames may be classified as:

(a) Miscellaneous isolated simple structural elements, e.g. beams and columns; or simple groups of elements, e.g. floors.
(b) Bridgeworks.
(c) Single storey factory units, e.g. portal frames.
(d) Multi-storey units, e.g. tower blocks.
(e) Oil rigs.

A real structure consists of a load bearing frame, cladding, and services as shown in Fig. 1.1(a). A load bearing frame is an assemblage of members (structural elements) arranged in a regular geometrical pattern in such a way that they interact through structural connections to support loads and maintain them in equilibrium without excessive deformation. Large deflections and distortions in structures are controlled by the use of bracing which stiffens the structure and may be in the form of diagonal structural elements, masonry walls or reinforced concrete lift shafts. A load bearing steel frame is idealized, for the purposes of structural design, as centre lines representing structural elements which intersect at joints, as shown in Fig. 1.1(b). Other shapes of load bearing frames are shown in Figs. 1.1(c) to (e).

Structural elements are required to resist forces in a variety of ways, and may act in tension, compression, flexure, shear, torsion, or in any combination of these forces. The structural behaviour of a steel element depends on the nature of the forces, the length and shape of the cross section of the member, and the magnitude of the yield stress. For

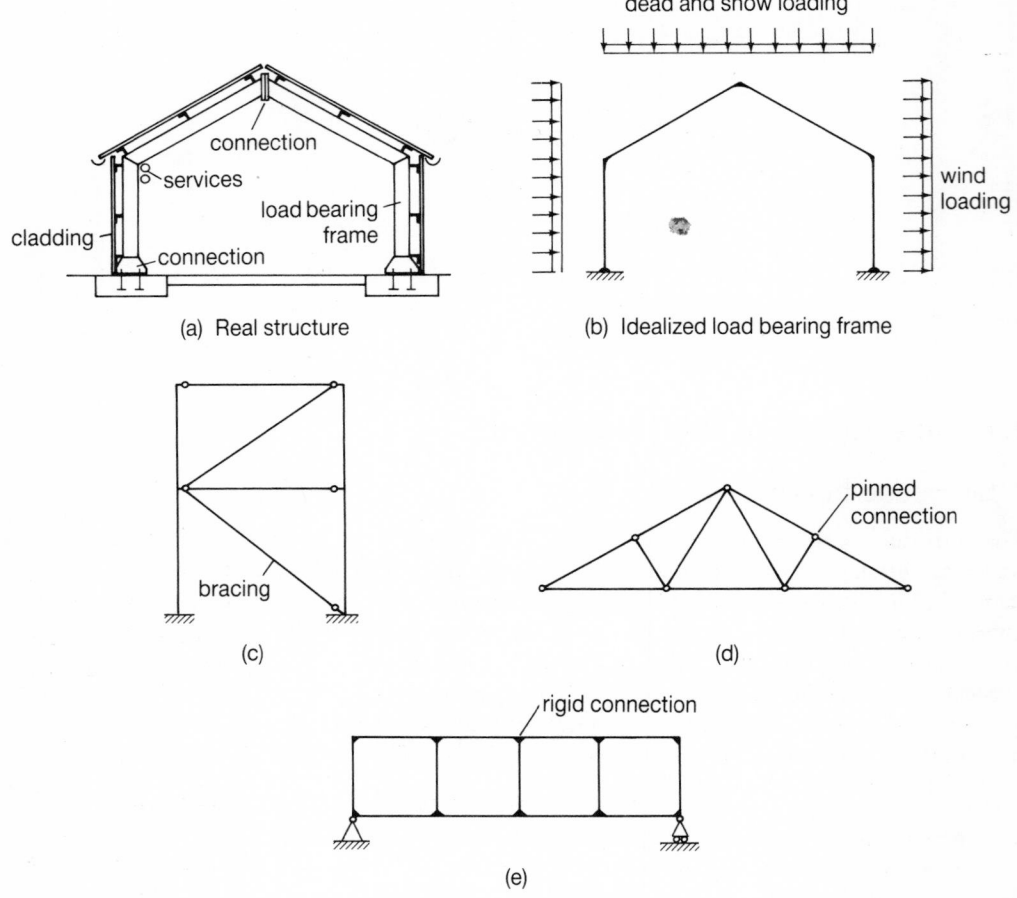

Fig. 1.1   Typical load bearing frames.

example a tie behaves in a linear elastic manner until yield is reached. A slender strut behaves in a non-linear elastic manner until first yield is attained, provided that local buckling does not occur first. A laterally supported beam behaves elastically until a plastic hinge forms, while an unbraced beam fails by elastic torsional buckling. These modes of behaviour are considered in detail in the following chapters.

The structural elements are made to act as a frame by connections. These are composed of plates, welds and bolts which are arranged to resist the forces involved. The connections are described for structural design purposes as 'pinned', 'semi-rigid' and 'rigid', depending on the amount of rotation, and are described and analysed in detail in Chapter 7.

1.1.2   Standard steel sections

The optimization of costs in steel construction favours the use of structural steel elements with standard cross sections and common bar lengths of 12 or 15 m. The billets of steel are hot rolled to form bars, flats, plates, angles, tees, channels, I sections and hollow

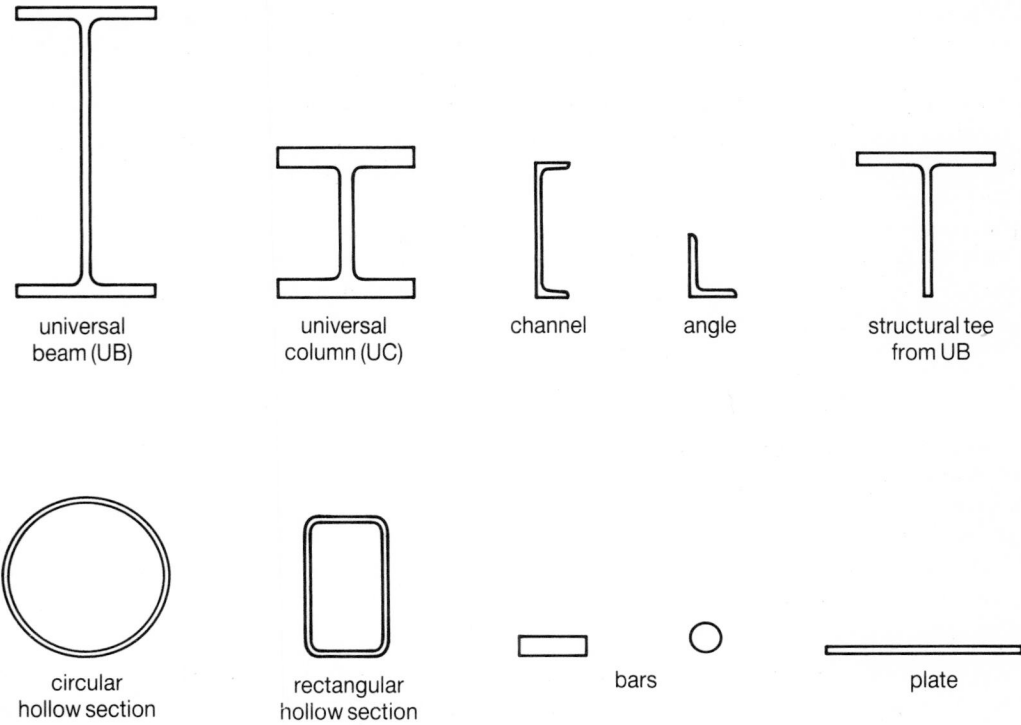

universal
beam (UB)   universal
column (UC)   channel   angle   structural tee
from UB

circular
hollow section   rectangular
hollow section   bars   plate

Fig. 1.2 Standard steel sections.

sections as shown in Fig. 1.2. The detailed dimensions of these sections are given in BS 4, Pt 1 and BS 4848, Pts 2 and 4.

Where thickness varies, e.g. Universal beams, columns and channels, sections are identified by the nominal size, i.e. 'depth × breadth × mass per unit length × shape'. Where thickness is constant, e.g. tees and angle sections, the identification is 'breadth × depth × thickness × shape'. In addition a section is identified by the grade of steel.

To optimize on costs steel plates should be selected from available stock sizes. Thicknesses are in the range 6, 8, 10, 12.5, 15 mm, and then in 5 mm increments. Thicknesses of less than 6 mm are available but because of lower strength and poorer corrosion resistance their use is limited to cold formed sections. Stock plate widths are in the range 1, 1.25, 1.5, 2, 2.5, and 3 metres, but narrow plate widths are also available. Stock plate lengths are in the range 2, 2.5, 3, 4, 5, 6, 10, and 12 metres. The adoption of stock widths and lengths avoids work in cutting to size and also reduces waste.

The application of some types of section is obvious, e.g. when a member is in tension a round or flat bar is the obvious choice. However a member in tension may be in compression under alternative loading and an angle, tee, or tube is often more appropriate. The connection at the end of a bar or tube, however, is more difficult to make.

If a structural element is in bending about one axis then the 'I' section is the most efficient because a large proportion of the material is in the flanges, i.e. at the extreme fibres. Alternatively, if a member is in bending about two axes at right angles and also supports an axial load then a tube, or rectangular hollow section, is more appropriate.

Other steel sections available are cold formed from steel plate into a variety of cross sections for uses such as lightweight lattice beams, glazing bars and shelf racks. Not all these sections are standardized because of the large variety of possible shapes and uses, however there is a wide range of sections listed in BS 2994. Local buckling can be a problem and edges are stiffened using lips. Also because of the relative thinness of the material when used as beams web crushing, shear buckling and lateral torsional buckling can be a problem. Although the thickness of the material (1 to 3 mm) is less than that of the standard sections the resistance to corrosion is good because of the surface finish obtained by pickling and oiling. After degreasing this surface can be protected by galvanizing, or painting, or plastic coating. The use in building of cold formed sections in light gauge plate, sheet and strip steel 6 mm thick and under is dealt with in BS 5950, Pt 5.

1.1.3   Structural classification of steel sections (cl 3.5.2, Pt 1)

A section, or element of a member, in compression due to an axial load may fail by local buckling. Local buckling can be avoided by limiting the width to thickness ratios (b/T or d/t) of each element of a cross section. The values are given in Table 7, Pt 1 and their use avoids tedious and complicated calculations.

Depending on the b/T or d/t ratios standard, or built-up sections, are classified for structural purposes as:

(a)   Plastic (low values of b/T or d/t) – a plastic hinge can be developed with sufficient rotation capacity to allow redistribution of moments within the structure.
(b)   Compact – full plastic moment capacity can be developed but local buckling may prevent development of a plastic hinge with sufficient rotation capacity to permit plastic design.
(c)   Semi-compact (high values of b/T and d/t) – stress at the extreme fibres can reach design strength but local buckling may prevent the development of the full plastic moment.
(d)   Slender – local buckling may prevent the stress from reaching the design strength.

Sections for simply supported beams are either plastic or compact, while sections for rigid jointed frames are plastic. Most grade 43 and grade 50 Universal Beam sections are plastic but there are some exceptions.

The classification described above is important in design calculations as shown in examples in this book.

1.2   Development, manufacture and types of steel

1.2.1   Outline of developments in design using ferrous metals

Prior to 1779, when the Iron Bridge at Coalbrookdale on the Severn was completed, the most important materials used for load bearing structures were masonry and timber. Ferrous materials were only used for fastenings, armaments and chains.

The earliest use of cast iron columns in factory buildings (circa 1780) enabled relatively large span floors to be constructed. Due to a large number of disastrous fires around 1795, timber beams were replaced by cast iron with the floors carried on brick jack arches between the beams. This mode of construction was pioneered by Strutt in an effort to attain a fireproof construction technique. Cast iron, however, is weak in tension and necessitates a tension flange larger than the compression flange and consequently cast iron was used mainly for compression members. Large span cast iron beams were impractical, and on occasions disastrous as in the collapse of the Dee bridge designed by Robert Stephenson in 1874. The last probable use of cast iron in bridge works was in the piers of the Tay bridge in 1879 when the bridge collapsed in high winds due to poor design and unsatisfactory supervision during construction.

In an effort to overcome the tensile weakness of cast iron, wrought iron was introduced in 1784 by Henry Cort. Wrought iron enabled the Victorian engineers to produce the following classic structures. Robert Stephenson's Brittania Bridge was the first box girder bridge and represented the first major collaboration between engineer, fabricator (Fairburn) and scientist (Hodgkinson), I. K. Brunel's Royal Albert Bridge at Saltash combined an arch and suspension bridge. Telford's Menai suspension bridge used wrought iron chains which have since been replaced by steel chains. Telford's Point Cysyllte is a canal aqueduct near Llangollen. The Brittania Bridge was replaced after a fire in 1970. The introduction of wrought iron also revolutionized ship building and enabled Brunel to produce the S.S. *Great Britain*.

Steel was first produced in 1740, but was not available in large quantities until Bessemer invented the converter in 1856. The first major structure to use the new steel exclusively was Fowler and Baker's railway bridge at the Firth of Forth. The first steel rail was rolled in 1857 and installed at Derby where it was still in use 10 years later. Cast iron rails in the same position lasted about 3 months. Steel rails were in regular production at Crewe under Ramsbottom from 1866.

By 1840 standard shapes in wrought iron, mainly rolled flats, tees and angles, were in regular production and were appearing in structures about 10 years later. Compound girders were fabricated by riveting together the standard sections. Wrought iron remained in use until around the end of the nineteenth century.

By 1880 the rolling of steel 'I' sections had become widespread under the influence of companies such as Dorman Long. Riveting continued in use as a fastening method until around 1950 when it was superseded by welding. Bessemer steel production in Britain ended in 1974 and the last open hearth furnace closed in 1980. Further information on the history of steel making can be found in Buchanan, Cossons, Derry, Pannel, and Rolt.

1.2.2 Manufacture of steel sections

The manufacture of standard steel sections, although now a continuous process, can be conveniently divided into three stages:

(a) Iron production.
(b) Steel production.
(c) Rolling.

Iron production is a continuous process and consists of chemically reducing iron ore in a blast furnace using coke and crushed limestone. The resulting material, called cast iron, is high in carbon, sulphur and phosphorus.

Steel production is a batch process and involves reducing the carbon, sulphur and phosphorus levels and adding, where necessary, manganese, chromium, nickel or vanadium. This process is now carried out using a Basic Oxygen Converter, which consists of a vessel charged with molten cast iron, scrap steel and limestone through which oxygen is passed under pressure to reduce the carbon content by oxidation. This is a batch process which typically produces about 250 to 300 tonnes every 40 minutes. The alternative electric arc furnace is in limited use (approximately 5% of the UK steel production), and is generally used for special steels such as stainless steel.

From the converter the steel is 'teemed' into ingots which are then passed to the rolling mills for successive reduction in size until the finished standard section is produced. The greater the reduction in size the greater the work hardening, which produces varying properties in a section. The variation in cooling rates for the different thicknesses introduces residual stresses which may be relieved by the subsequent straightening process. Steel plate is now produced using a continuous casting procedure which eliminates ingot casting, mould stripping, heating in soaking pits and primary rolling. Continuous casting permits better control, improved quality, reduced wastage and lower costs.

1.2.3   Types of steel

The steel used in structural engineering is a compound containing approximately 98% iron and small percentages of carbon, silicon, manganese, phosphorus, sulphur, niobium and vanadium as specified in BS 4360. Increasing the carbon content increases strength and hardness but reduces ductility and toughness. Carbon content therefore is restricted to between 0.25% to 0.2% to produce a steel that is weldable and not brittle. The niobium and vanadium are introduced to raise the yield strength of the steel, the manganese improves corrosion resistance, and the phosphorus and sulphur are impurities, BS 4360 also specifies tolerances, testing procedures and specific requirements for weldable structural steel.

1.3   Structural steel

1.3.1   Initiation of a design

The demand for a structure originates with the client who may be either a private person, a private or public company, local or national government, or a nationalized industry.

In the first stage preliminary drawings and estimates of costs are produced, followed by consideration of which structural materials to use, e.g. reinforced concrete, steel, timber, brickwork. If the structure is a building only an architect may be involved at this stage, but if the structure is a bridge or industrial building then a civil or structural engineer prepares the documents. When the client is satisfied with the layout and estimated costs then detailed design calculations, drawings and costs are prepared and incorporated in a legal contract document. The design documents should be adequate to detail, fabricate and erect the structure (cl 1.5, Pt 1).

The contract document is usually prepared by the consultant engineer and work is carried out by a contractor who is supervised by the consultant engineer. However, larger companies, local and national government, and nationalized industries generally employ their own consulting engineer. Alternatively direct labour may be used to carry out the work. A further alternative is for the contractor to produce a design and construct package, where the contractor is responsible for all parts and stages of the work. General recommendations for steelwork tenders and contracts are given in Appendix A, BS 5950, Pt 2.

1.3.2   The object of structural design (cl 1.0.1, Pt 1)

The object of structural design is to produce a structure that will not become unserviceable or collapse in its lifetime, and which fulfils the requirements of the client and user at reasonable cost.

The requirements of the client and user may include any or all of the following:

(a)   The structure should not collapse locally or overall.
(b)   It should not be so flexible that deformations under load are unsightly or alarming, or cause damage to the internal partitions and fixtures; neither should any movement due to live loads, such as wind, cause discomfort or alarm to the occupants or users.
(c)   It should not require excessive repair or maintenance due to accidental overload, or because of the action of weather.
(d)   In the case of a building, the structure should be sufficiently fire resistant to give the occupants time to escape, enable the fire brigade to fight the fire in safety, and to restrict the spread of fire to adjacent structures.

The designer should be conscious of the costs involved, which include:

(a)   The initial cost which includes fees, site preparation, materials and construction.
(b)   Maintenance costs, e.g. decoration and structural repair.
(c)   Insurance, chiefly against fire damage.
(d)   Eventual demolition.

It is the responsibility of the structural engineer to design a structure that is safe and which conforms to the requirements of the local bye-laws and building regulations. Information and methods of design are obtained from British Standards and Codes of Practice and these are 'deemed to satisfy' the local bye-laws and building regulations. In exceptional circumstances, e.g. the use of forms or methods validated by research or testing, a design may be accepted (cl 7, Pt 1).

A structural engineer is expected to keep up to date with the latest research information. In the event of a collapse or malfunction where it can be shown that the engineer has failed to reasonably anticipate the cause or action leading to collapse, or has failed to apply properly the information at his disposal, i.e. codes of practice, British Standards, Building Regulations, research or information supplied by the manufacturers, then he may be sued for professional negligence. Consultants and contractors carry liability insurance to mitigate the effects of such legal action.

1.3.3   Limit state design (cl 2.1.1, Pt 1)

It is self evident that a structure should be 'safe' during its lifetime, i.e. free from the risk of collapse. There are, however, other risks associated with a structure and the term safe is now replaced by the term 'serviceable'. A structure should not during its lifetime become 'unserviceable', i.e. it should be free from such risks as collapse, rapid deterioration, fire, cracking or excessive deflection.

Ideally it should be possible to calculate mathematically the risk involved in structural safety based on the variation in strengths of the material and variation in the loads. Reports, such as the Construction Industry Research and Information Association (CIRIA) Report 63, have introduced the designer to the elegant and powerful concept of 'structural reliability'. Methods have been devised whereby engineering judgement and experience can be combined with statistical analysis for the rational computation of partial safety factors in codes of practice. However, in the absence of complete understanding and data concerning aspects of structural behaviour, absolute values of reliability cannot be determined.

It is not practicable, nor is it economically possible, to design a structure that will never fail. It is always possible that the structure will contain material that is less than the required strength or that it will be subjected to loads greater than the design loads. It is therefore accepted that 5% of the material in a structure is below the design strength, and that 5% of the applied loads are greater than the design loads. This does not mean that collapse is inevitable however, because it is extremely unlikely that the weak material and overloading will combine simultaneously to produce collapse.

Philosophy and objectives must be translated into a tangible form using calculations. A structure should be designed to be safe under all conditions of its useful life and to ensure that this is accomplished certain distinct performance requirements, called 'limit states', have been identified. The method of limit state design recognizes the variability of loads, materials, construction methods and approximations in the theory and calculations. Limit states may apply at any stage of the life of a structure, or at any stage of loading. The limit states that are important for the design of steelwork are shown in Table 1, cl 2.1.1, Pt 1, and it should be noted that strength is a dominant theme throughout.

To reduce the number of load cases to be considered only serviceability and ultimate limit states are specified. Each of these sections is divided into four although some may not be crucial in every design. Calculations for limit states involve loads and load factors (see Chapter 3), and material factors and strengths (see Chapter 2).

Stability, an ultimate limit state, is the ability of a structure, or part of a structure, to resist overturning, overall failure and sway (cl 2.4.2, Pt 1). Calculations should consider the worst realistic combination of loads at all stages of construction (see Section 8.8.3).

All structures, and parts of structures, should be capable of resisting sway forces, e.g. by use of bracing, 'rigid' joints, or shear walls. Sway forces arise from horizontal loads, e.g. winds, and also from practical imperfections, e.g. lack of verticality. Sway forces from practical imperfections are difficult to quantify and notional forces are specified in cl 2.4.2.3, Pt 1.

Also involved in limit state design is the concept of structural integrity. Essentially this means that the structure should be tied together as a whole, but if damage does occur, it should be localized. Requirements for 'tying' are sometimes complex and examples are given in Chapter 8.

Deflection is a serviceable limit state. Deflections should not impair the efficiency of a structure, or its components, or cause damage to the finishes. Generally the worst realistic combination of unfactored imposed loads is used to calculate elastic deflections. These values are compared with allowable values related to the length of a member, span or height given in Table 5, cl 2.5.1, Pt 1.

Fortunately there are few structural failures, but when they do occur they are often associated with human error involved in design calculations, or construction, or in the use of the structure.

1.3.4   Methods of design (cl 2.1.2, Pt 1)

Methods of design are described in BS 5950 as;

(a)   simple,

(b)   semi-rigid,
(c)   rigid.

These titles refer to the types of connections and whether bracing is included.

Simple design assumes that 'pin joints' connect the members, and that joint rotations are prevented by bracing. Historically this method was popular because calculations could be done by hand. With the advent of the computer this justification has become less convincing, but the method is nevertheless still used.

Semi-rigid design acknowledges that end moments exist at the connections. Experimental and simple calculations show that for 'pin' joints there are small moments at the connections, and provided that they are not greater than 10% of the free bending moment for the member they may be taken into account to reduce the size of the beam. However they must be balanced by moments on the other members which may not prove to be cost effective. Bracing is generally used with this method of design.

Rigid design assumes that the connections between members are rigid and therefore the angles between members can be maintained without the use of bracing. This method often produces a more economical design but because calculations for the design of members and connections are more complicated a computer is generally used. Further information on methods of design is given in Section 8.9.

1.3.5   Errors

To err is human but the consequences of an error in structural design can lead to loss of life and damage to property, and it is necessary to appreciate where errors can occur. Small errors can occur in the rounding off of figures but these generally do not lead to failures. The common sense advice given in cl 1.0.3, Pt 1 is that the accuracy of the calculation should match the accuracy of the values given in BS 5950.

Errors that occur in structural design calculations and affect structural safety are:

1   Ignorance of the physical behaviour of the structure under load, which consequently introduces errors in the basic assumptions used in the theoretical analysis.
2   Errors in estimating the loads, especially the erection forces.
3   Numerical errors in the calculations. These should be eliminated by checking, but when speed in paramount checks are often ignored.
4   Ignorance of the significance of certain effects, e.g. residual stresses, fatigue.
5   Introduction of new materials, or methods, which have not been proved by tests.
6   Insufficient allowance for tolerances or temperature strains.
7   Insufficient information, e.g. erection procedures.

Errors that can occur in workshops or on construction sites are:

1   Using the wrong grade of steel, and when welding using the wrong type of electrode.
2   Using the wrong weight of section. A number of sections are the same nominal size but differ in web or flange thickness.
3   Errors in manufacture, e.g. holes in the wrong position.

Errors that occur in the life of a structure and also affect safety are:

1   Overloading.
2   Removal of structural material, e.g. to insert service ducts.
3   Poor maintenance.

1.4 Fabrication of steelwork

1.4.1 Drawings

Detailed design calculations are essential for any steelwork design but the sizes of the members, dimensions and geometrical arrangement are usually presented as drawings. Initially the drawings are used by the fabricator and eventually by the contractor on site. General arrangement drawings are often drawn to a scale of 1 : 100, while details are drawn to a scale of 1 : 20 or 1 : 10. Special details are drawn to larger scales where necessary.

Drawings should be easy to read and should not include superflous detail. Some important notes are:

(a)   Members and components should be identified by logically related mark numbers, e.g. related to the grid system used in the drawings.
(b)   The main members should be presented by a bold outline (0.4 mm wide) and dimension lines should be unobtrusive (0.1 mm wide).
(c)   Dimensions should be related to centre lines, or from one end; strings of dimensions should be avoided. Dimensions should appear once only so that ambiguity cannot arise when revisions occur. Fabricators should not be put in the position of having to do arithmetic in order to obtain an essential dimension.
(d)   Tolerances for erection purposes should be clearly shown.
(e)   The grade of steel to be used should be clearly indicated.
(f)   The size, weight and type of section to be used should be clearly stated.
(g)   Detailing should take account of possible variations due to rolling margins and fabrication variations (cl 1.6, Pt 1).

1.4.2 Tolerances (cl 7, Pt 2)

Tolerances are limits placed on unintentional inaccuracies that occur in dimensions which must be allowed for in design if structural elements and components are to fit together. In steelwork variations occur in the rolling process, marking out, cutting and drilling during fabrication, and in setting out during erection.

In the rolling process the allowable tolerances for length, width, thickness and flatness for plates are given in BS 4360. Length and width tolerances are positive while those for thickness and flatness are negative and positive. The dimensional and weight tolerances for sections are given in BS 4, Pt 1, or BS 4848, Pt 4, as appropriate.

Fabrication tolerances for built-up members are given in cl 7, Pt 2. During fabrication there is a tendency for members and components to increase rather than reduce, and the tolerance is therefore often specified as negative: it is often cheaper and simpler to insert packing rather than shorten a member, provided that the packing is not excessive. Where concrete work is associated with steelwork variations in dimensions are likely to be greater. When casting concrete, for example, errors in dimensions may arise from shrinkage or from warping of the shuttering, especially when it is re-used. Therefore, by virtue of the construction method, larger tolerances are specified for work involving concrete.

To facilitate erection all members and connections should be provided with the maximum tolerance that is acceptable from structural and architectural considerations. Values are given in cl 7.3, Pt 2. A typical example is a connection between a steel column and a reinforced concrete base. It would produce great difficulties if the base were set too high and a tolerance of approximately 50 mm is often included in the design, with provision for grouting under the base. Tolerances are also provided to allow lateral adjustment of the foundation bolts. Tolerances between concrete and steelwork are also important because two different contractors are involved.

1.4.3 Fabrication, assembly and erection of steelwork (cls 3 to 5, Pt 2)

The drawings produced by the structural designer are used first by the steel fabricator and later by the contractor on site.

The steel fabricator obtains the steel either direct from the rolling mills or from the steel stockist, and then cuts, drills and welds the steel components to form the structural elements as shown on the drawings. In general, for British practice, the welding is confined to the workshop and the connections on site are made using bolts. In America, however, site welding is common practice.

When marking out, the measurements of length for overall size, position of holes etc. can be done by hand, but if there are several identical components then wooden or cardboard templates are made, avoiding repeated measurement. Nowadays automatic machines controlled by computer or punched paper tape are used to cut and drill standard sections. When completed the steelwork should be marked clearly and manufactured to the tolerances given in cl 7, Pt 2.

When fabricated, parts of the structure are delivered to the site in the largest pieces that can be transported and erected. For example a lattice girder may be sent fully assembled to a site in this country, but sent in pieces to fit a standard transport container for erection abroad. All components should be assembled within the tolerances and cambers specified, and should not be bent or twisted or otherwise damaged. Further guidance on welding, bolting and protective treatment is given in cl 4, Pt 2.

On site the general contractor may be responsible for the assembly, erection, connections, alignment and levelling of the complete structure. Alternatively the erection work may be done by the steel fabricator, or sublet to a specialist steel erector. The objective of the erection process is to assemble the steelwork in the most cost effective way whilst maintaining the stability of individual members, and/or a part of the complete structure. To do this it may be necessary to introduce cranes and temporary bracing which must also be designed to resist the loads involved.

During assembly on site it is inevitable that some components will not fit, despite the tolerances that have been allowed. A typical example is that the faying surfaces for a friction grip joint are not in contact when the bolts are stressed. Other examples are given by Mann and Morris. The correction of some faults and the consequent litigation can be expensive.

1.4.4 Testing of steelwork (Section 7, Pt 1)

Steel is routinely sampled and tested during production to maintain quality. However, at times new methods of construction are suggested and there may be some doubt as to the validity of the assumptions of behaviour of the structure. Alternatively if the structure collapses there may be some dispute as to the strength of a component, or member, of the structure. In such cases testing of components, or part of the structure, may be necessary. However it is generally expensive because of the accuracy required, cost of material, cost of fabrication, necessity of repeating tests to allow for variations, and of reporting accurately.

BS 5950, Section 7, Pt 1 distinguishes between;

(a)   acceptance tests (cl 7.3.3, Pt 1) – non-destructive, for confirming structural performance,

(b)   strength tests (cl 7.3.4, Pt 1) – used to confirm the calculated capacity of a component or structure,

(c)   tests to failure (cl 7.3.5, Pt 1) – to determine the real mode of failure and the true capacity of a specimen,

(d)   check tests (cl 7.3.6, Pt 1) – where the assembly is designed on the basis of tests.

The size, shape, position of the gauges and method of testing of small sample pieces of steel is given in BS 4360 and BS 18. The tensile test is most frequently employed, and gives values of Young's modulus, limit of proportionality, yield stress or proof stress, percentage elongation and ultimate stress. Methods of destructive testing fusion welded joints and weld metal in steel are given in BS 709.

The Charpy V-notch test for impact resistance is used to measure toughness, i.e. the total energy, elastic and plastic, which can be absorbed by a specimen before fracture. The test specimen is a small beam of rectangular cross section with a 'V' notch at mid-length. The beam is fractured by a blow from a swinging pendulum, and the amount of energy absorbed is calculated from the loss of height of the pendulum swing after fracture. Details of the test specimen and procedure are given in BS 4360 and BS 131. The Charpy V-notch test is often used to determine the temperature at which transition from brittle to ductile behaviour occurs.

Structures which are unconventional, and/or methods of design which are unusual or not fully validated by research, should be subject to acceptance tests. Essentially these consist of loading the structure to ensure that it has adequate strength to support, 1 (test dead load) + 1.15 (remainder of dead load) + 1.25 (imposed load). Details of the tests are given in cl 7, Pt 1.

Where welds are of vital importance, e.g. in pressure vessels, they should be subjected to non-destructive tests. The defects that can occur in welds are slag inclusions, porosity, lack of penetration and sidewall fusion, liquation, solidification, hydrogen cracking, lamellar tearing and brittle fracture.

A surface crack in a weld may be detected visibly but alternatively a dye can be sprayed onto the joint which seeps into the cracks. After removing any surplus dye the weld is resprayed with a fine chalk suspension and the crack then shows as a coloured line on the white chalk background. A variant of this technique is to use fluorescent dye and a crack then shows as a bright green line in ultraviolet light. A surface crack may also be detected if the weld joint area is magnetized and sprayed with iron powder. The powder congregates along a crack, which shows as a black line.

Other weld defects cannot be detected on the surface and alternative methods must be used. Radiographic methods use an X-ray or gamma ray source on one side of the weld and a photographic film on the other. The rays are absorbed by the weld metal, but if there is a hole or crack there is less absorption which shows as a dark area on the film. Not all defects are detected by radiography since the method is sensitive to the orientation of the flaw, e.g. cracks at right angles to the X-ray beam are not detected. Radiography also requires access to both sides of the joint and the method is therefore most suitable for in-line butt welds for plates.

An alternative method of detecting hidden defects in welds uses ultrasonics. If a weld contains a flaw then high frequency vibrations are reflected. The presence of a flaw can therefore be indicated by monitoring the reduction of transmission of ultrasonic vibrations, or by monitoring the reflections. The reflection method is extremely useful for welds where access is only possible from one side. Further details can be obtained from Gourd.

References

Alexander S. J. and Lawson R. M. (1981). *Movement design in buildings*, Technical Note 107, Construction Industry Research and Information Association Publication.

BS 4 *Specification for hot rolled sections Pt 1*, British Standards Institution, London.

BS 18 *Methods for tensile testing of metals – Pt 1 steel general, Pt 4 steel tubes*, British Standards Institution, London.

BS 131 *The Charpy V-notch impact test on metals, Pt 2*, British Standards Institution, London.

BS 709 *Methods of destructive testing fusion welded joints and weld metal in steel*, British Standards Institution, London.

BS 2994 *Cold rolled sections*, British Standards Institution, London.

BS 4360 *Specification for weldable structural steels*, British Standards Institution, London.

BS 4848 *Pt 4, Hot rolled structural steel sections, equal and unequal angles, Pt 2 Hollow sections*, British Standards Institution, London.

BS 5950 *Structural use of steelwork in buildings Pts 2 and 5*, British Standards Institution, London.

Buchanan R. A. (1972). *Industrial Archeology in Britain*, Penguin.

CIRIA (1977). *Rationalisation of safety and serviceability factors in structural codes*, Report 63, Construction Industry Research and Information Association Publication.

Cossons N. (1975). *The BP Book of Industrial Archeology*, David and Charles.

Derry T. K. and Williams T. (1960). *A Short History of Technology*, Oxford University Press.

Gourd L. M. (1980). *Principles of welding technology*, Edward Arnold.

Mann A. P. and Morris L. J. (1981). *Lack of fit in steel structures*, Technical Report 87, Construction Industry Research and Information Association Publication.

Pannel J. P. M. (1964). *An Illustrated History of Civil Engineering*, Thames and Hudson.

Rolt L. T. C. (1970). *Victorian Engineering*, Penguin.

2

Mechanical Properties of Structural Steel

2.1 Variation of material properties

All manufactured material properties vary because the molecular structure of the material is not uniform and because of inconsistencies in the manufacturing process. The variations that occur in the manufacturing process are dependent on the degree of control. Variations in material properties must be recognized and incorporated into the design process. The material properties that are of most importance for structural design using steel are strength and Young's modulus. Other properties that are of lesser importance are hardness, impact resistance and melting point.

   If a number of samples are tested for a particular property, e.g. strength, and the number of specimens with the same strength (frequency) plotted against the strength, then the results approximately fit a normal distribution curve as shown in Fig. 2.1.

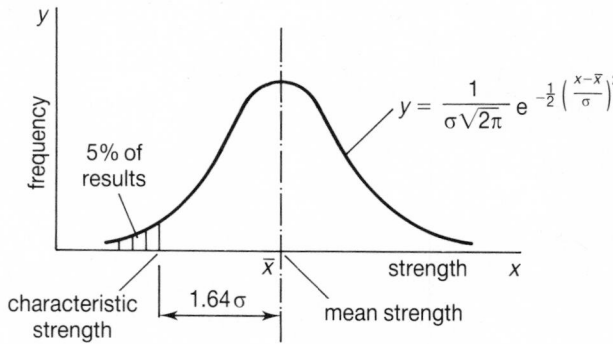

Fig. 2.1 Variation in material properties.

   This curve can be expressed mathematically by the equation shown in Fig. 2.1 which can be used to define 'safe' values for design purposes as follows.

2.2 Characteristic strength

A strength to be used as a basis for design must be selected from the variation in values shown in Fig. 2.1. This strength, when defined, is called the characteristic strength. If the characteristic strength is defined as the mean strength, then clearly from Fig. 2.1, 50% of the material is below this value. This is not acceptable. Ideally the characteristic strength should include 100% of the samples, but this also is impracticable because it is a low value and results in heavy and costly structures. A risk is therefore accepted and it is therefore recognized that 5% of the samples fall below the characteristic strength.

The characteristic strength is calculated from the equation

$$f_k = f_{mean} - 1.64\sigma$$

where the standard deviation for n samples

$$\sigma = [\Sigma(f_{mean} - f)^2/(n-1)]^{1/2}$$

Characteristic strengths are not given in BS 5950.

2.2 Design strength (cl 3.1.1, Pt 1)

The characteristic strength of steel is the value obtained from tests at the rolling mills, but by the time the steel becomes part of a finished structure this strength may be reduced, e.g. by corrosion or accidental damage. The strength to be used in design calculations is therefore the characteristic strength divided by a partial safety factor. The value of the partial safety factor adopted for steel in BS 5950 is not stated but a figure of approximately 1.05 is generally assumed.

The design strength (p_y) in BS 5950 is defined as the specified minimum yield strength (Y_s) but not greater than 0.84 (minimum ultimate tensile strength) as specified in BS 4360. The actual values of yield strength and ultimate tensile strength from tests are generally greater than these values.

Table 6, Pt 1 gives values of the design strength related to the thickness of the material. The relevant thickness for a rolled section is the flange thickness. Structural steel is described as grade 40, 43, 50, and 55, where the figures represent the strength in kgf/mm^2 approximately. The current use of various grades of steel are:

Grade 43 – lowest in cost and the most widely used.
Grade 50 – cost greater than grade 43 but economical where deflections are not critical; use increasing.
Grade 55 – rarely used.

2.4 Other properties for steel (cl 3.1.2, Pt 1)

Young's modulus for steel is obtained from the stress–strain relationship as shown in Fig. 2.2. This value is a material property and therefore values from a set of samples vary. However, the variation for steel is very small and BS 5950 assumes $E_s = 205 \, kN/mm^2$.

The elastic shear modulus (G_s) is related to Young's modulus by the expression

$$E_s = 2G_s(1 + v)$$

where Poisson's ratio $v = 0.3$, a value which is also used in calculations involving plate members.

The thermal coefficient of expansion for steel is given as $\alpha = 12E - 6/°C$ and is used in calculations for temperature changes which for the UK are in the range $-5°C$ to $+35°C$ (cl 2.3, Pt 1).

Hardness is a material property that is occasionally of importance in structural steel design. It is measured by the resistance the surface of the steel offers to indentation by

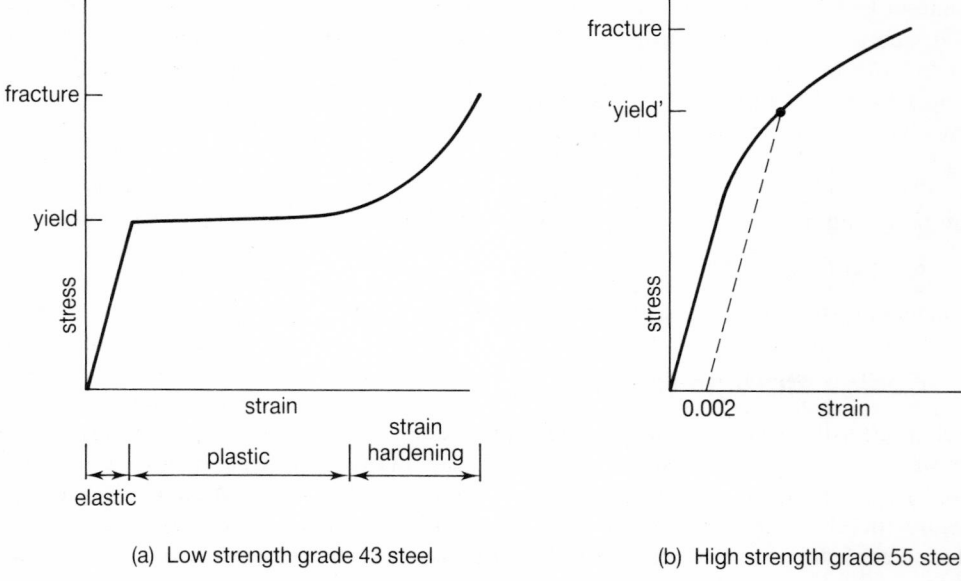

(a) Low strength grade 43 steel   (b) High strength grade 55 steel

Fig. 2.2   Tensile stress-strain relationships for steel.

either a hardened steel ball (Brinell test), a square based diamond pyramid (Vickers test) or a diamond cone (Rockwell test). Higher strength often correlates with greater hardness but this is not inevitable.

Ductility may be described as the ability of a material to change its shape without fracture. This is measured by the percentage elongation, i.e. $100 \times$ (change in length/ original length). Values of 20% can be obtained for mild steel but are less for high strength steel. A high value is advantageous because it allows the redistribution of stresses at ultimate load and the formation of plastic hinges.

2.5   Corrosion and durability of steel work (cl 2.5.2, Pt 1)

Durability is a service limit state and the following factors should be considered at the design stage;

(a)   environment,
(b)   degree of exposure,
(c)   shape of the members and details,
(d)   protective measure,
(e)   maintenance.

Methods of protecting steelwork are given in BS 5493 and the specification for weather resistant steel is given in BS 4360.

Corrosion of steel work reduces the cross section of members and thus affects safety. Corrosion, which occurs on the surface of steel, is a chemical reaction between iron, water and oxygen, which produces a hydrated iron oxide called rust. Electrons are liberated in the reaction and a small electrical current flows from the corroded area to the uncorroded area. The elimination of water, oxygen or the electrical current reduces the rate of corrosion.

In contrast pollutants in the air, e.g. sulphur dioxides from industrial atmospheres and salt from marine atmospheres, increase the electrical conductivity of water and accelerate the corrosion reaction.

Steel is particularly susceptible to atmospheric corrosion which is often severe in coastal or industrial environments and the corrosion may reduce the section size due to pitting or flaking of the surface. Modern rolling techniques and higher strength steels result in less material being used, e.g. the web of an 'I' section may be only 6 mm thick. Generally in structural engineering 8 mm is the minimum thickness used for exposed steel, and 6 mm for unexposed steel. For sealed hollow sections these limits are reduced to 4 mm and 3 mm respectively.

Corrosion of steel usually takes the form of rust, which is a complex hydrated oxide of iron. The rust builds up a deposit on the surface and may eventually flake off. The coating of rust does not inhibit corrosion, except in special steels, and corrosion progresses beneath the rust forming conical pits and the thickness of the metal is reduced. The conical pits can act as stress raisers, i.e. centres of high local stress, and in cases where there are cyclic reversals of stress, may become the initiating points of fatigue cracks, or brittle fracture.

The corrosion resistance of unprotected steel is dependent on its chemical composition, the degree of pollution in the atmosphere, and the frequency of wetting and drying. Low strength carbon steels are inexpensive but are particularly susceptible to atmospheric corrosion which is often greatest in industrial or coastal environments. High strength low alloy steels (Cr–Si–Cu–P) do not pit as severely as carbon steels and the rust that forms becomes a protective coating against further deterioration. These steels therefore have several times the corrosion resistance of carbon steels. The longer steel remains wet the greater the corrosion and therefore the detailing of steelwork should include drainage holes, avoid pockets, and allow the free flow of air for rapid drying.

The commonest and cheapest protection process is to clean the surface by sand or shot blasting, and then to paint with a lead primer, generally in the workshop prior to delivery on site. Joint contact surfaces need not be protected unless specified. On site the steel is erected and protection is completed with an undercoat and finishing coat, or coats, of paint. In the case of surfaces to be welded steel should not be painted, nor metal coated, within a suitable distance of any edges to be welded, if the paint or the metal coating specified is likely to be harmful to welders or impair the quality of the welds. Welds and adjacent parent metal should not be painted prior to de-slagging, inspection and approval.

Encasing steel in concrete provides an alkaline environment and no corrosion will take place unless water diffuses through the concrete carrying with it SO_2 and CO_2 gases from the air in the form of weak acids. The resulting corrosion of the steel and the increase in pressure spalls the concrete. Parts to be encased in concrete should neither be painted nor oiled, and where friction grip fasteners are used protective treatment should not be applied to the faying surfaces.

A more expensive protection is zinc, or aluminium spray coating, which is sometimes specified in corrosive atmospheres. Further improvements are hot dip zinc galvanizing, or the use of stainless steels. These and other forms of protection are described in BS 5493. Recently zinc coated highly stressed steel has been shown to be susceptible to hydrogen cracking.

2.6 Brittle fracture (cl 2.4.4, Pt 1)

This is critical at the ultimate limit state. The evidence of a brittle fracture is a small crack, which may or may not be visible, and which extends rapidly to produce a sudden failure with few signs of plastic deformation. This type of fracture is more likely to occur in welded structures as shown by Stout, Tor and Ruzek and the essential conditions leading to brittle fracture are:

(a)   There must be a tensile stress in the material though it need not be very high, and may be a residual stress from welding.
(b)   There must be a notch, or defect, or hole in the material which produces a stress concentration.
(c)   The temperature of the material must be below the transition temperature (generally below room temperature). At low temperatures crack initiation and propagation are more likely because of lower ductility.

The mechanism of failure is that the notch, defect, or hole raises the local tensile stresses to values as high as three times the average tensile stress. The material, which generally fails by a shearing mechanism, now tends to fail by a brittle fracture cleavage mechanism which exhibits considerably less plastic deformation. A drop in the temperature encourages the cleavage failure. A ductile material which has an extensive plastic range is more likely to resist brittle fracture and a test used as a guide to resistance to brittle fracture is the Charpy V-notch impact test as specified in BS 4360 (see Section 1.4.4).

Brittle fracture need not be considered except where tensile stresses exist, nor need it be considered for grade 43A base plates. Where tensile stresses exist then the maximum thickness is determined from Tables 3 and 4, Pt 1 in the UK provided that the service temperature does not fall below $-5°C$ (internal conditions) and $-15°C$ (external conditions). Where these conditions are not met then a Charpy impact test should be carried out to determine if $C_v > Y_s t/(710\ K)$ at the service temperature. Further information can be obtained from the Navy Department Advisory Committee (NDAC) on structural steels.

2.7   Residual stresses

Residual stresses are present in steel due to uneven heating and cooling. The stresses are induced in steel during rolling, welding which constrains the structure to a particular geometry, force fitting of individual components, lifting and transportation. These stresses may be relieved by subsequent reheating and slow cooling but the process is expensive. The presence of residual stresses adversely affects the buckling of columns, and introduces premature yielding, fatigue resistance and brittle fracture.

Welding raises the local temperature of the steel which expands relative to the surrounding metal. When it cools it contracts inducing tensile stresses in the weld and the immediately adjacent metal. These tensile stresses are balanced by compressive stresses in the metal on either side.

During rolling the whole of the steel section is initially at a uniform temperature, but as the rolling progresses some parts of the cross section become thinner than others and consequently cool more quickly. Thus, as in the welded joint, the parts which cool last have a residual tensile stress and the parts which cool first may be in compression. Since the cooling rate also affects the yield strength of the steel, the thinner sections tend to have a higher yield stress than the thicker sections. A tensile test piece cut from the thin web of a Universal beam will probably have a higher yield stress than one cut from the thicker flange. The residual stress and yield stress in rolled sections are also affected by the cold straightening which is necessary for many rolled sections before leaving the mills.

2.8 Fatigue (cl 2.4.3, Pt 1)

This is an ultimate limit state. The term fatigue is generally associated with metals and is the reduction in strength that occurs due to progressive development of existing small pits, grooves or cracks when subject to fluctuating loads. The rate of development of these cracks depends on the size of the crack and on the magnitude of the stress variation in the material and also the metallurgical properties. The number of stress variations or cycles of stress that a material will sustain before failure is called fatigue life and there is a linear experimental relationship between the log of the stress range and the log of the number of cycles. Welds are susceptible to a reduction in strength due to fatigue because of the presence of small cracks, local stress concentrations, and abrupt changes of geometry. Research into the fatigue strength of welded structures is described by Munse. Other references are Grundy and the European Convention for Structural Steelwork (ECCS).

All structures are subjected to varying loads but the variation may not be significant. Stress changes due to fluctuations in wind loading need not be considered, but wind induced oscillations must not be ignored. The variation in stress depends on the ratio of dead load to imposed load, or whether the load is cyclic in nature, e.g. where machinery is involved. For bridges and cranes fatigue effects are more likely to occur because of the cyclic nature of the loading which causes reversals of stress. In design calculations an estimate is made of the number of load cycles the structure is likely to experience during the life of the structure and the design stress is reduced accordingly. Guidance on this aspect is given in BS 5400 Pts 3 and 10 for bridges where design stresses are related to the stress range and the number of load cycles. Dynamic loads and impact effects for cranes are obtained from BS 6399, Pt 1.

2.9 Stress concentrations

Structural elements and connections often have abrupt changes in geometry and also contain holes for bolts. These features produce stress concentrations, which are localized stresses greater than the average stress in the element, e.g. tensile stresses adjacent to a hole are approximately three times the average tensile stress. If the average stress in a component is low then the stress concentration may be ignored, but if high then appropriate methods of structural analysis must be used to cater for this effect. In recent history, the effect of stress concentrations has been shown to be critical in plate web girders. Stress concentrations are also associated with fatigue as described in Section 2.8, and can also affect brittle fracture (see Section 2.6). Formulae for stress concentrations are given in Roark and Young.

2.10 Failure criteria for steel

The structural behaviour of a metal at or close to failure may be described as ductile or brittle. A typical brittle metal is cast iron which exhibits a linear load–displacement relationship until fracture occurs suddenly with little or no plastic deformation. In contrast mild steel is a ductile material which also exhibits a linear load–displacement relationship, but at yield large plastic deformations occur before fracture.

The minimum yield strength, or its equivalent, is the design strength in BS 5950 and is therefore an important failure criterion for steel. The tensile yield condition can be related to various stress situations, e.g. tension, compression, shear or various combinations of stresses.

There are four generally acceptable theoretical yield criteria as follows:

(a) The maximum stress theory, which states that yield occurs when the maximum principal stress reaches the uniaxial tensile stress.
(b) The maximum strain theory, which states that yield occurs when the maximum principal tensile strain reaches the uniaxial tensile strain at yield.
(c) The maximum shear stress theory, which states that yield occurs when the maximum shear stress reaches half of the yield stress in uniaxial tension.
(d) The distortion strain energy theory, or shear strain energy theory, which states that yielding occurs when the shear strain energy reaches the shear strain energy in simple tension. For a material subjected to principal stresses f_2, f_2, and f_3 it is shown by Timoshenko that this occurs when

$$(f_1 - f_2)^2 + (f_2 - f_3)^2 + (f_3 - f_1)^2 = 2Y_s^2 \qquad (2.1)$$

This theory was originally developed by Huber, Von-Mises and Hencky.

Alternatively Equation (2.1) can be expressed in terms of direct stresses f_b, f_{bc} and f_{bt}, and shear stress f_q on two mutually perpendicular planes. It can be shown from Mohr's circle of stress that the principal stresses

$$f_1 = (f_b + f_{bc})/2 - [(f_b - f_{bc})^2/4 + f_q^2]^{1/2} \qquad (2.2)$$

and

$$f_2 = (f_b - f_{bc})/2 + [(f_b - f_{bc})^2/4 + f_q^2]^{1/2} \qquad (2.3)$$

If equations (2.2) and (2.3) are inserted in Equation (2.1) with $f_3 = 0$ and Y_s is equal to the design stress p_y then

$$p_y^2 = f_{bc}^2 + f_b^2 - f_{bc}f_b + 3f_q^2 \qquad (2.4)$$

If f_{bc} is replaced by f_{bt} with a change in sign then

$$p_y^2 = f_{bc}^2 + 3f_q^2 \qquad (2.5a)$$

Generally this is expressed in the form

$$(f_{bc}/p_y)^2 + 3(f_q/p_y)^2 = 1 \qquad (2.5b)$$

This equation is a yield criterion which is applicable in some design situations but is not given in BS 5950.

References

BS 2573 *Permissible stresses in cranes and design rules.* Pt 1, British Standards Institution, London.

BS 4360 *Specification for weldable structural steels.* British Standards Institution, London.

BS 5135 *Specification for metal arc welding of carbon and carbon manganese steels,* British Standards Institution, London.

BS 5400 *Steel, concrete and composite bridges,* Pt 3, British Standards Institution, London.

BS 5400 *Code of practice for fatigue,* Pt 10, British Standards Institution, London.

BS 5493 *Code of practice for protection of iron and steel structures,* British Standards Institution, London.

BS 6399 Pt 1 *Design loads for buildings – Code of practice for dead and imposed loads,* British Standards Institution, London.

ECCS *European recommendations for steel construction*, European Convention for Structural Steelwork, Construction Press.

ECCS *Recommendations for fatigue design of steel structures*, European Convention for Structural Steelwork, Construction Press.

Grundy P. (1985). *Fatigue limit design for steel structures*, Civil Engineering Transactions, Institution of Civil Engineers, Australia, CE27, No. 1.

Munse W. H. (1984). *Fatigue of welded structures*, Welding Research Council.

NDAC (1970). *Brittle fracture in steel structures*, Navy Department Advisory Committee on Structural Steels, ed. G. M. Boyd, Butterworth.

Ogle M. H. (1982). *Residual stresses in a steel box-girder bridge*, Tech Note 110, Construction Industry Research and Information Association Publication.

Roark J. R. and Young W. C. (1975). *Formulae for Stress and Strain*, Fifth Edition, McGraw-Hill.

Stout, R. D., Tor S. S. and Ruzek J. M. (1951). *The effect of fabrication procedures on steels used in pressure vessels*, Welding Journal 30.

Timoshenko S. (1946). *Strength of Materials*, Pt 11, D. Van Nostrand.

3

Loads

3.1 Characteristic loads (cl 2.2, Pt 1)

Ideally, loads applied to a structure during its working life should be considered statistically and a characteristic load determined. The characteristic load might then be defined as the load which no more than 5% of the loads exceed, as shown in Fig. 3.1. However there are insufficient data to determine this and loads are defined and quantified as follows.

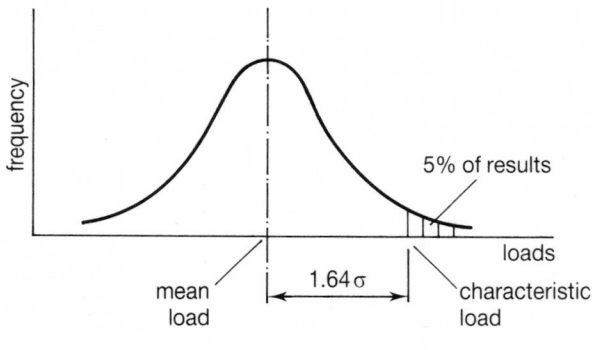

Fig. 3.1

The broad classification of loads for design purposes is 'dead loads' and 'imposed loads' (previously called live loads). Dead loads are those applied from permanent construction and imposed loads are from transient forces. Alternatively, loads may be classified as 'gravity' and 'non-gravity' loads as described in Section 8.2. These loads are now described in more detail.

3.1.1 Dead loads

These are due to the weight of the structure, i.e. walls, permanent partitions, floors, roofs, finishes and services. The actual weights of the materials should be used in the design calculations, but if these are unknown values may be obtained from BS 648. The density of steel is given as 7850 kg/m^3. The weights of tanks and other receptacles should be considered when full and empty. When wind loads are acting the empty condition often produces a worse effect.

3.1.2 Imposed floor and roof loads

These are due to moveable items such as furniture, occupants, machinery, vehicles, stored materials or snow. Imposed loads are generally expressed as static loads for convenience, although there may be minor dynamic forces involved. Imposed loads are subdivided into:

1 Imposed floor loads for various dwellings are given in BS 6399 Pt 1. These loads include a small allowance for impact and other dynamic effects that may occur in normal occupancy. They do not include forces resulting from the acceleration and braking of vehicles or movement of crowds. The loads are usually given in the form of a distributed load or an alternative concentrated load. The one that gives the most severe effect is used in design calculations. When designing a floor it is not necessary to consider the concentrated load if the floor is capable of distributing the load and for the design of the supporting beams the distributed load is always used. When it is known that mechanical stacking of materials is intended, or other abnormal loads are to be applied to the floor, then actual values of the loads should be used, not those obtained from BS 6399 Pt 1. In multi-storey buildings the probability that all the floors will simultaneously be required to support the maximum loads is remote and reductions to column loads are therefore allowed.

2 Imposed roof loads are related to snow load and access for maintenance. They are specified in BS 6399 Pt 3 and, as with floor loads, they are expressed as a uniformly distributed load on plan, or as an alternative concentrated load. In special situations where roof shapes are likely to result in drifting of snow then loads are increased and the magnitude of the load can be assessed from information given in BS 6399, Pt 3.

3.1.3 Wind loads

Wind loads are dynamic forces but for convenience they are expressed as static pressures in CP 3, Chapter V, Pt 2 (to be revised under BS 6399 Pt 2). The pressure at any point on a structure is related to the shape of the building, the basic wind speed, topography and ground roughness. These factors are incorporated in a single formula. The effects of vibration, such as resonance in tall buildings, must be considered separately.

3.1.4 Other loads and forces

1 Earth loads occur in situations such as retaining walls and foundations. Advice on earth pressures and allowable bearing pressures for different types of soil at service load conditions is given in CP 2004.

2 Dynamic forces (vibration, shock, acceleration, retardation and impact) are of importance in the design of cranes and information on loads and impact effects can be obtained from BS 6399 cl 7, Pt 1. Vertical static crane loads are increased by percentages from 10% to 25% depending on whether the crane is hand or electrically operated. When designing for earthquakes the inertial forces must be calculated but because this is not of importance in the UK there is no British Standard Code of Practice. Information in English can be obtained from American codes.

3 Erection forces are of great importance in steelwork construction because pre-fabrication is normal practice. Compression members which will be restrained in a completed structure may buckle during erection when subjected to relatively minor forces. Joints which are rigid when fully bolted may, during erection, act as a pin and induce collapse

of the structure. Suspension points for members or parts of structures may have to be specified to avoid damage to components. It is extremely difficult to anticipate all possible erecion forces and the contractor is responsible for erection, which should be carried out with due care and attention. Nevertheless a designer should have knowledge of the most likely method of erection and design accordingly. If necessary temporary stiffening or supports should be specified, and/or instructions given.

4   Temperature forces are introduced into steelwork structures which are restrained or highly redundant. Where it is necessary to allow for temperature change in the UK the range should be $-5°C$ to $+35°C$ (cl 2.3, Pt 1).

5   Accidental forces include explosions and fire and are described in Chapter 11.

3.2   Design loads (or factored loads) (cl 2.4.1, Pt 1)

The design load (or factored load), is obtained from the characteristic load by multiplying by a partial safety factor for loads (γ_f). This factor is often referred to as a load factor and the values recommended in Table 2, Pt 1 vary between 1 and 1.6. Load factors allow for the probability that there will be a variation in the value of the load, e.g. an imposed load is more likely to vary than a dead load. Also the values allow for inaccurate assessment of load effects, stress redistribution, variations in dimensional accuracy and the importance of the limit state under consideration. The load factor for water is generally taken as 1.4 as recommended in BS 8007. This appears to be conservative but if used for a tank for example, it allows for the tank overflowing, dimensional changes, and being filled with a denser liquid.

3.3   Pattern loading (cl 2.2.1, 5.1.2, Pt 1)

All possible design loads appropriate to a structure should be considered in design calculations. The loads should be considered separately and in realistic combinations to determine which is most critical for strength and stability of the structure. The magnitude and frequency of fluctuating loads, erection forces, temperature change and settlement at the supports should also be considered.

For continuous structures, connected by rigid joints or continuous over the supports, vertical loads should be arranged in the most unfavourable but realistic pattern for each element. Dead load factors need not be varied when considering such pattern loading, but should be varied when considering stability against overturning. Where horizontal loads are being considered pattern loading of vertical loads need not be considered.

To check the sway stability of a structure notional horizontal forces should be applied. These notional forces may arise from practical imperfections, such as lack of verticality, and should be taken as (cl 5.1.2.3, Pt 1):

0.5% of the factored dead plus imposed load applied horizontally.

These notional forces should be assumed to act in any one direction at a time and should be applied at each roof and floor level or their equivalent. The notional forces should not:

(a)   be applied when considering overturning;
(b)   be combined with the applied horizontal loads;
(c)   be combined with temperature effects;
(d)   be taken to contribute to the reactions at the foundations.

For the design of a simply supported beam it is obvious that the critical condition for strength is when the beam supports the maximum dead load plus the maximum imposed load at the ultimate limit state. The size of the beam is then determined from this loading condition and checked for deflection at the serviceability limit state.

A more complicated structure is a simply supported beam with a cantilever as shown in Fig. 3.2(a).

Assuming that the beam is of uniform section and that the dead load is uniformly applied over the full length of the beam, it is now necessary to consider various combinations of the imposed load as shown in Figs. 3.2(b) to (d). Although partial loading of spans is possible this is not generally considered except in special cases of rolling loads, e.g. a train on a bridge span.

For a particular section it is not immediately apparent which load combination is most critical because it depends on the relative span dimensions and loads. Therefore calculations are necessary to determine the conditions and section for maximum bending moment and shear force at the ultimate limit state.

In other situations, e.g. when checking the overturning of a structure, the critical combination of loads may be the minimum dead load, minimum imposed load, and maximum wind load (see also Chapter 8).

3.4 Tutorial Problems

Problem 3.4.1

The plan of part of a college classroom floor supported by beams and columns is shown in Fig. 3.3. The floor consists of floor finish on screed on precast prestressed concrete units

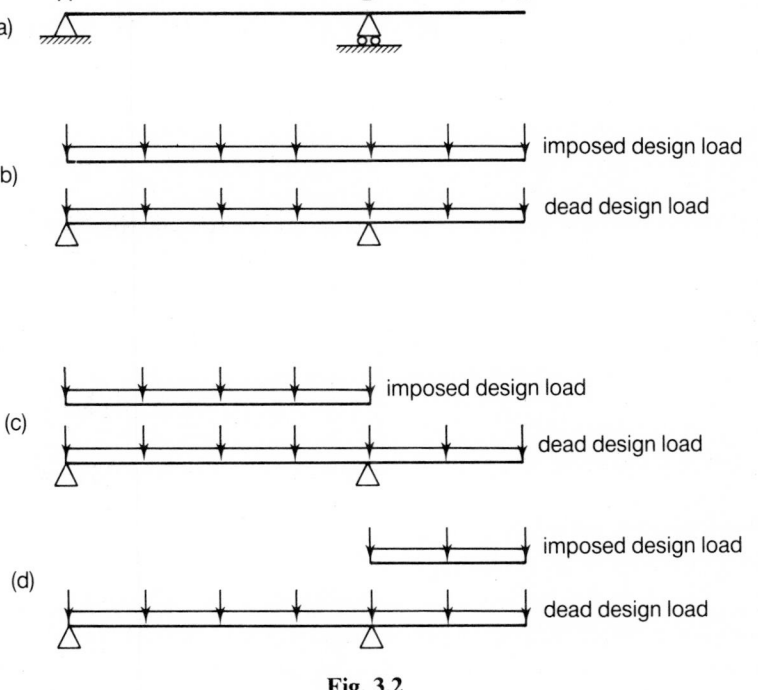

Fig. 3.2

supported by steel beams. Assume that dead load is equivalent to 150 mm of reinforced concrete, the beams are simply supported, and allow 1 kN/m² for permanent partitions. Determine (a) the maximum axial load on a column, (b) the maximum shear force for the longitudinal beams, (c) the maximum bending moment for the longitudinal beams.

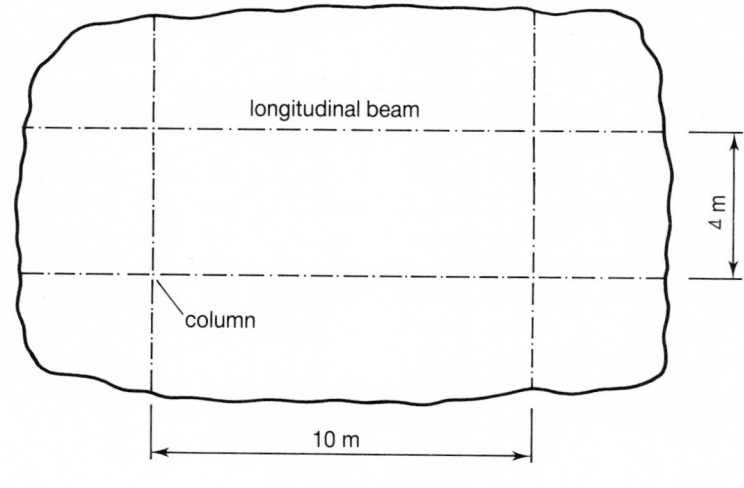

Fig. 3.3

Answers

(a)   445.4 kN
(b)   222.7 kN
(c)   556.75 kNm.

Problem 3.4.2

A water tower composed of steel plates and sections is shown in Fig. 3.4. Assume that the tank is built from 8 mm thick steel plate and that the dead load of the supporting steel work is 30 kN. The basic wind speed is 43 metres per second, the topography factor is unity, the ground roughness factor is 0.93 and the statistical probability factor is unity. To allow for wind pressure on the supporting steelwork assume a force factor of 0.5. Determine the critical reactions (N, H, and M) at the centre-line of the underside of the base for:

(a)   maximum vertical load
(b)   overturning and sliding
(c)   combined dead, imposed and wind loads.

Answers

(a) $N = 856.0$ kN, $H = 0$ kN, $M = 0$ kNm: (b) $N = 297.5$ kN, $H = 27.45$ kN, $M = 138.6$ kNm: (c) $N = 686.6$ kN, $H = 23.52$ kN, $M = 118.8$ kNm.

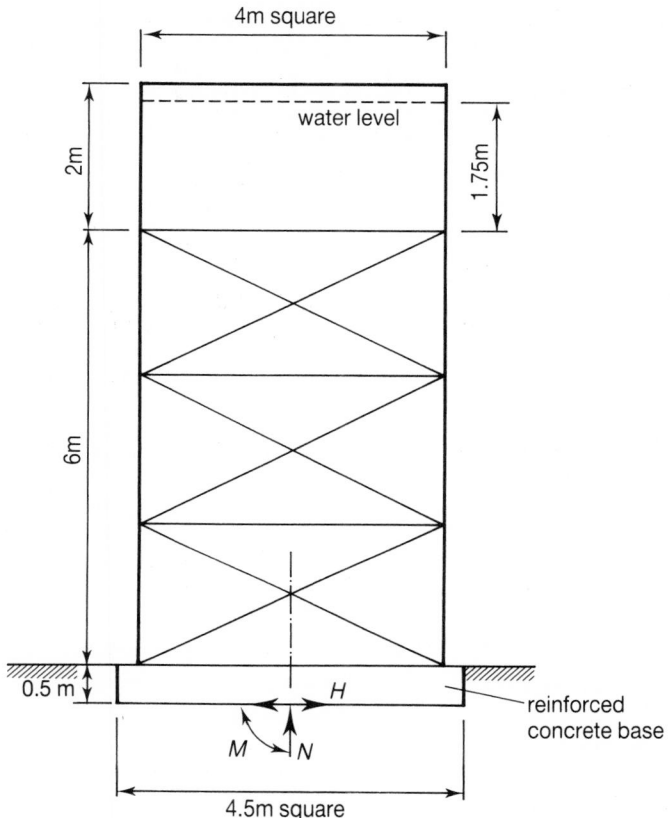

4m square

water level

2m

1.75m

6m

0.5 m

H

M *N*

reinforced
concrete base

4.5m square

Fig. 3.4

References

BS 648 *Schedule of weights of building materials*, British Standards Institution, London.

BS 2573 *Pt 1 Rules for the design of cranes – Specification for classification, stress calculations and design criteria for structures*, British Standards Institution, London.

BS 6399 *Pt 1 Design loads for buildings – Code of practice for dead and imposed loads*, British Standards Institution, London.

BS 6399 *Pt 3 Loadings for buildings – Code of practice for imposed roof loads*, British Standards Institution, London.

BS 8007 *Code of practice for the structural use of concrete for retaining aqueous liquids*, British Standards Institution, London.

CP 3 *Basic data for the design of buildings, Chapter V Loading*, British Standards Institution, London.

CP 2004 *Foundations*, British Standards Institution, London.

4

Laterally Restrained Beams

4.1 Introduction and structural classification of sections (cl 3.5.2, Pt 1)

Chapters 4 and 5 are concerned with the design of members that are predominantly in bending, i.e. where axial loads, if any, are small and transverse shear forces are not excessive. Chapter 4 contains basic theoretical work on section properties and the design of laterally restrained beams using compact standard sections and plastic methods of analysis. Chapter 5 introduces the theory for laterally unrestrained beams and applies this to standard sections, plate girders, castellated beams and tapered beams.

Sections of steel beams in common use are shown in Fig. 4.1. The rolled sections shown at (a) are used most often and of these the 'I' section is used widely. Some sections are of uniform thickness while others are of different thickness for the web and flange. The rolled sections are generally in stock, are lowest in cost, require less design and connections are straightforward. Hollow sections are not as efficient in bending but corrosion resistance is better and aesthetically they may be more acceptable. Cold formed sections are thinner and are therefore more susceptible to corrosion unless protected, however they are very economical for use as purlins. Fabricated sections are used when a suitable rolled section is not available, but costs are higher and delivery times are longer. Castellated sections are used for large spans with relatively low loads and where transverse shear forces are not excessive. Tapered beams are efficient in resisting bending moments but must be checked for shear forces. Composite steel–concrete sections are used for floors.

Cross sections of steel beams are classified as plastic, compact, semi-compact, or slender as described in Section 1.1.3. Classification is determined by calculating the width/thickness ratio of flange or web elements and comparing them with the limiting values given in Table 7, Pt 1. The design stresses in Table 6, Pt 1 for slender elements are reduced by a factor obtained from Table 8, Pt 1.

All members subjected to bending should be checked for the following at critical sections (cl 4.2.1.3, Pt 1):

(a) a combination of bending and shear force
(b) deflection
(c) lateral restraint
(d) local buckling
(e) web bearing and buckling.

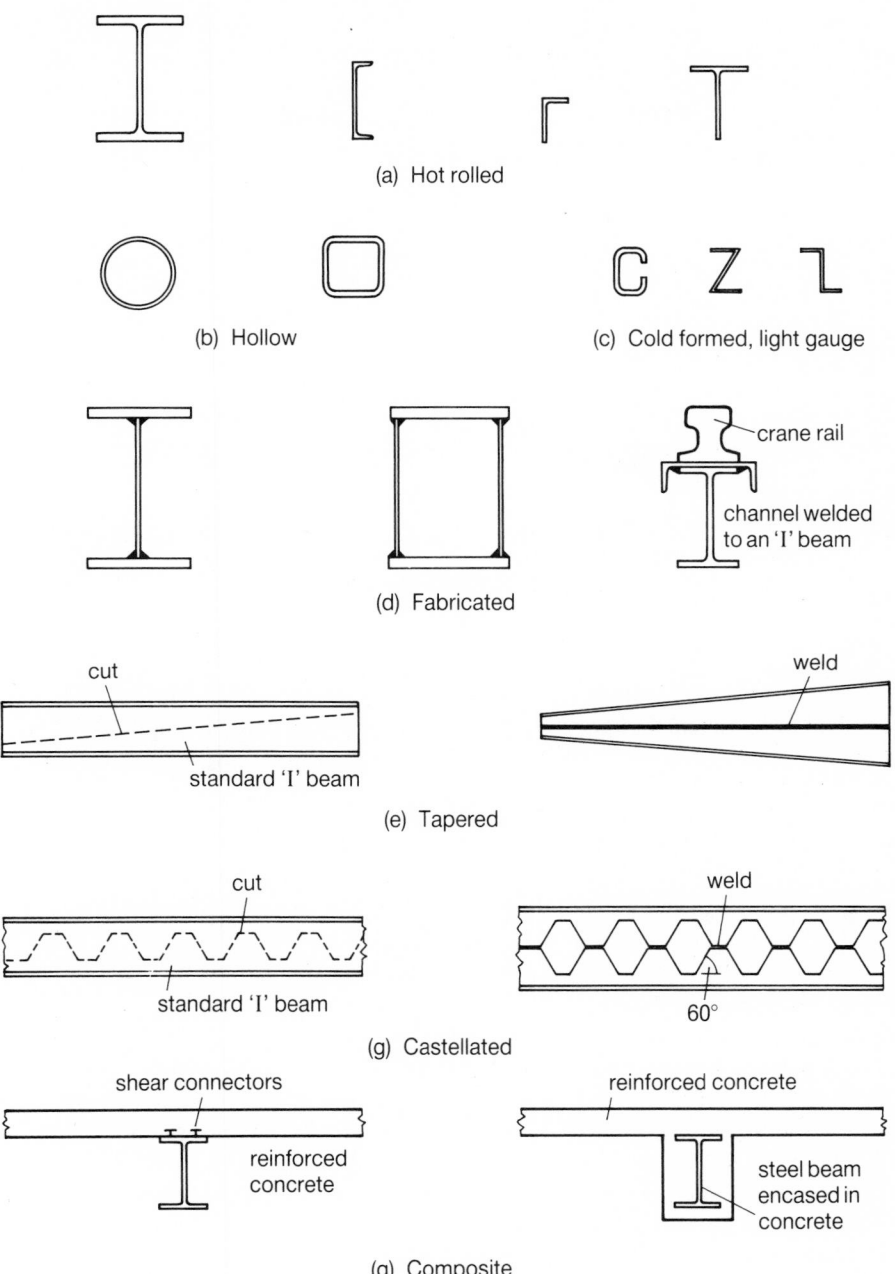

Fig. 4.1   Types of steel beams.

This chapter is concerned with members that are predominantly subjected to bending and where lateral torsional buckling and local buckling of the compression flange are prevented. It is important to recognize the characteristics of these two forms of buckling as shown in Fig. 4.2. Lateral torsional buckling exhibits vertical movement (bending about the x–x axis), lateral displacement (bending about the y–y axis), and torsional rotation (rotation about the z–z axis). Local buckling exhibits local deformation of an outstand, e.g. a flange of an 'I' beam.

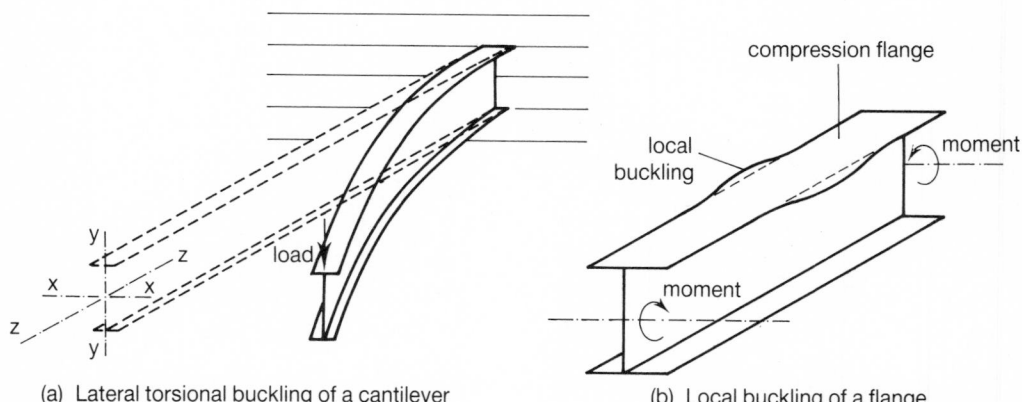

(a) Lateral torsional buckling of a cantilever   (b) Local buckling of a flange

Fig. 4.2   Buckling of beams.

Lateral torsional buckling occurs when the buckling resistance about the y–y axis and the torsional resistance about the z–z axis are low. The buckling resistance about the y–y axis can be improved by lateral restraints, e.g. transverse members which prevent lateral movement of the compression flange. Local buckling occurs when the flange outstand to thickness ratio (b/T) is high, and is avoided by choosing 'plastic' sections (see Section 1.1.3).

Where both types of buckling are prevented then the section can be stressed to the maximum design stresses in bending, i.e. yield, and plastic methods of analysis and design can be used (cls 2.1.2.3 and 5.3, Pt 1). The steel must comply with BS 4360 or have plastic properties as described in cl 5.3.3, Pt 1. If semi-compact or slender sections are used the plastic moment capacity is reduced to $M_c = p_y z$ (cl 4.2.5, Pt 1), where p_y is the design strength reduced by a factor obtained from Table 8, Pt 1.

If a steel 'I' section is used as a simply supported beam and loaded with a uniformly distributed load then the bending moment distribution varies parabolically. If the section is bent about a major axis then the stress distribution at centre span at various stages of loading is shown in Fig. 4.3(c). In the early stages of loading the stress distribution is elastic, then elastic–plastic and finally fully plastic. The corresponding moment curvature relationship is shown in Fig. 4.3(b).

The fully plastic stage corresponds to the fully plastic condition for the tensile stress strain relationship for the steel shown in Fig. 4.3(a). Theoretically the load cannot be increased beyond the fully plastic condition but strain hardening occurs and this increases the resistance. Note that for full plasticity large strains occur, of the order of 20%, which makes mild steel ideal, while other steels with less plastic strain behave in a more brittle fashion.

Although bending is the predominant design criterion checks must be made for the magnitude of the shear stresses. The shear stresses are introduced from vertical shear forces, or torsion forces if present. Where $d/t > 63\varepsilon$, e.g. a plate girder, the web should be checked for shear buckling (cl 4.2.5, Pt 1).

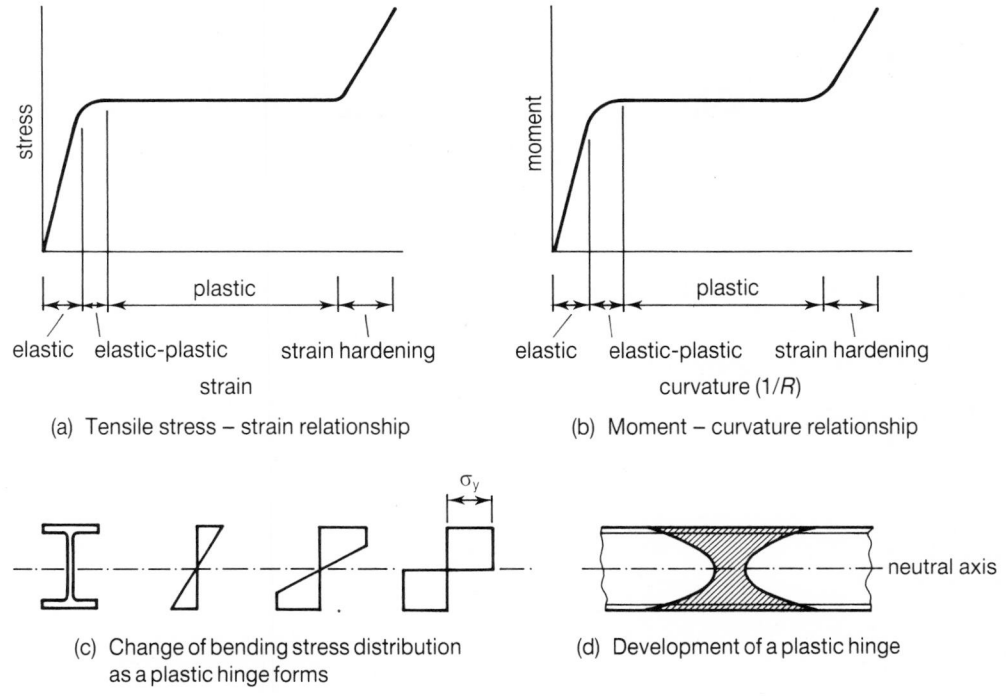

Fig. 4.3 Development of a plastic hinge.

In the design of beams calculations are required in the elastic stage of behaviour, e.g. stresses and deflections, and also at the fully plastic stage e.g. collapse load. The calculations involve certain basic section properties.

4.2 Elastic section properties and analysis in bending

4.2.1 Sectional axes and sign conventions

For all standard sections rectangular centroidal axes x–x and y–y are defined parallel to the main faces of the section, as shown in Fig. 4.4. The position of these axes is given in the Standard Section Tables. In angles, and other sections where the rectangular and principal axes do not coincide, the principal axes are denoted by u–u and v–v. The major axis u–u is conventionally inclined to the x–x axis by an angle α, as shown in Fig. 4.4(e) and (f). For equal angles, $\alpha = 45°$.

In problems involving simple uniaxial or biaxial bending of symmetrical sections a strict sign convention is not necessary, but for the solution of complex problems it is desirable. In this book the positive conventions of sagging curvature and downward deflections are adopted; and the direction of the angle α is anti-clockwise, consistent with the Standard Section Tables. Fig. 4.5(a) shows the coordinates of a point P in the positive quadrant of a section and the positive directions for the externally applied forces and couples. The positive directions of the corresponding stress resultants (shear forces and bending moments) are shown for the horizontal and vertical planes in Fig. 4.5(b). Positive directions

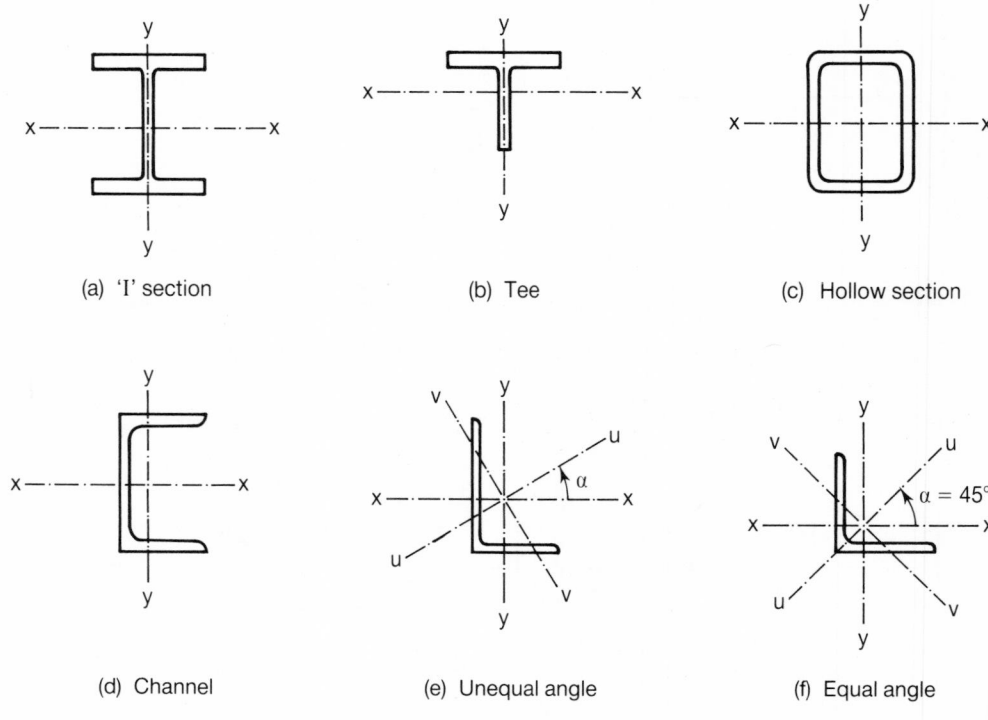

Fig. 4.4 Sectional axes.

relative to the u–u and v–v axes can be inferred. The convention for the moments has been chosen so that positive moments give tensile stresses in the positive quadrant of the section.

Normally the coordinates of points in a section relative to the rectangular axes are known, or can be easily obtained. The coordinates relative to the principal axes are given by

$$u = x \cos \alpha + y \sin \alpha$$
$$v = y \cos \alpha - x \sin \alpha$$

$$(4.1)$$

External forces and shear forces transform in exactly the same way, thus

$$W_u = W_x \cos \alpha + W_y \sin \alpha$$
$$W_v = W_y \cos \alpha - W_x \sin \alpha$$

$$(4.2)$$

However the directions chosen for the moments are consistent with the rules for a right hand set of axes, which gives rise to changes in sign, thus

$$M_u = M_x \cos \alpha - M_y \sin \alpha$$
$$M_v = M_y \cos \alpha + M_x \sin \alpha$$

$$(4.3)$$

4.2.2 Elastic second moments of area (I)

This property is derived from the simple theory of elastic bending (see Croxton and Martin Vol 1). In design it is used to calculate stresses and deflections in the elastic stage of

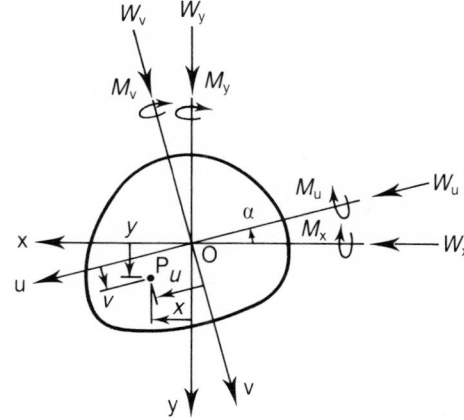

(a) Coordinates and external loads

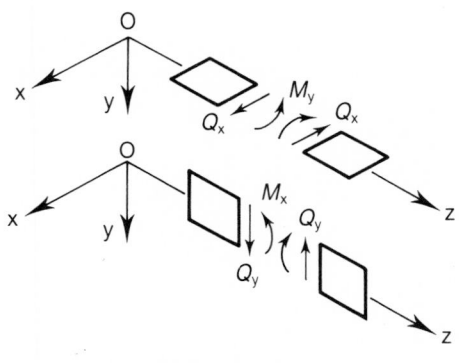

(b) Stress resultants

Fig. 4.5   Sign conventions.

behaviour, i.e. at service loads. Second moments of area for all standard sections are given in the Section Tables but for fabricated sections they must be calculated. The procedure involves application of the theorems of parallel axes which, for the single element of area A in Fig. 4.6, can be stated as follows

$$I_x = I_a + Ay^2$$
$$I_y = I_b + Ax^2 \qquad\qquad (4.4)$$
$$I_{xy} = I_{ab} + Axy$$

where

I_x = elemental second moment of area about x–x
I_y = elemental second moment of area about y–y
I_{xy} = elemental product moment of area about x–x and y–y
a–a and b–b are centroidal axes through the element, parallel to x–x and y–y respectively.

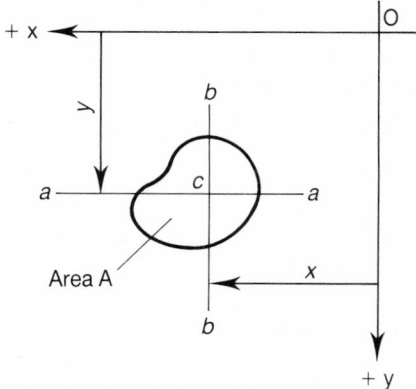

Fig. 4.6   Parallel axes for an element.

For the determination of I_{xy}, which can be either positive or negative, the correct signs must be allocated to the coordinates x and y. The positive directions are indicated by arrows in Fig. 4.6.

When second moments of area about the rectangular axes have been computed, the direction of the principal axes can be obtained from

$$\tan 2\alpha = 2I_{xy}/(I_y - I_x) \tag{4.5}$$

The principal second moments of area are then given by

$$\begin{aligned}
I_u &= I_x \cos^2 \alpha + I_y \sin^2 \alpha - I_{xy} \sin 2\alpha \\
I_v &= I_x \sin^2 \alpha + I_y \cos^2 \alpha + I_{xy} \sin 2\alpha
\end{aligned} \tag{4.6}$$

If I_x is arranged to be greater than I_y, then α will be less than 45° and I_u will be the major principal second moment of area. A negative result for Equation (4.5) indicates that α is to be measured clockwise from x–x.

4.2.3   Elastic section moduli

These values are derived from the second moments of area by dividing by the distance to the extreme fibres, i.e. $Z = I/y$. Values of section moduli are given in Standard Section Tables. For structural tees two values of Z are given, referring to the extreme fibres in the table and the stalk.

4.2.4   Elastic bending of symmetrical sections

When either of the rectangular axes is an axis of symmetry the normal bending stress at any point in the section is given by

$$f = M_x y/I_x + M_y x/I_y \tag{4.7}$$

If the directions of the bending moments and the coordinates are in accordance with the sign convention of Fig. 4.5, a positive result indicates that the stress is tensile.

For simple bending about the x–x axis, which is the most common event, the stress in the extreme fibres

$$f_{max} = M_x y_{max}/I_x = M_x/Z_x \tag{4.8}$$

Similarly for bending about the y–y axis

$$f_{max} = M_x y_{max}/I_x = M_x/Z_x \tag{4.9}$$

4.2.5   Elastic bending of unsymmetrical sections

When a section is subjected to bending about an axis which is not a principal axis the effect is the same as if the section were subject to the components of the bending moment acting about the principal axis. In other words the bending is biaxial.

For standard rolled angles the principal second moments of area and the directions of the principal axes are given in the Section Tables. Transforming bending moments and coordinates to the principal axes by means of Equations (4.3) and (4.1). Bending stress

$$f = M_u v/I_u + M_v u/I_v \tag{4.10}$$

This is the same as Equation (4.7) but with all the same terms related to the principal axes. If the sign convention of Fig. 4.5 is observed, a positive result indicates tension.

In other cases the additional calculations required for the solution of problems by principal axes can be avoided by the use of 'effective bending moments'. These are modified bending moments which can be considered to act about the rectangular axes of the section. The bending stress is then given by an expression having exactly the same form as Equation (4.7)

$$f = M_{ex} y/I_x + M_{ey} x/I_y \tag{4.11}$$

where M_{ex} and M_{ey} are effective bending moments about x–x and y–y axes respectively and are given by

$$\begin{aligned} M_{ex} &= (M_x - M_y I_{xy}/I_y)/[1 - I_{xy}^2/(I_x I_y)] \\ M_{ey} &= (M_y - M_x I_{xy}/I_x)/[1 - I_{xy}^2/(I_x I_y)] \end{aligned} \tag{4.12}$$

These expressions are derived from the application of conventional elastic bending theory to curvature in both the xz and yz planes. One such derivation is given by Megson.

By successive differentiation with respect to z, the longitudinal dimension, similar expressions for the effective shear force and effective load intensity can be obtained, thus

$$\begin{aligned} Q_{ex} &= (Q_x - Q_y I_{xy}/I_x)/[1 - I_{xy}^2/(I_x I_y)] \\ Q_{ey} &= (Q_y - Q_x I_{xy}/I_y)/[1 - I_{xy}^2/(I_x I_y)] \end{aligned} \tag{4.13}$$

and

$$\begin{aligned} w_{ex} &= (w_x - w_y I_{xy}/I_x)/[1 - I_{xy}^2/(I_x I_y)] \\ w_{ey} &= (w_y - w_x I_{xy}/I_y)/[1 - I_{xy}^2/(I_x I_y)] \end{aligned} \tag{4.14}$$

It should be noted that the quantities I_x and I_y are interchanged in Equations (4.13) and (4.14). This is because the expressions for the shear force and the load intensity in the x direction are obtained by successive differentiation of bending moments along the y–y axis and vice versa.

All bending moment problems with unsymmetrical sections can be solved simply by replacing ordinary loads, shears, and bending moments by their effective counterparts.

Note however that these effective counterparts have values related to both the x–x and y–y axes, even if the section is only loaded in the direction of one of the rectangular axes.

EXAMPLE 4.1 Principal axes for an unequal angle section

Find the directions of the principal axes and the values of the principal second moments of area for the angle section in Fig. 4.7(a).

For the calculation of section properties the work is simplified considerably, with insignificant loss of accuracy, by using the dimensions of the section profile, i.e. the shape formed by the centre line of the elements, as shown in Fig. 4.7(b).

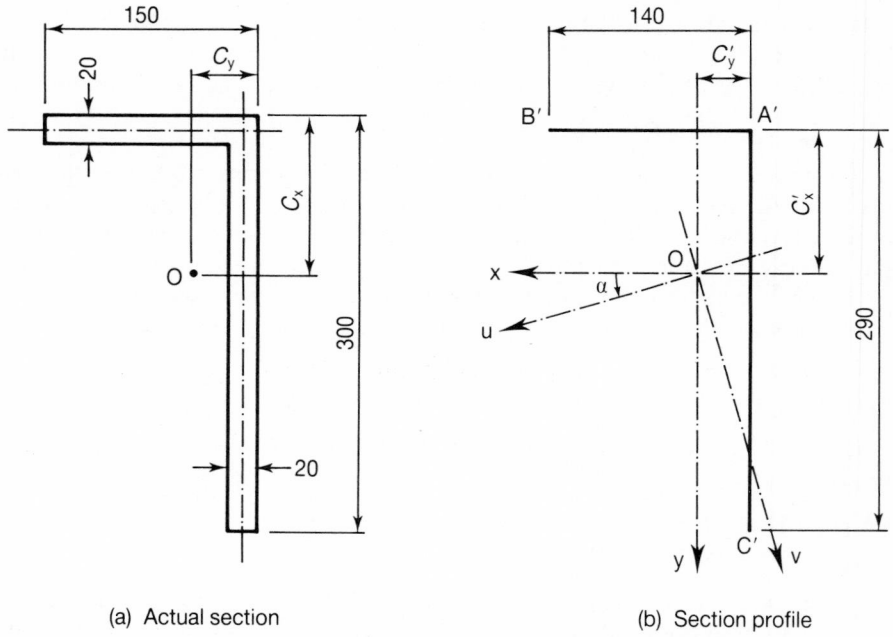

(a) Actual section   (b) Section profile

Fig. 4.7 Principal axes for an unequal angle.

The position of the centroid O is found by taking moments of the area about the centre lines of each leg in turn

Areas	mm^2
A′B′	$140 \times 20 = 2800$
A′C′	$290 \times 20 = 5800$
	$\overline{8600}$

Taking moments about A′B′
$$8600 C'_x = 5800 \times 290/2, \quad \text{hence} \quad C'_x = 97.8 \text{ mm}$$

Taking moments about A′C′
$$8600 C'_y = 2800 \times 140/2, \quad \text{hence} \quad C'_y = 22.8 \text{ mm}$$

Hence for the full section (see Fig. 4.7(a))

$C_x = 107.8$ mm and $C_y = 32.8$ mm

Coordinates of the centroids of the legs AB and AC are therefore given by

Leg A'B' $x = 140/2 - C'_y = 47.2$ mm

$y = -C'_x = -97.8$ mm

Leg A'C' $x = -C'_y = -22.8$ mm

$y = 290/2 - C'_x = 47.2$ mm

The second moments of area about the rectangular axes are obtained in the usual way by applying the parallel axes formula to each leg.

$I_x(\text{leg}) = bd^3/12 + A(\text{leg})y^2$

where b and d are dimensions of the leg parallel to the x–x and y–y axes respectively.

$$\text{mm}^4$$

Leg A'B' $140 \times 20^3/12 + 2800(-97.8)^2 = 26.87\text{E}6$

Leg A'C' $20 \times 290^3/12 + 5800 \times 47.2^2 = 53.57\text{E}6$

$$I_x = 80.44\text{E}6$$

$I_y(\text{leg}) = db^3/12 + A(\text{leg})x^2$

$$\text{mm}^4$$

Leg A'B' $20 \times 140^3/12 + 2800 \times 47.2^2 = 10.81\text{E}6$

Leg A'C' $290 \times 20^3/12 + 5800(-22.8)^2 = 3.21\text{E}6$

$$I_y = 14.02\text{E}6$$

The product moment of area I_{xy} is obtained by applying the parallel axis formula to each leg

$I_{xy}(\text{leg}) = I_{ab} + A(\text{leg})xy$

For each leg the term I_{ab} is equal to zero, because the parallel axes through the centroid of the leg are principal axes.

$$\text{mm}^4$$

Leg A'B' $2800 \times 42.7 \times (-97.8) = -12.93\text{E}6$

Leg A'C' $5800 \times (-22.8) \times 47.2 = -6.24\text{E}6$

$$I_{xy} = -19.17\text{E}6$$

Direction of the principal axes from Equation (4.5)

$\tan 2\alpha = 2I_{xy}/(I_y - I_x)$

$2\alpha = \arctan\left[2(-19.17)/(14.02 - 80.44)\right],$ hence $\alpha = 15°$

Principal second moments of area from Equations (4.6)

$I_u = I_x \cos^2 \alpha + I_y \sin^2 \alpha - I_{xy} \sin 2\alpha$

$I_v = I_x \sin^2 \alpha + I_y \cos^2 \alpha + I_{xy} \sin 2\alpha$

Substituting values

$$I_u = 85.58E6 \text{ mm}^4 \text{ and } I_v = 8.88E6 \text{ mm}^4.$$

As a check on the transformation

$$I_x + I_y = I_u + I_v \text{ which is correct.}$$

EXAMPLE 4.2   *Structural tee in biaxial bending*

Calculate the maximum extreme fibre stresses in a standard $292 \times 419 \times 113$ kg structural tee cut from a Universal beam. The tee is loaded by two moments as shown in Fig. 4.8.

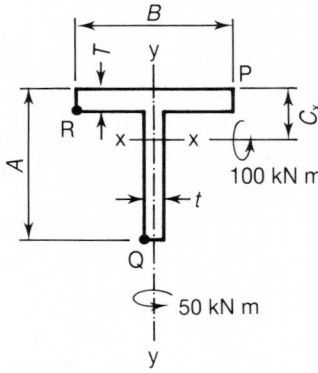

Fig. 4.8   Structural tee in biaxial bending.

From section tables $B = 293.8$ mm, $A = 425.5$ mm, $t = 16.1$ mm, $T = 26.8$ mm, $C_x = 108$ mm, $I_x = 246.6E6 \text{ mm}^4$, $I_y = 56.76E6 \text{ mm}^4$, $Z_x(\text{flange}) = 2.277E6 \text{ mm}^3$, $Z_x(\text{toe}) = 0.7776E6 \text{ mm}^3$, $Z_y = 0.3865E6 \text{ mm}^3$.

By inspection the maximum compressive stress occurs at a point P because the stresses from both moments are compressive.

$$f_P = M_x/Z_x(\text{flange}) + M_y/Z_y = 100E6/2.277E6 + 50E6/0.3865E6 = 173 \text{ N/mm}^2$$

The maximum tensile stress can occur at point Q or point R, depending on the relative magnitude of the bending moments. It is necessary to check both points.

Using the sign convention of Fig. 4.5 both bending moments are positive and the coordinates of the points are given by

Point Q, $x = t/2 = 16.1/2 = 8.05$ mm, $y = A - C_x = 425.5 - 108 = 317.5$ mm
Point R, $x = B/2 = 293.8/2 = 146.9$ mm, $y = T - C_x = 26.8 - 108 = -81.2$ mm.

From Equation (4.7) the stresses

$$f = M_x y/I_x + M_y x/I_y$$
$$f_Q = 100 \times 317.5/246.6 + 50 \times 8.05/56.76 = 135.8 \text{ N/mm}^2$$
$$f_R = 100(-81.2)/246.6 + 50 \times 146.9/56.76 = 96.5 \text{ N/mm}^2.$$

Summarizing, the maximum stresses are:

At Q, $f_Q = +135.8$ N/mm^2 (tension)

At P, $f_P = -173$ N/mm^2 (compression).

EXAMPLE 4.3 Bending stresses in an unequal angle section

Calculate the bending stresses in the angle section shown in Fig. 4.9 where $I_x = 80.44E6$, $I_y = 14.02E6$, $I_{xy} = -19.17E6$ mm^4.

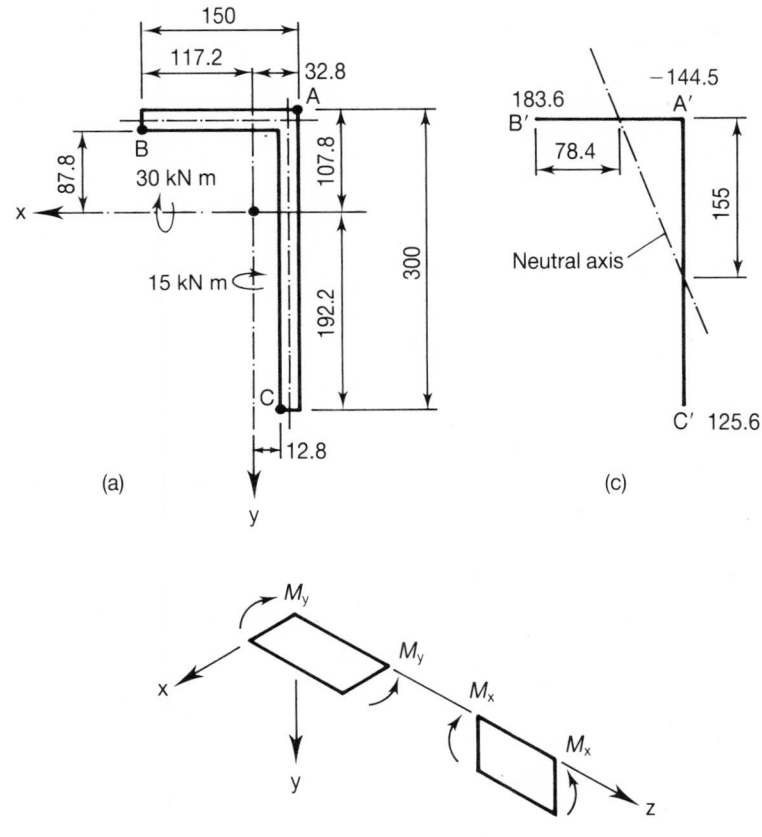

(b) positive bending moments

Fig. 4.9 Bending stresses in an unequal angle.

Effective moments from Equations (4.12)

$$M_{ex} = (M_x - M_y I_{xy}/I_y)/(1 - I_{xy}^2/I_x I_y) = 74.92 \text{ kNm}$$
$$M_{ey} = (M_y - M_x I_{xy}/I_x)/(1 - I_{xy}^2/I_x I_y) = 32.86 \text{ kNm}$$

Maximum compressive stress at A

$$f_A = M_{ex} y_A/I_x + M_{ey} x_A/I_y$$
$$= 74.92E6(-107.8)/80.44E6 + 32.86E6(-32.8)/14.02E6$$

$$= -177.3 \text{ N/mm}^2 \text{ (compression)}.$$

Check for maximum tensile stress at B, and at C

$$f_B = M_{ex}y_B/I_x + M_{ey}x_B/I_y$$
$$= 74.92\text{E}6(-87.8)/80.44\text{E}6 + 32.86\text{E}6 \times 117.2/14.02\text{E}6$$
$$= +192.9 \text{ N/mm}^2 \text{ (tension)}.$$
$$f_C = M_{ex}y_C/I_x + M_{ey}x_C/I_y$$
$$= 74.92\text{E}6 \times 192.2/80.44\text{E}6 + 32.86\text{E}6(-12.8)/14.02\text{E}6$$
$$= +149.0 \text{ N/mm}^2 \text{ (tension)}.$$

Summarizing, the maximum stresses are:

At B, $f_B = +192.9 \text{ N/mm}^2$ (tension)

At A, $f_A = -177.3 \text{ N/mm}^2$ (compression).

Note that although the position of the centroid and the values of the second moment of area can be calculated without significant error from the profile dimensions of the section, the same is not true of the stresses.

EXAMPLE 4.4   Bending about principal axes of an angle section

Recalculate the stresses at points A, B and C in the previous example considering bending about the principal axes.

The coordinates of the points are transformed in accordance with Equations (4.1), $u = x \cos \alpha + y \sin \alpha$ and $v = y \cos \alpha - x \sin \alpha$, $\sin \alpha = 0.2588$ and $\cos \alpha = 0.9659$.

Point	x/y axes		u/v axes	
	x	y	u	v
A	−32.8	−107.8	−59.6	−95.6
B	117.2	−87.8	90.5	−115.1
C	−12.8	192.2	37.5	189.0

Bending moments transform in accordance with Equations (4.3)

$$M_u = M_x \cos \alpha - M_y \sin \alpha = 25.10 \text{ kNm}, \quad M_v = M_y \cos \alpha + M_x \sin \alpha = 22.25 \text{ kNm}.$$

Bending stresses from Equation (4.10)

$$f = M_u v/I_u + M_v u/I_v, f_A = -177.3, f_B = 192.9, f_C = 149.0 \text{ N/mm}^2$$

which are the same as obtained previously.

4.2.6   Elastic analysis of beams

The elastic analysis of simply supported beams with examples of shear force and bending moment diagrams and deflection calculations are given in many introductory books on structural analysis, e.g. Croxton and Martin Vol. 1. The analysis of continuous beams is more complicated but there is a choice of methods such as area–moment, moment distribution, slope deflection and matrix methods. These methods are also covered by Croxton and Martin Vol. 2 with a computer program for analysis using matrix methods.

4.2.7 Elastic deflections of beams (cl 2.5.1, Pt 1)

The deflections under serviceability loads of a building or part should not impair the strength or efficiency of the structure or its components or cause damage to the finishes. When checking for deflections the most adverse realistic combination and arrangement of serviceability loads should be assumed, and the structure may be assumed to be elastic.

The theory and methods of calculating deflections for static and hyperstatic structures are given in Croxton and Martin Vols. 1 and 2. For simple beams standard cases can be superimposed and some useful cases are shown in Figs. 4.10 and 4.11.

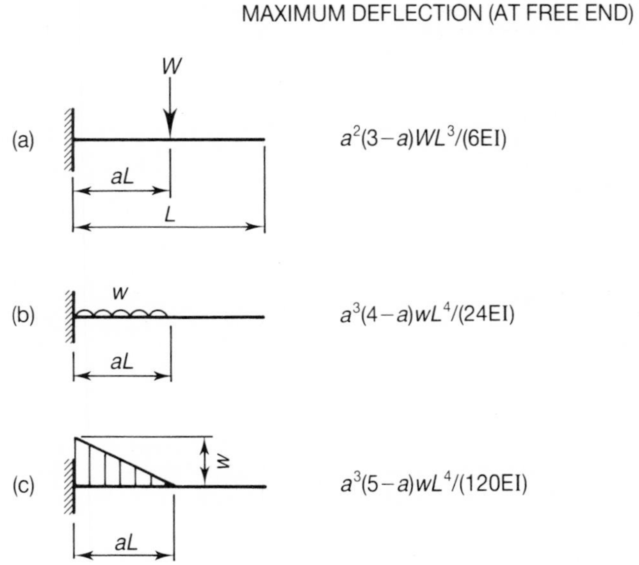

MAXIMUM DEFLECTION (AT FREE END)

(a) $a^2(3-a)WL^3/(6EI)$

(b) $a^3(4-a)wL^4/(24EI)$

(c) $a^3(5-a)wL^4/(120EI)$

Fig. 4.10 Deflections of cantilevers.

For simply supported beams the central deflection rather than the maximum is given, so that deflections from individual cases can be added. For most loading cases the central deflection only differs by a small percentage from the maximum. In case (a) of Fig. 4.11, for example, the difference is always within 2.5%. A notable exception is the case of equal end moments acting in the same direction, when the central deflection is zero. However in such a case the deflection at other points along the beam is likely to be small. A more accurate analysis can be formed if it is suspected that the deflection is likely to exceed the limit.

EXAMPLE 4.5   Deflections for a hyperstatic structure

The sizes of the members for the symmetrical structure shown in Fig. 4.12 have been determined and the structure requires to be checked for deflections in the elastic range of behaviour. The imposed characteristic loads are shown on Fig. 4.12 and the second moment of area is $I_x = 127.56E6$ mm^4. The bending moments (positive clockwise) at the joints are given in the following table.

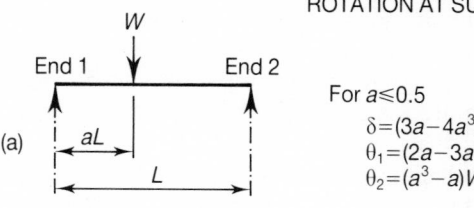

CENTRAL DEFLECTION, AND
ROTATION AT SUPPORTS

(a)

For $a \leq 0.5$

$$\delta = (3a - 4a^3)WL^3/(48EI)$$
$$\theta_1 = (2a - 3a^2 + a^3)WL^2/(6EI)$$
$$\theta_2 = (a^3 - a)WL^2/(6EI)$$

(b)

$$\delta = 5wL^4/(384EI)$$
$$\theta_1 = wL^3/(24EI)$$
$$\theta_2 = -\theta_1$$

(c)

For $a \leq 0.5$

$$\delta = (3a^2 - 2a^4)wL^4/(96EI)$$
$$\theta_1 = (a^4 - 4a^3 + 4a^2)wL^3/(24EI)$$
$$\theta_2 = (a^4 - 2a^2)wL^3/(24EI)$$

(d)

$$\delta = wL^4/(120EI)$$
$$\theta_1 = 5wL^3/(192EI)$$
$$\theta_2 = -\theta_1$$

(e)

$$\delta = (M_1 - M_2)L^2/(16EI)$$
$$\theta_1 = (2M_1 - M_2)L/(6EI)$$
$$\theta_2 = (2M_2 - M_1)L/(6EI)$$

Fig. 4.11 Displacements of simply supported beams.

Joint	Span	Moment (kNm)
B	AB	+60
B	BC	−60
C	BC	−6
C	CG	+72
C	CD	−66
G	CG	+36

For all beams

$$E_s I = 205E3 \times 127.56E6 = 26.15E12 \text{ N mm}^2.$$

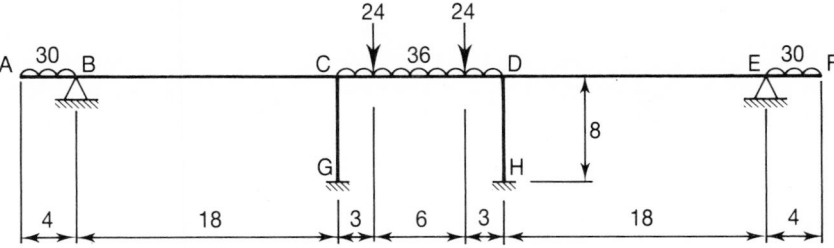

Dimensions in m; loads in kN.

Fig. 4.12   Symmetrical continuous structure.

Span CD

1. Uniform load (see Fig. 4.11(b))

$$\delta_1 = 5wL^4/(384E_s I) = 5WL^3/(384E_s I)$$
$$= 5 \times 36E3 \times 12E3^3/(384 \times 26.15E12) = 30.9 \text{ mm}.$$

2. Concentrated loads (see Fig. 4.11(a))

$$\delta_2 = 2[3a - 4a^3]WL^3/(48E_s I)$$
$$= 2[3 \times (3/12) - 4 \times (3/12)^3] \times 24E3 \times 12E3^3/(48 \times 26.15E12) = 45.4 \text{ mm}$$

3. End moments (see Fig. 4.11(e))

$$\delta_3 = L^2(M_1 - M_2)/(16E_s I)$$
$$= 12E3^2 \times (-66 - 66)E6/(16 \times 26.15E12) = -45.4 \text{ mm (upwards)}$$

Total deflection $= \delta_1 + \delta_2 + \delta_3 = 30.9 + 45.4 + (-45.4) = 30.9$ mm.

Limit for beams with plaster finish, Table 5, Pt 1

$$= L/360 = 12E3/360 = 33.3 \text{ mm}.$$

Span BC

End moments (see Fig. 4.11(e))

$$\delta_3 = L^2(M_1 - M_2)/(16E_s I)$$
$$= 18E3^2 \times (-60 + 6)E6/(16 \times 26.15E12) = -41.8 \text{ mm (upwards)}$$

An accurate analysis gives -43.4 mm at 7.35 mm from B.

Limit for beams with plaster finish, Table 5, Pt 1

$$= L/360 = 18E3/360 = 50 \text{ mm}.$$

Cantilever span AB

For this span the deflection is due to the flexure of the cantilever, assuming the beam is horizontal at B, plus the effect of the anti-clockwise rotation of the beam at B, i.e. $-\theta_1$ for span BC.

End moments for span BC (see Fig. 4.11(e))

$$\theta_1 = L(2M_1 - M_2)/(6E_s I)$$
$$= 18\text{E}3 \times (-2 \times 60 + 6)\text{E}6/(6 \times 26.15\text{E}12) = -0.01308 \text{ radians.}$$

1. Deflection at A due to rotation

$$\delta_1 = -L\theta_1 = -4\text{E}3 \times (-0.01308) = 52.32 \text{ mm}$$

2. Deflection due to load (see Fig. 4.10(b))

$$\delta_2 = a^3(4-a)wL^4/(24E_s I) = WL^3/(8EI)$$
$$= 30\text{E}3 \times 4\text{E}3^3/(8 \times 26.15\text{E}12) = 9.18 \text{ mm}$$

Total deflection $= \delta_1 + \delta_2 = 52.3 + 9.2 = 61.5$ mm.

Limit for a cantilever beam with plaster finish, Table 5, Pt 1

$$= L/180 = 4\text{E}3/180 = 22.2 \text{ mm.}$$

The deflection of the cantilever exceeds the limit and stiffening is required, e.g. increase the size of the section, or add flange plates. It is also necessary to stiffen spans AB and BC because the deflection is dependent on both.

4.2.8   Span/depth ratios for simply supported beams

An initial estimate for the depth of a simply supported beam carrying a uniformly disributed load can be obtained by using the deflection limit. If f_{max} is the maximum elastic bending stress at service load then from elastic bending theory

$$f_{max} = My/I = (WL/8) \times (D/2)/I$$

rearranging

$$W = 16f_{max}I/(LD) \tag{i}$$

Limit for beams with plaster finish, Table 5, Pt 1

$$L/360 = 5WL^3/(384E_s I) \tag{ii}$$

Eliminating W by combining Equations (i) and (ii) and putting $E_s = 205\text{E}3 \text{ N/mm}^2$, the span depth ratio

$$L/D = 2733/f_{max}$$

If a beam is laterally restrained so that lateral torsional buckling does not occur, then the maximum bending stress at service load is $p_y/1.6$, and the span/depth ratios for different grades of steel and different limits are

Grade	Design stress	Span/depth ratio	
		$L/360$	$L/200$
43	275	15.9	28.6
50	355	12.3	22.2
55	450	9.71	17.5

Note that since Young's modulus (E_s) is a constant for all grades of steel, the stiffness of a beam does not increase with a higher grade of steel. If a design is governed by deflection then there is no advantage in using a higher grade of steel.

4.3 Elastic shear stresses

4.3.1 Elastic shear stress distribution in a symmetrical section

When a beam is bent elastically by a system of transverse loads, plane sections no longer remain plane after bending, but are warped by shear strains. In most cases the effect is small and the errors introduced in the use of conventional bending theory are negligible. Important exceptions are discussed briefly in Section 4.3.2. Formulae for the calculation of shear stresses in an elastic beam are derived by considering the variation in bending stresses along a short length of beam.

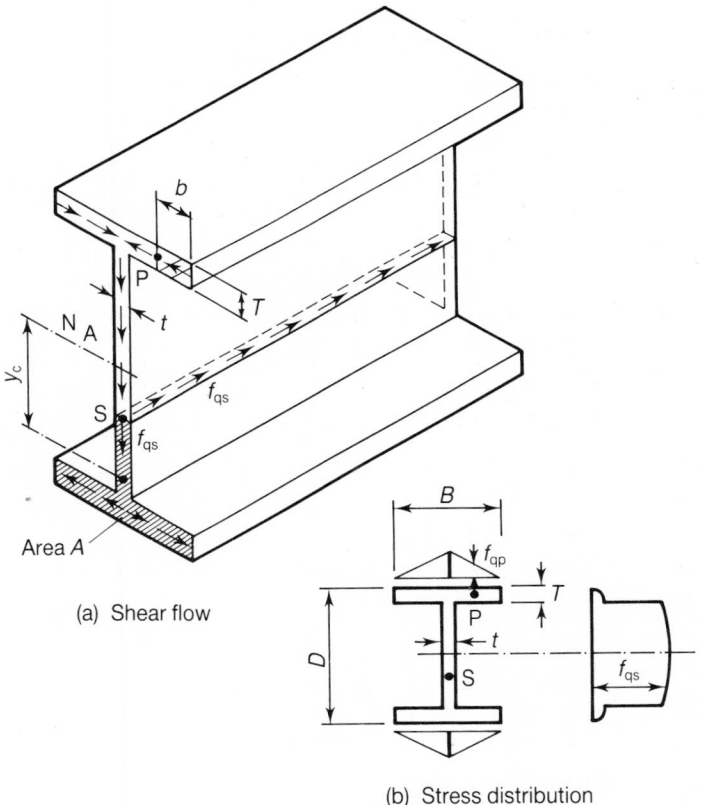

(a) Shear flow

(b) Stress distribution

Fig. 4.13 Shear stresses in an 'I' beam.

Consider the very short length of beam in Fig. 4.13(a). At a point S in the web the shear stresses on the vertical and the horizontal section are complementary and are given by the established formula

$$f_{qs} = QAy_c/(It) \qquad (4.15)$$

where

Q = the vertical shear force on the section

A = the hatched area, i.e. the part of the section between point S and the extreme fibres

y_c = the distance from the centroid of the area A to the neutral axis

I = the second moment of area of the whole section about the neutral axis

t = the thickness of the section at the point S

The formula cannot be used to obtain the vertical shear stress in the outstanding parts of the flange. However as this must be equal to zero at the top and bottom faces, it must be very small. In fact the resistance of the section to vertical shear is provided almost entirely by the web.

The resultant of the longitudinal shear stress in the web is in equilibrium with the change in the normal tensile force on the area A due to the variation in bending moment along the beam. Similar longitudinal stresses exist in the flanges and give rise to horizontal complementary stresses in the directions shown. For example, at point P in the top flange:

$A = BT$, $t = T$, $y_c = (D - T)/2$, and Equation (4.15) becomes

$$f_{qp} = Qb(D - T)/(2I) \tag{4.16}$$

This expression is linear with respect to the variable b, and f_{qp} has a maximum value at the centre of the flange where $b = B/2$, i.e.

$$f_{qp(max)} = QB(D - T)/(4I) \tag{4.17}$$

The complete distribution of shear stress on the cross section is shown in Fig. 4.13(b).

Equation (4.15) can be expressed in terms of the shear flow, which is the product of the shear stress and the thickness of the section, thus

$$q_s = tf_{qs} = QAy_c/I \tag{4.18}$$

In the longitudinal sense the shear flow is equal to the shear force per unit length of beam, and is a convenient quantity for the calculation of the shear force to be resisted by bolts or welds in a fabricated section.

4.3.2   Elastic shear stresses in thin walled open sections

The shear stress distribution for a rectangular cross section subjected to a transverse shear force is shown in Fig. 4.14(a) and can be calculated using Equation (4.15). The shear stress distribution for the open cross section is different as shown in Fig. 4.14(b). The distribution for an 'I' section is shown in Fig. 4.14(c). Steel sections are usually composed of relatively thin elements, for which the analysis can be simplified by:

(a)   referring all dimensions to the profile of the section;

(b)   assuming that the shear stress does not vary across the thickness;

(c)   ignoring any shear stresses acting at right angles to the section profile. As these are equal to zero at each outside surface they must always be very small in a thin walled section.

It it is further assumed that the load is applied in such a way that no twisting of the beam occurs, the shear flow at a point S on the profile of the section is given by

$$q_s = q_o - (Q_{ey}/I_x) \int_o^s ty:ds - (Q_{ex}/I_y) \int_o^s tx:ds \tag{4.19}$$

where Q_{ex} and Q_{ey} are the effective shear forces obtained from Equation (4.13) or by applying the effective loads obtained from Equation (4.14). The variable s is the distance

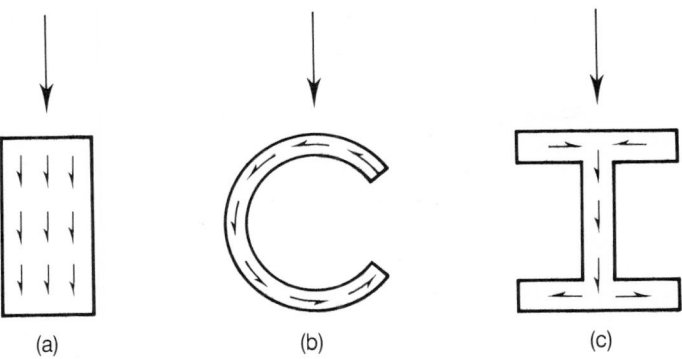

Fig. 4.14 Distribution of shear stresses.

around the profile to the point of interest, starting from any point at which the shear flow q_o is known. At any open end, such as the end of a flange, the value of q_o is zero. The direction of s can be chosen arbitrarily and, provided that the sign convention of Fig. 4.5 is adopted, a positive sign for q_s indicates that the shear flow is in the direction chosen for s.

4.3.3 Elastic shear stresses in thin walled closed sections

The shear stress and shear flow in a symmetrical closed section can be obtained directly from Equations (4.15) and (4.18) respectively. For an unsymmetrical section Equation (4.19) can be used, but analysis is complicated by the fact that q_o is not known at any point. The problem can be solved by first cutting the section at some point and finding the position of the shear centre in the resulting open section. The shear flow in the closed section then results from the combined action of the applied shear loads transferred to the shear centre of the cut (open) section, and the torque on the closed section due to the transference of loads. Examples of shear stress distribution are given in Fig. 4.14. A similar approach is used to find the position of the shear centre of the closed section.

4.3.4 Elastic shear lag

The simple theory of bending is based on the assumption that plane sections remain plane after bending. In reality shear strains cause the section to warp. The effect in the flanges is to modify the bending stresses obtained by the simple theory, producing higher stresses near the junction of a web and lower stresses at points remote from it as shown in Fig. 4.15. This effect is described as 'shear lag'. The discrepancies produced by shear lag are minimal in rolled sections, which have relatively narrow and thick flanges. However in plate girders, or box sections, having wide thin flanges the effects can be significant when they are subjected to high shear forces, especially in the vicinity of concentrated loads where the sudden change in shear force produces highly incompatible warping distortions.

4.3.5 Elastic shear centre

Equation (4.19) is only valid if no twisting of the beam occurs at the section considered. Torsion in a section can be generated by a transverse load if the resultants of the shear

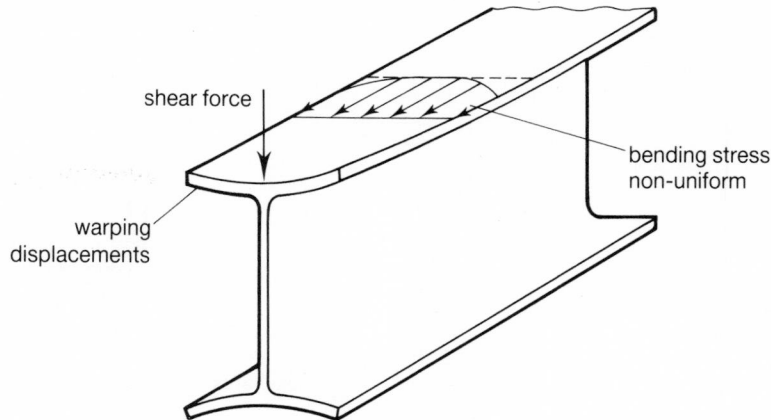

Fig. 4.15   Shear lag effects for an 'I' section.

stresses in the elements of the section produce a torque. To counteract this the line of action of the applied load must pass through the shear centre. In a symmetrical section the shear centre lies on an axis of symmetry, and loads applied along such an axis do not cause twisting. In some sections the position of the shear centre can be inferred from the direction of the shear flow (see examples Fig. 4.16). In Fig. 4.16(a) the shear centre lies at the intersection of the two axes of symmetry and is coincident with the centroid; in (b) and (c) it lies at the intersection of lines of shear flow; in (d), if the flanges are of the same size, the shear stresses in them set up opposing torques about the centroid which is therefore the shear centre.

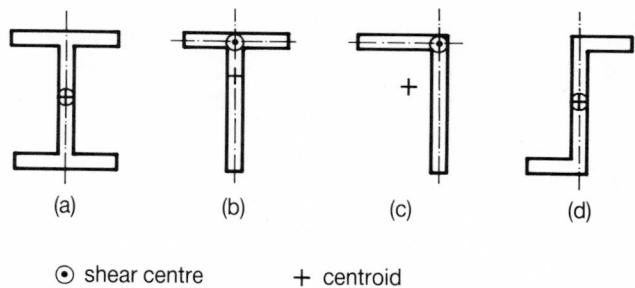

Fig. 4.16   Position of shear centre.

For the general case of an unsymmetrical thin walled open section subject to biaxial bending, with shear forces Q_x and Q_y, the position of the shear centre can be found by determination of the shear flow from Equation (4.18) or (4.19), applying Q_x and Q_y in turn, assuming that they pass through the shear centre. Consider, for example, the section profile in Fig. 4.17. If point B is chosen as the fulcrum it is only necessary to find the resultant shear forces in the leg CD due to Q_x and Q_y in turn. These forces produce torques equal to $Q_x y'$ and $Q_y x'$ respectively. By taking moments about B the values of y' and x' can be obtained. There is no need to calculate the shear stresses in AB or BC because their lines of action pass through the point B, and generate no moment. The resultant shear forces

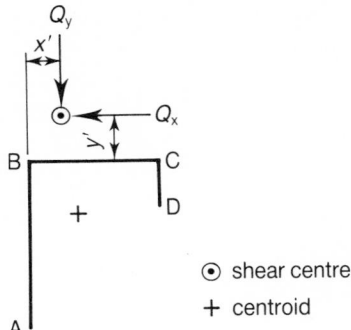

Fig. 4.17 Shear centre–unsymmetrical section.

in CD are obtained by integrating the shear stresses obtained by Equation (4.19) along the leg.

The above process is tedious since, for each value of Q_x and Q_y, the corresponding effective shear forces Q_{ex} and Q_{ey} must be calculated and applied. If there is an axis of symmetry Equation (4.18) can be used and the analysis is simplified.

EXAMPLE 4.6 Distribution of shear stresses for an angle section

Calculate the shear stresses in the simply supported angle section shown in Fig. 4.18(a). $I_x = 80.44E6$, $I_y = 14.02E6$, $I_{xy} = -19.17E6$ mm^4.

To the left of mid-span the shear forces are

$$Q_x = 5 \text{ and } Q_y = 10 \text{ kN}$$

Effective shear forces from Equation (4.13)

$$Q_{ex} = (Q_x - Q_y I_{xy}/I_x)/[1 - I_{xy}^2/(I_x I_y)] = 10.95 \text{ kN}$$
$$Q_{ey} = (Q_y - Q_x I_{xy}/I_y)/[1 - I_{xy}^2/(I_x I_y)] = 24.97 \text{ kN}$$

Shear flow from Equation (4.19)

$$q_s = q_o - (Q_{ey}/I_x) \int_o^s ty:ds - (Q_{ex}I_y) \int_o^s tx:ds \tag{i}$$

For the horizontal leg starting from the left hand end

$$q_o = 0, \quad s = s_1, \quad x = (117.2 - s_1), \quad \text{and} \quad y = -97.8 \text{ mm}$$

Substituting these values in (i) and integrating

$$q_{s1} = 0.00781 s_1^2 - 1.224 s_1 \tag{ii}$$

This equation shows that $q_{s1} = 0$ only when $s_1 = 0$. Differentiating with respect to s_1 and equating to zero gives a turning point at $s_1 = 78.4$ mm. Hence from (ii) $q_{s1(max)} = -47.96$ N/mm, and $s_1 = 140$ mm, $q_{s1} = -18.28$ N/mm. The negative signs indicate that the shear flow is in the opposite direction to s_1.

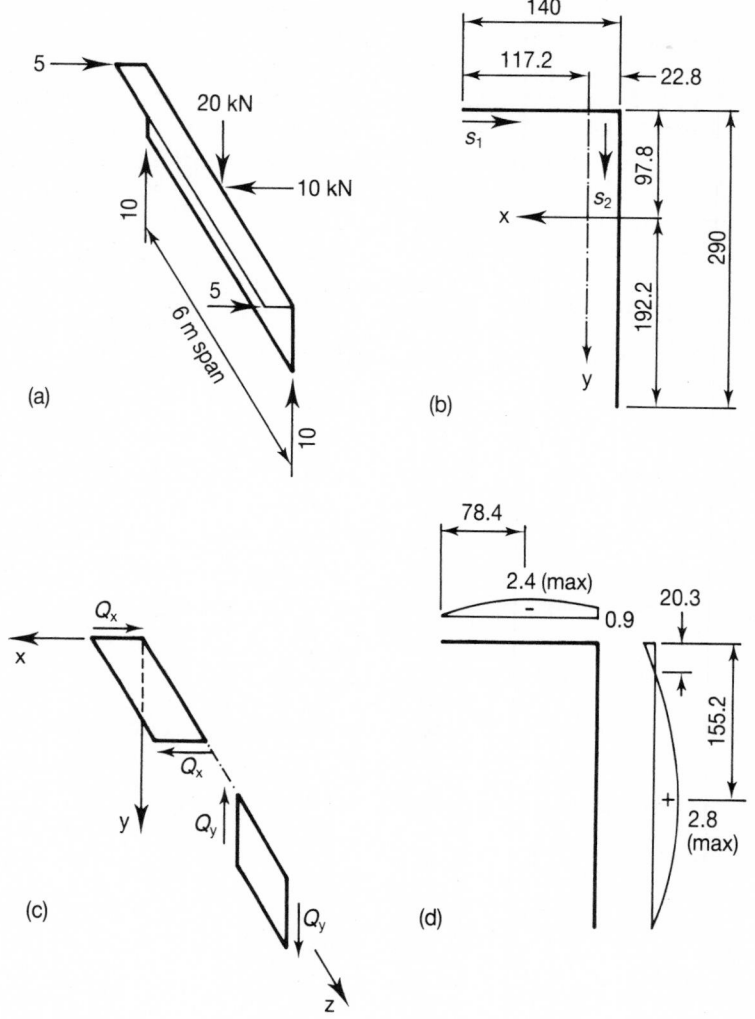

Fig. 4.18 Distribution of shear stresses for an unsymmetrical section.

For the vertical leg

$$s = s_2, \quad x = -22.8 \text{ mm}, \quad y = (s_2 - 97.8) \text{ mm}, \quad q_o = q_A = -18.28 \text{ N/mm}$$

Substitution of these values into (i) gives

$$q_{s2} = -18.28 + 0.9633s_2 - 0.003104s_2^2 \tag{iii}$$

Solving (iii) for s_2 shows that when $q_{s2} = 0$, $s_2 = 20.3$ mm. There is also a turning point at $s_2 = 155.2$ mm. Hence from (iii) $q_{s2(max)} = 56.46$ N/mm. The positive sign indicates that the shear flow is in the direction of s_2. As a check, putting $s_2 = 290$ mm gives $q_{s2} = 0$, which is correct.

The shear stresses in N/mm^2 are obtained by dividing the shear flows by the thickness, i.e. 20 mm, and are plotted for the whole section in Fig. 4.18(d).

This example demonstrates the method of analysis but the shear stresses are very low and do not justify such a detailed treatment. As a rough guide as to whether detailed analysis is required the shear forces are divided by the area of the appropriate leg.

If the shear stress is only required at particular points in the section Equation (4.15) can be used with the effective shear forces, taking each axis in turn and superimposing the results. Integration is avoided but the directions of the stresses have to be found by inspection. It can be seen from the position of the neutral axis in Example 4.3 that the maximum shear stresses occur where the neutral axis intersects the profile of the section, as in a symmetrical section. If these points have previously been found, then the maximum shear stress can be calculated directly as above, using Equation (4.15).

EXAMPLE 4.7   Shear centre for a channel section

Find the position for the shear centre for the channel shown in Fig. 4.19 which has a uniform thickness *T.*

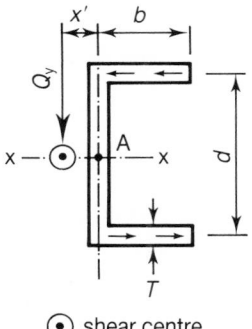

● shear centre

Fig. 4.19   Shear centre of a channel.

As the shear centre lies on the axis of symmetry x–x, there is no need to consider Q_x. If point A at the intersection of axis x–x with the centre line of the web is taken as the fulcrum, then only the shear force in the flanges need be considered since the resultant shear force in the web produces no moment about A. Equation (4.18) is

$$q_s = QAy_c/I$$

The distribution of shear flow in the flanges is linear with zero at the ends and maximum at the web centre-line. Hence, for maximum shear flow $A = bT$, $Q = Q_y$, $y_c = d/2$, and $I = I_x$, which gives

$$q_{s(max)} = Q_y bTd/(2I_x)$$

The resultant shear force is equal to half the maximum shear flow multiplied by the flange width, i.e.

$$\text{Force} = Q_y b^2 Td/(4I_x)$$

The torque about point A from both flanges is equivalent to the torque produced by the applied shear force when it passes through the shear centre, thus

$$Q_y x' = Q_y b^2 T d^2 / (4I_x)$$

from which

$$x' = b^2 d^2 T / (4I_x) \qquad (4.20)$$

4.4   Plastic section properties and analysis

4.4.1   Plastic section modulus

Plastic methods may be used in the design of structures provided that cls. 5.3.1 to 5.3.7, Pt 1 are satisfied. These clauses cover types of loading, grades of steel, geometric properties, restraints, stiffeners and fabrication restrictions. For the general case of a steel section symmetrical about the plane of bending, the stress distributions in the elastic and fully plastic state are shown in Fig. 4.3(c). For equilibrium of normal forces, the tensile and compressive forces must be equal. In the elastic state, when the bending stress varies from zero at the neutral axis (NA) to a maximum at the extreme fibres, this condition is achieved when the neutral axis passes through the centroid of the section. In the fully plastic state, because the stress is uniformly equal to the yield stress, equilibrium is obtained when the neutral axis divides the section into two equal areas.

$$M_p = (\text{first moment of area about the plastic NA}) Y_s \qquad (4.21)$$

EXAMPLE 4.8   Section moduli for an 'I' section

Determine the plastic section moduli about the x–x and y–y axes for the 'I' section shown in Fig. 4.20(a). The section is for a 914 × 419 × 388 kg Universal Beam with the root radius omitted.

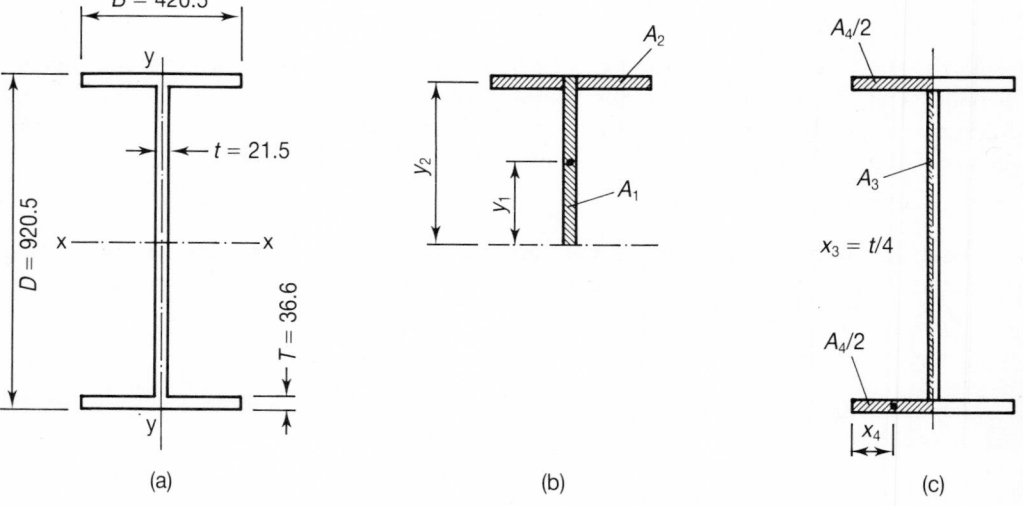

(a)   (b)   (c)

Fig. 4.20   Plastic section modulus for an 'I' section.

To determine the plastic section modulus about the x–x axis divide the section into A_1 and A_2 as shown in Fig. 4.20(b) where

$$A_1 = (D/2)t = (920.5/2)21.5 = 9895.375 \text{ mm}^2$$
$$A_2 = (B - t)T = (420.5 - 21.5)36.6 = 14603.5 \text{ mm}^2$$

and

$$y_1 = D/4 = 920.5/4 = 230.125 \text{ mm}$$
$$y_2 = D/2 - T/2 = (920.5 - 36.6)/2 = 441.95 \text{ mm}$$

Plastic section modulus

$$\begin{aligned} S_x &= 2(A_1 y_1 + A_2 y_2) \\ &= 2(9895.375 \times 230.125 + 14603.4 \times 441.95) \\ &= 17.462\text{E6 mm}^3. \end{aligned}$$

The value obtained from the Section Tables is 17.657E6 mm³ which is slightly greater because of the additional material at the root radius.

Similarly for the plastic section modulus about the y–y axis divide the section into areas A_3 and A_4 as shown in Fig. 4.20(c) where

$$A_3 = (D - 2T)t/2 = (920.5 - 2 \times 36.6)21.5/2 = 9108.475 \text{ mm}^2$$
$$A_4 = 2(B/2)T = 2(420.5/2)36.6 = 15390.3 \text{ mm}^2$$

and

$$x_3 = t/4 = 21.5/4 = 5.375 \text{ mm}$$
$$x_4 = B/4 = 420.5/4 = 105.125 \text{ mm}$$

Plastic section modulus

$$\begin{aligned} S_y &= 2(A_3 x_3 + A_4 x_4) \\ &= 2(9108.475 \times 5.375 + 15390.3 \times 105.125) \\ &= 3.334\text{E6 mm}^3. \end{aligned}$$

The value obtained from the Section Tables is 3.339E6 mm³ which is slightly greater because of the additional material at the root radius.

From the Section Tables the ratio of plastic/elastic section modulus (shape factor) for this section about the x–x axis is

$$S_x/Z_x = 17.657\text{E6}/15.616\text{E6} = 1.1307.$$

The value of the shape factor for bending about the x–x axis generally quoted for 'I' sections in current use in design is 1.15. This value together with the stress diagrams shown in Fig. 4.4(c) explain the use of the limit for moment capacity of $M_o < 1.2p_y Z$ which prevents plasticity at service load (cl 4.2.5, Pt 1).

The corresponding value of the shape factor for bending about the y–y axis is $S_y/Z_y = 3.339\text{E6}/2.160\text{E6} = 1.55$, which is typical for an 'I' section.

EXAMPLE 4.9   Plastic section moduli for a channel section

Determine the plastic section moduli about the x–x and y–y axes for the channel section shown in Fig. 4.21(a). The section is for a $432 \times 102 \times 65.54$ kg channel with the root radius omitted.

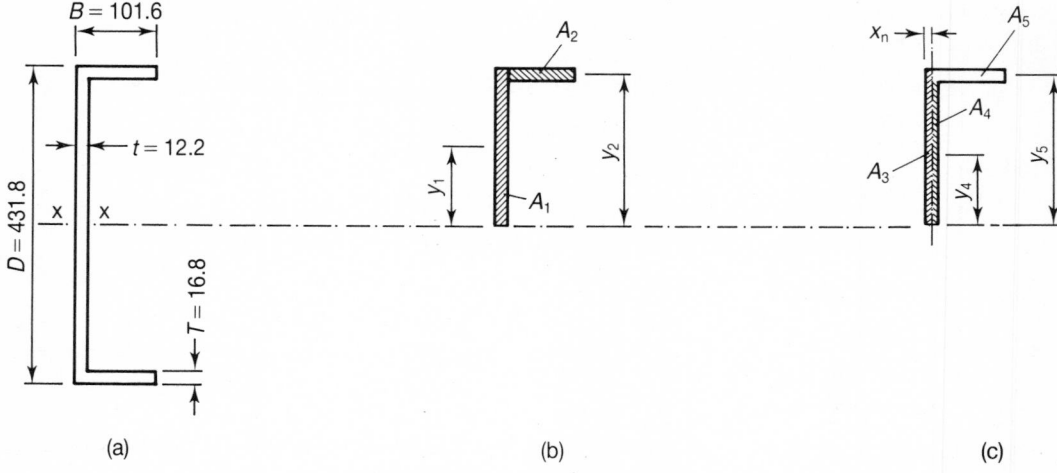

Fig. 4.21   Plastic section modulus for a channel section.

To determine the plastic section modulus about the x–x axis divide the section into A_1 and A_2 as shown in Fig. 4.21(b) where

$$A_1 = (D/2)t = (431.8/2)12.2 = 2633.98 \text{ mm}^2$$
$$A_2 = (B - t)T = (101.6 - 12.2)16.8 = 1501.92 \text{ mm}^2$$

and

$$y_1 = D/4 = 431.8/4 = 107.95 \text{ mm}$$
$$y_2 = D/2 - T/2 = (431.8 - 16.8)/2 = 207.5 \text{ mm}$$

Plastic section modulus

$$\begin{aligned}
S_x &= 2(A_1 y_1 + A_2 y_2) \\
&= 2(2633.98 \times 107.95 + 1501.92 \times 207.5) \\
&= 1.192\text{E6 mm}^3.
\end{aligned}$$

The value obtained from the Section Tables is 1.207E6 mm^3 which is slightly different because of the additional material at the root radius and the fact that the flanges taper.

Similarly for the plastic section modulus about the y–y axis divide the section into areas A_3, A_4 and A_5 as shown in Fig. 4.21(c). It is first necessary to determine the position of the neutral axis y–y.

Since the axis y–y divides the total area into two equal parts

$$x_n = (A_1 + A_2)/D = (2633.98 + 1501.92)/431.8 = 9.578 \text{ mm}.$$

Areas

$$A_3 = (D/2)x_n = (431.8/2)9.578 = 2067.89 \text{ mm}^2$$
$$A_4 = (D/2 - T)(t - x_n) = (431.8/2 - 16.8)(12.2 - 9.578) = 522.04 \text{ mm}^2$$
$$A_5 = (B - x_n)T = (101.6 - 9.578)16.8 = 1545.97 \text{ mm}^2$$

and

$$x_3 = x_n/2 = 9.578/2 = 4.789 \text{ mm}$$
$$x_4 = (t - x_n)/2 = (12.2 - 9.578)/2 = 1.311 \text{ mm}$$
$$x_5 = (B - x_n)/2 = (101.6 - 9.578)/2 = 46.011 \text{ mm}$$

Plastic section modulus

$$S_y = 2(A_3 x_3 + A_4 x_4 + A_5 x_5)$$
$$= 2(2067.89 \times 4.789 + 522.04 \times 1.311 + 1545.97 \times 46.011)$$
$$= 0.1634E6 \text{ mm}^3.$$

The value obtained from the Section Tables is $0.1531E6 \text{ mm}^3$ which is slightly different because of the additional material at the root radius and the fact that the flanges taper.

4.4.2   Plastic methods of analysis

A plastic collapse mechanism depends on the formation of a plastic hinge(s) and this will now be considered in detail. The tensile stress–strain curve for mild steel is shown in Fig. 4.3(a). The curve is idealized into three stages namely elastic, plastic, and strain hardening stages of deformation. The moment–curvature for a beam made of the same material is shown in Fig. 4.3(b) with the corresponding distribution of stress at various loading stages shown in Fig. 4.3(c). The spread of the plastic hinge along the length of the beam is shown in Fig. 4.3(d).

The amount of rotation that can take place at a plastic hinge is determined by the length of the yield plateau shown in Fig. 4.3(b). For mild steel the length is considerable and the work hardening stage is ignored. For higher grade steels work hardening occurs immediately after yielding and there is no plateau. A plastic hinge does form but with an increasing moment of resistance. In design this increase in resistance due to work hardening is ignored which errs on the side of safety.

Plastic methods of analysis and design consider a structure at collapse when sufficient plastic hinges have formed to produce a mechanism. Examples of collapse mechanisms are shown in Fig. 4.22.

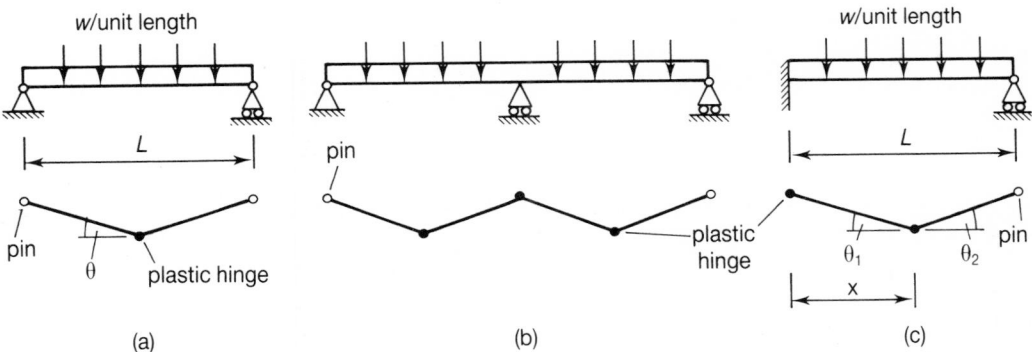

Fig. 4.22   Examples of plastic collapse mechanisms.

In simple situations, as shown in Fig. 4.22(a), the position of the plastic hinge is obvious and it is simple to calculate the collapse load using the method of virtual work.

The method must only be applied to structures where the material becomes plastic at the yield stress and is capable of accommodating large plastic deformations. It must therefore not be applied to brittle materials such as cast iron. However it can be applied to reinforced concrete because the steel reinforcement behaves plastically at collapse but care must be taken to check the rotation at the hinge for heavily reinforced sections.

The plastic method can be seen as a more rational method for design because all parts of the structure can be given the same safety factor against collapse. In contrast for elastic methods the safety factor varies. Intrinsically the plastic method of analysis is simpler than the elastic method because there is no need to satisfy elastic strain compatibility conditions. However calculations for instability and elastic deflections require careful consideration when using the plastic method, but nevertheless it is very popular for the design of some structures, e.g. beams and portal frames.

The method of analysis demonstrated in this chapter is based on the principle of virtual work. This states that if a structure, which is in equilibrium, is given a set of small displacements then the work done by the external loads on the external displacements is equal to the work done by the internal forces on the internal displacements. More concisely, external work equals internal work. The displacements need not be real, they can be arbitrary, which explains the use of the word 'virtual'. However the external and internal geometry must be compatible.

It is tacitly assumed that collapse is due to the formation of plastic hinges at certain locations and that other possible causes of failure, e.g. local or general instability, axial or shear forces, are prevented from occurring. It is also important to understand that at collapse:

(a)   the structure is in equilibrium, i.e. the forces and moments, externally and internally, balance,
(b)   no bending moment exceeds the plastic moment of resistance of a member,
(c)   there are sufficient hinges to form a collapse mechanism.

These three conditions lead to three theorems for plastic analysis.

Lower bound theorem: if only conditions (a) and (b) are satisfied then the solution is less than or equal to the collapse load.
Upper bound theorem: if only conditions (a) and (c) are satisfied then the solution is greater than or equal to the collapse load.
Uniqueness theorem: if conditions (a), (b) and (c) are satisfied then the solution is equal to the collapse load.

Settlement of the supports has no effect on the solution at collapse because the only effect is to change the amount of rotation required. This is in contrast to elastic methods of analysis where settlement calculations must be included.

Plastic hinges form in a member at the maximum bending moment. However at the intersection of two members where the bending moment is the same the hinge forms in the weaker member. Generally the locations of hinges are at restrained ends, intersection of members and at point loads. The hinges may not form simultaneously as loading increases but this is not important for calculating the final collapse load. Generally the number of plastic hinges

$$n = r + 1$$

where

r is the number of redundancies.

However there are exceptions, e.g. partial collapse of a beam in a structure.

Equating external to internal work for the member shown in Fig. 4.22(a).

$$wL(L/4)\theta = m_p(2\theta); \quad \text{hence} \quad m_p = wL^2/8$$

In other conditions, as shown in Fig. 4.22(c), the position of the plastic hinge for minimum collapse load is not obvious and the calculations are more complicated.

Equating external to internal work for the member shown in Fig. 4.22(c).

$$wx(x\theta_1/2) + w(L-x)^2\theta_2/2 = m_p(2\theta_1 + \theta_2) \tag{i}$$

From geometry

$$x\theta_1 = (L-x)\theta_2 \tag{ii}$$

Combining (i) and (ii)

$$wL = 2m_p[2/x - 1/(L-x)] \tag{iii}$$

Differentiating (iii) to determine the value of x for which w is a minimum.

$$x^2 - 4Lx + 2L^2 = 0, \quad \text{hence} \quad x = L(2 - \sqrt{2}) \tag{iv}$$

Combining (iii) and (iv)

$$m_p = (1.5 - \sqrt{2})wL^2 \tag{4.22}$$

The method can be applied to a variety of structures. Further explanation and examples are given in Moy, and Croxton and Martin Vol. 2.

4.5   Effect of a shear force on the plastic moment of resistance (cls 4.2.5 and 4.2.6, Pt 1)

In general the effect of a shear force (F_v) is to reduce the plastic moment of resistance but the reduction for an 'I' section is small for $F_v/P_v < 0.6$, where $P_v = 0.6 A_v Y_s$. When the ratio $d/t > 63\sqrt{(275/p_y)}$ the web should be checked for shear buckling in accordance with cl 4.4.5, Pt 1. The limiting value of d/t is based on experimental work by Longbottom and Heyman, and later work by Horne (1958).

A theory that describes the effect of shear on the plastic moment of resistance is given by Horne (1971). If a shear force is applied to an 'I' section most of the shear force is resisted by the web. If it is assumed that all of the shear force is resisted by the web the effect on the plastic moment of resistance is shown in Fig. 4.23(a). The outer fibres in bending are at yield while the inner fibres are linear elastic. It is not possible to resist shear forces on the outer fibres and thus the inner fibres resist all of the shear force.

Plastic moment of resistance of the web

$$M_{pw} = [td^2/4 - th^2/4 + th^2/6]Y_s = [td^2/4 - th^2/12]Y_s \tag{i}$$

From the parabolic distribution of shear stress

$$F_v = (2/3)th\tau_y \tag{ii}$$

Defining

$$P_v = tD\tau_y \tag{iii}$$

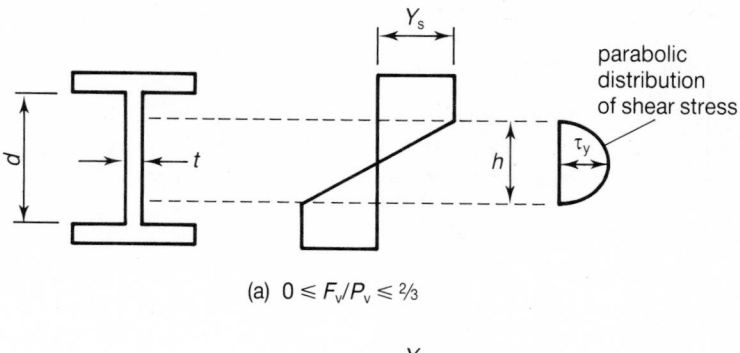

(a) $0 \leqslant F_v/P_v \leqslant \frac{2}{3}$

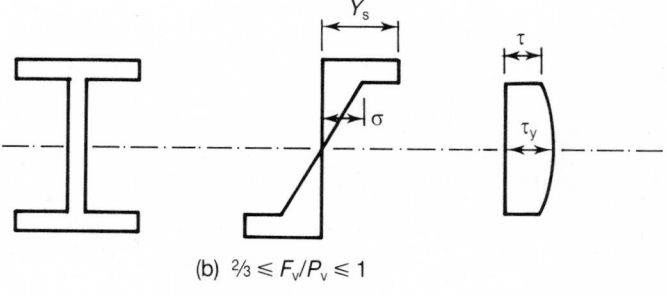

(b) $\frac{2}{3} \leqslant F_v/P_v \leqslant 1$

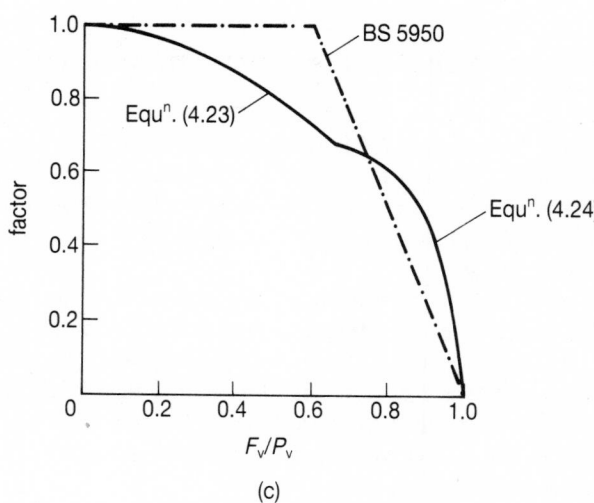

(c)

Fig. 4.23   Effect of a shear force on the plastic section modulus of the web of an 'I' section.

Combining (i) to (iii)

$$M_{pw} = (td^2 \, Y_s/4)[1 - (3/4)(F_v/P_v)^2] \tag{4.23}$$

This expression is valid for $0 \leqslant F_v/P_v \leqslant 2/3$.

When $2/3 \leqslant F_v/P_v \leqslant 1$ then assuming the stress distributions shown Fig. 4.23(b) the plastic moment of resistance of the web

$$M_{pw} = td^2\sigma/6 \qquad\qquad\qquad (i)$$

Applied shear force

$$F_v = dt\tau + (2/3)dt(\tau_y - \tau) \qquad\qquad\qquad (ii)$$

If $P = dt\tau_y$, ratio

$$F_v/P_v = (1/3)(2 + \tau/\tau_y) \qquad\qquad\qquad (iii)$$

Adopting a failure criterion of the form

$$(\sigma/Y_s)^2 + (\tau/\tau_y)^2 = 1 \qquad\qquad\qquad (iv)$$

Combining (i) to (iv)

$$M_{pw} = (td^2 Y_s/4) \times [(2/3)\sqrt{\{1 - (3F_v/P_v - 2)^2\}}] \qquad\qquad\qquad (4.24)$$

The expressions in the square brackets in Equations (4.23) and (4.24) are plotted in Fig. 4.23(c) expressed as $M_{pw} = $ factor $(td^2 Y_s/4)$. This theory is approximate and conservative but it does give a general appreciation of the effect of a shear force on the plastic moment of resistance. A fuller description and less conservative theories are given in Horne (1971).

Also shown in Fig. 4.23 are the recommendations in BS 5950. According to cl 4.2.6, Pt 1 where $F_v/P_v < 0.6$ there is no reduction in the plastic moment of resistance for plastic or compact sections. This recommendation is based on the work of Morris and Randall who stated that shear can be ignored unless the average shear stress in the web exceeds $Y_s/3$, or $Y_s/4$ when the ratio of the overall depth to flange width (D/B ratio) exceeds 2.5.

Where $F_v/P_v > 0.6$, cl 4.2.6, Pt 1 recommends that the reduced plastic moment of resistance for plastic or compact sections

$$M_p = p_y(S - S_v\rho_1) \qquad\qquad\qquad (4.25)$$

where

$$\rho_1 = 2.5(F_v/P_v) - 1.5$$

EXAMPLE 4.10   Plastic section modulus reduced by shear

Determine the plastic modulus for a $762 \times 267 \times 197$ kg UB grade 43 when subjected to a shear force of 1145 kN.

$$F_v/P_v = F_v/(0.6Dtp_y) = 1145E3/(0.6 \times 769.6 \times 15.6 \times 265) = 0.6$$

According to cl 4.2.6, Pt 1 there is no reduction in the plastic modulus.

According to Horne, dividing Equation (4.23) by Y_s

$$S_{p(web)} = (td^2/4)[1 - (3/4)(F_v/P_v)^2] = 0.73(td^2/4),\text{ i.e. } 27\% \text{ loss in the web.}$$

Reduced plastic section modulus of the complete section

$$S_{p(reduced)} = S_{p(whole)} - S_{p(web)}(1 - \text{factor})$$
$$= 7.167E6 - (15.6 \times 685.8^2/4)(1 - 0.73) = 6.672E6 \text{ mm}^3$$

Percentage reduction in plastic modulus for the whole section

$$= 100(7.167\text{E}6 - 6.672\text{E}6)/7.167\text{E}6 = 6.91\%$$

Although the ratio of $F/P = 0.6$ is relatively high the reduction of the value of the plastic modulus for the whole section is small.

4.6   Lateral restraint (cls 4.2.2, 4.3.2, 5.3.5, Pt 1)

The full rotation required at a plastic hinge in a beam may not be realized unless lateral support is provided at the hinge position. It may also be necessary to provide lateral support at other points along the span to ensure that lateral torsional buckling does not occur. Lateral torsional buckling is considered to be prevented if the compression flange is prevented from moving laterally, either by an intersecting member, or by frictional restraint from intersecting floor units.

Full restraint exists if the friction or positive connection of a floor or other construction to the compression flange of the member is capable of resisting a lateral force of not less than 2.5% of the maximum factored force in the compression flange of the member, under factored loading. This load should be considered as distributed uniformly along the flange, provided that the dead load of the floor and the imposed load it supports together constitute the dominant loading on the beam. The floor construction should be capable of resisting this lateral force   (cl 4.2.2, Pt 1).

Where one or more lateral restraints are required at intervals within the span of the beam, these intermediate lateral restraints should be capable of resisting a total force of not less than 2.5% of the maximum factored force in the compression flange, divided between the intermediate lateral restraints in proportion to their spacing   (cl 4.3.2.1, Pt 1). Restraints must prevent lateral movement, e.g. by bracing, or be connected to a robust part of the structure.

In addition.   Where three or more intermediate lateral restraints are provided, each intermediate lateral restraint should be capable of resisting a force of not less than 1% of the maximum factored force in the compression flange   (cl 4.3.2.2, Pt 1).

Where practicable, torsional restraint should be provided at a plastic hinge. Where impracticable, restraint should be provided within a distance of $D/2$ of the plastic hinge (cl 5.3.5, Pt 1). Within a member containing a plastic hinge the maximum distance from the hinge restraint to an adjacent restraint

$$L_{\mathrm{m}} = 38r_{\mathrm{y}}/\sqrt{[\,f_{\mathrm{c}}/130 + (p_{\mathrm{y}}/275)^2(x/36)^2]} \tag{4.26}$$

where

f_{c} is the average compressive stress due to axial load in N/mm^2
x is the torsional index.

Web stiffeners at hinge locations should be provided where a load is applied within $D/2$ of a plastic hinge location which exceeds 10% of the shear capacity of the member. The stiffener should be provided within a distance of half the depth of the member either side of the hinge location. If the stiffeners are plates $b_{\mathrm{s}}/t_{\mathrm{s}} < 9$, or if sections are used $\sqrt{(I_{\mathrm{so}}/J_{\mathrm{s}})} < 9$ (cl 5.3.6, Pt 1)
where

I_{so} is the second moment of area of the stiffener about the face of the element

J_s is the torsion constant of the stiffener.

The above restraint at a plastic hinge need not be met at the last hinge to form provided that it can be clearly identified (cl 5.5.3.1, Pt 1).

4.7 Web strength of 'I' sections

4.7.1 Web shear strength (cl 4.2.3, Pt 1)

Most of the shear force (90 to 95%) is resisted by the web of an 'I' section when bending about the x–x axis. The maximum shear resistance is $P_v = 0.6A_v p_y = 0.6Dt p_y$, but if $d/t > 63\sqrt{(275/p_y)}$ then the web should be checked for shear buckling as described in Section 5.7.5. Values of A_v for other sections are given in cl 4.2.3, Pt 1.

4.7.2 Web buckling strength (cls 4.5.1.3 and 4.5.2.1, Pt 1)

Where point loads, or reactions from supports or other members, are applied to a beam as shown in Fig. 4.24 then the web should be checked for buckling as a strut with a slenderness ratio of $\lambda = 2.5d/t$. The effective width of the strut is determined by assuming a 45° dispersion angle from the edge of the load as shown in the examples in Fig. 4.24. The buckling resistance

$$P_w = (b_1 + n_1)t p_c \qquad (4.27)$$

where

b_1 is the stiff bearing length (see Fig. 4.24)
n_1 is the length obtained by the 45° dispersion (see Fig. 4.24)
p_c is the compressive strength obtained from Table 27(c), Pt 1 with a slenderness
ratio $\lambda = 2.5d/t$

The slenderness ratio ($\lambda = 1/r$) is obtained from considering the buckling of the web about an axis parallel to the web. The radius of gyration

$$r_y = \sqrt{(I_y/A)} = \sqrt{[(bt^3/12)/(bt)]} = t/(2\sqrt{3})$$

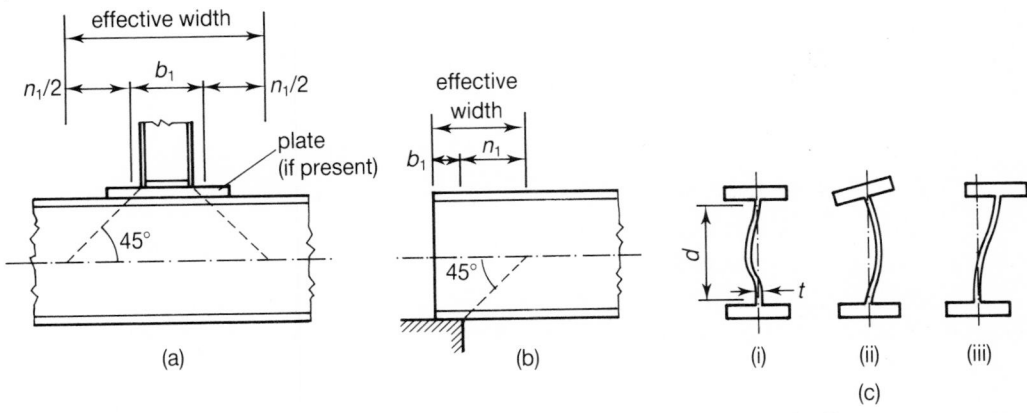

Fig. 4.24 Effective width of a strut for buckling.

If the effective length is taken as $0.72d$ the slenderness ratio

$$\lambda = 1/r_y = 0.72d/[t/(2\sqrt{3})] = 2.5d/t \qquad (4.28)$$

The value of the effective length of $0.72d$ is empirical but it gives reasonable accuracy in tests where the flanges are effectively fixed ended (see Fig. 4.24(c)(i)) as shown by Astill *et al.* This is the minimum practical effective length of the equivalent strut, and if lateral movement or rotation of the ends is possible the effective length must be increased (see Figs. 4.24(c)(ii) and (iii)). Alternatively stiffeners can be introduced to prevent these movements and resist part of the load.

In the Section Tables Equation (4.28) is expressed as

$$P_w = C_1 + L_b C_2 + t C_3$$

where

C_1 is the beam component
C_2 is the stiff bearing component
C_3 is the flange plate component

Where the web strength is inadequate then stiffeners must be introduced and designed as shown in Section 5.7.6.

4.7.3   Web bearing strength (cls 4.5.1.3 and 4.5.3, Pt 1)

Where point loads, or reactions from supports or other members, are applied to a beam as shown in Fig. 4.25 then the web should be checked for bearing stresses. The web bearing strength

$$P_{wbg} = (b_1 + n_2)tp_{yw} \qquad (4.29)$$

where

b_1 is the stiff bearing length (see Fig. 4.25)
n_2 is the length obtained by the 1:2.5 dispersion through the flange (see Fig. 4.25)
p_{yw} is the design strength of the web obtained from Table 6, Pt 1

In the Section Tables Equation (4.29) is expressed as

$$P_{wbg} = C_1 + L_b C_2 + t C_3$$

where

C_1 is the beam component
C_2 is the stiff bearing component
C_3 is the flange plate component

Where the strength of the web is inadequate then stiffeners must be introduced and designed as shown in Section 5.7.6.

EXAMPLE 4.11   Simply supported beam

The floor of an office building consists of 125 mm pre-cast concrete units, with a mass of 205 kg/m² , topped with a 40 mm concrete screed and 20 mm wood blocks. Lightweight partitions supported by the floor are equivalent to a superficial load of 1.0 kN/m² and the

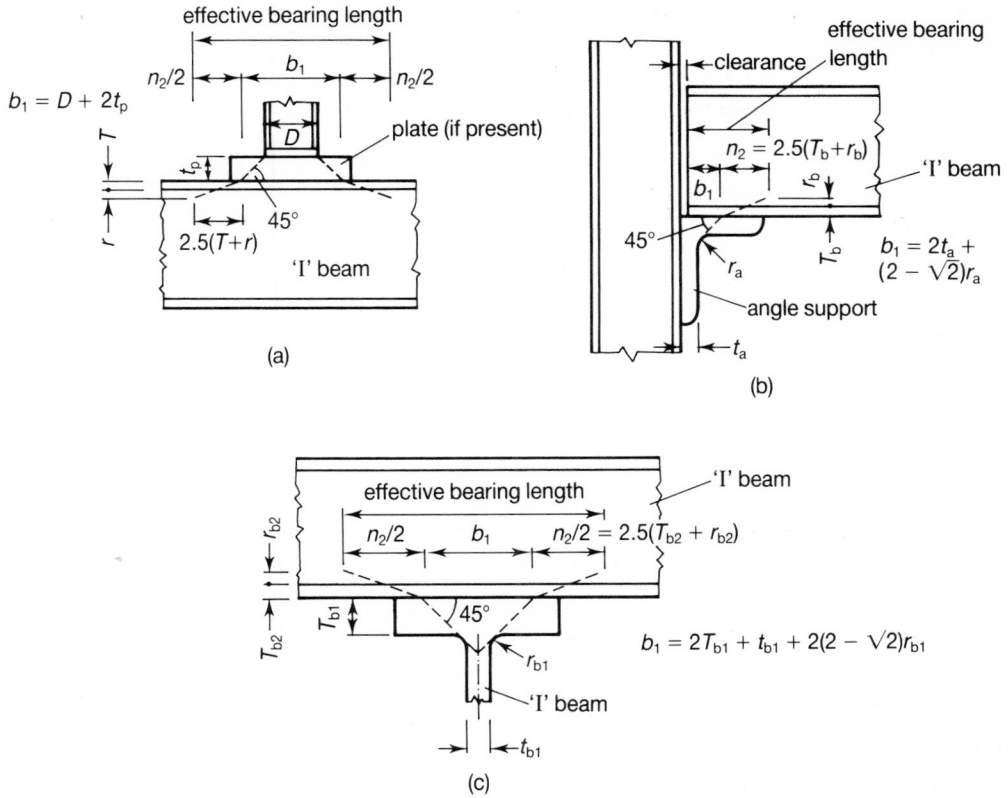

$b_1 = D + 2t_p$

$n_2 = 2.5(T_b + r_b)$

$b_1 = 2t_a + (2 - \sqrt{2})r_a$

(a)

(b)

$n_2/2 = 2.5(T_{b2} + r_{b2})$

$b_1 = 2T_{b1} + t_{b1} + 2(2 - \sqrt{2})r_{b1}$

(c)

Fig. 4.25   Effective bearing length.

suspended ceiling has a mass of $40 \, \text{kg/m}^2$. The floor rests on the top flanges of simply supported steel beams spanning 8 m and at a pitch of 3.75 m.

Characteristic loads	kg/m²	kN/m²
Dead load		
self weight of steel beam (assumed)	20	
125 mm precast units	205	
40 mm concrete screed $2400 \times 0.04 =$	96	
20 mm wood blocks $900 \times 0.02 =$	18	
suspended ceiling	40	
	$379 \times 9.81/1E3 =$	3.72
lightweight partitions		1.00
Total dead load =		4.72
Imposed load (office general)		
BS 6399 Pt 1		2.50

Maximum design bending moment at mid-span with load factors taken from Table 2, Pt 1

$$\text{kNm}$$

Dead load BM $= \gamma_f WL/8 = 1.4 \times 4.72 \times 8 \times 3.75/8 =$ 198.2
Imposed load BM $= \gamma_f WL/8 = 1.6 \times 2.5 \times 8 \times 3.75/8 =$ 120.0

Total BM $=$ 318.2

Using Grade 43 steel and assuming $p_y = 275 \text{ N/mm}^2$ the plastic section modulus required

$S_x = M/p_y = 318.2\text{E}6/275 = 1.157\text{E}6 \text{ mm}^3$.

From the Section Tables, try $457 \times 152 \times 60$ kg UB, $S_x = 1.248\text{E}6 \text{ mm}^3$,
  $T = 13.3$ mm and from Table 6, Pt 1 $p_y = 275 \text{ N/mm}^2$.
  Check local buckling of the flange from Table 7, Pt 1
  $b/T = (152.9/2)/13.3 = 5.75 < 8.5\sqrt{(275/275)}$
  Check for buckling of web
  $d/t = 407.78/8 = 50.96$
  From Table 7, Pt 1
  $79\varepsilon = 79\sqrt{(275/275)} = 79 > 50.96$, therefore plastic section.

Check for plasticity at service load (cl 4.2.5, Pt 1)
  $M_c = p_y S_x = 275 \times 1.284\text{E}6/\text{E}6 = 353.1 \text{ kNm}$
  $1.2p_y Z_x = 1.2 \times 275 \times 1.120\text{E}6/\text{E}6 = 369.6 > 353.1 \text{ kNm}$, therefore satisfactory.

Check lateral restraint at the ultimate limit state (cl 4.2.2, Pt 1)
  Total load on span $= (1.4 \times 4.72 + 1.6 \times 2.5) \times 8 \times 3.75 = 318.24 \text{ kN}$
  Frictional force $= \mu W = 0.3 \times 318.24 = 95.47 \text{ kN}$
  Disturbing force in compression flange
  $= (2.5/100)[M/(D - T)] = (2.5/100)[318.2\text{E}6/(454.7 - 13.3)]/1\text{E}3$
  $= 18.02 < 95.47 \text{ kN}$, therefore laterally restrained.

Check deflections at service load (cl 2.5.1, Pt 1)
  Imposed load on span $= 2.5 \times 8 \times 3.75 = 75 \text{ kN}$
  Maximum deflection $= 5WL^3/(384EI)$
             $= 5 \times 75 \times (8\text{E}3)^3/(384 \times 205 \times 254.64\text{E}6) = 9.58 \text{ mm}$
  From Table 5, Pt 1, if no brittle finishes, deflection limit $= L/200 = 8\text{E}3/200 =$
  $40 > 9.58$ mm, therefore satisfactory.

Check web strength at the ultimate limit state (cl 4.5, Pt 1).
  Check shear buckling of the web (cl 4.2.3, Pt 1)
  $d/t = 407.7/8 = 50.96$
  $63\varepsilon = 63\sqrt{(275/275)} = 63 > 50.96$, therefore satisfactory.

The load is distributed so that the only concentrated loads are at the end supports. Shear force at the end
  $= WL/2 = (1.4 \times 4.72 + 1.6 \times 2.5)8 \times 3.75/2 = 159.12 \text{ kN}$.

Maximum shear strength of web
  $P_v = 0.6Dtp_y = 0.6 \times 454.7 \times 8 \times 275/1\text{E}3$
      $= 600.2 > 159.12 \text{ kN}$, therefore satisfactory.

Web buckling strength (cl 4.5.2.1, Pt 1).
  Length of stiff bearing for type of angle seating as shown in Fig. 4.25(b) assuming
  an $150 \times 75 \times 10$ mm angle and 3 mm clearance
  $b_1 = 2t_a + (2 - \sqrt{2})r_a - \text{clearance} = 2 \times 10 + (2 - \sqrt{2})11 - 3 = 23.44 \text{ mm}$
  Slenderness ratio of web $\lambda = 2.5d/t = 2.5 \times 407.7/8 = 127.4$

From Table 27(c) for $p_y = 275$ N/mm², $p_c = 88$ N/mm².
From Equation (4.27) buckling resistance
$P_w = (b_1 + n_1)tp_c = (23.44 + D/2)tp_c$
$\quad = (23.44 + 454.7/2)8 \times 88/1E3 = 176.6 > 159.12$ kN therefore satisfactory.

Web bearing strength (cls 4.5.1.3 and 4.5.3, Pt 1) from Equation (4.29)
$P_{wbg} = (b_1 + n_2)tp_{yw} = [23.44 + 2.5(T + r)]tp_{yw}$
$\quad = [23.44 + 2.5 \times (13.3 + 10.2)]8 \times 275/1E3 = 180.8 > 159.12$ kN
$\quad$ therefore satisfactory.

Web strength satisfactory therefore no web stiffeners required. No torsional restraint required at the plastic hinge because it is the last hinge to form (cl 5.5.3.1, Pt 1).

Check self weight of steel beam $= 60/3.75 = 16 < 20$ kg/m² assumed in loading calculations, therefore acceptable.

EXAMPLE 4.12 Support for a conveyor

Part of the support for a conveyor consists of a pair of identical beams as shown in Fig. 4.26. Each beam is connected to a stanchion at end A by a cleat and is supported on a cross beam at D by bolting through the connecting flanges. Lateral restraint is provided by transverse beams at A, B and E connected to rigid supports.

Assuming 'simple design' as described in cl 2.1.2.2, Pt 1.

Reactions and shears are determined by taking moments about A, then about D.
$\quad R_D = (2 \times 225 + 5 \times 450 + 10 \times 150)/8 = 525$ kN
$\quad R_A = (6 \times 225 + 3 \times 450 - 2 \times 150)/8 = 300$ kN.

Important bending moments are
$\quad M_B = R_A \times 2 = 300 \times 2 = 600$ kNm
$\quad M_C = R_A \times 5 - 225 \times 3 = 300 \times 5 - 225 \times 3 = 825$ kNm
$\quad M_D = -150 \times 2 = -300$ kNm

The shear force and bending moment diagrams are shown in Fig. 4.26.
$\quad$ Using Grade 50 steel and assuming $p_y = 355$ N/mm² the plastic section modulus required

$\quad S_x = M_{max}/p_y = 825E6/355 = 2.342E6$ mm³.

From the Section Tables try $533 \times 210 \times 92$ kg UB, $S_x = 2.366E6$ mm³, $T = 15.6$ mm and from Table 6, Pt 1 $p_y = 355$ N/mm².

Check local buckling of the flange from Table 7, Pt 1
$\quad b/T = (B/2)/T = (209.3/2)/15.6 = 6.71$
$\quad 8.5\sqrt{(275/p_y)} = 8.5\sqrt{(275/355)} = 7.48 > 6.71$
$\quad$ Check for buckling of web
$\quad d/t = 476.5/10.2 = 46.72$
$\quad$ From Table 7, Pt 1
$\quad 79\varepsilon = 79\sqrt{(275/355)} = 69.53 > 46.72$, therefore plastic section.

Check on plasticity at service load (cl 4.2.5, Pt 1)
$\quad p_y S_x = 355 \times 2.366E6/E6 = 839.9$ kNm
$\quad 1.2p_y Z_x = 1.2 \times 355 \times 2.076E6/E6 = 884.4 > 839.9$ kNm, therefore satisfactory.

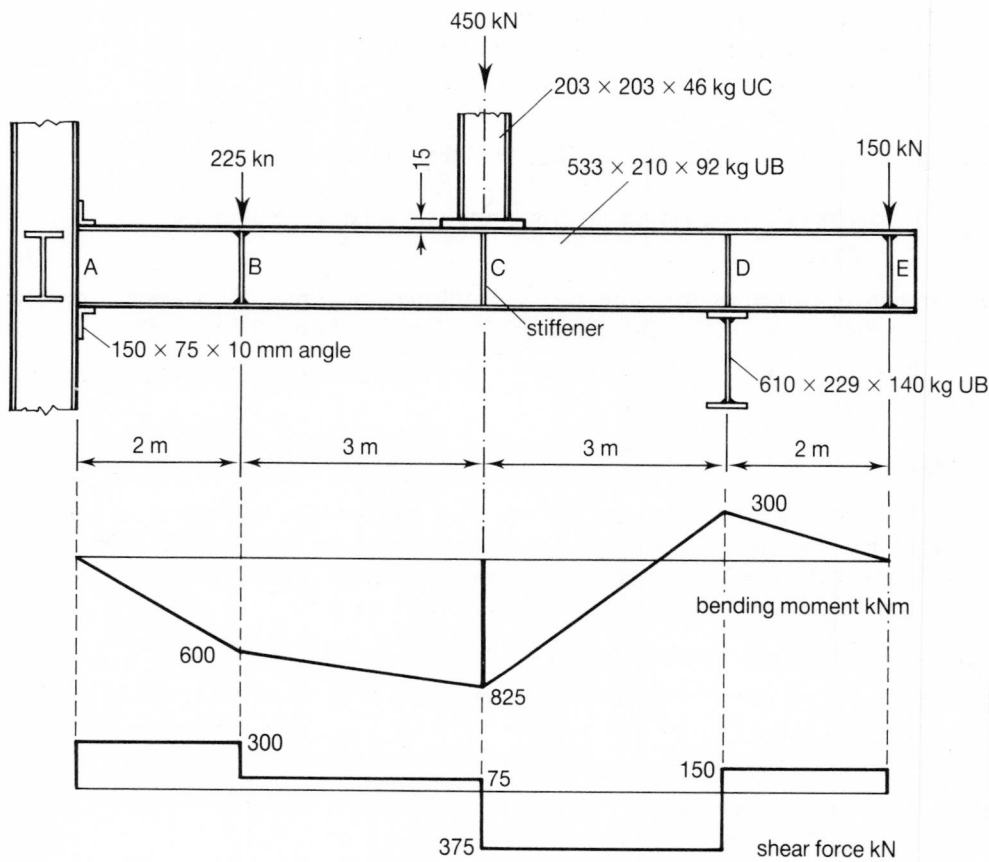

Fig. 4.26   Support for a conveyor.

Check deflections at service load (cl 2.5.1, Pt 1)

Using area and area moment methods the deflections due to the imposed loads are
midspan in ABCD 11.3 mm (downwards)
end of cantilever DE 6.5 mm (upwards)
From Table 5, Pt 1, because there are no brittle finishes, deflection limit for
ABCD = $L/200 = 8E3/200 = 40 > 11.3$ mm, therefore satisfactory
DE = $L/180 = 2E3/180 = 11.1 > 6.5$ mm, therefore satisfactory.

Check web strength at the ultimate limit state (cl 4.5, Pt 1).

Check shear buckling of the web (cl 4.2.3, Pt 1)
$d/t = 476.5/10.2 = 46.72$
$63\varepsilon = 63\sqrt{(275/p_y)} = 63\sqrt{(275/355)} = 55.45 > 46.72$, therefore satisfactory.

From Fig. 4.26 shear force in span CD, $F_v = 375$ kN.

Maximum shear strength of web
$P_v = 0.6p_yDt = 0.6 \times 355 \times 533.1 \times 10.2/1E3 = 1158.2 > 375$ kN,
therefore satisfactory.

Web buckling strength at A (cl 4.5.2.1, Pt 1).
    Length of stiff bearing for the angle seating shown at A
    For a $150 \times 75 \times 10$ mm angle (see Fig. 4.25(b))
    $b_1 = 2t_a + (2 - \sqrt{2})r_a - \text{clearance} = 2 \times 10 + (2 - \sqrt{2})11 - 3 = 23.44$ mm
    Slenderness ratio of web $\lambda = 2.5d/t = 2.5 \times 476.5/10.2 = 116.8$
    From Table 27(c) for $p_y = 355$ N/mm^2, $p_c = 110$ N/mm^2.
    From Equation (4.27) buckling resistance
    $P_w = (b_1 + n_1)tp_c = (23.44 + D/2)tp_c$
        $= (23.44 + 533.1/2)10.2 \times 110/1E3 = 325.4 > 300$ kN, therefore satisfactory.

Web bearing strength at A (cls 4.5.1.3 and 4.5.3, Pt 1) from Equation (4.29)
    $P_{wbg} = (b_1 + n_2)tp_{yw} = [23.44 + 2.5(T + r)]tp_{yw}$
        $= [23.44 + 2.5 \times (15.6 + 12.7)]10.2 \times 355/1E3 = 341.1 > 300$ kN,
        therefore satisfactory.

Web buckling strength at D (cl 4.5.2.1, Pt 1).
    Length of stiff bearing for support on a $610 \times 229 \times 140$ kg transverse UB (see Fig. 4.25(c))
    $b_1 = 2T + t + 2(2 - \sqrt{2})r_b = 2 \times 22.1 + 13.1 + 2(2 - \sqrt{2}) \times 12.7 = 64.74$ mm
    Slenderness ratio of web $\lambda = 2.5d/t = 2.5 \times 476.5/10.2 = 116.8$
    From Table 27(c) for $p_y = 355$ N/mm^2 $p_c = 110$ N/mm^2.
    From Equation (4.27) buckling resistance
    $P_w = (b_1 + n_1)tp_c = (69.6 + D)tp_c$
        $= (64.74 + 533.1)10.2 \times 110/1E3 = 670.8 > 525(R_D)$ kN, therefore satisfactory.

Web bearing strength at D (cls 4.5.1.3 and 4.5.3, Pt 1) from Equation (4.29)
    $P_{wbg} = (b_1 + n_2)tp_{yw} = [b_1 + 2.5 \times 2(T + r)]tp_{yw}$
        $= [64.74 + 2.5 \times 2(15.6 + 12.7)]10.2 \times 355/1E3 = 746.8 < 525(R_D)$ kN,
        therefore satisfactory.

Web buckling strength at C (cl 4.5.2.1, Pt 1).
    Length of stiff bearing
    $b_1 = D + 2t_p = 203.2 + 2 \times 15 = 233.2$ mm (see Figs. 4.24(a) and 4.25(a))
    Slenderness ratio of web $\lambda = 2.5d/t = 2.5 \times 476.5/10.2 = 116.8$
    From Table 27(c) for $p_y = 355$ N/mm^2, $p_c = 110$ N/mm^2.
    From Equation (4.27) buckling resistance
    $P_w = (b_1 + n_1)tp_c = (233.2 + 533.1) \times 10.2 \times 110/1E3 = 859.8 > 450$ kN,
        therefore satisfactory.

Web bearing strength at C (cls 4.5.1.3 and 4.5.3, Pt 1) from Equation (4.29)
    $P_{wbg} = (b_1 + n_2)tp_{yw} = [b_1 + 2.5 \times 2(T + r)]tp_{yw}$
        $= [233.2 + 2.5 \times 2(15.6 + 12.7)]10.2 \times 355/1E3 = 356.8 < 450$ kN,
        therefore satisfactory.

Stiffener at C (plastic hinge location (cl 5.3.6, Pt 1))
    Applied force $F = 450$ kN should not exceed 10% of web capacity.
    Maximum shear strength of web
    $P_v = 0.6Dtp_y = 0.6 \times 533.1 \times 10.2 \times 355/1E3 = 1158.2$ kN
    $0.1P_v = 0.1 \times 1158.2 = 115.8 < 450$ kN, therefore stiffeners required.
    If flat plates, $t_s = b_s/9 = (B - t)/(2 \times 9) = (209.3 - 10.2)/18 = 11.06$ mm.
    Use 12 mm plate.

Check at C if shear force reduces M_p (cl 4.2.5, Pt 1)

$F_v/P_v = 375/1158.2 = 0.324 < 0.6$, therefore no reduction.

The plastic hinge at C is the last to form and therefore no lateral restraint is required (cl 5.5.3.1, Pt 1).

EXAMPLE 4.13   Two span beam

Determine the size of Universal Beam required to support the design loads shown in Fig. 4.27. Assume that the compression flange is fully restrained and that lateral torsional buckling does not occur.

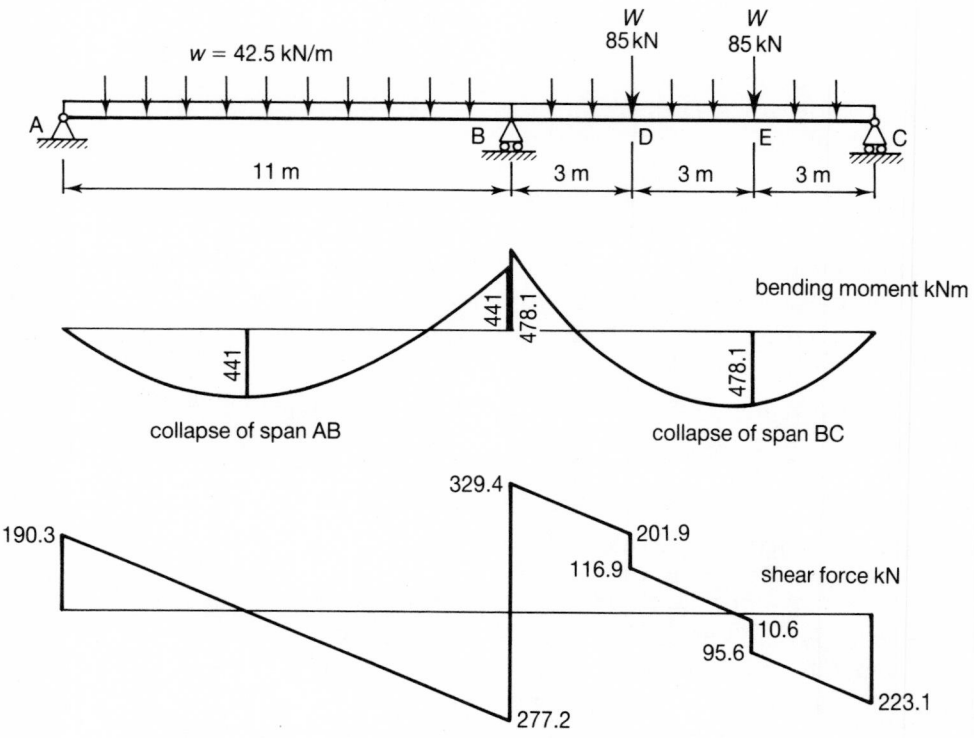

Fig. 4.27   Two span beam.

Plastic analysis of the beam produces the following.

Collapse of span AB considered as a propped cantilever (see Equation (4.22))

$$M_p = (1.5 - \sqrt{2})wL^2 = (1.5 - \sqrt{2})42.5 \times 11^2 = 441 \text{ kNm.}$$

Collapse of span BC with plastic hinges assumed to be at E and B as shown in Fig. 4.27

External work = internal work

$$W(2L/3)\theta_1 + W(2L/3)\theta_1/2 + w(2L/3)(L/3)\theta_1$$
$$+ w(L/3)(L/6)\theta_2 = m_p[\theta_1 + (\theta_1 + \theta_2)] \tag{i}$$

and from geometry $(2L/3)\theta_1 = (L/3)\theta_2$; hence $\theta_2 = 2\theta_1$ (ii)

Combining equations (i) and (ii) and rearranging

$$M_p = WL/4 + wL^2/12 = 85 \times 9/4 + 42.5 \times 9^2/12 = 478.1 \text{ kNm.}$$

This value of M_p is greater than the value for span AB and therefore is used to determine the size of a section which is continuous for two spans. The assumption that a plastic hinge is at E is not correct, it actually occurs at 3.25 m from C and $M_p = 479.1$ kNm. However the error is small and is ignored.

Using Grade 50 steel and assuming $p_y = 355$ N/mm^2 the plastic section modulus required
$S_x = M_{max}/p_y = 478.1E6/355 = 1.347E6$ mm^3.

From the Section Tables try $457 \times 191 \times 67$ kg UB
$S_x = 1.471E6$ mm^3, $T = 12.7$ therefore from Table 6, Pt 1 $p_y = 355$ N/mm^2.
Check local buckling of the flange from Table 7
$b/T = (B/2)/T = (189.9/2)/12.7 = 7.48$
$8.5\sqrt{(275/p_y)} = 8.5\sqrt{(275/355)} = 7.48$, therefore acceptable.

Check buckling of the web
$d/t = 407.9/8.5 = 48.0$
From Table 7, Pt 1
$79\sqrt{(275/p_y)} = 79\sqrt{(275/355)} = 69.53 > 48.0$, therefore plastic section.

Check to prevent plasticity at service load (cl 4.2.5, Pt 1)
$p_y S_x = 355 \times 1.471E6/E6 = 522.2$ kNm
$1.2p_y Z_x = 1.2 \times 355 \times 1.296E6/E6 = 552.1 > 522.2$ kNm, therefore satisfactory.

Check deflections at service load (cl 2.5.1, Pt 1)
Using area and area moment methods the deflections due to the imposed load are 20.7 mm for AB and 25 mm for BC.
From Table 5 Pt 1, if there are no brittle finishes, deflection limit for
AB is $L/200 = 11E3/200 = 55 > 20.7$ mm, therefore satisfactory
$BC = L/200 = 9E3/200 = 45 > 25$ mm, therefore satisfactory.

Sketch the shear force diagram (see Fig. 4.27) at the ultimate limit state for the collapse of span BC.
For EC
$E\rangle + 478.1 + 42.5 \times 3 \times 1.5 - 3R_c = 0$; hence $R_c = 223.1$ kN
For AB
$B\rangle + 478.1 - 42.5 \times 11 \times 11/2 + 11R_A = 0$; hence $R_A = 190.3$ kN
For ABC
$\uparrow + R_A + R_B + R_C - \Sigma wL - \Sigma W = 0$
$+ 190.3 + R_B + 223.1 - 42.5 \times 20 - 2 \times 85 = 0$; hence $R_B = 606.6$ kN.

Check web strength at the ultimate limit state (cl 4.5, Pt 1).
Check shear buckling of the web (cl 4.2.3, Pt 1)
$d/t = 407.9/8.5 = 47.99$
$63\varepsilon = 63\sqrt{(275/p_y)} = 63\sqrt{(275/355)}$
$= 55.45 > 47.99$, therefore satisfactory.

From Fig. 4.27 maximum shear force at B in span BC $= 329.4$ kN.

Maximum shear strength of web
$P_v = 0.6Dtp_y = 0.6 \times 453.6 \times 8.5 \times 355/1E3$
$= 821.24 > 329.4$ kN therefore satisfactory.
Check at B is shear force reduces M_p (cl 4.2.5, Pt 1)
$F_v/P_v = 329.4/821.24 = 0.401 < 0.6$, therefore no reduction.

Web buckling strength at C (cl 4.5.2.1, Pt 1)

Length of stiff bearing for a $150 \times 75 \times 10$ mm angle seating (see Fig. 4.25(b))

$b_1 = 2t_a + (2 - \sqrt{2})r_a - \text{clearance} = 2 \times 10 + (2 - \sqrt{2})11 - 3 = 23.44$ mm

Slenderness ratio of web $\lambda = 2.5d/t = 2.5 \times 407.9/8.5 = 120.0$

From Table 27(c) for $p_y = 355$ N/mm^2, $p_c = 106$ N/mm^2.

From Equation (4.27) buckling resistance

$P_w = (b_1 + n_1)tp_c = (23.44 + D/2)tp_c$

$\quad = (23.44 + 453.6/2)8.5 \times 106/1\text{E}3 = 225.5 > 223.1$ kN, therefore satisfactory.

Web bearing strength at C (cls 4.5.1.3 and 4.5.3, Pt 1) from Equation (4.29)

$P_{wbg} = (b_1 + n_2)tp_{yw} = [23.44 + 2.5(T + r)]tp_{yw}$

$\quad = [23.44 + 2.5 \times (12.7 + 10.2)]8.5 \times 355/1\text{E}3$

$\quad = 243.5 > 223.1$ kN, therefore satisfactory.

Web buckling strength at B (cl 4.5.2, Pt 1).

Length of stiff bearing for support on a $914 \times 305 \times 289$ kg transverse UB at B (see Fig. 4.25(c))

$b_1 = 2T + t + 2(2 - \sqrt{2})r_b = 2 \times 32 + 19.6 + 2(2 - \sqrt{2})19.1 = 106.0$ mm

Slenderness ratio of web $\lambda = 2.5d/t = 2.5 \times 407.9/8.5 = 120$

From Table 27(c) for $p_y = 355$ N/mm^2, $p_c = 106$ N/mm^2.

From Equation (4.27) buckling resistance

$P_w = (b_1 + n_1)tp_c = (b_1 + D)tp_c$

$\quad = (106 + 453.6) \times 8.5 \times 106/1\text{E}3 = 504.2 < 606.6(R_B)$ kN, therefore not satisfactory and stiffeners required.

Web bearing strength at B (cls 4.5.1.3 and 4.5.3, Pt 1) from Equation (4.29)

$P_{wbg} = (b_1 + n_2)tp_{yw} = [b_1 + 2.5 \times 2(T + r)]tp_{yw}$

$\quad = [106 + 2.5 \times 2(12.7 + 10.2)]8.5 \times 355/1\text{E}3$

$\quad = 665.4 > 606.6(R_B)$ kN, therefore satisfactory.

Stiffener at plastic hinge location B (cl 5.3.6, Pt 1)

Reaction force $R_B = 606.6$ kN should not exceed 10% of web capacity.

Maximum shear strength of web

$P_v = 0.6Dtp_y = 0.6 \times 453.6 \times 8.5 \times 355/1\text{E}3 = 821.24$ kN

$0.1P_v = 0.1 \times 821.24 = 82.12 < 606.6$ kN, therefore stiffeners required.

If flat plates, $t_s = b_s/9 = (B - t)/(2 \times 9) = (189.9 - 8.5)/18 = 10.08$ mm.

Use 10 mm plate. Use the same stiffeners at E.

A continuous $457 \times 191 \times 67$ kg UB can be used for both spans. If a minimum weight design is required then:

(a) a smaller section can be used for span AB but there would have to be a splice which would increase fabrication costs.

(b) A smaller section could be used for span AB, with flange plates added to increase the moment of resistance for span BC, but this would also increase fabrication costs.

4.8   Tutorial problems

4.8.1   Properties of an angle section

Determine the position of the principal axes and the values of the principal second moments of area for a $200 \times 150 \times 18$ mm angle. Use the centre line of the elements and ignore the radii.

Answers

$A = 5976$ mm^2; $c_x = 63.94$; $c_y = 38.94$ mm; $I_x = 23.837$E6; $I_y = 11.555$E6; $I_{xy} = -9.831$E6 mm^4; $\alpha = 29°$; $I_u = 29.287$E6; $I_v = 6.105$E6 mm^4

4.8.2   Bending about principal axes

Calculate the bending stresses for positions A, B and C for the moments as shown in Fig. 4.9, for a $200 \times 150 \times 18$ mm angle. Consider bending about the principal axes.

Answers

$u_A = -65.06$; $v_A = -37.04$; $u_B = +74.86$; $v_B = -94.02$; $u_C = +47.65$; $v_C = 129.15$ mm; $M_u = 18.97$; $M_v = 27.66$ kNm; $f_A = -318.76$; $f_B = +278.27$; $f_C = 299.54$ N/mm^2

4.8.3   Elastic displacements

Calculate the elastic central deflection of the column CG shown in Fig. 4.12 if the end moments are $M_{CG} = +36$, and $M_{GC} = +18$ kNm (positive clockwise). Also find the rotation of joint C.

Answers

$\delta = 2.75$ mm (right to left); $\theta_C = +0.00275$ radians

4.8.4   Position of shear centre

Determine the position of the shear centre for a $432 \times 102 \times 65.54$ kg channel ignoring the tapers and radii.

Answers

$x' = 30.83$ mm

4.8.5   Section properties of a tee section

Determine the elastic and plastic section properties of a $419 \times 457 \times 194$ kg structural tee.

Answers

$A = 24.50$E3 mm^2; $y_e = 356.4$ mm   from   base; $I_e = 440.84$E6 mm^4; $z_{e1} = 4.245$E6; $z_{e2} = 1.237$E6 mm^3; $y_p = 29.13$ mm from top of section; $S = 2.187$E6 mm^3

4.8.6   Plastic collapse of a propped cantilever

A beam, of span L, is built-in at A and simply supported at B and a point load is applied at C at a distance of $L/4$ from B. Determine the plastic collapse load (W_p) if the plastic moment of resistance of the section is m_p. If $L = 6$ m, $W_p = 100$ kN and $p_y = 275$ N/mm^2 determine the minimum size of Universal Beam section assuming the beam is restrained laterally and ignoring self weight. Finally if the load factor is 1.5 calculate the maximum bending stress at service load.

Answers

$W_p = 20m_p/(3L)$; $S = 327.3E3$ mm³; $254 \times 102 \times 28$ kg   UB; $b/T = 5.105 < 8.5$; $d/t = 35.2 < 63$, therefore plastic; $M_C = 81WL/512 = 63.28$ kNm; $f_{max} = 205.5$ N/mm²

4.8.7   Web strength of a beam

Determine the web shear, buckling and bearing strengths of a beam for an angle support of the type shown in Fig. 4.25(b). A $200 \times 100 \times 15$ mm angle supports a $457 \times 152 \times 82$ kg UB of Grade 43 steel and the clearance is 5 mm.

Answers

$p_y = 265$ N/mm²; $d/t = 38 < 64.2$; $P_v = 791.3$; $P_w = 373.3$; $P_{wbg} = 302.1$ kN

References

Astill A. W., Holmes M. and Martin L. H. (1980). *Web buckling of steel 'I' beams*, CIRIA Tech. note 102.
Croxton P. C. L. and Martin L. H. (1987 and 1989). *Solving Problems in Structures Vols. 1 and 2*, Longman Scientific and Technical.
Horne, M. R. (1971). *Plastic Theory of Structures*, Nelson.
Horne, M. R. (1958). *The full plastic moment of sections subjected to shear force and axial loads*, British Welding Journal, **5**, 170.
Longbottom, E. and Heyman J. (1956). *Experimental verification of the strength of plate girders designed in accordance with the revised BS153 tests on full size and on model plate girders*, Proc. ICE, **5**(III), 462.
Megson, T. H. G. (1980). *Strength of Materials for Civil Engineers*, Nelson.
Morris, L. J. and Randall, A. L. (1979). *Plastic Design*, Constrado.
Moy, S. S. (1981). *Plastic Methods for Steel and Concrete Structures*, Macmillan Press.
Trahair N. S. and Bradford M. A. (1988). *The Behaviour and Design of Steel Structures*, Chapman and Hall.
Zbirohowski-Koscia K. (1967). *Thin Walled Beams*, Crosby Lockwood.

5

Laterally Unrestrained Beams

5.1 Introduction

In the previous chapter it was assumed that under flexure a beam could attain its full plastic moment under load. A prerequisite to this assumption is that the compression flange should be unable to deflect in a horizontal plane. If this condition is not satisfied then a phenomenon known as lateral buckling or lateral torsional buckling can occur. This phenomenon is analogous to the premature failure of a strut by Euler buckling before the squash load is attained.

The chapter will initially consider the theory of lateral torsional buckling for sections symmetric about both axes. It will then consider how the problem is handled in design for rolled sections. Finally the chapter will consider design methods used for fabricated beams such as plate girders, and castellated beams.

5.2 Lateral torsional buckling for a beam symmetric about both axes

Consider the beam sketched in Fig. 5.1(a). A moment of M is applied at both ends in such directions as to produce single curvature bending with the top flange in compression. If the ends are supported (Fig. 5.1(b)) in such a manner as to allow rotation about a vertical axis but to prohibit relative displacement of the top and bottom flange, and thereby prohibit twisting at the support, then the buckled shape will be as Fig. 5.1(c).

The equation governing bending about the minor axis is

$$EI_y \, d^2u/dz^2 = -M\phi \qquad (5.1)$$

where EI_y is the flexural rigidity about the minor axis.

The remaining symbols are defined in Fig. 5.1.

The torsion equation has two components; that due to uniform torsion

$$GJ \, d\phi/dz,$$

where GJ is the torsional rigidity and the warping torsion

$$EH \, d^3\phi/dz^3,$$

where EH is the warping rigidity, and the disturbing torque of $M \, du/dz$.

The two torsion terms act in opposite directions and thus the final equation is

$$GJ \, d\phi/dz - EH \, d\phi^3/dz^3 = M \, du/dz \qquad (5.2)$$

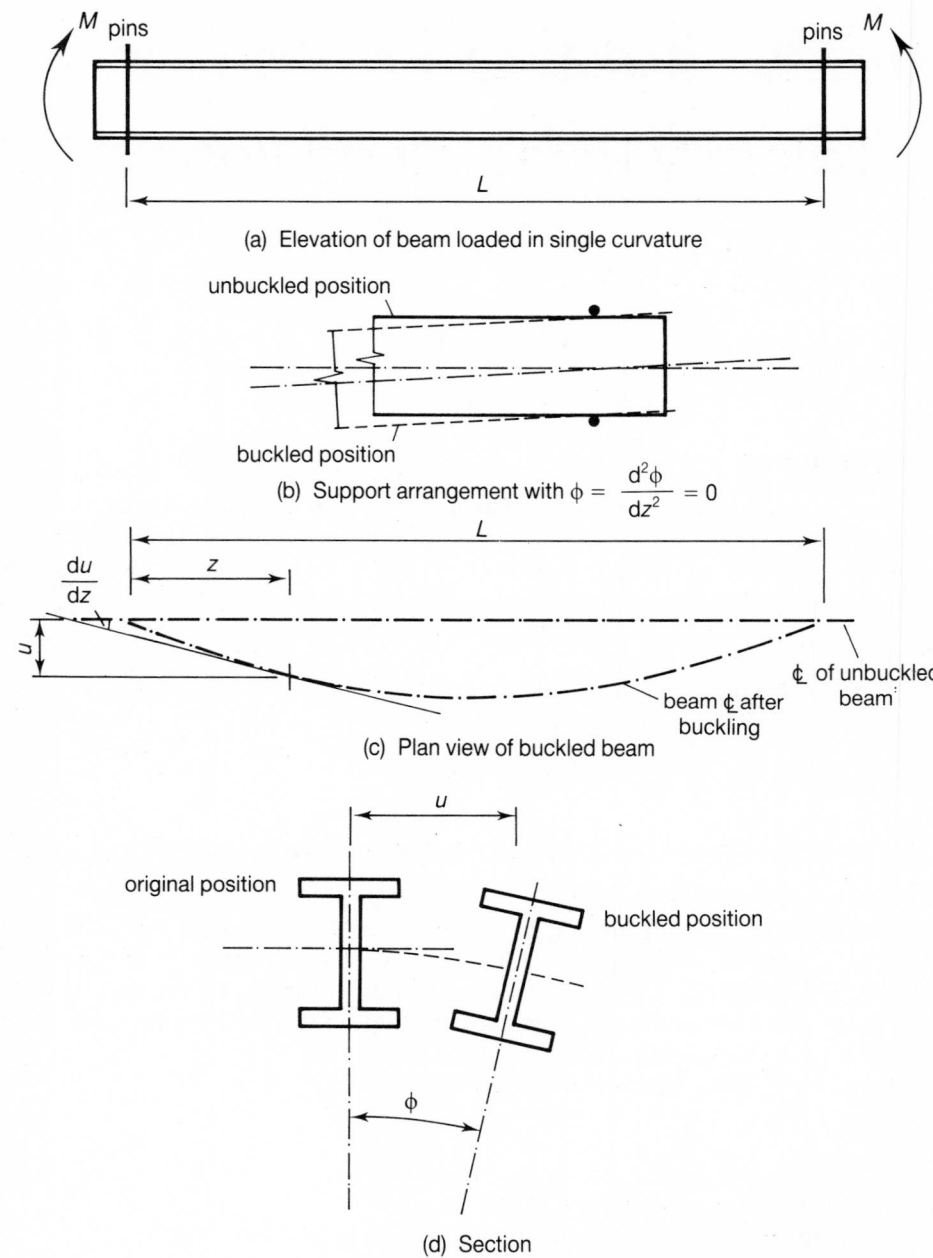

(a) Elevation of beam loaded in single curvature

(b) Support arrangement with $\phi = \dfrac{d^2\phi}{dz^2} = 0$

(c) Plan view of buckled beam

(d) Section

Fig. 5.1   Lateral torsional buckling of beams.

For the full derivation of Equations (5.1) and (5.2) reference should be made to Trahair and Bradford, or Kirby and Nethercot.

If Equation (5.2) is differentiated with respect to z, and d^2u/dz^2 is substituted from Equation (5.1), the following governing equation is obtained,

$$GJ\,d^2\phi/dz^2 - EH\,d^4\phi/dz^4 = -M^2/EI_y \qquad (5.3)$$

The solution to Equation (5.3) is given by the first term of a Fourier series for ϕ,

$$\phi = \phi_0 \sin(\pi z/L) \tag{5.4}$$

where ϕ_0 is the angle of twist at mid-span.

Substituting Equation (5.4) into Equation (5.3) together with the boundary condition $\phi = 0$ at $z = L$, gives with M_E, the elastic critical buckling moment, replacing the generalized moment, M,

$$M_E = (\pi/L)(EI_y GJ)^{0.5}(1 + \pi^2 EH/(GJL^2))^{0.5} \tag{5.5}$$

Equation (5.5) ignores the effect of major axis bending which may be allowed for by introducing a factor, γ, defined as

$$\gamma = 1 - I_y/I_x \tag{5.6}$$

with the result that the equation for the elastic critical moment may be written as

$$M_E = (\pi/L)(EI_y GJ/\gamma)^{0.5}(1 + \pi^2 EH/(GJL^2))^{0.5} \tag{5.7}$$

If the behaviour of a set of beams is considered which were geometrically perfect, i.e. had no imperfections due to lack of straightness, and had no residual stresses due to the manufacturing process, then at low slenderness ratios the beam achieves its full plastic capacity, whereas at high slenderness ratios the behaviour closely approximates to that predicted by Equation (5.7). At intermediate slenderness ratios the behaviour is dependent on the buckling moment and the plastic collapse moment. This behaviour is indicated by the chain dotted line in Fig. 5.2. If the results from a series of tests on real beams which are geometrically imperfect and have residual stresses are plotted (Fig. 5.2) (from Kirby and Nethercot), it will be seen that the results for real beams lie below those for ideal beams except at low slenderness ratios where the beams behave plastically. A more detailed discussion of residual stresses is presented later in this section.

It was observed in experimental work that the effect of the strength of the beam, i.e. its plastic moment capacity, and any variation in elastic properties could be eliminated by plotting the results non-dimensionally with the abscissa plotted as $(M_p/M_E)^{0.5}$ and the ordinate as M_b/M_p, where M_p is the plastic moment capacity of the section and M_b is the failure moment.

By analogy with strut behaviour, for which reference should be made to Section 6.5.2, the slenderness ratio, λ, may be defined by

$$\lambda = (\pi^2 E/p_y)^{0.5}(P_y/P_E)^{0.5} \tag{5.8}$$

where P_E is the Euler buckling load and P_y is the squash load, the Code defines an equivalent slenderness ratio for beam buckling, λ_{LT}, by

$$\lambda_{LT} = (\pi^2 E/p_y)^{0.5}(M_p/M_E)^{0.5} \tag{5.9}$$

Equation (5.9) is not convenient for design purposes as the calculation of M_E is cumbersome. Further, note that in Equation (5.9) all variables must be in consistent units, e.g. Newtons and metres.

To simplify the problem two further section properties are required and are defined by either of the following Equations, (5.10) or (5.11), for sections symmetric about both axes.

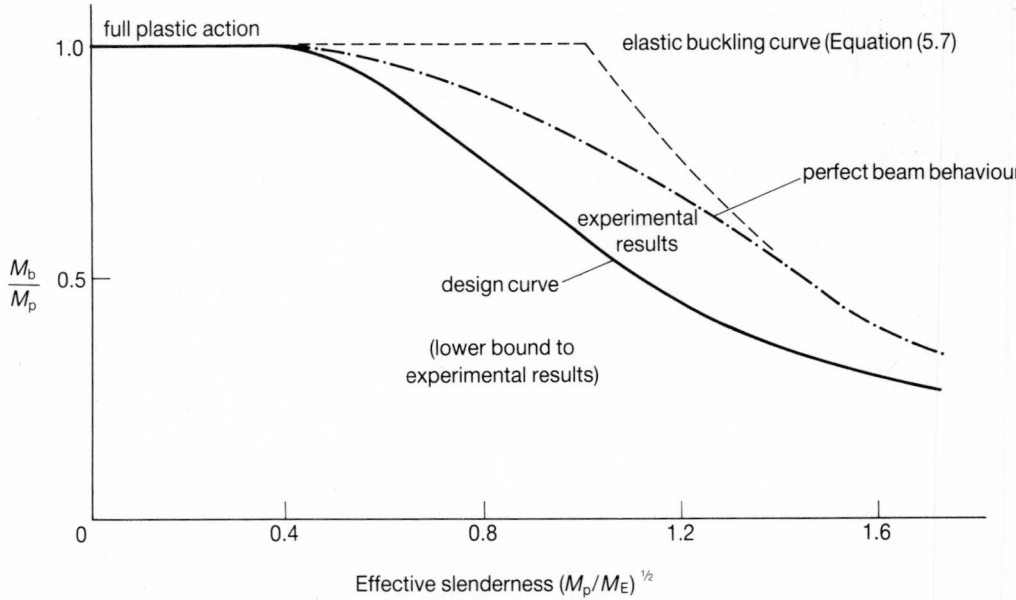

Fig. 5.2   Beam behaviour with lateral torsional buckling.

For sections symmetric about the major axis Equation (5.10) must be used, whereas for sections symmetric about the minor axis Equation (5.11) is to be used.

$$x = 1.132(AH/I_y J)^{0.5}$$
$$u = (I_y S_x^2 \gamma/AH)^{0.25} \tag{5.10}$$

and

$$x = 0.566h_s(A/J)^{0.5}$$
$$u = (4S_x^2 \gamma/A^2 h_s^2)^{0.25} \tag{5.11}$$

where h_s is the distance between the shear centres of the flanges.

Given that

$$r_y^2 = I_y/A \tag{5.12}$$
$$G = E/2(1 + \nu) = E/2.6 \tag{5.13}$$

since $\nu = 0.3$ (cl 3.1.2, Pt 1)

and defining

$$\lambda = L/r_y \tag{5.14}$$

Equation (5.5) may be written as

$$M_E = (\pi/L)(EI_y GJ/\gamma)^{0.5}(1 + \pi^2 \times 2.6x^2 I_y J/A(1.132\lambda r_y)^2))^{0.5} \tag{5.15}$$

which reduces to

$$M_E = (\pi/L)(EI_y GJ/\gamma)^{0.5}(1 + 20(x/\lambda)^2))^{0.5} \tag{5.16}$$

or

$$M_E = (\pi/L)(EI_y GJ/\gamma)^{0.5}(20)^{0.5}(x/\lambda)(1 + (1/20)(\lambda/x)^2))^{0.5} \tag{5.17}$$

Defining

$$v = (1 + (\lambda/x)^2/20)^{-0.25} \tag{5.18}$$

and substituting the values of u and x defined in Equation (5.10) together with Equation (5.18) into Equation (5.16) gives

$$M_E = \pi^2 ES_x/(\lambda uv)^2 \tag{5.19}$$

Note Equation (5.18) defining v only applies to sections whose shear centre lies on the major axis but is not necessarily coincident with the minor axis. For sections whose shear centre lies on the minor axis but which is not coincident with the major axis reference should be made to Section 5.6.

Substituting Equation (5.19) into the definition of λ_{LT} given by Equation (5.9), i.e.,

$$\lambda_{LT} = (\pi^2 E/p_y)^{0.5}(M_p/M_E)^{0.5}$$

gives

$$\lambda_{LT} = uv\lambda \tag{5.20}$$

To deal with the intermediate cases between buckling and full plastic moment a 'Perry-Robertson' approach is used (see Section 6.5.2), in that the moment that a beam can sustain, M_b, is given as the least root of

$$(M_b - M_E)(M_b - M_p) = \eta_{LT} M_E M_b \tag{5.21}$$

where η_{LT} is a coefficient to allow for initial imperfections and residual stresses.

It should be noted that unlike the situation for struts, dealt with in Chapter 6, there is no theoretical justification for the use of a 'Perry-Robertson' approach to the calculation of the moment capacity of a beam. It is entirely an empirical approach derived from experimental results. It is also to be noted that this approach for struts is strictly only applicable for dealing with initial curvature or non-concentric loading. The problem of residual stresses caused by non-uniform cooling of the section after rolling is much more complex. It has been found from experimental work that these residual stresses are often very high, in some cases of the order of the yield strength of the material. Typical results for the magnitude of residual stresses are given in Fig. 5.3 (Kirby and Nethercot). These residual stresses do not affect the plastic capacity of the section, as plastic behaviour is independent of any imposed self-equilibrating residual stress patterns, but they reduce the capacity of the section when any elastic state of stress is considered and therefore reduce the moment capacity when buckling takes place (Nethercot (1974a)). Since it is very difficult to predict either the magnitude of these residual stresses or their effect, it was decided that in the Code a convenient way of dealing with this was to increase the coefficient η_{LT} in the Perry-Robertson equation beyond that strictly required to deal with initial imperfections due to curvature and non-concentric loading. In the case of sections fabricated from plates by welding, the residual stresses are due to deformations induced during the welding process and are much more severe than the stresses induced in rolled section (Nethercot (1974b)).

It was noticed from tests (Fig. 5.2) that buckling does not occur at values of $(M_p/M_E)^{0.5}$ of less than 0.4, thus a limiting slenderness ratio, λ_{LO}, is defined by

$$\lambda_{LO} = 0.4(\pi^2 E/p_y)^{0.5} \tag{5.22}$$

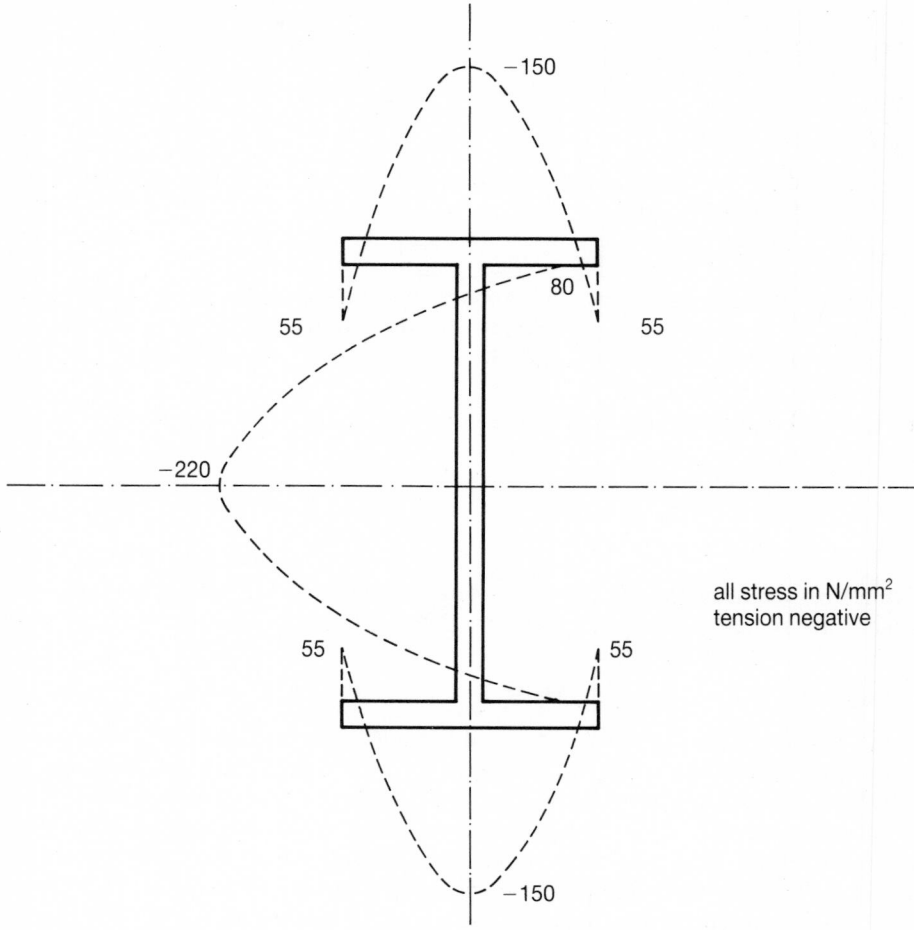

Fig. 5.3 Typical residual stresses in a grade 43 'I' section.

The imperfection constant, η_{LT}, is defined, depending on the type of section, whether rolled or welded, in terms of a constant, α_b, which has a value of 0.007, the slenderness ratio λ_{LT} and the limiting slenderness ratio λ_{LO}.
For rolled sections

$$\eta_{LT} = \alpha_b (\lambda_{LT} - \lambda_{LO}) \geqslant 0 \tag{5.23}$$

for welded sections

$$\eta_{LT} = 2\alpha_b \lambda_{LO} \tag{5.24}$$

subject to the conditions

$$\eta_{LT} \leqslant 2\alpha_b (\lambda_{LT} - \lambda_{LO})$$
$$\eta_{LT} \geqslant \alpha_b (\lambda_{LT} - \lambda_{LO}) \tag{5.25}$$
$$\eta_{LT} \geqslant 0$$

In order to demonstrate the principles involved a full calculation will be done to determine the flexural capacity of a beam loaded in single curvature. It is to be emphasized that this is not the approach that would be used in design.

EXAMPLE 5.1   Calculation of moment capacity

Determine the maximum moment that can be carried by a 254 × 146 × 43 UB in Grade 50 steel loaded by end moments in single curvature and simply supported over a span of 8 m in such a way as to avoid twisting of the section at the supports.

Section properties:

$$T = 12.7 \text{ mm}, \quad J = 24.1 \text{ cm}^4, \quad H = 0.103 \text{ dm}^6, \quad I_x = 6560 \text{ cm}^4, \quad I_y = 677 \text{ cm}^4,$$
$$S_x = 568 \text{ cm}^3, \ D = 259.6 \text{ mm and } r_y = 3.51 \text{ cm}.$$

From Table 6, $p_y = 355 \text{ N/mm}^2$

$$E = 205 \text{ kN/mm}^2$$
$$G = E/2(1 + v) = 205/2(1 + 0.3) = 78.85 \text{ kN/mm}^2$$
$$\gamma = 1 - I_y/I_x = 1 - 677/6560 = 0.897$$

Use Equation (5.7) to evaluate M_E

$$M_E = (\pi/L)(EI_y GJ/\gamma)^{0.5}(1 + \pi^2 EH/(GJL^2))^{0.5}$$

working in kNm units,

$$EI_y = 205E6 \times 677E\text{-}8 = 1387.5 \text{ kNm}^2$$
$$GJ = 78.85E6 \times 24.1E\text{-}8 = 19.00 \text{ kNm}^2$$
$$EH = 205E6 \times 0.103E\text{-}6 = 21.12 \text{ kNm}^4$$
$$L = 8 \text{ m}$$
$$M_E = (\pi/8)(1387.85 \times 19.00/0.897)^{0.5}(1 + \pi^2 \times 21.12/19.00 \times 8^2))^{0.5}$$
$$= 72.87 \text{ kNm}$$
$$M_p = p_y S_x = 355 \times 568/1E3 = 201.64 \text{ kNm}$$

Use Equation (5.9) to calculate λ_{LT}

$$\lambda_{LT} = (\pi^2 E/p_y)^{0.5}(M_p/M_E)^{0.5}$$
$$= (\pi^2 \times 205E3/355)^{0.5}(201.64/72.87)^{0.5}$$
$$= 125.6$$

Use Equation (5.22) to calculate λ_{LO}

$$\lambda_{LO} = 0.4(\pi^2 E/p_y)^{0.5}$$
$$= 0.4(\pi^2 \times 205E3/355)^{0.5}$$
$$= 30.2$$

Use Equation (5.23) to calculate η_{LT}

$$\eta_{LT} = \alpha_b(\lambda_{LT} - \lambda_{LO})$$
$$= 0.007(125.6 - 30.2)$$
$$= 0.668$$

Use Equation (5.21) to calculate M_b

$$(M_b - M_E)(M_b - M_p) = \eta_{LT} M_E M_b$$
$$(M_b - 72.87)(M_b - 201.64) = 0.668 \times 72.87 M_b$$

from which M_b equals either 54.74 or 268.5 kNm.

The latter value is inadmissible since it exceeds the plastic moment capacity, thus

$$M_b = 54.74 \text{ kNm}.$$

The theoretical approach outlined above needs modifying to deal with the real design situation in that it is very rare for a beam to be loaded in single curvature, i.e. to have moments applied only to its ends, and to have the end conditions specified when deriving the basic theory. It is thus necessary to modify the above approach to allow for loading along the span of the beam either in the form of distributed loading or in the form of discrete point loading and to allow for the effects of support conditions where for example twisting may occur or where warping may be suppressed.

BS 5950 Part 1 is presented in terms of an allowable bending strength, p_b, defined by

$$p_b = M_b / S_x \tag{5.26}$$

Dividing both sides of Equation (5.21) by S_x^2, and defining an elastic critical buckling stress, p_E, given by $p_E = M_E / S_x$, gives

$$(p_b - p_E)(p_b - p_y) = \eta_{LT} p_E p_b \tag{5.27}$$

The elastic critical buckling stress is obtained by substituting Equation (5.20) into (5.19), i.e.

$$p_E = \pi^2 E / \lambda_{LT}^2 \tag{5.28}$$

The imperfection constant η_{LT} is only dependent on λ_{LT} and λ_{LO}. The latter, the limiting slenderness ratio, given by Equation (5.22) is dependent only on p_y. Thus the allowable bending stress, p_b, is a function only of the design strength p_y and the equivalent slenderness ratio λ_{LT}. Values of p_b for rolled sections are given in Table 11 and for welded sections in Table 12.

Before the problems of non-uniform moments and varying end conditions can be discussed it is necessary to define a series of terms:

Lateral restraint (cl 4.3.2, Pt 1).

A lateral restraint is one which prevents the lateral movement (in a direction normal to the span) of the beam as a whole. This restraint should have adequate stiffness to prevent such movement and should be designed to take 2.5% of the force in the compression flange of the beam at ultimate limit state. Further discussion of this point is provided in Section 8.7.2 of this text.

Torsional restraint (cl 4.3.3, Pt 1).

A torsional restraint is where relative movement in a vertical plane between the flanges of the beam is prevented. Such restraints should be capable of resisting a couple determined by multiplying the distance between the centroids of the flanges by a force equal to 1% of the maximum compression force in the flange at ultimate loading.

Effective length of a beam.

This can be defined in a similar manner to that for a strut (Chapter 6) in that it is the length L in the elastic critical moment Equation (5.5) which will give the same critical moment as the standard case for an alternative support condition, e.g. a cantilever in which one end is rigidly held (encastré) and the other end free.

5.3   Design modifications to the basic procedure

5.3.1   Effective lengths

It is convenient to consider simply supported beams, which themselves can be divided into two categories–those with end restraints only and those with additional intermediate restraints–and cantilevers separately.

5.3.1.1   *Destabilizing loads (cl 4.3.4, Pt 1)*

The basic theory assumes that any applied loading on the top flange is not free to move laterally with the flange relative to the beam centroid as the beam buckles, i.e. its line of action is fixed relative to the beam centroid. Clearly when this condition does not hold there will be an additional torsion term in the buckling Equation (5.2) caused by the lateral movement of the load with the resultant effect that the elastic critical moment is reduced. Although for certain loading cases the calculation of the elastic critical moment can be made (Trahair and Bradford, Anderson and Trahair), this is of limited use and the Code allows for the case of destabilizing loads by making increases in the effective lengths for the normal case in which the load is not destabilizing of about 20% for simply supported beams and substantially higher increases for cantilevers which are more prone to the effects of destabilizing loads owing to the presence of a free end.

Bradford (1989) indicates that for beams on a seat support (where there is no restraint in any form to the top, compression, flange), with the beam subjected to destabilizing loads the use of an effective length of $1.2L + 2D$ can produce unsafe estimates of the buckling capacity.

5.3.1.2   *Simply supported beams–end restraints only (cl 4.3.5, Pt 1)*

In this case the effective lengths should be taken from Table 9. It is often difficult to relate the theoretical support condition in Table 9 with that achieved by practical connections. Some guidance on this is given in Section 5.3.3.

5.3.1.3   *Simply supported beams–intermediate restraints (cl 4.3.5, Pt 1)*

In this case each segment of the beam whether between intermediate restraints or an intermediate restraint and an end restraint should be assessed using Table 9 with L being taken as the distance between restraints for each segment of the beam. In the case of destabilizing loads $1.2L$ should be used.

5.3.1.4   *Cantilevers (cl 4.3.6, Pt 1)*

Where a cantilever has intermediate restraints to the compression flange, or has a moment applied to the tip of the cantilever it should be treated in line with the provisions of a simply supported beam (Sections 5.3.1.2 and 3). Where, however, it is unrestrained along its length and subject to a loading system other than a moment at the tip, Table 10 should be used.

5.3.2   Effects due to loading

Here two cases must be considered. The first is when loading is applied only at positions of lateral restraint and the second is where loading is applied between positions of lateral restraint. Note that in general the effect of the self weight of the beam can be ignored when assessing whether the load is applied at or between lateral restraints (see Example 5.3).

5.3.2.1   *Loading applied at points of lateral restraint*

Consider the beam in Fig. 5.4 which is loaded through steel cross beams bolted to the top flange. The bending moment diagram is also given in Fig. 5.4(b). Since the beam has lateral restraints at A, B, C and D each section of the beam can be considered separately for lateral torsional buckling (Fig. 5.4(c)). It will then be observed that each section has a moment gradient along its length, i.e. the moments at each end are unequal. From experimental work (Nethercot and Rockey) it was observed that the imposition of a moment gradient increased the load carrying capacity of the beam when failure involving buckling took place.

This phenomenon is handled by calculating the beam capacity, M_b, as if under a uniform moment, but comparing this value with a value of the design moment $\bar{M}$ which is given by

$$\bar{M} = mM_A \tag{5.29}$$

where M_A is the numerical maximum applied moment in the beam segment being considered and m is a moment reduction factor given by

$$m = 0.57 + 0.33\beta + 0.10\beta^2 \geqslant 0.43 \tag{5.30}$$

where β is the ratio of the value of the smaller moment at one end of the beam segment to the larger value at the other end such that $1.0 \geqslant \beta \geqslant -1.0$ (Fig. 5.5)

5.3.2.2   *Beam with loading applied between lateral restraints*

The method described above for loading applied only at restraints will overestimate the moment capacity of the beam if applied to the case where loading is applied between restraints, and will thus be unsafe (Trahair and Bradford, Kirby and Nethercot) and so an alternative empirical approach has been adopted whereby the design moment $\bar{M}$ is set equal to M_A and the beam slenderness Equation (5.20) is modified with an additional factor n to become

$$\lambda_{LT} = nuv\lambda \tag{5.31}$$

Values of n are taken from Tables 15 and 16 of BS 5950 Pt 1. It will be seen that Tables 15 and 16 are dependent on two moment ratios β and γ. The signs of these are important. Table 17 gives typical configurations and also the signs of the two moment ratios. The use of Tables 15 and 16 is best illustrated by examples (see Example 5.4).

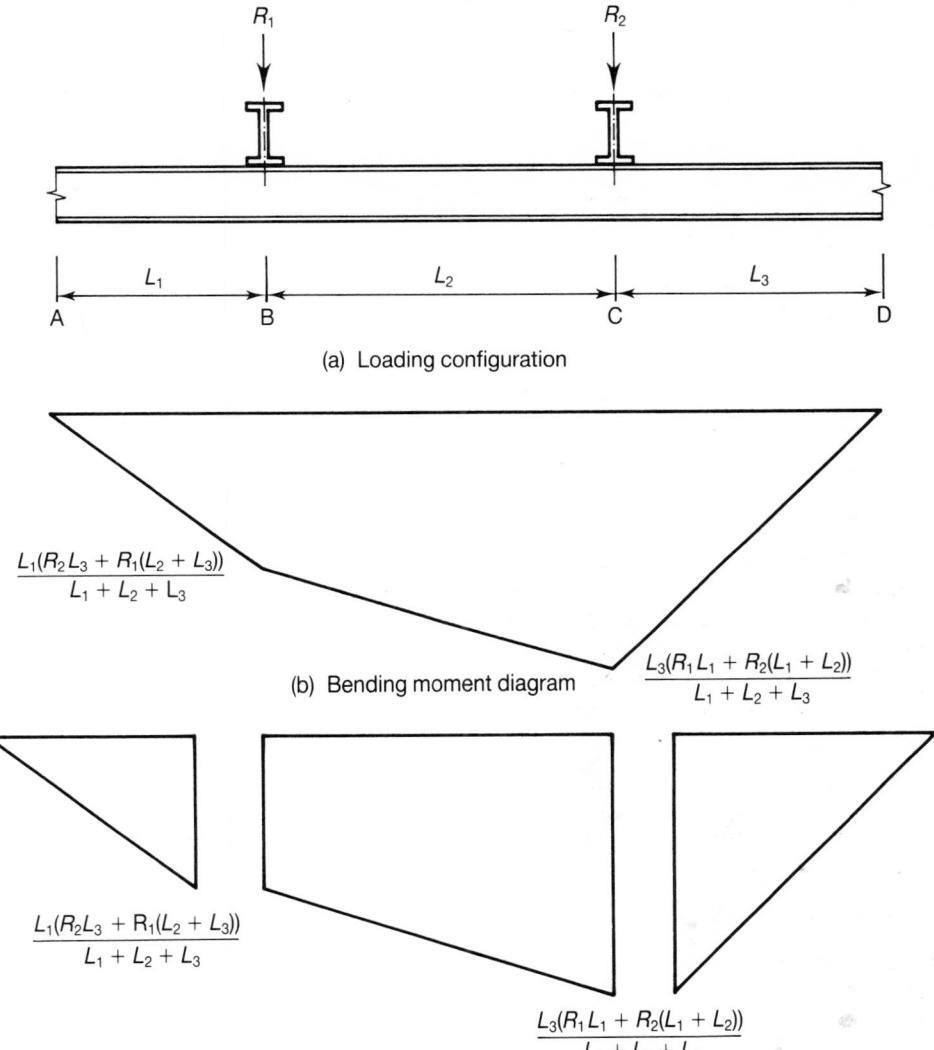

(a) Loading configuration

$$\frac{L_1(R_2 L_3 + R_1(L_2 + L_3))}{L_1 + L_2 + L_3}$$

(b) Bending moment diagram

$$\frac{L_3(R_1 L_1 + R_2(L_1 + L_2))}{L_1 + L_2 + L_3}$$

$$\frac{L_1(R_2 L_3 + R_1(L_2 + L_3))}{L_1 + L_2 + L_3}$$

$$\frac{L_3(R_1 L_1 + R_2(L_1 + L_2))}{L_1 + L_2 + L_3}$$

(c) Bending moment diagram for each segment

Fig. 5.4   Loading applied at lateral restraints.

5.3.3   Effect of connection detailing on effective lengths

Examination of Table 9 leads to the observation that the effective length of a beam is dependent on whether there is torsional restraint to the end of the beam and whether there is lateral restraint to the compression flange at the support and on the relative rotation between the tension and compression flanges at the support. The existence of torsion restraint, or lateral restraint to the compression flange or the suppression of relative rotation between the tension and compression flanges clearly depend on the connection detail or support detail at the ends of the beam. Figure 5.6 gives some typical connection

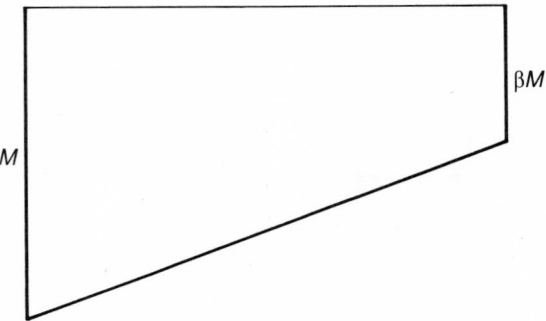

(a) Moments at the end of a beam segment having
    the same sign ($0 \leqslant \beta \leqslant 1.0$)

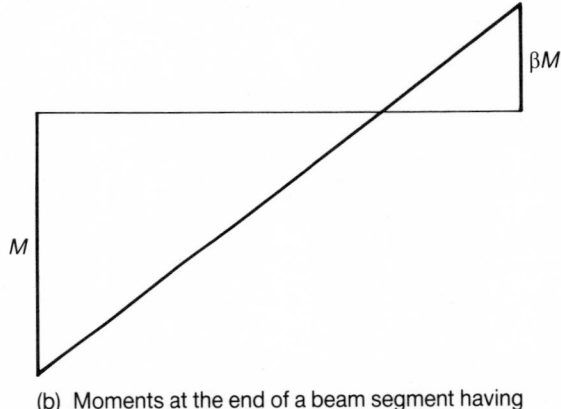

(b) Moments at the end of a beam segment having
    opposite sign ($|M| > |\beta M|$, $-1 \geqslant \beta \geqslant 0$)

Fig. 5.5   Determination of the sign of the moment ratio β.

and support details with notes on the restraint offered at the end of the beam (after Pillinger).

5.3.4   Use of member capacity tables for flexural design

In Volume 1 of the Steelwork Design Guide to BS 5950 Part 1 produced by SCI and the BCSA a set of flexural capacity tables for rolled steel members in both Grade 43 and 50 are provided. For each section size the values of M_b are tabulated for a range of effective lengths L_E which depends on the section size, and for values of n, the slenderness correction factor for loading applied between restraints, between 0.4 and 1.0 in increments of 0.2. For each section, the section classification is also given. The tabulated value of M_b should be compared with the design value $\bar{M}$. The plastic moment capacity of the beam M_p must, of course, be greater than the maximum value of the applied moment M_A.

Linear interpolation for intermediate values of n may be used, but it should be noted that the value of M_b will be conservative and that this could in extreme cases lead to the selection of a larger beam than strictly necessary. Note, it is still necessary to carry out the remaining design checks both under ultimate and service load conditions.

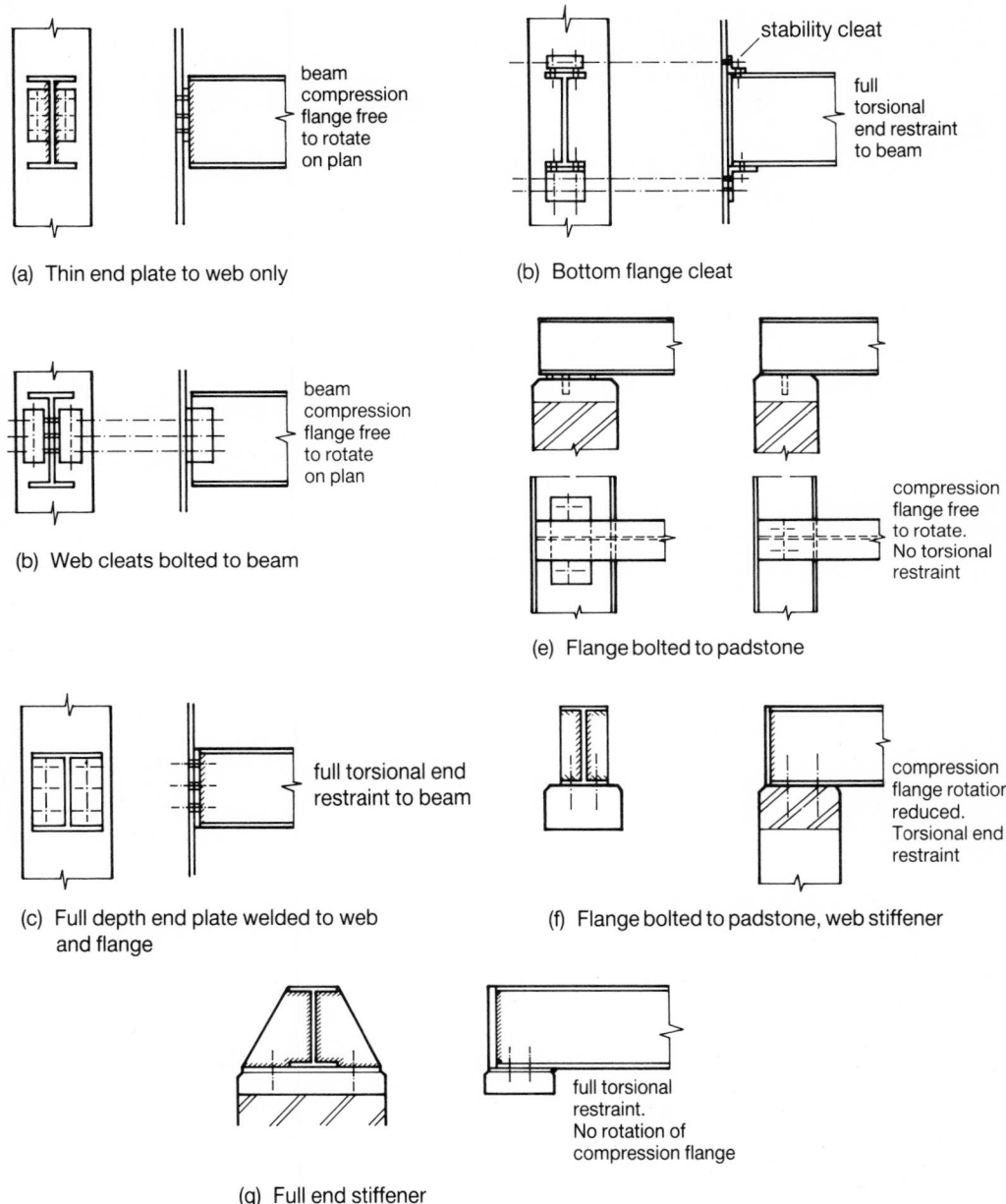

(a) Thin end plate to web only

beam compression flange free to rotate on plan

(b) Web cleats bolted to beam

beam compression flange free to rotate on plan

(c) Full depth end plate welded to web and flange

full torsional end restraint to beam

(b) Bottom flange cleat

stability cleat

full torsional end restraint to beam

(e) Flange bolted to padstone

compression flange free to rotate. No torsional restraint

(f) Flange bolted to padstone, web stiffener

compression flange rotation reduced. Torsional end restraint

(g) Full end stiffener

full torsional restraint. No rotation of compression flange

Fig. 5.6 Typical connection and support detail.

Before continuing by giving a series of design examples to illustrate the design of beams subjected to lateral torsional buckling, it will be convenient to summarize the design process to check for lateral torsional buckling. The fact that a check needs to be made in a particular design for lateral torsional buckling does not override the basic design principle that at any point in the beam the magnitude of the applied moment must be less than the capacity of the beam at ultimate limit state. The design summary is best done as an annotated flow chart (Fig. 5.7).

EXAMPLE 5.2   *Calculation of moment capacity–code approach*

Recalculate the carrying capacity of the beam of Example 5.1 using the code approach first using exact values for u and x and secondly approximate values.

(a)   Exact values

$n = 1.0$ (Loading at restraints only)

$u = 0.889$ (Appendix B)

$\lambda = L_E/r_y$

$\quad = 8000/35.1 = 227.9$

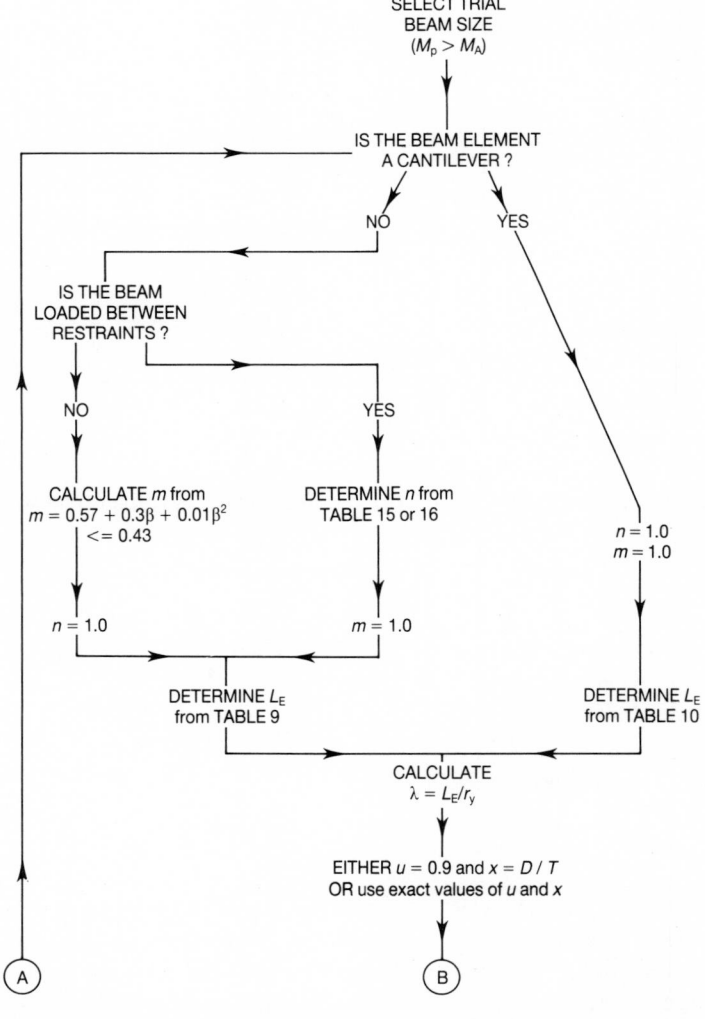

Fig. 5.7   Flow chart for rolled section beam design with lateral torsional buckling.

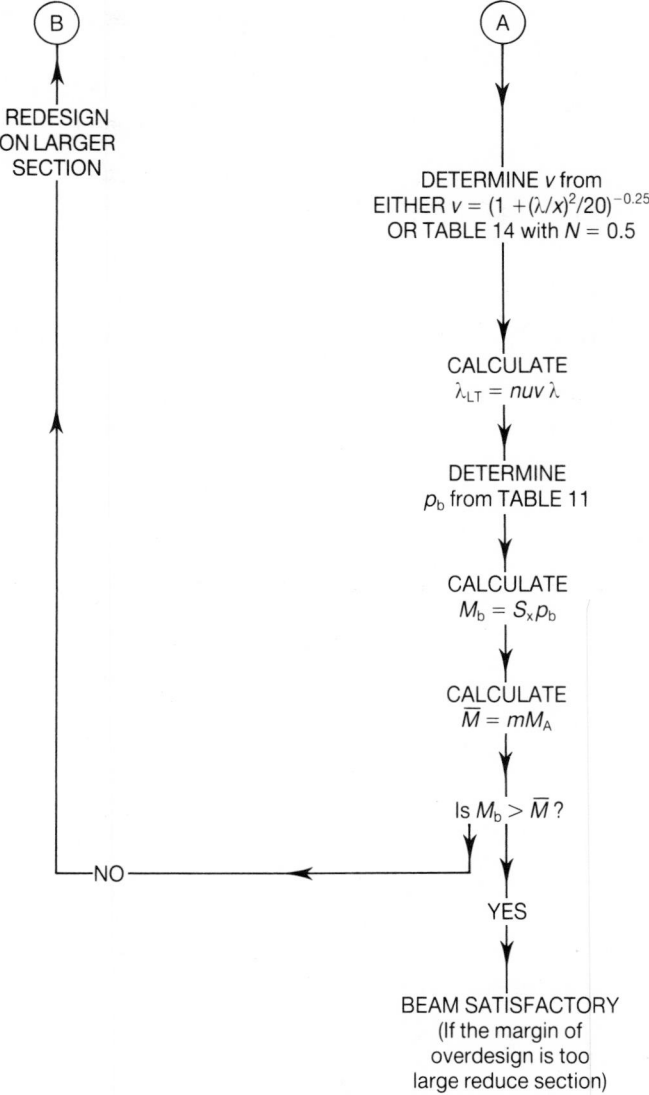

Note: For the design of any beam element it is permissible to set $m = n = 1.0$
for a conservative design

Fig. 5.7   continued

$$x = 21.1$$
$$\lambda/x = 227.9/21.1 = 10.8$$
$$v = 0.619 \text{ (either Table 14, with } N = 0.5 \text{ or Equation (5.18))}$$
$$\lambda_{LT} = nuv\lambda$$
$$= 1.0 \times 0.889 \times 0.619 \times 227.9$$
$$= 125.4$$
$$p_b = 96 \text{ N/mm}^2 \text{ (Table 11, with } p_y = 355),$$

$$M_b = p_b S_x$$
$$= 96 \times 568/1E3 = 54.53 \text{ kNm}$$

(b)   Approximate calculations

$$u = 0.9$$
$$x = D/T = 259.6/12.7 = 20.4$$
$$\lambda/x = 227.9/20.44 = 11.15$$
$$v = 0.61$$
$$\lambda_{LT} = 1.0 \times 0.9 \times 0.61 \times 227.9 = 125.1$$

The value of λ_{LT} is virtually identical to that calculated above and the calculation need be taken no further.

The difference in the two approaches will not always give as close an answer as this. However the use of $u = 0.9$ and $x = D/T$ will alway give an answer less than the exact approach.

EXAMPLE 5.3   Beam with loading applied at restraints

Prepare a design in Grade 50 steel for the beam for which the data are given in Fig. 5.8.

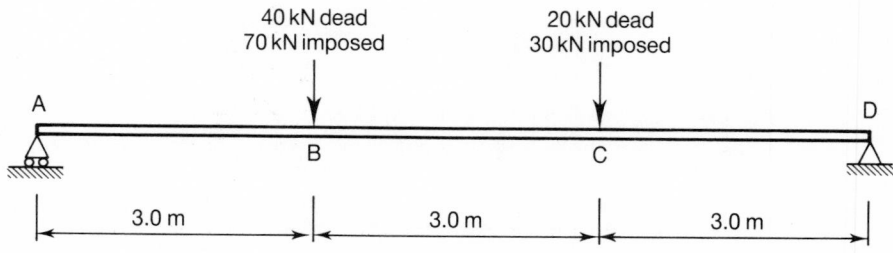

Notes: (1) All loads are service loads
(2) Lateral torsional restraints exist at A, B, C and D

Fig. 5.8   Design data for Example 5.3.

Factored loading at ULS:

at B. $1.4 \times 40 + 1.6 \times 70 = 168$ kN
at C. $1.4 \times 20 + 1.6 \times 30 = 76$ kN
            Total load $= 244$ kN

The self weight of the beam will be allowed for by slightly overdesigning and then finally checking.

Reactions: at A 137 kN, at D 107 kN

The bending moment and shear force diagrams are given in Fig. 5.9.

It is not usual for shear to be critical on a simply supported beam using a rolled section, therefore first design for lateral torsional buckling. It is clear from the bending moment

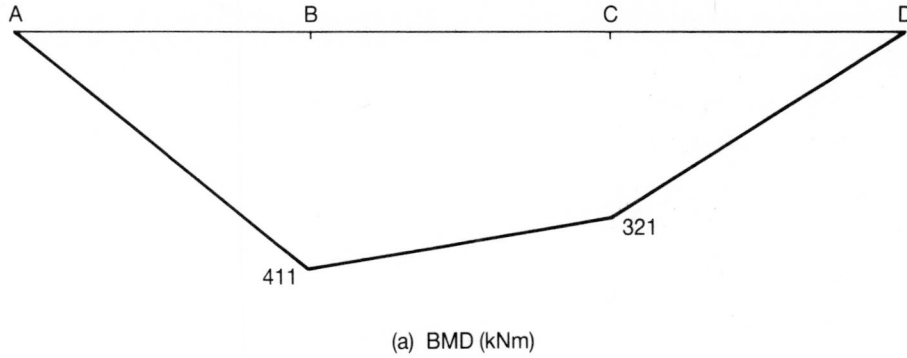

(a) BMD (kNm)

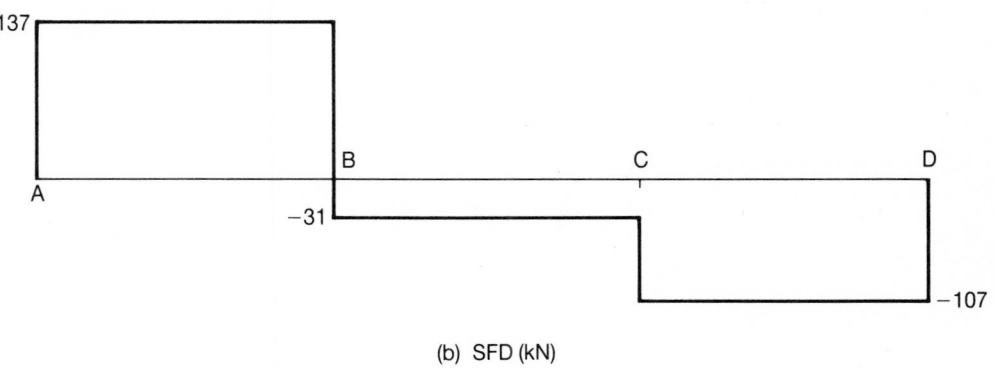

(b) SFD (kN)

Fig. 5.9 BMD and SFD for Example 5.3.

diagram that the smallest moment gradient is in the middle segment of the beam (BC) since the effective lengths will be the same on each section as the segment lengths and the end conditions are identical.

The loads are not destabilizing, since the loads though applied on the top flange are not free to move as they are applied at points of lateral torsional restraint.

Since the loading is at restraint points, $n = 1$, and m, therefore, will need calculating.

Beam segment BC.

Ratio of moments at the ends of BC,

$$\beta = 321/411 = 0.78$$

$$m = 0.57 + 0.33\beta + 0.10\beta^2 \qquad (5.30)$$
$$= 0.57 + 0.33 \times 0.78 + 0.10 \times 0.78^2$$
$$= 0.89$$

Calculate the design moment, $\bar{M}$

$$\bar{M} = mM_A \qquad (5.29)$$

M_A is the maximum moment at the end of the segment, i.e. the moment at B, so

$$\bar{M} = 0.89 \times 411$$
$$= 365.8 \text{ kNm}$$

Try a $406 \times 178 \times 74$ UB ($M_p = 532.5$ kNm)

First check allowable ratios of b/T and d/t for plastic action (Table 7)

$$\varepsilon = (275/p_y)^{0.5}$$

$p_y = 355 \text{ N/mm}^2$ as $T = 16$ mm, so

$$\varepsilon = (275/355)^{0.5} = 0.89$$

Flange outstand:

Maximum value of $b/T = 8.5\varepsilon = 8.5 \times 0.89 = 7.57$
Actual value $= 5.62$

Web slenderness:

Maximum value of $d/t = 79\varepsilon = 79 \times 0.89 = 70.3$
Actual value $= 37.2$

Both are satisfactory, therefore the section is plastic.

Section properties:

$D = 412.8 \text{ mm}$, $t = 9.7 \text{ mm}$, $I_x = 27300 \text{ cm}^4$, $r_y = 4.03 \text{ cm}$, $S_x = 1500 \text{ cm}^3$,
$u = 0.881$, $x = 27.6$

Effective length, L_E
Since the beam is restrained laterally and torsionally at B and C,

$$L_E = 1.0L = 1.0 \times 3000 = 3000 \text{ mm}$$

$$\lambda = L_E/r_y = 3000/40.3$$
$$= 74.4$$

$$\lambda/x = 74.4/27.6 = 2.70$$

$v = 0.925$ (using either Equation (5.18) or Table 14 with $N = 0.5$)

$$\lambda_{LT} = nuv\lambda$$
$$= 1.0 \times 0.881 \times 0.925 \times 74.4$$
$$= 60.6$$

(5.31)

From Table 11 with $p_y = 355$

$$p_b = 252 \text{ N/mm}^2$$

$$M_b = p_b S_x$$
$$= 252 \times 1500/1E3$$
$$= 378 \text{ kNm.}$$
$$\bar{M} = 366 \text{ kNm.}$$

Thus the beam is adequate.

However check for the effects of self weight of the beam:

Factored self weight $= 1.4 \times 74 \times 9.81/1\text{E}3 = 1.01$ kN/m.

Maximum moment due to self weight $= 1.01 \times 9^2/8 = 10.2$ kNm.

Effective design moment due to self weight $= 0.89 \times 10.2 = 9.1$ kNm, thus, total design moments, $\bar{M} = 365.8 + 9.1 = 374.9$ kNm.

This is still less than the moment capacity.

Strictly the end portions of the span ought to be checked:

$$L_E = 3000,$$

thus $M_b = 378$ kNm, as before

$$\beta = 0, \quad \text{so} \quad m = 0.57,$$

and $\bar{M} = mM_A = 0.57 \times 411 = 234.2$ kNm.

Thus the end sections of the beam are also satisfactory.

Shear check:

Shear capacity, $P_v = 0.6p_y Dt$
$$= 0.6 \times 355 \times 412.8 \times 9.7/1\text{E}3$$
$$= 852.9 \text{ kN}.$$

This is greater than the maximum reaction at A of 137 kN.

Also given the large margin by which the shear capacity exceeds the applied shear, there is no need to check for reduction in the plastic moment capacity due to shear.

Web buckling under the two point loads:

This will be found to be satisfactory as the web capacity exceeds the applied loading.

Deflection:

Midspan deflection, δ, due to a point load, W, is given by

$$\delta = (WL^3/48EI)(3(b/L) - 4(b/L)^3)$$

where W is the load, L the span, EI the flexural rigidity and b the least distance from the load to either support.

load at B:

$W = 70$ kN (imposed load only), $L = 9$ m, $b = 3$ m, $EI = 205\text{E}6 \times 27300\text{E}{-8} = 55965$ kNm^2, thus

$$\delta = (70 \times 9^3/(48 \times 55965))(3(3/9) - 4(3/9)^3)$$
$$= 0.016 \text{ m}$$

Load at C:

      $W = 30$ kN, all other data as at B

      $\delta = 0.007$ m

Total deflection $= 0.016 + 0.007 = 0.023$ m
allowable deflection is span/200 (Table 5)
span/200 $= 9/200 = 0.045$ m

This is greater than the actual deflection and is therefore satisfactory.

EXAMPLE 5.4   Beam design with loading applied between lateral torsional restraints

Prepare a design in Grade 50 steel for the beam whose data are given in Fig. 5.10.

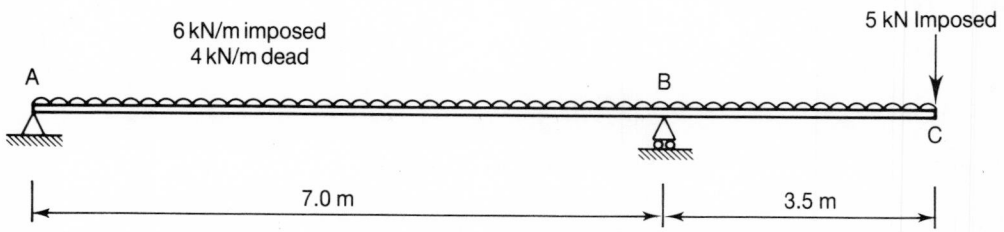

Notes: (1) All loading is service loading
       (2) Lateral torsional restraints exist at A and B
       (3) The TOP flange is restrained at C

Fig. 5.10   Design data for Example 5.4.

Factored loading:

      UDL: $1.4 \times 4 + 1.6 \times 6 = 15.2$ kN/m

      Point load: $1.6 \times 5 = 8$ kN

Figure 5.11 shows the beam under factored loading together with the bending moment and shear force diagram.

For this example it is necessary to check both span AB and BC.

Span AB.

Since the beam is torsionally restrained and laterally restrained at both supports, the effective length factor is 1.0 (Table 9). It is not permissible to take the effective length as the distance between the points of contraflexure calculated from the applied loading as these will not form nodal points in the post-buckled shape of the beam. Nodal points in the post-buckled shape can only occur at points of lateral restraint, since at a nodal point the lateral deflection in the post-buckled shape must be zero (Kirkby and Nethercot).

    It will be assumed that although the loading is applied to the top flange it is not free to move laterally and is therefore not destabilizing.

    Since the loading is applied between restraints $m = 1$, and n, therefore, will need calculating

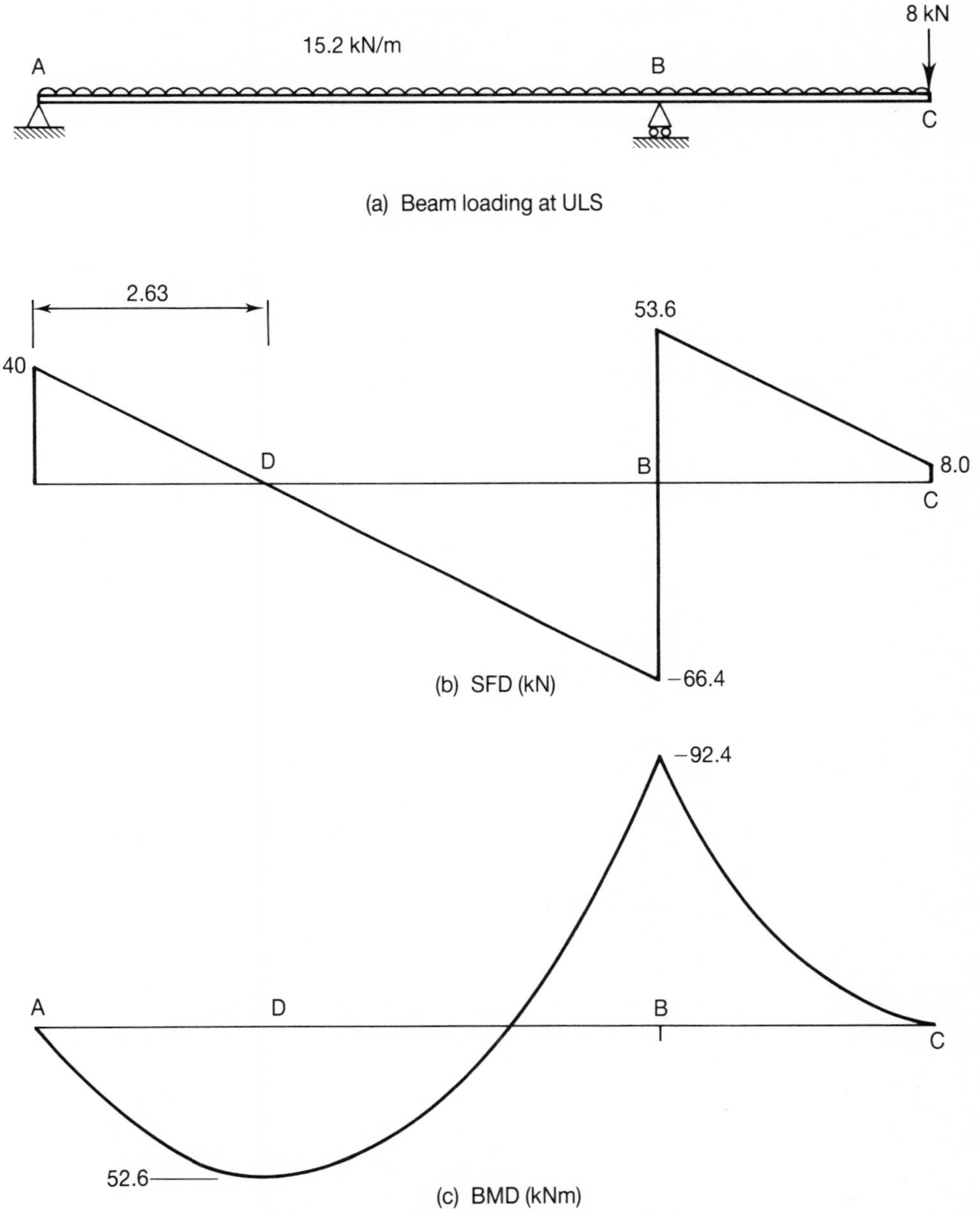

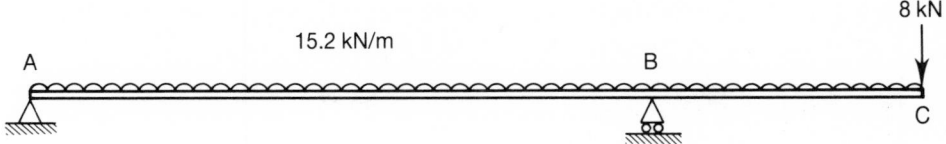

8 kN
15.2 kN/m
A
B
C
(a) Beam loading at ULS
2.63
53.6
40
D
B
8.0
C
−66.4
(b) SFD (kN)
−92.4
A
D
B
C
52.6
(c) BMD (kNm)

Fig. 5.11 BMD and SFD for Example 5.4.

Try a 305 × 127 × 48 Grade 50 UB ($M_p = 148$ kNm)

The checks for b/T and d/t indicate that the section is plastic.

$S_x = 706$ cm^3, $r_y = 2.75$ cm, $D = 310.4$ mm, $t = 8.9$ mm, $T = 14.0$ mm, $x = 23.3$, $u = 0.874$, $I_x = 9500$ cm^4

$$\bar{M} = mM_A = 1.0 \times 92.4 = 92.4 \text{ kNm}$$

Note that M_A is calculated at the point of the maximum absolute valued moment irrespective of whether that moment is hogging or sagging.

$$L_E = 7000 \text{ mm}$$

$$\lambda = L_E/r_y = 7000/27.5$$
$$= 255$$

$$\lambda/x = 255/23.3 = 10.94$$

$$v = 0.615$$

Calculation of n:

The BMD for span AB is replotted in Fig. 5.12(a) and below is drawn the reference diagram from Table 16 (Fig. 5.12(b)).

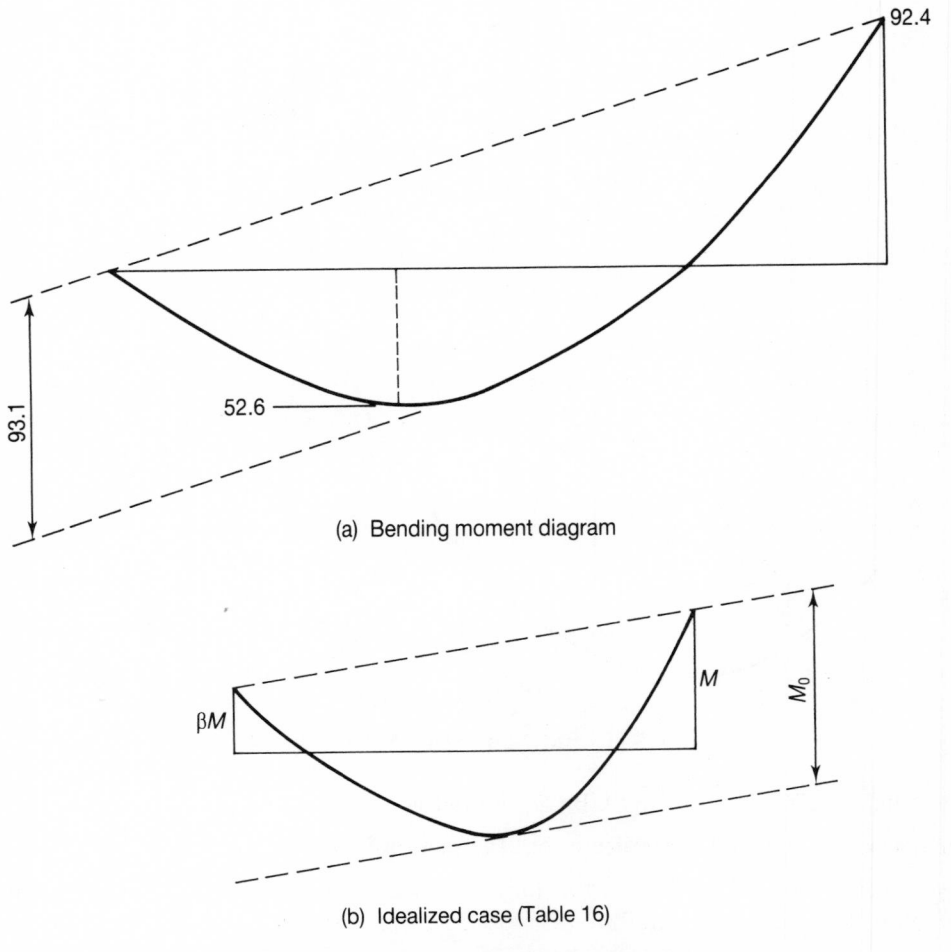

(a) Bending moment diagram

(b) Idealized case (Table 16)

Fig. 5.12   Calculation of n for Example 5.4.

Table 16 is to be used since the loading is not substantially concentrated in the middle fifth.

Comparison between Figs. 5.12(a) and (b) indicates that values of β and γ are needed.

Calculation of β:

Maximum moment, M, is 92.4 kNm at B and minimum moment, βM, is 0 at A, thus

$$\beta = 0$$

Calculation of γ:

Bending moment at midspan (with the span taken as the distance between restraints), $M_0 = 15.2 \times 7^2/8 = 93.1$ kNm.

From Table 17, the third example in that Table is equivalent to the case under consideration here and indicates that γ is negative, thus

$$\gamma = -M/M_0 = -92.4/91.4 = -0.99.$$

From Table 16 with $\beta = 0$ and $\gamma = -0.99$, $n = 0.66$

$$\lambda_{LT} = nuv\lambda$$
$$= 0.66 \times 0.615 \times 0.874 \times 255$$
$$= 91$$

From Table 11 with $p_y = 355$, $p_b = 160$ N/mm^2

$$M_b = p_b S_x$$
$$= 160 \times 700/1E3 = 112.0 \text{ kNm}$$

This is satisfactory.

Span BC

This span acts as a cantilever, so $m = n = 1.0$ (Fig. 5.7).

The effective length factor is 0.9 (Table 10), since the cantilever is continuous over the support with lateral and torsional restraint and has top flange restraint at the free end, and the loading is not destabilizing.

$$\lambda = 0.9 \times L_E/r_y = 0.9 \times 3000/27.5$$
$$= 98$$

$$\lambda/x = 98/23.3 = 4.21$$

$$v = 0.853$$

$$\lambda_{LT} = nuv\lambda$$
$$= 1.0 \times 0.874 \times 0.853 \times 98$$
$$= 73$$

From Table 11, $p_b = 199$ N/mm^2

$$M_b = p_b S_x = 199 \times 706/1E3 = 140 \text{ kNm}$$

$$\bar{M} = mM_A = 1.0 \times 92.4 = 92.4 \text{ kNm}$$

The beam is therefore satisfactory for buckling in both spans.

The shear check will be found to be satisfactory as will the web bearing check at B.

Deflection check:

Since the loading configuration cannot easily be decomposed into the standard cases given in Section 4.2.7, the relevant formulae and the definitions of symbols are given in Fig. 5.13(a) for a UDL and Fig. 5.13(b) for the point load.

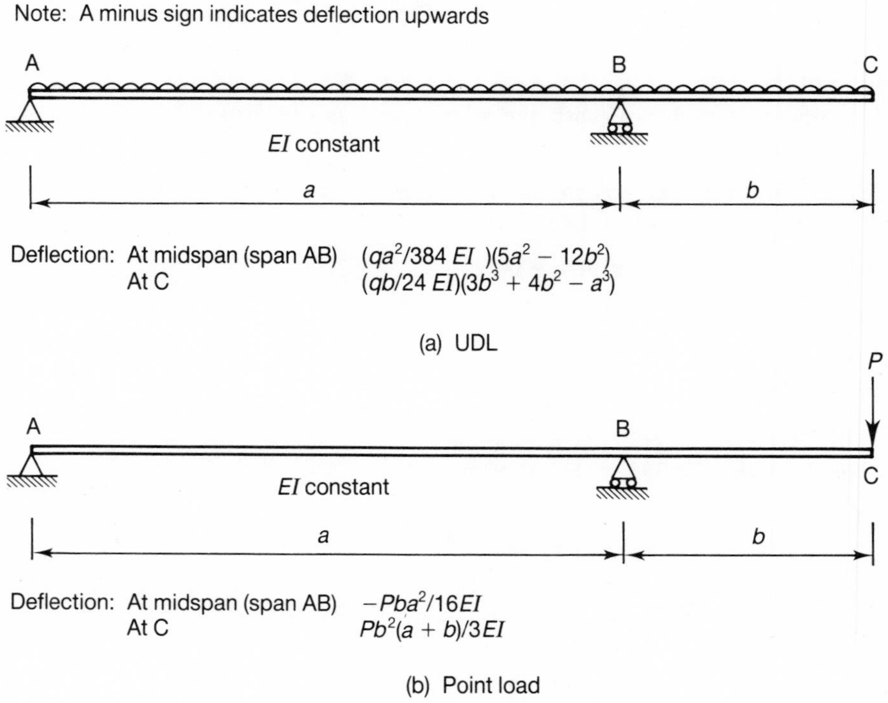
Note: A minus sign indicates deflection upwards
A
B
C
EI constant
a
b
Deflection: At midspan (span AB) $(qa^2/384\,EI)(5a^2 - 12b^2)$
At C $(qb/24\,EI)(3b^3 + 4b^2 - a^3)$
(a) UDL
P
A
B
C
EI constant
a
b
Deflection: At midspan (span AB) $-Pba^2/16EI$
At C $Pb^2(a + b)/3EI$
(b) Point load

Fig. 5.13 Deflection formulae for Example 5.4.

Working in kN and m as basic units,

$$EI = 205\text{E}6 \times 9500\text{E-}8 = 19475 \text{ kNm}^2$$

Span AB:

$$\begin{aligned}\delta &= (6 \times 7^2/384 \times 19475)(5 \times 7^2 - 12 \times 3^2) - 5 \times 3 \times 7^2/(16 \times 19475)\\ &= 0.005 - 0.002\\ &= 0.003 \text{ m}\end{aligned}$$

Allowable deflection = span/200 (Table 5) span/200 = 7/200 = 0.035, therefore deflection is satisfactory.

Span BC:

$$\begin{aligned}\delta &= (6 \times 3/(24 \times 19475))(3 \times 3^3 + 4 \times 3^2 - 7^3) + 5 \times 3^2(7 + 3)/(3 \times 19475)\\ &= -0.009 + 0.008\\ &= -0.001 \text{ m}\end{aligned}$$

This deflection is upwards, so strictly no check is required on its magnitude.

5.4 Conservative design approaches

There are two methods to be considered here.

5.4.1 Method one

This is given in the Code in the last line of Table 13, and will be found as a note to Fig. 5.7. It quite simply sets $m = n = 1.0$, and then uses the methods detailed above.

5.4.2 Method two (cl 4.3.7.7, Pt 1)

This method is only applicable to rolled universal sections or joists and to simply supported beams, and is based on the following equations:

$$p_b/p_y = (1 + (1 + \eta)(p_e/p_y))/2 + (((1 + (1 + \eta)(p_e/p_y)/2)^2 - p_e/p)^{0.5}$$
$$\eta = 0.007 \, (\pi^2 E/p_y)^{0.5}((p_y/p_e)^{0.5} - 0.4)$$

and

$$p_e = (2.5E6/\lambda^2)(1 + (\lambda/x)^2/20)^{0.5}$$

The first equation is the solution of the Perry-Robertson equation for beam buckling (Equation (5.27)), the second equation is Equation (5.23) written in terms of stresses, and the third equation is a simplified version of the critical elastic buckling moment equation (Equation (5.5)) obtained taking approximations for the section properties similar to those used to obtain Table 7 of BS 449. These equations form the basis of Table 19.

This approximate method cannot be used for cantilevers, or for simply supported beams which cantilever over either one or both supports.

The method is best presented as a series of steps:

Step 1. Calculate $\lambda = nL_E/r_y$

where λ is the slenderness ratio

L_E the effective length between restraints determined in accordance with Sections 5.3.1.2 and 3 of this text

n is a correction factor for the loading pattern and is taken from Table 20

Note, this value of n is NOT to be confused with the value of n from Section 5.3.2.2.

Step 2. Determine the torsional index, x, for the section.

It is sufficiently accurate to take $x = D/T$.

Step 3. Using the appropriate section of Table 19, determine p_b.

Step 4. Calculate $M_b = p_b S_x$ and compare with the maximum value of the applied moment in the beam segment being considered.

EXAMPLE 5.5   Beam capacity using the conservative approach

Calculate the carrying capacity of the beam in Examples 5.1 and 5.2 using the methods of Section 5.4.2.

Step 1. Calculation of λ

$n = 1.0$ (from top diagram of Table 20)

$\lambda = 1.0 \times 8000/35.1 = 227.9$

Step 2. Calculation of x

$x = D/T = 259.6/12.7 = 20.44$

Step 3. Determination of p_b

from Table 19(d), $p_b = 98$ N/mm^2

Step 4. Calculation of M_b

$M_b = p_b S_x = 98 \times 568/1\text{E}3 = 55.7$ kNm.

This value is very slightly higher than the value obtained using the exact methods, but the difference is not significant.

5.5   Lateral torsional buckling for other rolled sections

5.5.1   Channel sections

These are treated exactly the same as rolled 'I' sections since they have equal top and bottom flanges, even though the shear centre is not at the centroid of the section. It does however lie on the major axis, thus the value of v may either be calculated from Equation (5.18) or found from Table 14 with $N = 0.5$.

5.5.2   Box sections (including rectangular hollow sections (RHS)) (cl B2.6, Pt 1)

5.5.2.1   *Sections of uniform wall thickness (including RHS)*

If the values of the slenderness ratio do not exceed the values given in Table 38 Pt 1, then the sections may be designed to give full plastic moment capacity. If these conditions are not satisfied then the procedure given in Section 5.5.2.2 should be followed.

5.5.2.2   *Sections of non-uniform wall thickness*

The modified slenderness ratio, λ_{LT}, is given by

$$\lambda_{LT} = 2.25n(\phi_b \lambda)^{0.5} \tag{5.32}$$

where ϕ_b is the buckling index and is defined by

$$\phi_b = ((S_x^2/AJ)(1 - I_y/I_x)(1 - J/2.6I_x))^{0.5} \tag{5.33}$$

where A is the area, I_x and I_y are the second moments of area for major and minor axes respectively, and J is the torsional constant, which for a Box section is given by

$$J = 4A_h^2 / \sum (s/t)$$

where A_h is the area enclosed by the mean perimeter and s and t are the breadth and thickness of each enclosing element respectively.

For an RHS section a more accurate value for J is given by

$$J = t^3 h/3 + 4A_h^2 t/h$$

where h is the mean perimeter of the section.

5.5.3 Plates and flats (cl B2.7, Pt 1)

For a flat or plate subjected to moment about the major axis, the beam slenderness ratio, λ_{LT}, is given by

$$\lambda_{LT} = 2.8n(L_E d/t^2)^{0.5} \tag{5.34}$$

where d is the depth and t the thickness of the plate.

Equation (5.34) is derived directly from the basic elastic critical moment equation (Equation (5.5)) with the warping constant H set equal to zero, and the definition of λ_{LT} given in Equation (5.9).

5.5.4 Angles

Here the approach is different in that the maximum stress in the angle must remain elastic under all loading conditions imposing flexure. The allowable stress is dependent on the slenderness ratio in that the higher the slenderness ratio, the lower the permissible stress,

$$
\begin{aligned}
&\text{for} \quad L/r_w \leqslant 100, \ M_b = 0.8p_y Z \\
&\text{for} \quad L/r_w \leqslant 180, \ M_b = 0.7p_y Z \\
&\text{for} \quad L/r_w \leqslant 300, \ M_b = 0.6p_y Z
\end{aligned}
\tag{5.35}
$$

where L is the unrestrained length, r_w the radius of gyration about the weakest axis, and Z the elastic section modulus about the relevant axis.

Having considered the behaviour of rolled sections with respect to lateral torsional buckling, it is now useful to continue by considering fabricated sections, i.e. compound beams, plate girders and castellated beams, both with and without lateral torsional buckling. For these varieties of beam it will also be necessary to consider additional design criteria that were not covered in the previous chapter on the design of restrained beams. These additional design criteria include further studies of the behaviour of web buckling and the design of web stiffeners, neither of which are generally required for rolled sections.

5.6 Compound beams

5.6.1 Introduction

Compound beams are beams fabricated from rolled sections by either welding additional flange plates on one or both flanges or by welding an inverted channel on to the top flange of an 'I' beam. Typical cross sections of compound beams are given in Fig. 5.14.

Compound beams are used either where there is restricted construction depth and it is necessary therefore to increase the flexural capacity of the basic rolled section (Fig. 5.14(a)), or where it is necessary to stiffen the top flange, and increase flexural capacity, where substantial horizontal loads are applied at the level of the top flange in, say, crane beams (Fig. 5.14(b) and (c)).

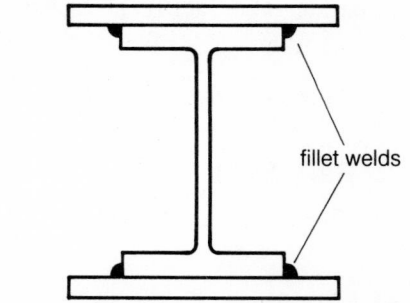

fillet welds

(a) Both flanges strengthened with cover plates

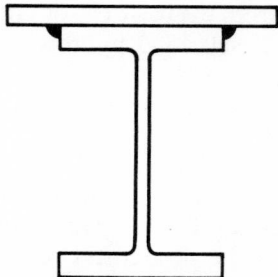

(b) Compression flange only strengthened with a cover plate

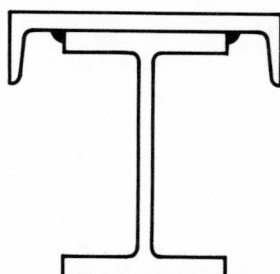

(c) Section strengthened with an inverted channel section

Fig. 5.14 Typical compound beam cross sections.

Where the beams are fully restrained against lateral torsional buckling the design follows the same procedure as that of Chapter 4, except that the section properties required must be calculated from first principles. Where, however, the beams are not fully restrained lateral torsional buckling will need consideration. The basic effect on the buckling behaviour is as follows: when the compression flange is larger than the tension flange, as is the case in most compound beams, the elastic critical moment is enhanced, and when the tension flange is larger, the critical moment is decreased. This effect is due to an additional disturbing torque caused by the effect of unequal flanges. The additional torque is negative when the compression flange is larger, hence giving more stability, and is positive when the tension flange is larger, hence giving less stability. This additional torque is caused by the shear centre and elastic centre not being coincident. Where equal strengthening is applied to both flanges, as in Fig. 5.14(a), this effect is not present as the shear centre is still coincident with the elastic centre.

Most compound beams are usually symmetric about the minor axis, and thus only the effect of the lack of symmetry about the major axis must be taken into account when determining the buckling behaviour. A beam which is symmetric about only one axis is known as monosymmetric. This has the effect that the basic buckling theory presented in Section 5.2 needs modifying to allow for the effect of monosymmetry.

5.6.2 Lateral torsional buckling of monosymmetric beams

The effect of unequal flanges is to introduce an additional torque, T_{add}, acting on the section given by,

$$T_{add} = M\beta_x \, d\phi/dz \tag{5.36}$$

where β_x is a parameter which is a function of the degree of monosymmetry of the section and is defined as,

$$\beta_x = 2y_0 - \int_A (y(x^2 + y^2) \, dA)/I_x \tag{5.37}$$

where y_0 is the distance from the shear centre to the centroid of the compression flange, and the co-ordinate system is such that the x axis corresponds to the major axis of the section and the y axis to the minor axis. The integral is evaluated over the whole cross section of the beam.

For the method used to calculate the position of the shear centre reference should be made to Section 4.3.5.

The governing differential equation for lateral torsional buckling is obtained by modifying the equation for sections symmetric about both axes (Equation (5.3)) to take account of the additional torque from Equation (5.36) thus giving

$$(GJ + \beta_x M) \, d^2\phi/dz^2 - EH \, d^4\phi/dz^4 = - M^2/EI_y \tag{5.38}$$

By following a similar procedure to that for the case of symmetry about both axes, the elastic critical moment, M_E, is given by

$$M_E = (\pi/L)(EI_y \, GJ/\gamma)^{0.5}((1 + \pi^2 EH/(GJL^2) + (\gamma_M/2)^2)^{0.5} + \gamma_M/2) \tag{5.39}$$

where

$$\gamma_M = (\beta_x/L)(EI_y/GJ)^{0.5} \tag{5.40}$$

The warping constant, H, can no longer be calculated as

$$H = I_y h_s^2/4 \tag{5.41}$$

which is only applicable to doubly symmetric beams, but is calculated from

$$H = N(1 - N)I_y h_s^2 \tag{5.42}$$

where N is defined as the ratio of the second moment of area of the compression flange about the minor axis of the beam (I_{cf}) to the sum of the second moments of area of the tension and compression flanges about the minor axis of the beam $(I_{cf} + I_{tf})$, or

$$N = I_{cf}/(I_{cf} + I_{tf}) \tag{5.43}$$

Note, for equal flange beams, $N = 0.5$.

The same definition of λ_{LT} as for doubly symmetric beams, is used from monosymmetric beams, i.e.

$$\lambda_{LT} = (\pi^2 E/p_y)^{0.5}(M_p/M_E)^{0.5} \tag{5.9}$$

λ_{LT} may also be written as

$$\lambda_{LT} = uv\lambda \tag{5.20}$$

but with v calculated from

$$v = ((4N(1 - N) + (\lambda/x)^2/20 + \psi^2)^{0.5} + \psi)^{-0.5} \tag{5.44}$$

where $\psi = \beta_x/h_s$ $\tag{5.45}$

and with u and x calculated from Equation (5.11).

Note that with $N = 0.5$ and $\psi = 0$, Equation (5.44) reduces to Equation (5.18).

The exact solution for calculating β_x for common monosymmetric sections is given in Fig. 5.15 (after Anderson and Trahair).

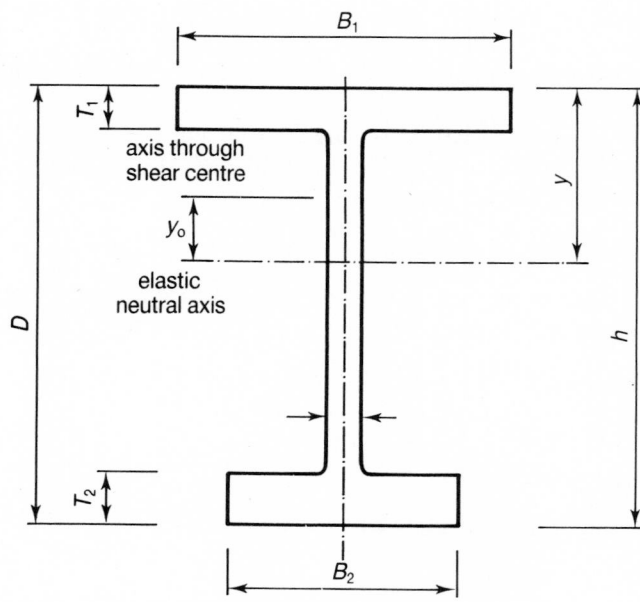

$$y_0 = h/[1 + (B_1/B_2)^3(T_1/T_2)] - \bar{y}$$
$$\beta_x = (h - \bar{y})(B_2^3 T_2/12 + B_2 T_2(h - \bar{y})^2 + (h - \bar{y})^3 t/4)/I_x$$
$$\quad - \bar{y}(B_1^3 T_1/12 + B_1 T_1(\bar{y})^2 + (\bar{y})^3 t/4)/I_x - 2y_0$$

Fig. 5.15 Exact calculation of monosymmetry index.

However, the Code does allow a simplification to the calculation of ψ, in that, when

$$N > 0.5, \ \psi = 0.8(2N - 1)(1 + D_L/D)$$

$$N < 0.5, \ \psi = 1.0(2N - 1)(1 + D_L/D)$$

(5.46)

where D is the overall depth of the section and D_L is the depth of any lip to the top flange.

For a full discussion of the behaviour of monosymmetric beams reference should be made to Anderson and Trahair, or Trahair and Bradford.

5.6.3 Other design criteria

A compound beam will need checking for shear, web bearing, web buckling and deflection in exactly the same manner as for normal rolled sections. However, in addition, the size of the fillet weld between the parent beam and the additional strengthening will need to be determined. This is performed using the theory given in Section 4.9.

5.6.4 Crane beams (cl 4.11, Pt 1)

For crane beams the effect of dynamic loads to allow for 'snatch' and inertia must be considered. The Crane design Code (BS 2573) categorizes cranes into four groups, Classes Q1 to Q4, with Q1 being the lightest duty crane and Q4 the heaviest. For cranes of Class Q1 and Q2 any dynamic effect may be allowed for by applying a percentage increase to the static load (BS 6399 Pt 1, cl 7). It should be noted that for Classes Q3 and Q4, BS 2573 should be consulted, and that in addition the partial safety factors given in BS 5950 Pt 1 Table 2 will not be applicable to these cases. Also crane girders carrying Class Q3 and Q4 cranes need designing for the effect of horizontal forces induced by crabbing. This is where the two ends of the main crane unit do not remain at right angles to the crane girders, but rotate slightly about a vertical axis and attempt to jam against the crane girder.

The load factors to be used for Classes Q1 and Q2 are:

Dead load:

1.4

Imposed load:

Vertical and horizontal forces acting independently—1.6

Vertical and horizontal forces acting in combination—1.4

For external crane girders not in operation CP3 Ch V Pt 2 may be used for calculating wind loads, but when the crane girders are in operation BS 2573 must be used to calculate the wind forces.

EXAMPLE 5.6   Design of a crane beam

Design a beam in Grade 43 steel to support the crane bridge (Class Q2) carrying an electric crane. Design data are given in Fig. 5.16. The crane beam has a span of 7.5 m and is to be taken as simply supported.

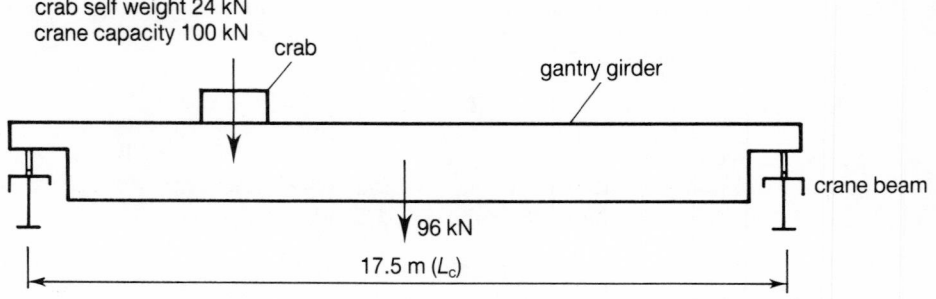

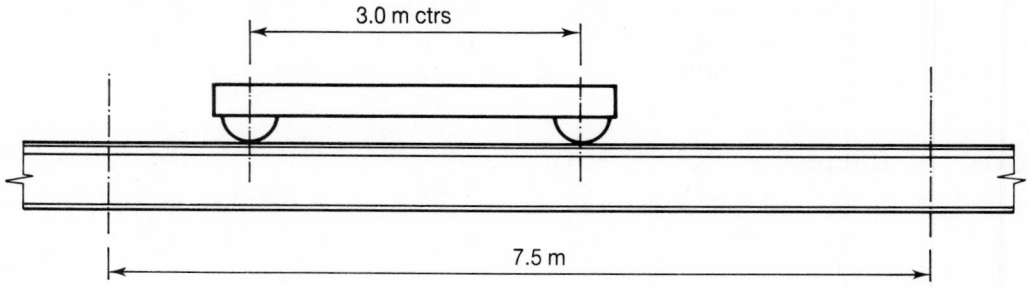

Fig. 5.16   Design data for Example 5.6.

Calculation of maximum static wheel load at A:

Due to weight of crane: $96/4 = 24$ kN

Due to crane load: $W(1 - L_h/L_c)/2 = (100 + 24)(1 - 1.1/17.5)/2$
$$= 58.1 \text{ kN}$$

Total load due to weight of crane and crane load:

$24 + 58.1 = 82.1$ kN

To allow for impact etc., the static load should be increased by 25% (BS 6399 Pt 1, cl 7).

Design load $= 82.1 \times 1.25 = 102.6$ kN

The horizontal load should be taken as 10% of the sum of the weight of the crab and the load to be lifted (BS 6399 Pt 1, cl 7)

Design horizontal load $= 0.1 \times (100 + 24) = 12.4$ kN

This is carried by four wheels, so the load per wheel $= 12.4/4 = 3.1$ kN.

Calculation of maximum forces.

To calculate the maximum effect of moving forces influence lines must be used. The reader is referred to Coates *et al.*, Neville and Ghali, or Marshall and Nelson for the basic theory, as only the results will be quoted herein.

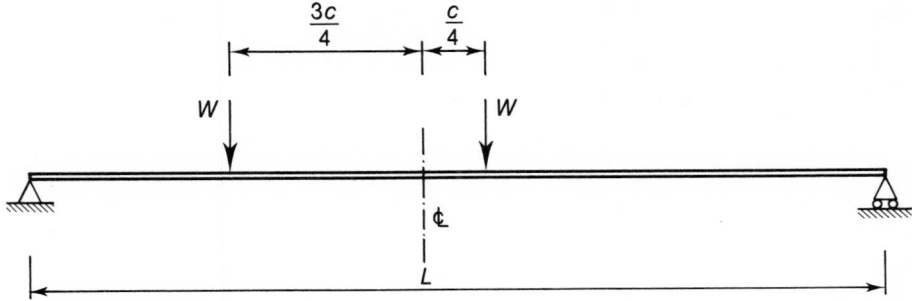

Fig. 5.17 Load train position for maximum bending moment.

(a) Maximum moment.

It may be shown that the maximum moment caused by a train of loads occurs when the mid-point of the beam is coincident with a position midway between the centre of gravity of the load train and the nearest load, and that the maximum moment occurs beneath the load (Fig. 5.17)

$$M_{max} = (W/(2L))(L - c/2)^2 \qquad (5.47)$$

Due to crane loads:

$$M_{max} = (102.6/(2 \times 7.5))(7.5 - 3/2)^2$$
$$= 246.2 \text{ kNm for UNFACTORED vertical loads.}$$

Due to UNFACTORED surge loads:

$$M_{max} = (3.1(7.5 \times 2))(7.5 - 3/2)^2$$
$$= 7.4 \text{ kNm.}$$

Due to self weight:

The maximum UNFACTORED moment due to self weight will be calculated from $WL/8$, where W is the total self weight. This is not strictly consistent with the calculation of the moments due to the applied loading as these are calculated under the wheel load, whereas the moment due to the dead load has been calculated at centre span. The result will be marginally conservative since the moment due to the dead load has been overestimated, and in any case the moment due to the self weight of the crane beam is small compared with that due to the reaction from the crane beam gantry.

Moment due to self weight $= WL/8 = 10 \times 7.5/8 = 9.4 \text{ kNm.}$

(b) Maximum shear (Fig. 5.18)

$$V_{max} = W(2 - c/L) \qquad (5.48)$$

Due to crane loads:

$$V_{max} = 102.6(2 - 3/7.5)$$
$$= 164.1 \text{ kn UNFACTORED}$$

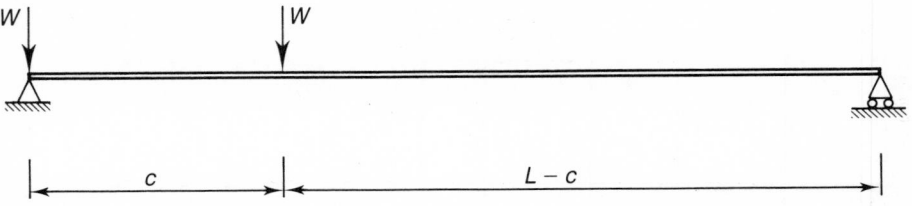

Fig. 5.18   Load train position for maximum shear force.

Due to surge loads:

$$V_{max} = 3.1(2 - 3/7.5)$$
$$= 5.0 \text{ kN UNFACTORED}$$

Due to self weight:

$$V_{max} = W/2$$
$$= 10/2 = 5.0 \text{ kN UNFACTORED.}$$

Load Combinations:

Vertical loading only:

Moment

$$1.4 \times 9.4 + 1.6 \times 246.2 = 407.1 \text{ kNm}$$

Shear

$$1.4 \times 5.0 + 1.6 \times 164.1 = 270 \text{ kN}$$

Vertical and horizontal loading in combination:

Due to vertical loading:

Moment

$$1.4(9.4 + 246.2) = 357.8 \text{ kNm}$$

due to horizontal loading:

Moment

$$1.4 \times 7.4 = 10.4 \text{ kNm}$$

In this combination the shear will be less and need not be checked.

Thus the beam must be checked for bending under both combinations and shear from the first only.

Trial Section:

Try a $610 \times 229 \times 101$ UB and a 305×89 channel, both in grade 43. The principal dimensions and properties are given in Fig. 5.19(a) to (c).

It is now necessary to calculate the section properties required. To do this the idealized section in Fig. 5.20 will be used. (The channel is assumed to have parallel flanges of thickness equal to the tabulated thickness at half depth and all fillets are ignored.)

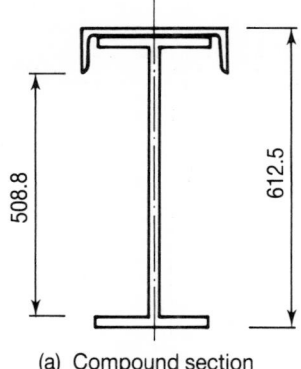

(a) Compound section

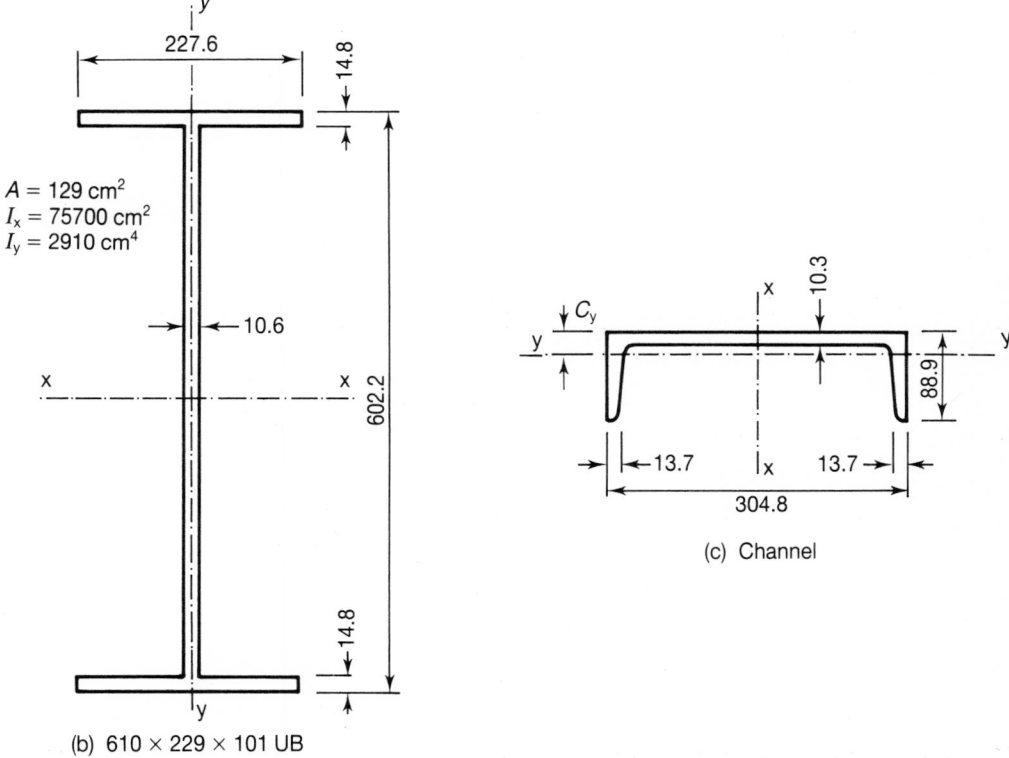

(b) 610 × 229 × 101 UB

(c) Channel

Fig. 5.19   Principal dimensions and section properties.

Equal Area Axis:

> Total area = 2 × 227.6 × 14.8 + 10.6 × 572.6 + 2 × 88.9 × 13.7 + 277.4 × 10.3
>
> = 18100 mm^2

Half area   = 9050 mm^2

Area of bottom flange of UB = 227.6 × 14.8

= 3368 mm^2

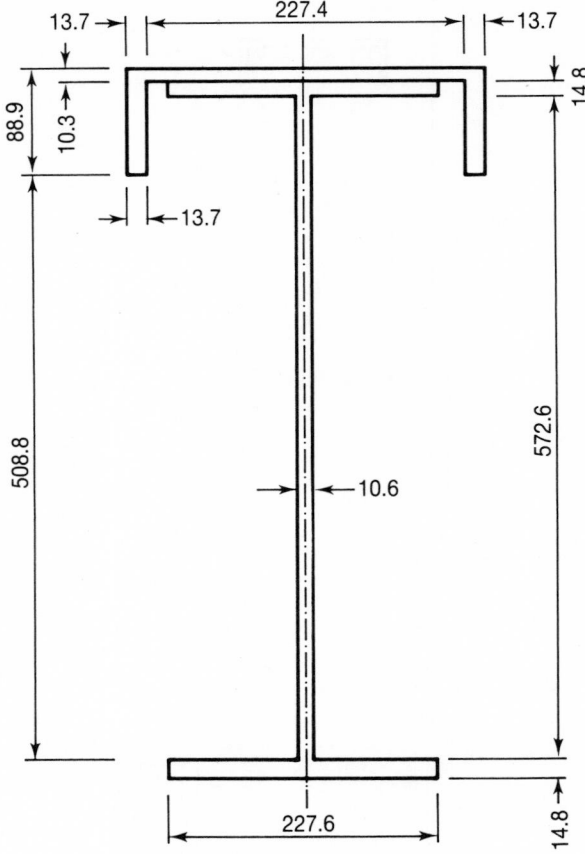

Fig. 5.20   Idealized section.

Area of web of UB to underside of downstand $= 508.8 \times 10.6$
$$= 5393 \text{ mm}^2$$

Area of bottom flange and web $= 3368 + 5393$
$$= 8761$$

Additional area required to equal area axis $= 9050 - 8761$
$$= 289 \text{ mm}^2$$

This will take in the downstands of the channel and the UB web, so additional height, x, required is given by

$$x(2 \times 13.7 + 10.6) = 289$$

or $x = 7.6$ mm

Plastic Section Modulus:

(a)   below equal area axis (Fig. 5.21)

$$(\Sigma A \bar{y})_T = 10.6 \times 516.4^2/2 + 227.6 \times 14.8(516.4 + 14.8/2) + 2 \times 13.7 \times 7.6^2/2$$
$$= 3.176E6 \text{ mm}^3$$

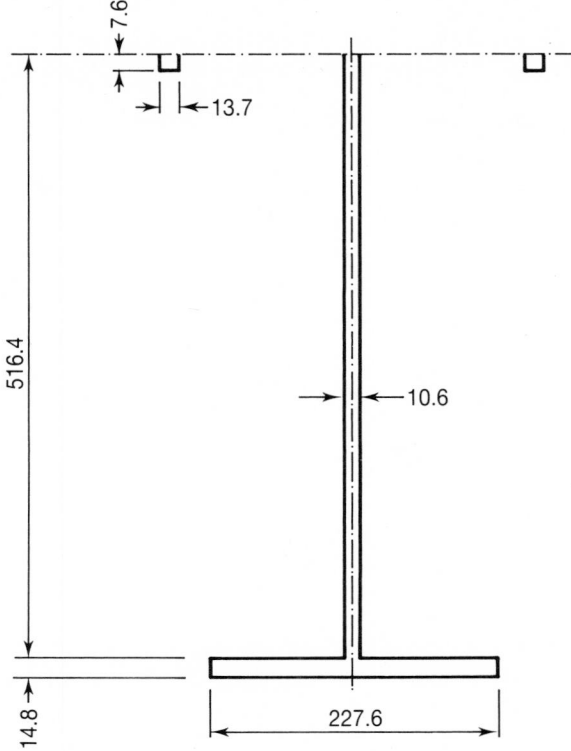

Fig. 5.21 Section below plastic equal area axis.

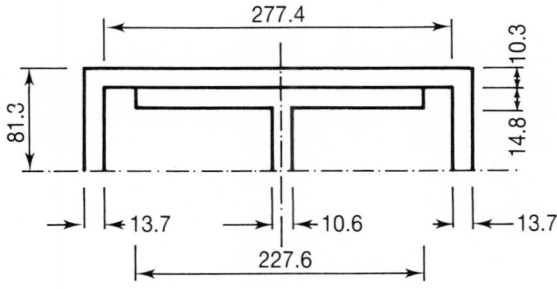

Fig. 5.22 Section above plastic equal area axis.

(b) above equal area axis (Fig. 5.22)

$$(\Sigma A \bar{y})_{\mathrm{C}} = 2 \times 13.7 \times 81.3^2/2 + 10.6 \times 56.2^2/2 + 227.6 \times 14.8(56.2 + 14.8/2)$$
$$+ 277.4 \times 10.3 \times (81.3 - 10.3/2)$$
$$= 0.539\mathrm{E}6 \ \mathrm{mm}^3$$

$$S_{\mathrm{x}} = (\Sigma A \bar{y})_{\mathrm{T}} + (\Sigma A \bar{y})_{\mathrm{C}}$$
$$= 3.72\mathrm{E}6 \ \mathrm{mm}^3$$

Elastic section properties.

The subscript UB refers to the Universal beam and CH to the channel section.

Cross-sectional area (from tabulated values)

$$A = A_{UB} + A_{CH}$$
$$= 129 + 53.1$$
$$= 182.1 \text{ cm}^2$$

which compares with the value of 181.0 cm², determined using the approximate section.

$$I_y = (I_y)_{UB} + (I_x)_{CH}$$
$$= 2910 + 7060$$
$$= 9970 \text{ cm}^4$$

$$r_y = (I_y/A)^{0.5}$$
$$= (9970/182.1)^{0.5}$$
$$= 7.4 \text{ cm.}$$

Position of elastic neutral axis, $\bar{x}$:

$$182.1\bar{x} = 129 \times 602.2/2 + 53.1 \times (602.2 + 10.3 - 21.8)$$

or,

$$\bar{x} = 385.5 \text{ mm}$$

$$I_x = (I_x)_{UB} + A_{UB}h_1^2 + (I_y)_{CH} + A_{CH}h_2^2$$

$$h_1 = \bar{x} - D_{UB}/2$$
$$= 385.5 - 602.2/2$$
$$= 84.4 \text{ mm}$$
$$= 8.44 \text{ cm}$$

$$h_2 = (D_{UB} + t_{CH}) - \bar{x} - C_x$$
$$= (602.2 + 10.3) - 385.5 - 21.8$$
$$= 205.2 \text{ mm}$$
$$= 20.52 \text{ cm}$$

$$I_x = 75700 + 129 \times 8.44^2 + 325 + 53.1 \times 20.52^2$$
$$= 1.076\text{E}5 \text{ cm}^4$$

Before going on to determine the bending capacity of the section it is necessary to determine its classification.

$$\varepsilon = (275/p_y)^{0.5}$$

For $T < 16$, $p_y = 275$ (Table 2),

$$\varepsilon = 1.0$$

Flange outstand (b/T) (cl 3.5.5, Pt 1):

Two cases need checking (See Fig. 3 Pt 1)

(a)   Total flange outstand to inside of lip (Fig. 5.23(a))

$$b = (304.8 - 2 \times 13.7)/2 = 138.7$$

$$T = 14.8 + 10.3 = 25.1$$

$$b/T = 138.7/25.1 = 5.5$$

Allowable value $= 7.5\varepsilon = 7.5$

(b)   Outstand of secondary flange (Fig. 5.23(b))

$$b = (304.8 - 2 \times 13.7 - 227.6)/2 = 24.9$$

$$T = 10.3$$

$$b/T = 24.9/10.3 = 2.4$$

both b/T ratios are below the allowable values and are thus satisfactory.

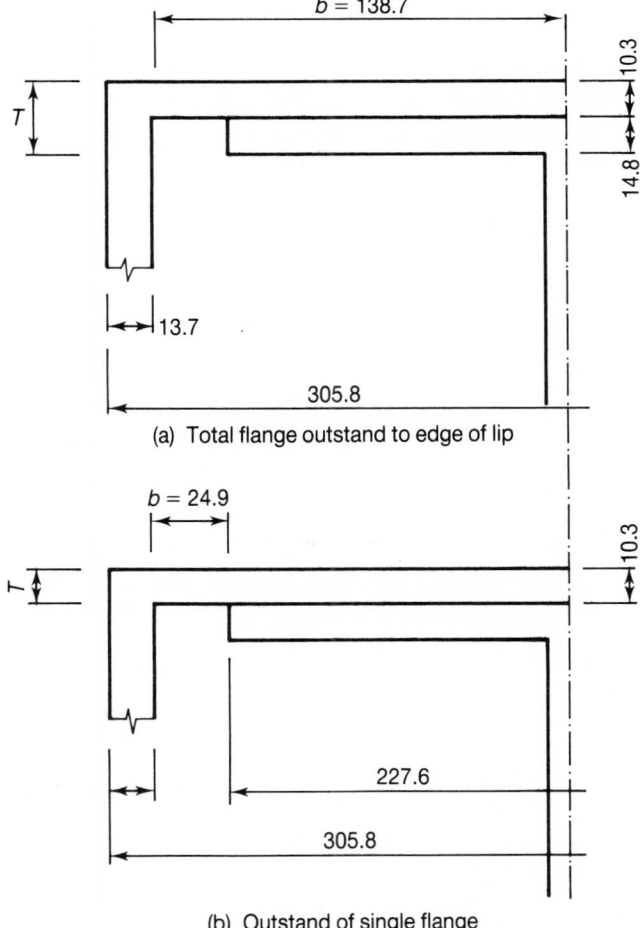

(a) Total flange outstand to edge of lip

(b) Outstand of single flange

Fig. 5.23   Flange instability check.

Web slenderness (d/t)

actual $d/t = 51.6$

allowable $d/t = 79\varepsilon = 79$

All checks indicate the section is plastic.

Plastic moment capacity.

Minimum elastic section modulus, Z:

$$Z = 1.076\text{E}9/385.5$$
$$= 2.79\text{E}6 \text{ mm}^3$$

$$S_x = 3.72\text{E}6 \quad \text{and} \quad 1.2Z = 3.35\text{E}6,$$

but since $S_x > 1.2Z$, the factor 1.2 may be replaced by the load factor (cl 4.2.5), which is in excess of 1.4. The plastic moment capacity is, therefore, given by $S_x p_y = 3.72\text{E}6 \times 275/1\text{E}6 = 1023$ kNm which is in excess of the moment induced by the applied loading.

Lateral torsional buckling:

Use Equation (5.11) to calculate the buckling parameters u and x.

$$\gamma = 1 - I_y/I_x$$
$$= 1 - 9970\text{E}4/1.076\text{E}9$$
$$= 0.907$$

Distance between shear centres of the flanges, h_s:

First determine the effective centroid, $\bar{x}$, of the combined top flange ignoring the down-stands (Fig. 5.24)

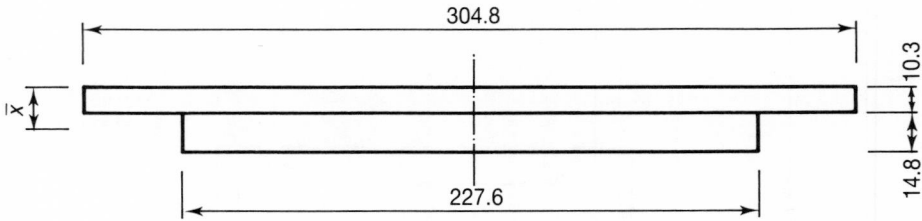

Fig. 5.24 Calculation of flange centroid ignoring downstand.

$$\bar{x}(227.6 \times 14.8 + 304.8 \times 10.3) = 304.8 \times 10.3^2/2 + 277.6 \times 14.8(14.8/2 + 10.3)$$

$$\bar{x} = 12.3 \text{ mm}$$

$$h_s = (D_{\text{UB}} + t_{\text{CH}}) - \bar{x} - T_{\text{UB}}/2$$
$$= (602.2 + 10.3) - 14.8/2 - 12.3$$
$$= 592.8 \text{ mm}$$

Torsional second moment of area, J:

Replace top flange by an equivalent plate of thickness equal to the combined thickness of the flange of the UB and the web of the channel, i.e. a thickness of $14.8 + 10.3 = 25.1$ and a width, b, given by

$$b = (14.8 \times 227.6 + 277.4 \times 10.3)/25.1$$
$$= 248 \text{ mm}$$

$$J = (1/3)\Sigma bt^3 \tag{5.49}$$

$$J = (227.6 \times 14.8^3 + 572.6 \times 10.6^3 + 248 \times 25.1^3 + 2 \times 88.9 \times 13.7^3)/3$$
$$= 1.933\text{E}6 \text{ mm}^4$$

$$u = (4S_x^2\gamma/(A \times h_s)^2)^{0.25}$$
$$= (4 \times (3.72\text{E}6)^2 \times 0.907/(18100 \times 592.8)^2)^{0.25}$$
$$= 0.813$$

$$x = 0.566h_s(A/J)^{0.5}$$
$$= 0.566 \times 592.8 \times (18100/3.72\text{E}6)^{0.5}$$
$$= 23.4$$

Effective length, L_E:

A typical support detail for light and medium weight crane girders is shown in Fig. 5.25.

This type of connection will give little torsional restraint to the beam and no lateral restraint to the compression flange which will also be free to rotate on plane. This coupled with the fact that the loading is destabilizing gives, from Table 9,

$$L_E = 1.2L + 2D$$

where D is taken as the overall depth of the compound section.

$$L_E = 1.2 \times 7500 + 2 \times 612.5$$
$$= 10255 \text{ mm}$$

$$\lambda = L_E/r_y$$
$$= 10225/74$$
$$= 138$$

$$\lambda/x = 138/23.4 = 5.9$$

calculation of v:

This is given by Equation (5.44).

To evaluate this two further parameters are needed:

N:

$$N = I_{cf}/(I_{cf} + I_{tf}) \tag{5.43}$$

$$I_{tf} = (I_y)_{UB}/2 = 2910/2 = 1455$$

Note, the error involved in taking the second moment of area of the flange equal to half

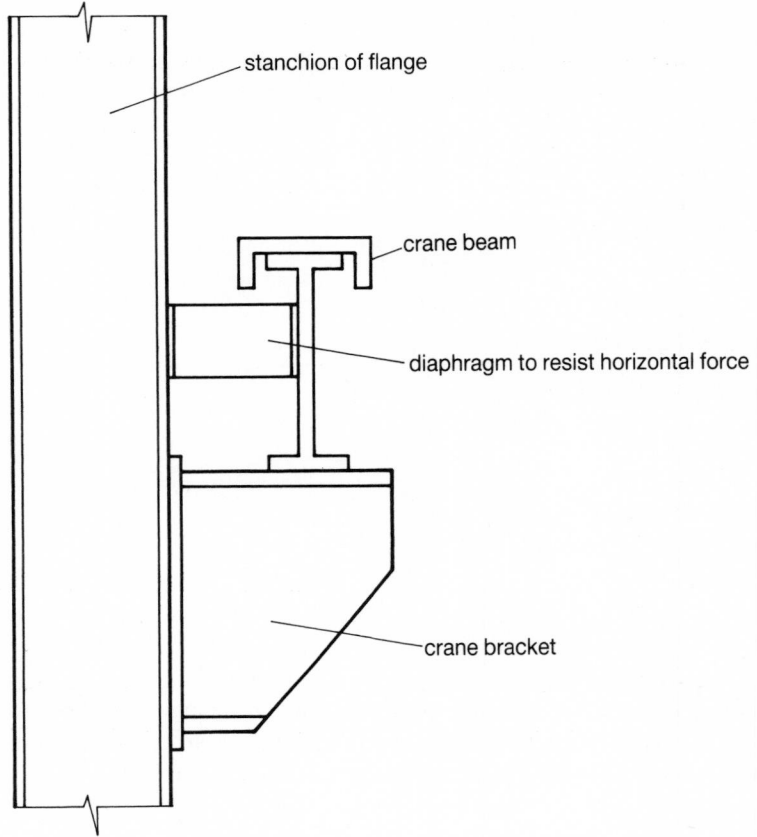

Fig. 5.25 Typical support details.

the minor axis second moment of area is negligible, since the contribution of the web to the value of I_y is very small.

$$I_{cf} = (I_x)_{CH} + (I_y)_{UB}/2 = 7060 + 2910/2 = 8515$$
$$N = 8515/(8515 + 1455)$$
$$= 0.854$$

ψ:

for $N > 0.5$, $\psi = 0.8(2N - 1)(1 + D_L/(2D))$ (5.46)

$$D_L = 88.9$$
$$\psi = 0.8(2 \times 0.854 - 1)(1 + 88.9/(2 \times 612.5))$$
$$= 0.608$$

$$v = ((4N(N - 1) + (1/20)(\lambda/x)^2 + \psi^2)^{0.5} + \psi)^{-0.5}$$
$$= ((4 \times 0.854(1 - 0.854) + 5.9^2/20 + (0.608^2)^{0.5} + 0.608)^{-0.5} \quad (5.44)$$
$$= 0.671$$

Calculation of λ_{LT}:

Since the loads are destabilizing, $n = m = 1.0$ (also refer to Cl 4.11.3 Pt 1)

$$\lambda_{LT} = nuv\lambda$$
$$= 1.0 \times 0.813 \times 0.671 \times 138$$
$$= 75$$

from Table 12 with $p_y = 275$, $p_b = 151$ N/mm^2

$$M_b = S_x p_b$$
$$= 3.72\text{E}6 \times 151/1\text{E}6$$
$$= 566 \text{ kNm}$$

The design moment $\bar{M}$ is given by,

$$\bar{M} = mM_a$$
$$= 1.0 \times 407.1$$

Thus the beam is satisfactory under vertical loading only.

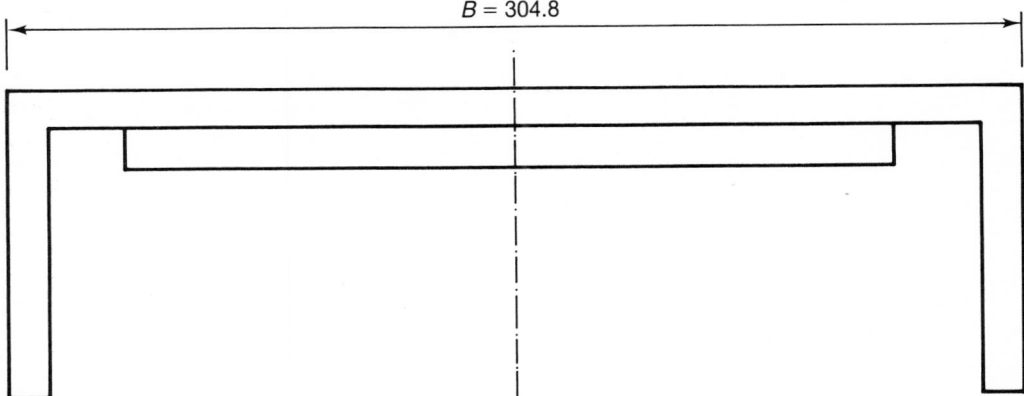

$B = 304.8$

Fig. 5.26   Section resisting horizontal loads.

It is now necessary to check under biaxial bending. The section resisting horizontal loading is given in Fig. 5.26

$$Z_y = (Z_x)_{CH} + I_F/(B/2))$$

where I_F is the second moment of area of the flange of the UB about the vertical centroidal axis

$$Z_y = 0.463\text{E}6 + (14.8 \times 227.6^3/12)/(304.8/2)$$
$$= 0.463\text{E}6 + 0.095\text{E}6$$
$$= 0.558\text{E}6 \text{ mm}^3$$

$$M_{cy} = Z_y p_y$$
$$= 0.558 \times 275$$
$$= 154 \text{ kNm}$$

$$M_y/M_{cy} = 10.4/148 = 0.07$$
$$M_x/M_{cx} = 357.8/1023 = 0.35$$

The following interaction equation needs to be satisfied:

$$M_x/M_{cx} + M_y/M_{cy} \leqslant 1.0$$

This is identical to Equation (6.4), except $F = 0$,

thus calculating the left hand side,

$$0.35 + 0.07 = 0.42.$$

This is less than unity and is therefore satisfactory.

Shear.

Shear capacity, P_v:

$$P_v = 0.6p_y Dt$$
$$= 0.6 \times 275 \times 10.6 \times 602.2/1E3$$
$$= 1053 \text{ kN}$$

This is well in excess of the maximum applied shear of 270 kN, and thus no reduction need be made in the plastic moment capacity of the section.

Note the shear is carried by the original section only.

Weld design.

The required shear capacity of the weld, q_s, is given by,

$$q_s = QAy_c/I_x$$
$$y_c = h_2 = 205.2 \text{ mm}$$
$$A = 5310 \text{ mm}^2$$
$$Q = 270 \text{ kN}$$
$$I_x = 1.076\text{E}9 \text{ mm}^4$$
$$q_s = 270 \times 5310 \times 205.2/1.076\text{E}9$$
$$= 0.28 \text{ kN/mm}$$

This shear is taken by two welds, so use a minimum weld of 6 mm (0.9 kN/mm per weld).

Web bearing and web buckling at the support (Fig. 5.27)

Web bearing

$$b = 73, \quad \text{and} \quad n_2 = 27.5 \times 2.5 = 68.8 \text{ mm}$$
$$\text{Capacity} = (b_1 + n_2)tp_y$$
$$= (73 + 68.8) \times 10.8 \times 275/1E3$$
$$= 421 \text{ kN}$$

This is greater than the applied reaction of 270 kN, and is therefore satisfactory.

Web buckling:

$$\lambda = 2.5(d/t) = 2.5 \times (547.3/10.6) = 129$$

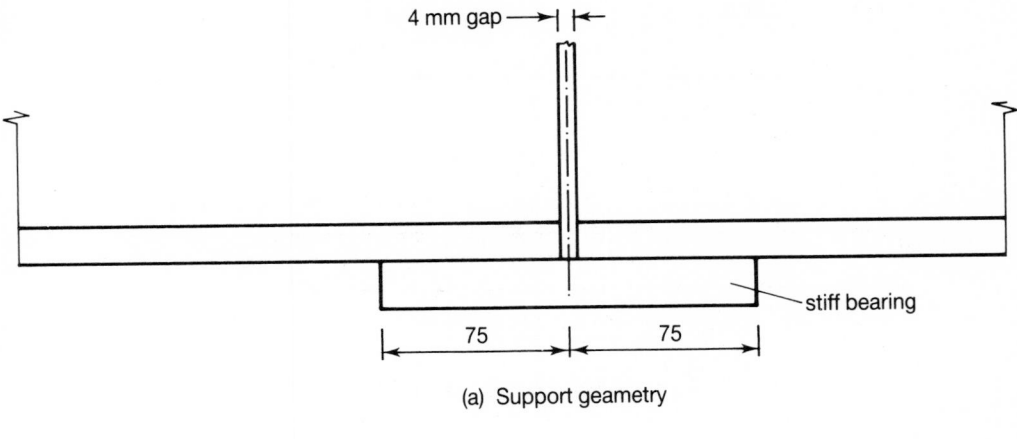

(a) Support geametry

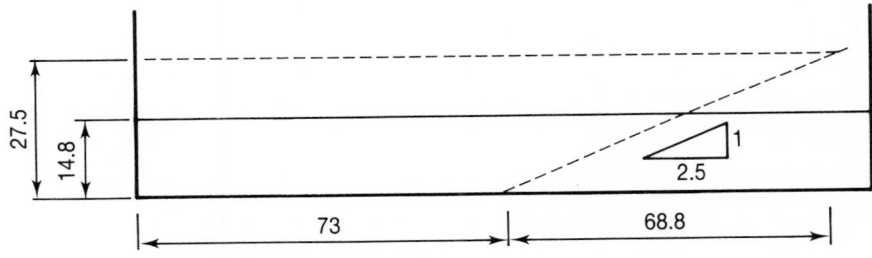

(b) Geometry for web bearing check

Fig. 5.27 Support details.

From Table 27(c) with $p_y = 275$, $p_c = 87$ N/mm^2, $b_1 = 73$, and $n_1 = D/2 = 602.2/2 = 301.1$ mm

$$\text{Capacity} = (b_1 + n_1)tp_c$$
$$= (73 + 201.1) \times 10.6 \times 87/1\text{E}3$$
$$= 345 \text{ kN}$$

This is also satisfactory.

Local compression below the wheel (cl 4.11.5) (Fig. 5.28)

Dispersion length x_R is given by,

$$x_R = 2(H_R + T)$$
$$= 2(65 + 10.3 + 14.8)$$
$$= 180.4 \text{ mm.}$$

$$\text{Capacity} = x_R tp_y$$
$$= 180.4 \times 10.6 \times 275/1\text{E}3$$
$$= 526 \text{ kN}$$

$$\text{Actual wheel load} = 1.6 \times 102.6$$
$$= 164.2 \text{ kN.}$$

This too is satisfactory.

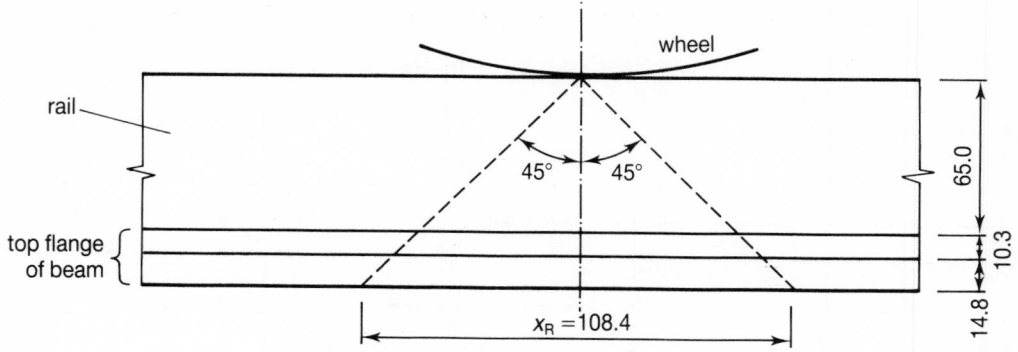

Fig. 5.28   Local compression check beneath wheel.

Deflection.

The midspan deflection will be at a maximum when the wheel loads are symmetrically disposed about the beam centre line.

Vertical Deflection:

For ONE wheel the deflection δ is given by

$$\delta = (WL^3/(48EI))(3(a/L) - 4(a/L)^3)$$

with $a < L/2$

For one wheel,

$$L = 7.5 \text{ m}$$

$$a = 2.25 \text{ m}$$

$$I = 1.076\text{E} - 3 \text{ m}^4$$

$$W = 82.1 \text{ kN}, \quad \text{and}$$

$$E = 205 \text{ GN/m}^2$$

Thus the deflection for one wheel is 2.9E-3 m, and so for both wheels, 5.8E-3 m.

The allowable deflection, from Table 5 is span/600 = 7.5/600 = 12.5E-3 m.

Thus the vertical deflection is satisfactory.

Horizontal deflection.

Only the compound top flange will be assumed to resist the applied surge loads as in the bending check.

$$I = (I_x)_{CH} + I_F$$

The notation is as Fig. 5.26.

$$I = 0.0706\text{E}9 + 0.0145\text{E}9 = 0.0851\text{E}9 \text{ mm}^4$$

The horizontal deflection, δ_H, may be determined pro rata from the vertical deflection, so

$$\delta_H = 2.9\text{E-}3(1.076/0.0851)(3.1/82.1)$$
$$= 1.4\text{E-}3 \text{ m}$$

Total deflection $= 2.8\text{E-}3$ m

Allowable deflection $= \text{span}/500 = 7.5/500 = 0.015$ m

The beam is therefore satisfactory for horizontal deflection.

This now completes all the checks required.

5.7 Plate girders

5.7.1 Introduction

Plate girders are used either on long spans where a rolled section would need to be spliced and would therefore probably be inefficient, or to support heavy loads, such as in a bridge structure, where a rolled section would have insufficient load carrying capacity. Plate girders are used rather than compound beams as they will generally have a lower self weight. It should be recognized, however, that plate girders are expensive because of the large amount of fabrication required.

Plate girders are built up from two flange plates and a web plate, generally from the same grade steel. They are fabricated using continuous automatic electric welding to produce fillet welds between the web and flanges. The welds are initially performed on one face of the web, the beam is then inverted and the welding is carried out on the other face on the return pass through the automatic welder. To ensure that in the final state the flanges are parallel, the initial set of welds is carried out with the flanges slightly inclined (Fig. 5.29). On the return pass the flanges are held vertical as the welding is carried out. The double pass welding process will build in a pattern of high residual stresses partly due to the necessity of restraining the flanges into position during welding and partly due to the large amount of heat required during the welding process and the subsequent cooling of the section. There are also residual stresses built into the flange plates during

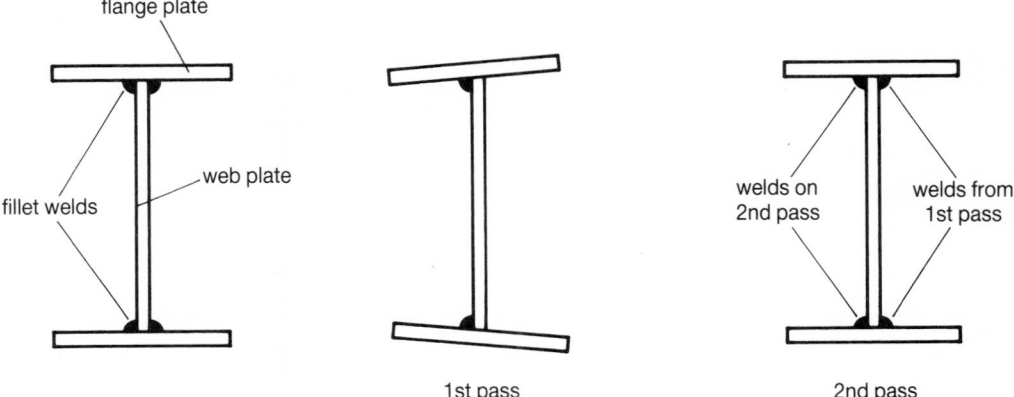

Fig. 5.29 Plate girder fabrication.

preparation. The magnitude of these stresses will depend on whether the plates were flame cut or sheared to size (Nethercot (1974b)).

The stiffeners are then generally welded manually. The stiffeners are needed either to provide support to the web which is very thin, or to resist point loads applied to the top flange. Generally the top and bottom flanges are the same size, thus giving a section symmetrical about both axes. Only plate girders with vertical stiffeners are dealt with in this text. It is possible to use horizontal stiffeners, although these are more common in bridge beams, and thus BS 5950 Part 1 does not specifically deal with horizontal stiffeners but refers the designer to Part 4 of BS 5400 (the Bridge Code).

5.7.2   Minimum web thickness

Cl 4.4.2.2, Pt 1 gives minimum web thicknesses for serviceability:

Without intermediate stiffeners, $t \geqslant d/250$

with transverse stiffeners only,

$$a > d : t \geqslant d/250$$
$$a \leqslant d : t \geqslant (d/250)(a/d)^{0.5}$$

where t is the web thickness, d the depth of the web and a the intermediate stiffener spacing.

These criteria are to ensure that the web will not buckle under normal service conditions. It should be noted that it is possible to use a thinner minimum size web when the web panels are shorter than the beam depth. This recognizes that such a web panel has a higher strength.

It is also necessary to ensure that the web is sufficiently strong to ensure that the flange will not buckle into the web (cl 4.4.2.3, Pt 1)

Without intermediate stiffeners: $t \geqslant (d/250)(p_{yt}/345)$

with intermediate stiffeners:

$$a > 1.5d : t \geqslant (d/250)(p_{yt}/345)$$
$$a \leqslant 1.5d : t \geqslant (d/250)(p_{yt}/455)^{0.5}$$

where p_{yt} is the design strength of the compression flange.

It will be noted that when these two sets of criteria are compared that if Grade 43 steel is used the serviceability criteria will be more critical, but if Grade 50 is used the buckling criteria become more important. This is because buckling strengths are dependent on Young's modulus which is the same for both steel grades.

5.7.3   Bending resistance

The methods that are to be used to calculate the bending capacity of a plate girder depend on the classification of both the flanges and the web calculated from the b/T and d/t ratios. The limiting b/T and d/t ratios for a particular classification are given in Table 7.

If the web slenderness (d/t) is less than 63ε then the beam should be designed as a rolled section (cl 4.4.4.1, Pt 1).

If the web is slender, i.e. $d/t > 63\varepsilon$, and the flanges are plastic, compact or semi-compact, then there are three ways of determining the moment capacity of the section (cl 4.4.4.2, Pt 1):

1   The flanges are assumed to resist the applied moment with each flange being subjected to an uniform stress of p_y and the web to resist the applied shear.

   This is a useful assumption when initially sizing a plate girder to obtain preliminary sizes, but it might be slightly unecomonic for the final design. The design procedure is, however, much simpler than that needed for either of the remaining methods. There is however the implication in the Code that this method may not be used for beams subjected to lateral torsional buckling as it is implied that the full value of the design stress is to be taken, i.e. for beams that are fully restrained.

2   The whole section may be used to resist the effects of axial force and moment and the web used to resist combined shear and longitudinal stresses. For sections whose flanges are semi-compact, the moment in the web should be determined using simple elastic theory. For sections with compact or plastic flanges, simple plastic theory may be used.

   Where the loading is a UDL over part or whole of the span, the web must be designed to resist the inplane effects of this load (Section 5.7.5.5) and that the resultant capacity of the web to resist bending stress is reduced (Equation 5.68). It should be noted that in any case the maximum capacity of the web to resist bending is given by $0.664t^3$ kNm, where t is the web thickness in millimetres. This maximum capacity is obtained from Equations 5.69 and 5.73. For most webs the full moment capacity is low, and it will be found for most beams difficult to satisfy the design interaction equation (Equation 5.68) if edge loading is applied to the web. It is thus recommended that the web should only be used to resist bending where the loading is discrete and applied at load bearing stiffeners.

   The design procedures for this method are in any case more complex, but may result in smaller flanges and a thinner web, although the stiffener requirement may be increased.

3   The last method is to allow part of the loading to be resisted by method two and the remainder by method one.

   There seems little merit in this approach at it appears to produce neither a simple design procedure nor an economic section. It will, therefore, not be discussed further.

For sections with slender flanges the moment capacity is to be calculated using a similar approach to that of slender rolled section (Section 4.1) (cl 4.4.4.3, Pt 1).

   In all plate girder design there is the inherent problem that if a minimum weight solution to a problem is adopted by using, say, thin webs, the solution may not be economic in terms of fabrication costs as the stiffener requirement will be increased. Only experience will determine the most economic solution to a given problem in terms of both material and fabrication costs.

5.7.4   Lateral torsional buckling of plate girders

The procedure adopted is the same as that for rolled sections, except that as was pointed out in Section 5.2 different values of the imperfection coefficient, η_{LT}, must be used owing to the higher levels of residual stresses. This means that Table 12 must be used to determine P_b.

All section properties required for calculating η_{LT} must be determined from first principles for each individual case. However, the following approximate formulae may be found useful:

$$J = (1/3)\Sigma bt^3 \tag{5.49}$$

where b and t are the width and thickness of the web and flange plates

$$H = I_y h_s^2/4 \tag{5.41}$$

Either Equation (5.10) or (5.11) may be used to calculate u and x for a girder symmetric about both axes.

Note, for a conservative design, $u = 1.0$ (cl 4.3.7.5, Pt 1).

5.7.5   Web design

Before considering the methods that may be used to design webs, it will be convenient to consider how the webs of plate girders behave experimentally.

5.7.5.1   *Experimental behaviour of plate girder webs*

It is only intended to give a brief review in this section, fuller details may be found in Porter *et al.* and Rockey *et al.* (1978).

As the shear load is increased on a stiffened web panel no undue deformation occurs until the web panel buckles in shear (Fig. 5.30(a)). This point, however, does not mark the point at which the maximum load capacity of the web occurs as the load can still be increased with the effect of part of the web that has buckled taking load in tension. This 'tension' member acts obliquely across the web panel although it is not coincident with the web panel diagonal (Fig. 5.30(b)). The girder now acts in a similar manner to a truss with the compression forces being taken by the flanges and the intermediate stiffeners. This additional reserve of strength is known as tension field action.

The stress at which the panel first buckles can be predicted with very good accuracy from classical plate buckling theory (Bulson). For a plate with simply supported edges, this elastic critical shear buckling stress, τ_{cr}, is given by

$$\tau_{cr} = k_{cr}(\pi^2 E/(1 - v^2)(d/t)^2 \tag{5.50}$$

where,

for $a/d > 1.0 : k_{cr} = 5.35 + 4(a/d)^2$

for $a/d < 1.0 : k_{cr} = 5.35(a/d)^2 + 4$ \qquad\qquad (5.51)

and d/t is the web slenderness.

By analyzing the collapse mechanism of Fig. 5.31, and using the Huber-von Mises' yield criterion, it may be shown (Porter *et al.*, Rockey *et al.* (1978)), that the ultimate shear load, V_s is given by,

$$V_s/V_{yw} = \tau_{cr}/\tau_{yw} + 3\sin^2\theta(\cot\theta - a/d)(\sigma_{ty}/\sigma_{yv}) + 4\sqrt{3}\sin\theta(M_p^*(\sigma_{ty}/\sigma_{yw})) \tag{5.52}$$

where

$$\sigma_{ty} = -(\sqrt{3}/2)(\tau_{cr}/\tau_{yw})\sin 2\theta + (1 + (\tau_{cr}/\tau_{yw})^2(0.75\sin^2 2\theta - 1))^{0.5} \tag{5.53}$$

$$M_p^* = M_{pf}/d^2 t\sigma_{yw} \tag{5.54}$$

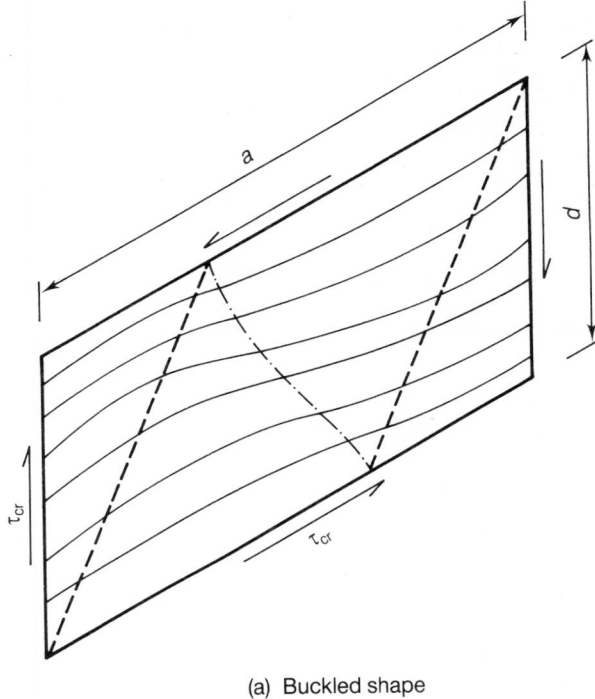

(a) Buckled shape

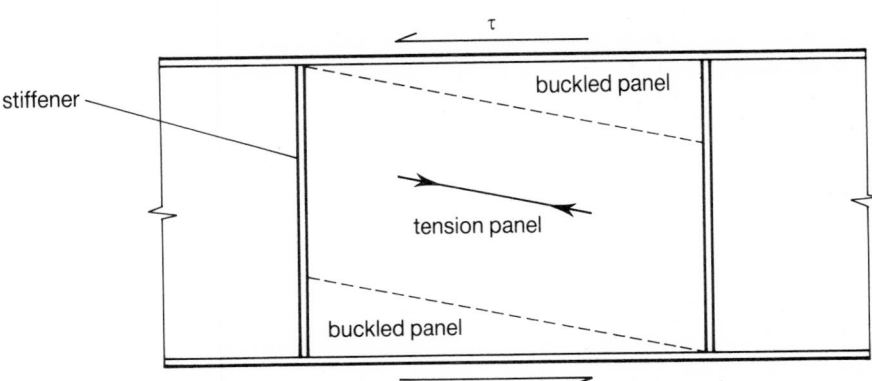

(b) Panel showing tension field

Fig. 5.30 Shear failure of a plate girder.

and

V_{yw} is the shear capacity of the web, d and t the depth and thickness of the web respectively, τ_{yw} the yield strength of the web in shear and σ_{yw} in tension, a/d the aspect ratio of the web and M_{pf} the plastic moment capacity of the flange.

The first term of Equation (5.52) is simply the panel shear buckling strength, albeit non-dimensionalized, the second term is due to the membrane effect of the panel supported by the vertical stiffeners and the third term is the component due to the flanges.

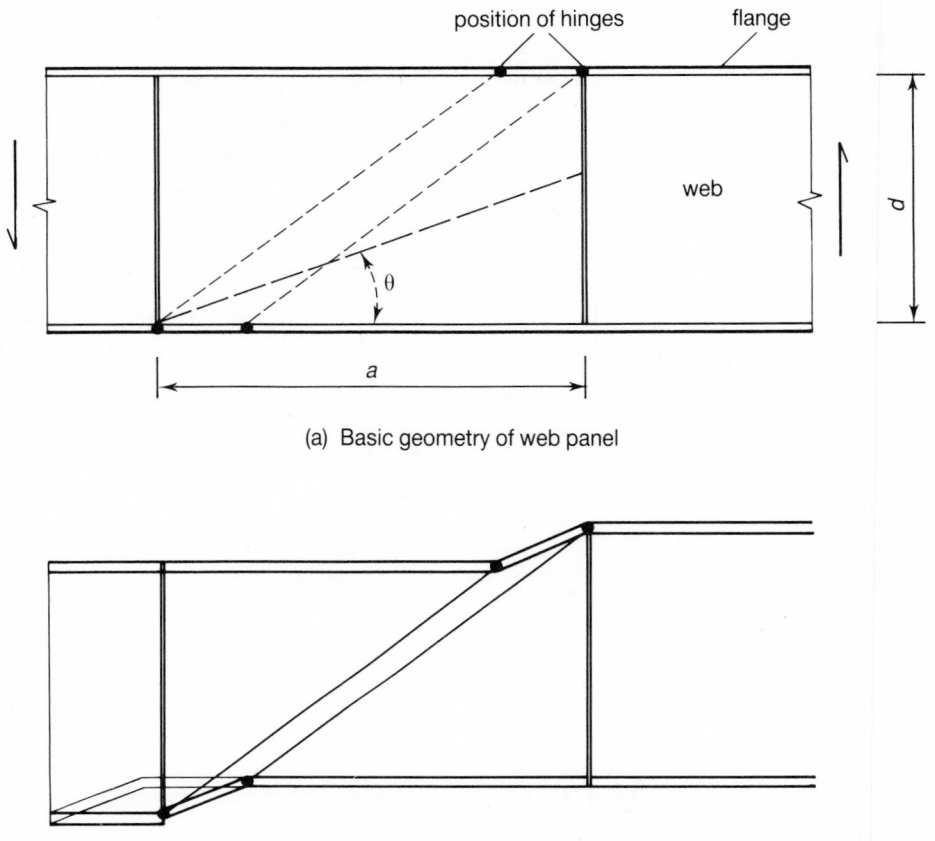

(a) Basic geometry of web panel

(b) Failure mechanism used to derive post buckling strength of web

Fig. 5.31   Overall behaviour of web at failure.

It was found that an optimum solution was given by Equation (5.53) when θ was equal to $(2/3)\arctan(d/a)$, although the solution is not very sensitive to slight changes in θ.

BS 5950 allows webs to be designed with or without tension field action. Before discussing the design implications of either of these alternatives, it is convenient to present the code design equations for both these methods.

5.7.5.2   *Web strength without tension field action (cl 4.4.5.3 and Appendix H1)*

To demonstrate how BS 5950 arrives at the relevant design equations consider the case where $a/d < 1.0$.

If Equations (5.50) and (5.51) are combined, then

$$\tau_{cr} = (5.35(d/a)^2 + 4.0)(\pi^2 E/(12(1 - v^2))(t/d)^2 \tag{5.55}$$

If the numerical values of v ($=0.3$) and E ($=205\text{E}3\ \text{N/mm}^2$) are substituted into Equation

(5.55), then

$$\tau_{cr} = ((d/a)^2 + 4.0/5.35)(\pi^2 \times 205E3/(12(1-0.3^2))(t/d)^2 \quad (5.56)$$
$$= ((d/a)^2 + 0.75) \times (996)^2 (t/d)^2 \quad (5.57)$$

The Code replaces the coefficient of 996 in Equation (5.57) with 1000, and with a slight change in notation, the elastic critical stress, q_e, is given by

$$\text{for} \quad a/d \leqslant 1.0 : q_e = (0.75 + 1/(a/d)^2)(1000/(d/t))^2 \quad (5.58a)$$

$$\text{for} \quad a/d > 1.0 : q_e = (1 + 0.75/(a/d)^2)(1000/(d/t))^2 \quad (5.58b)$$

The upper limit on web strength clearly must be the ultimate strength capacity of the web, i.e. $0.6p_{yw}$. The Code recognizes this by identifying three modes of behaviour. The first is where the web strength is governed by its ultimate capacity, the third where the capacity is governed solely by the elastic critical stress and the second, intermediate, stage where a degree of interaction between the first and third stage behaviour occurs. To quantify the divisions between the three modes a web slenderness ratio, λ_w, is defined by,

$$\lambda_w = (0.6p_{yw}/q_e)^{0.5} \quad (5.59)$$

and the web design strength, q_{cr} by

$$\text{for} \quad \lambda_w \leqslant 0.8, \ q_{cr} = 0.6p_{yw}$$

$$\text{for} \quad 0.8 < \lambda_w < 1.25, \ q_{cr} = 0.6p_{yw}(1 - 0.8(\lambda_w - 0.8))$$

$$\text{for} \quad \lambda_w > 1.25, \ q_{cr} = q_e \quad (5.60)$$

Equations (5.60) form the basis of Table 21.

The concept of using a linear transformation between stocky webs where yield governs and slender webs where buckling governs was first suggested by Bassler, and Rockey and Skaloud.

5.7.5.3   *Web strength using tension field theory (cl 4.4.54 and Appendix H2, Pt 1)*

It will be recalled from the theory of tension field action (Section 5.7.5) that there are three components to the predicted shear strength: the first was the basic buckling strength, the second due to tension action of the web, and the third due to the contribution of the flanges (Equation (5.52)). In the tests used to derive that equation, loading was in pure shear thus allowing the flanges to contribute full flexural capacity to the web. Clearly if the flanges are fully stressed in flexure they cannot contribute to the web strength. This is allowed for, as will be seen later, by applying a reduction factor to the third term of Equation (5.52).

The Code gives the strength of the web using tension field theory as,

$$V_b/dt = q_b + q_f\sqrt{K_f} \leqslant 0.6p_{yw} \quad (5.61)$$

where V_b is the maximum shear capacity, q_b, the basic web capacity and $q_f\sqrt{K_f}$, the flange component.

Consider the derivation of each of the components in turn,

(a) q_b:

$$q_b = q_{cr} + y_b/(2(a/d + (1 + (a/d)^2)^{0.5})) \quad (5.62)$$

with,

$$y_b = (p_{yw}^2 - 3q_{cr}^2 + \phi_t^2)^{0.5} - \phi_t$$

and

$$\phi_t = 1.5q_{cr}/(1 + (a/d)^2)^{0.5}$$

Equation (5.62) corresponds to the first two terms of Equation (5.52) with two differences, the basic shear buckling strength, τ_{yw}, is replaced by the basic web strength without tension field action, q_{cr}, and the resultant tension force is assumed to act at an angle of $\theta = 0.5 \times \arctan(d/a)$

(b) $q_f \sqrt{K_f}$:

$$q_f = 0.6p_{yw}(4\sqrt{3} \sin (\theta/2) (y_b/p_{yw})^{0.5}) \tag{5.63}$$

$$K_f = (M_{pf}/4M_{pw}) (1 - f/p_{yf}) \tag{5.64}$$

where M_{pf} is the plastic moment capacity of the smallest flange about its own equal area axis parallel to the flange, f is the mean longitudinal stress in the smaller flange due to the applied loading (flexural and axial), p_{yf} the design strength of the flange, and M_{pw} the plastic moment capacity of the web about its own equal area axis normal to the web.

Equation (5.63) corresponds to the third term of Equation (5.52), with the additional factor (Equation (5.64)) allowing for part of the capacity of the flanges being not available to resist the reactions from the tension field owing to flexural stresses in the flanges. It is also subject to the same general comments as those for q_b.

Values of q_b and q_f are to be found in Tables 22 and 23.

5.7.5.4   *End panels (cl 4.4.5.4.2 et seq., Pt 1)*

Where the webs are designed without tension field theory, no special considerations need to be made. Where webs are designed using tension field theory, then special consideration is required as there is a need to provide a sufficiently strong anchorage to resist the reactions set up by the tension field. These reactions comprise a moment, M_{tf}, and a shear, R_{tf}, which must be resisted by the stiffener system at the end of the beam. This enhanced stiffener system is referred to as an end post.

There are three basic configurations that may be used for the design of the end panels (Fig. 5.32)

(a)   End panel not designed using tension field theory (Fig. 5.32(a)).

In this case, the end panel (Panel B) is designed without tension field action and must resist the applied shear, the horizontal shear force R_{tf}, and the moment M_{tf}. The end stiffener is designed to resist the reaction from the external loading together with the compression due to the moment M_{tf}.

(b)   End panel designed using tension field theory–single stiffener (Fig. 5.32(b))

Panel B is designed using tension field theory. The combined bearing stiffener and end post is designed to resist the compressive stresses due to the external reaction and due to a moment of $(2/3)M_{tf}$.

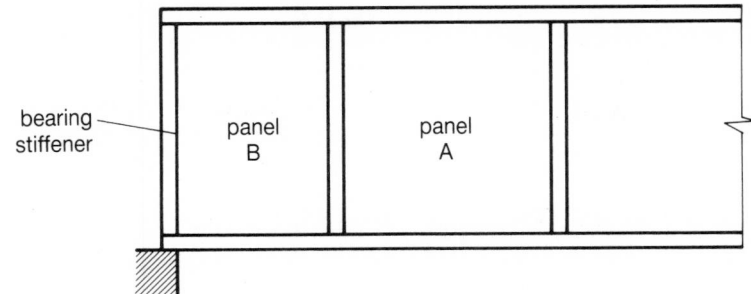

(a) End panel (panel B) designed without tension field theory

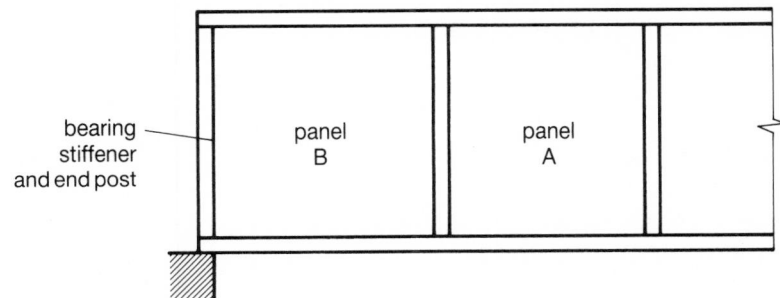

(b) Single end post – end panel (panel B) designed using tension field theory

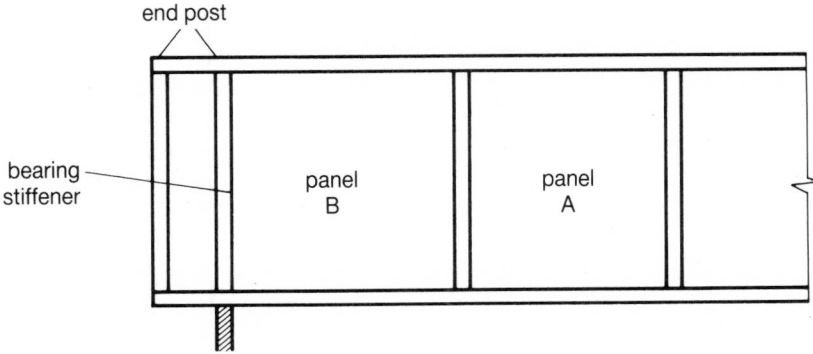

(c) Double end post – end panel (panel B) designed using tension field theory

Fig. 5.32 Configurations for web design.

(c)   End panel designed using tension field theory–double stiffener (Fig. 5.32(c))

Here the end panel (Panel B) can still be designed to utilize tension field theory. The bearing stiffener is designed to take only the compressive stresses from the external reaction, whilst the end post is designed to resist the horizontal shear, R_{tf}, and the moment, M_{tf}.

Whilst configuration (a) is simple it will give a substantially heavier web section at the end of the beam, which, unless it is proposed to use different web thicknesses along the length of the beam, will produce a heavy beam that could prove uneconomical. It should be pointed out that changing the web thickness along the length of the beam will increase fabrication costs. This increase may not offset the decrease in basic costs by lightening the beam. An alternative to altering the web size is to decrease the size of the end panel by reducing the distance between stiffeners. This will have the effect of increasing the basic shear capacity of the web. Configuration (b) will produce a very heavy end post but will allow a lighter web section to be used. A very thick end post may give problems with achieving a satisfactory weld. Configuration (c) will not be acceptable if an overhang at the end of the beam cannot be accommodated in the final structure for whatever reason. It is however likely to produce the most economic solution even allowing for the extra length of girder and increased fabrication costs as the end post and bearing stiffeners will be lighter than those of configuration (b). The centre-line distance between the end post and bearing stiffener must be sufficient for the fillet welds between the individual elements to be made satisfactorily. This distance should be not less than about 300 mm.

The values of R_{tf} and M_{tf} are calculated from the following formulae (cl 4.4.5.4.4, Pt 1):

$$R_{tf} = 0.5H_q,$$
$$M_{tf} = 0.1H_{qd}$$

(5.65)

where,

$$H_q = 0.75dtp_{yw}(1 - q_{cr}/(0.6p_{yw}))^{0.5}$$

(5.66)

Note that if $q_b > f_v$, then H_q may be multiplied by the ratio α where,

$$\alpha = (f_{cr} - q_{cr})/(q_b - q_{cr})$$

(5.66a)

where f_v is the shear stress in the panel utilizing tension field action (Panel A in Fig. 5.32(a) and Panel B in Figs 5.32(b) and (c)), q_b is the basic shear strength in the panel utilizing tension field action, and q_{cr} is the critical strength in the same panel.

5.7.5.5   Design of webs subjected to edge loading (cl 4.5.2.2, Pt 1)

Where loading, either as a UDL or as isolated point loads, is applied to the top flange of the beam other than at load bearing stiffeners, the compressive stresses due to these applied loads, f_{ed}, must not exceed the design strength, p_{ed}.

The design values are taken from the results of research work by Rockey and El-gaaly, and Rockey *et al.* (1972), and are determined on the basis that the applied load is concentrated over a very short bearing length on the top flange, thus giving an upper bound solution.

The design values are,

for the case where the compression flange is restrained from rotating relative to the web,

$$p_{ed} = (2.75 + 2/(a/d)^2)E/(d/t)^2$$

(5.67a)

or in the case where the compression flanges are free to rotate with respect to the web,

$$p_{ed} = (1.0 + 2/(a/d)^2)E/(d/t)^2$$

(5.67b)

The first case corresponds to where there are stiffeners between the flanges, and the second case to the situation, albeit rare, when there are no stiffeners between the flanges.

The applied stress, f_{ed}, is calculated by:

1 dividing any point loads or distributed loads shorter than the least panel dimension by the least panel dimension (a or d as appropriate)
2 adding the intensity (force/unit length) of any distributed loads over the whole length of the panel, and
3 dividing the sum obtained in (2) by the web thickness, t.

5.7.56 *Design of webs subjected to bending, shear and axial force (H3, Pt 1)*

This case arises when the girder is designed as a whole to take bending (Method (b), Section 5.7.3), and the web, therefore, is subject to bending, shear and any axial force. The web so designed must satisfy the following interaction equation (Allen and Bulson, Rockey (1971)),

$$(M_w/M_{cr})^2 + f_c/p_{c,cr} + (f_v/q_b)^2 \leqslant 1.0 \tag{5.68}$$

where M_w is the maximum moment in the web panel, M_{cr} is the buckling resistance moment in the web panel, f_c is the axial stress in the web, $p_{c,cr}$ is the critical axial strength of the web panel.

The axial stress has two components; the first is due to any applied axial force on the section (if the load is tensile, f_c is negative) and the second occurs when the mid-depth of the panel does not coincide with the centroidal axis of the beam. In this case, which only applies to girders with different sized flanges, the bending moment must be decomposed into two components, an axial force component and a moment about the web mid-depth (Fig. 5.33).

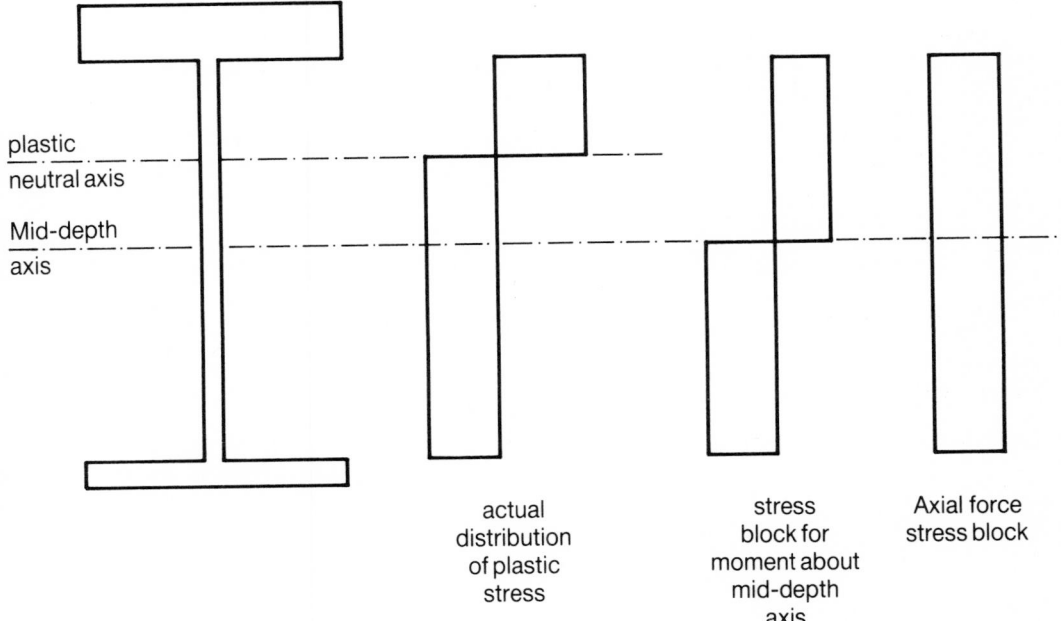

plastic
neutral axis

Mid-depth
axis

actual
distribution
of plastic
stress

stress
block for
moment about
mid-depth
axis

Axial force
stress block

Fig. 5.33 Unequal flange 'I' beam.

It should be noted that the critical design condition for plate girders where the web has been designed to take part of the bending moment will be Equation (5.68), especially when there is edge loading on any of the web panels, since in this case the term $f_c/p_{c,cr}$ is modified to take account of the edge stress as follows.

The stress ratio $f_c/p_{c,cr}$ is replaced by $((f_c/f_{c,cr})^2 + (f_{ed}/p_{ed})^2)^{0.5}$ when f_c is compressive, by $((f_{ed}/p_{ed})^2 - (f_c/p_{c,cr})^2)^{0.5}$ when f_c is tensile and the absolute value of f_{ed}/p_{ed} exceeds the absolute value of $f_c/p_{c,cr}$ or by $-((f_c/f_{c,cr})^2 - (f_{ed}/p_{ed})^2)^{0.5}$ when f_c is tensile and the absolute value of $f_c/p_{c,cr}$ exceeds the absolute value of f_{ed}/p_{ed}.

The critical bending moment M_{cr} on the web is given by,

$$M_{cr} = p_{b,cr}(td^2/4) \tag{5.69}$$

Derivation of the critical bending and axial strengths:

Bulson gives theoretical results for thin plates simply supported on all four edges for a variety of load cases. These results are reproduced in Fig. 5.34, where it will be seen that for any load case a minimum value of the coefficient K is found. In any particular case, the critical stress σ_{cr} is given by,

$$\sigma_{cr} = K\pi^2 E(t/d)^2/(12(1 - v^2)) \tag{5.70}$$

For uniform compression, the minimum value of K is 4, giving

$$\sigma_{cr} = 4\pi^2 E/(12(1 - v^2)) \tag{5.71}$$

If the numerical values of E and v are substituted into Equation (5.71), then

$$\sigma_{cr} = (861)^2/(d/t)^2$$

The code replaces the coefficient of 861 in the above equation by a lower value of 815, thus giving the critical stress for uniaxial strength $p_{c,cr}$ as

$$p_{c,cr} = (815/(d/t))^2 \tag{5.72}$$

Equation (5.72) only applies if one flange is in tension. Where both flanges are in compression the value of $p_{c,cr}$ is still calculated from Equation (5.72) but is subject to a maximum limit of $43p_y/(15 + d/(t\varepsilon))$ for welded sections and $43p_y/(4 + d/(t\varepsilon))$ for rolled sections.

For a triangular distribution of bending stress, (i.e. the classical elastic distribution), the minimum value of K is 23.9. In order to allow for the fact that a plastic distribution of bending stress is rectangular, this coefficient has been reduced by a factor of 0.6, giving

$$\sigma_{cr} = (0.6 \times 23.9)\pi^2 E(t/d)^2/(12(1 - v^2)),$$

or,

$$\sigma_{cr} = (1630/(d/t))^2$$

The Code thus defines the critical bending strength $p_{b,cr}$ as

$$p_{b,cr} = (1630/(d/t))^2 \tag{5.73}$$

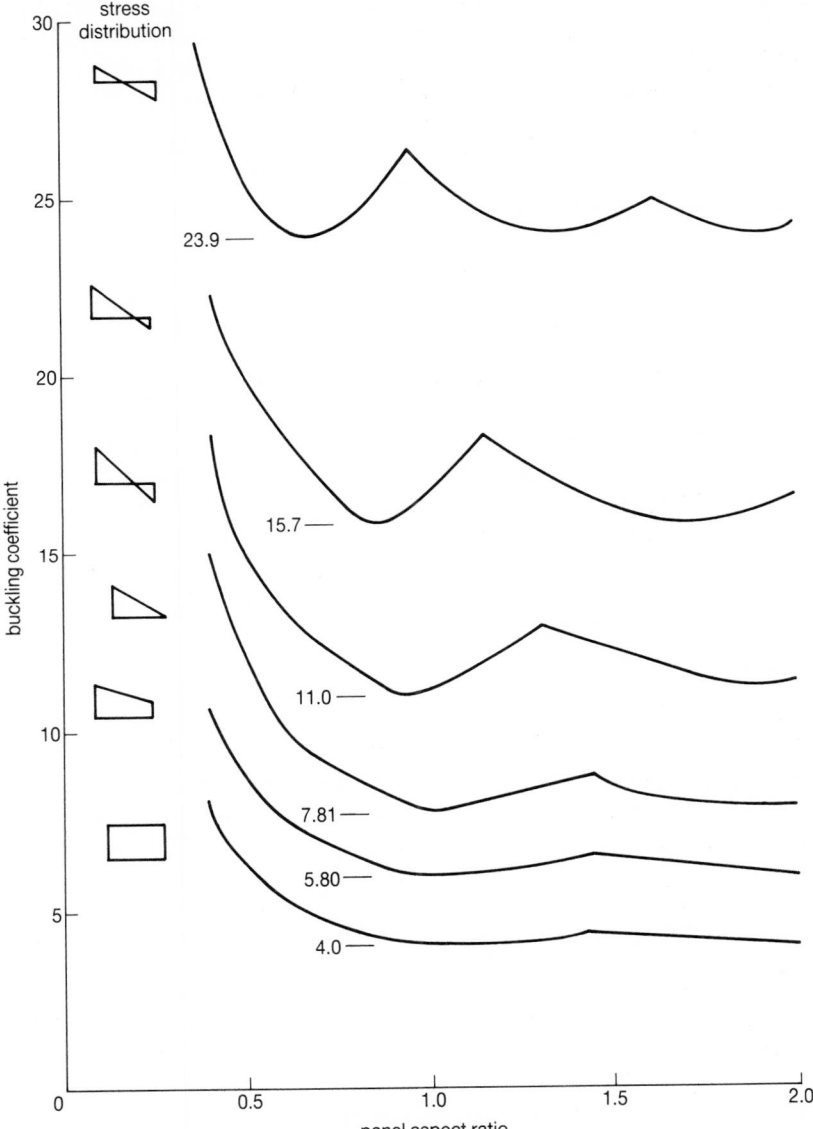

Fig. 5.34 Buckling strength coefficients.

5.7.6 Stiffeners

These are subdivided into three categories.

1 Intermediate stiffeners. These are required to stiffen the web and control the buckling of the web. The importance of these stiffeners will have become apparent when discussing web design as the web design stress is usually a function of the stiffener spacing.

2   Load bearing stiffeners. These are required under heavy point loads applied to the top flange in order not to place the web under high local stresses, and at the points of support to carry heavy reactions. In general where load bearing stiffeners are required the centres of intermediate stiffeners are chosen such that load bearing stiffeners also function as intermediate stiffeners.

3   Torsion stiffeners. These are a special category of load bearing stiffener, in that as well as resisting the axial effect of the end reaction, they are also required to give torsional restraint to the top flange of the beam.

5.7.6.1   *Design of intermediate stiffeners (cl 4.4.6, Pt 1)*

The function of intermediate stiffeners is to stiffen the web and to control the buckling of the web, and, where the web has been designed using tension field theory, it must be strong enough to supply the reaction to the tension force exerted in the web. There are thus two design criteria that must be considered: minimum stiffness and strength.

The minimum stiffness criterion is based on work by Rockey *et al.* (1981) whose experimental work indicates that at low panel aspect ratios (a/d) the minimum stiffness required from the stiffener is very high, whereas at low aspect ratios this requirement is much reduced. From the analysis of these results it can be demonstrated that the critical second moment of area I_{cr} for an intermediate stiffener, assuming all edges of the web plate to be simply supported, is given by

$$I_{cr} = 1.28dt^3/a^2$$

where a is the stiffener spacing, d the depth of the web and t the web thickness. In the Code the coefficient in the above equation has been replaced by 1.5, the web thickness t defined as the minimum thickness of web required utilizing tension field action to resist the applied shear, and a limit placed on its use of $a < \sqrt{2}d$, i.e.

$$\text{for} \quad a < \sqrt{2}d : I_s = 1.5d^3t^3/a^2 \tag{5.74}$$

This latter limit is arrived at since it can be demonstrated that for compression buckling of a panel a change from 1 to 2 half sine waves occurs at $a/d = \sqrt{2}$, and that thereafter the buckling coefficient is sensibly independent of the aspect ratio of the panel. Thus, for values of a greater than $\sqrt{2}d$, the value of $a = \sqrt{2}d$ has been substituted into Equation (5.74), giving

$$\text{for} \quad a > \sqrt{2}d : I_s = 0.75dt^3 \tag{5.75}$$

This latter equation is very conservative when compared to experimental results as no allowance is made for the effect of the panel aspect ratio.

Where the stiffener is subjected to either lateral loads or moments due to the effect of loading applied transversely to the web, the values of I_s, in Equations (5.74) and (5.75) should be increased by $2FD^3/Et$ for lateral loads and M_sD^2/Et for moments, where D is the overall depth of the section, t the web thickness and F the factored lateral load deemed to be applied at the level of the compression flange. M_s is the moment on the stiffener due to eccentrically applied loading.

Intermediate stiffeners not subjected to applied loading must be checked for a force F_q,

$$F_q = V - V_s \leqslant P_q \tag{5.76}$$

where V is the maximum shear force adjacent to the stiffener, V_s is the shear buckling strength of the web without the contribution of tension field action, and P_q is the buckling resistance of the stiffener, the calculation of which is dealt with in the next section.

Equation (5.76) is noted by Rockey *et al.* (1981) to be a safe lower bound solution to the exact equation for calculating the force in the stiffener.

In addition, intermediate stiffeners which carry loading must satisfy both the requirements for load bearing stiffeners and the following interaction equation,

$$(F_q - F_x)/P_q + F_x/P_x + M_s/M_{ys} \leqslant 1.0 \tag{5.77}$$

where F_x is the external load or reaction, P_x is the buckling resistance of a load bearing stiffener, and M_{ys} is the moment capacity of the stiffener based on its elastic modulus. Note that if in Equation (5.77) $F_q - F_x$ is negative, then it is set equal to zero.

5.7.6.2   *Load bearing stiffeners (cl 4.5, Pt 1)*

These are required at points where there are high external loads, either applied loads or reactions, and they may also function as intermediate stiffeners. In the latter case, the stiffener must satisfy the design criteria for both intermediate stiffeners (Equations (5.74) to (5.77)) and load bearing stiffeners.

The outstand of a load bearing stiffener should not exceed $19t_s\varepsilon$, where t_s is the stiffener thickness and $\varepsilon = (275/p_y)^{0.5}$. When the outstand is between $19t_s\varepsilon$ and $13t_s\varepsilon$, only a core section having an outstand of $13t_s\varepsilon$ should be considered when calculating the stiffener load capacity.

The buckling resistance of a load bearing stiffener is calculated on a section which includes part of the web, as it was observed (Rockey *et al.* (1981)) that part of the web also acted with the stiffener to resist the effects of stiffener buckling, and it was found when the results were analyzed that there was a good fit between calculation and experimental results if a length of web equal to twenty times the web thickness was taken. The Code, thus, allows a section of web to be equal to 20 times its thickness to be taken into account when calculating the section properties of a load bearing stiffener, i.e. the effective cross-sectional area, second moment of area of the stiffener about the centre line of the web, and the corresponding radius of gyration of a stiffener can be calculated as (Fig. 5.35),

$$A_e = b_{es}t_s + 40t_w^2 \tag{5.78}$$

$$I_e = b_{es}^3 t_s/12 + 40t_w^4/12 \tag{5.79}$$

for a stiffener with the web extending on both sides of the stiffener. In the case of a stiffener at the end of the beam the web will only extend on one side and, therefore, the coefficient 40 that appears in Equations (5.78) and (5.79) should be replaced by 20.

The effective length of an intermediate stiffener shall be taken as 0.7 times the stiffener length and for a load bearing stiffener 0.7 times the stiffener length if the flange through which the load is applied is prevented from rotating by other structural elements, otherwise a factor of 0.7 should be used.

The value of p_c used to calculate the buckling resistance should be that obtained from Table 27(c), with an effective design strength p_y' taken as 20 N/mm^2 less than the actual design strength of the stiffener, p_y. The reason for this reduction is an attempt to make a simple allowance for the higher residual stresses in welded sections than those in rolled sections without having to set up additional tables for strut buckling.

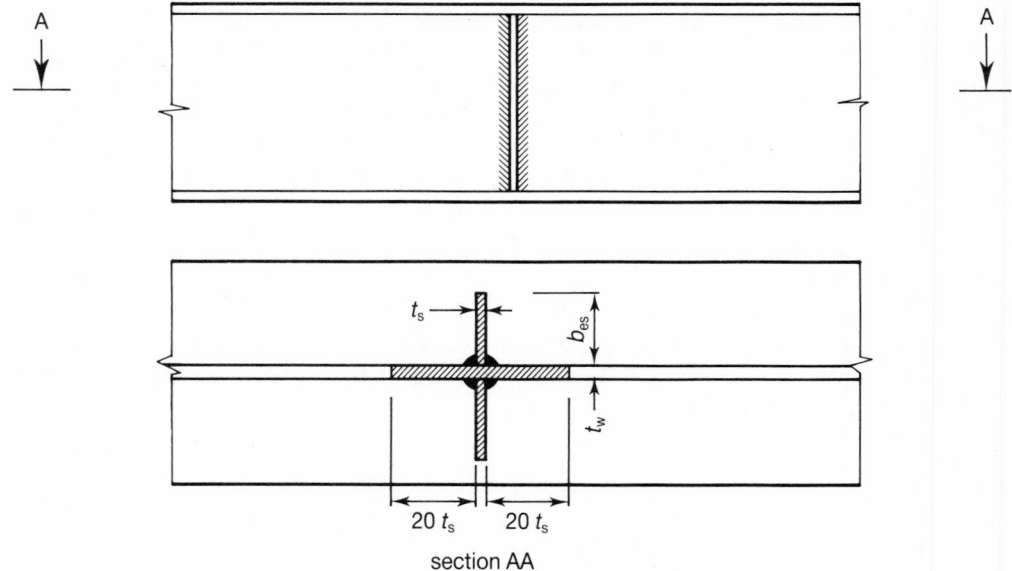

Fig. 5.35   Geometry for intermediate stiffeners.

The load bearing stiffener should satisfy the following relationship,

$$P_x > F_x \qquad (5.80)$$

where P_x is the buckling resistance of the stiffener and F_x is the applied load or reaction.

In addition the following relationship checking the bearing at the contact between the stiffener and the flange must be satisfied,

$$A_{min} > 0.8F_x/p_{ys} \qquad (5.81)$$

where A_{min} is the minimum area required to be in contact with the flange and p_{ys} is the design strength of the stiffener. The actual area A is generally less than the full cross-sectional area of the stiffener as the corners of the stiffener are coped to clear the web-to-flange fillet weld (Fig. 5.36)

5.7.6.3   Torsion stiffeners (cl 4.5.3, Pt 1)

These are required at the ends of a beam where torsional restraint is not otherwise provided. They are designed as load bearing stiffeners carrying the beam reaction, but are also subject to the specification of a minimum second moment of area I_s.

$$I_s = 0.34\alpha_s D^3 T_c \qquad (5.82)$$

where

for $\lambda \leqslant 50 : \alpha_s = 0.006$

for $50 < \lambda \leqslant 100 : \alpha_s = 0.3/\lambda$

and, $\lambda > 100 : \alpha_s = 30/\lambda^2$

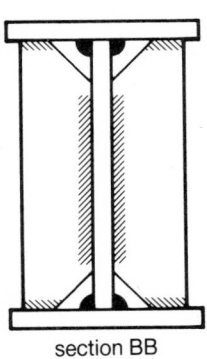

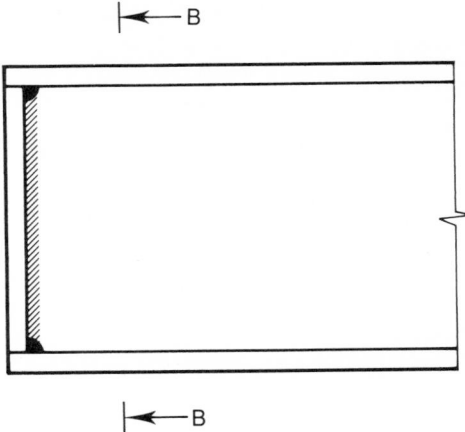

section BB

Fig. 5.36 Geometry for end stiffeners.

D is the overall depth of the beam at the support, T_c the maximum thickness of the compression flange and λ the slenderness ratio of the beam defined as $\lambda = L_E/r_y$

5.7.6.4 *Design of the stiffener to beam connection*

Where the intermediate stiffener carries no external loading then the shear s to be carried by the weld between the stiffener and the web is given by

$$s = t^2/5b_s \tag{5.83}$$

It should be noted that Equation (5.83) is not dimensionless, as the web thickness t, and the outstand of the stiffener b_s, are both in mm and the final result is in kN/mm.

Stiffeners not subjected to external loads may terminate clear of the tension flange, but must extend up to the compression flange, though they need not be connected to it.

The connection between load bearing stiffeners and the web must be designed for the lesser of:

1   the tension capacity of the stiffener, and
2   the sum of the forces where they act in the same direction or the largest when they do not.

Stiffeners transmitting compression should either be fitted against the loaded flange or connected by continuous welds. In addition, where the stiffener carries a support reaction or it forms the end stiffener of a stiffened web or acts as a torsion stiffener, the stiffener should be fitted or continuously welded to both flanges.

In order to illustrate the design process, particularly that for webs using the possible options of including tension field theory, a single design will be undertaken in which for simplicity the compression flange will be taken as continuously restrained against lateral torsional buckling and that the flanges will be designed to resist the bending moment and the web to resist shear. The overall depth of the beam will be kept constant as will the size of the flange plates throughout the possible permutations of web design.

EXAMPLE 5.7   *Plate girder design*

Design a plate girder in Grade 50 steel to carry an imposed service load of 150 kN/m over a span of 20 m. The compression flange is continuously restrained against lateral torsional buckling.

For a plate girder the usual span/depth ratios are in the order of between 8 and 12 to 1. The higher the ratio the lower the flange size will be but the web will need to be thicker to overcome web buckling.

One method of determining initial dimensions is to use a minimum weight analysis. It is to be emphasized that a minimum weight design is not synonymous with a minimum cost design as no account has been taken of fabrication problems.

If the moment M is entirely resisted by the flanges, then

$$M = p_y BTd \tag{5.84}$$

where p_y is the design strength of the flanges, B and T the width and thickness, respectively, of the flanges, and d is the depth of the web.

The cross-sectional area A of the beam is given by

$$A = 2BT + dt \tag{5.85}$$

where t is the web thickness.

Eliminating BT between Equations (5.84) and (5.85) gives

$$A = 2M/(dp_y) + dt \tag{5.86}$$

Setting $d = kt$
where k is the web slenderness,
d may be eliminated from Equation (5.86) to give

$$A = 2M/(ktp_y) + kt^2 \tag{5.87}$$

Differentiating Equation (5.87) with respect to t and setting the result equal to zero gives the optimum value of t as

$$t = (M/(p_y k^2))^{0.33} \tag{5.88}$$

and the optimum value of the depth d as

$$d = (Mk/p_y)^{0.33} \tag{5.89}$$

Note that these optimum values produce a beam in which the areas of a single flange and the web are equal. If the optimization procedure is carried out with the web also carrying bending, the flange plates will be calculated as being a quarter of the size of the web.

To calculate the maximum bending moment an initial assumption will need to be made of the self weight, so assume that the self weight is 60 kN. This will need checking in due course, although it is only a small proportion of the total loading.

Total ultimate load on the beam W is given by:

$$W = 1.6 \times 150 \times 20 + 1.4 \times 60 = 4884 \text{ kN}$$

Maximum moment $= WL/8$
$$= 4884 \times 20/8$$
$$= 12.21 \text{ MNm}$$

The moment is to be taken by the flanges, so the optimum depth between the flanges is given by Equation (5.89),

$$d = (Mk/p_{yf})^{0.33} \tag{5.89}$$

The flange thickness is likely to be greater than 16 mm, thus take p_y as 345 N/mm². The web is to be made as slender as possible, so for the initial design take $k = 220$, thus

$$d = (12.21\text{E}9 \times 220/345)^{0.33}$$
$$= 1982 \text{ mm}$$

and

from Equation (5.88), $t = 9$ mm.

It should be noted that 9 mm (and in a later section of the example, 8 mm) plate is a non-standard thickness although it would be available at extra cost as a special order.

Round the value of d to a slightly more convenient figure and set, therefore, the overall depth to 2000 mm.

For a plastic section the maximum flange outstand b is given by $7.5\varepsilon T$ (Table 7 for a welded section)

$$\varepsilon = (275/p_y)^{0.5} = (275/345)^{0.5} = 0.89$$

Thus the maximum outstand b is given by

$$7.5 \times 0.89T = 6.68T$$

The area of the flange is given by $(2 \times 6.68T)T = 13.36T^2$

The moment capacity is given, approximately, by

$$M = DA_f p_y$$
$$= 2000 \times 13.36T^2 \times 345 = 9.22\text{E}6 \times T^2 \text{ Nm}$$

or,

$$T = (12.21\text{E}9/9.33\text{E}6)^{0.5} = 36.4 \text{ mm}$$

Use flange plates 500×40 mm (actual b/T ratio 6.25).

Note that this size of flange plate is very close to that predicted by the optimization procedure which gives 36.6 by 489.

Actual moment of resistance $= hA_f p_{yf}$
$$= (2000 - 40)(40 \times 500) \times 345 \text{ Nm}$$
$$= 13.52 \text{ MNm}.$$

This is greater than the applied moment and therefore allows a margin if the self weight has been underestimated. The optimum cross-sectional area is 53750 mm² which is equivalent to a self weight of 82.5 kN (an increase of 22.5 kN over the original estimate, or, an increase in applied moment of 79 kNm compared with the moment used in the design of 12.21 MNm).

To ensure that there will be no problems with the deflection, calculate the second moment of area of the section without the web and then estimate the deflection.

The second moment of area of the flanges alone about the neutral axis of the beam, neglecting the I value of the flanges, is given by

$$I = 2A_f \times ((D - T)/2)^2$$
$$= 2 \times 40 \times 500 \times ((2000 - 40)/2)^2$$
$$= 3.84\text{E}10 \text{ mm}^4$$

$$\delta = (5/384)(qL^4/EI)$$
$$= (5/384)(150 \times 20^4/(205\text{E}6 \times 3.84 - 2))$$
$$= 0.040 \text{ m}$$

Allowable deflection = span/200 = 20/200 = 0.10 m

Thus the estimated deflection is below the allowable. The actual deflection will be lower than the calculated value of 40 mm, and thus no further check will be made.

Web design

The web will be designed using all the possible options allowed under BS 5950. The maximum value of d/t that may be used is 250 (Section 5.7.2)

(a)   *Design without tension field theory*

The minimum web thickness that can be used is 9 mm ($d/t = 213$) as there is no combination of d/t higher than this and a/d that will give a high enough value of q_{cr}.

$$V = W/2$$
$$= 4884/2 = 2442 \text{ kN}$$

$$d = D - 2T = 2000 - 2 \times 40 = 1920$$

Required shear strength $= 2442\text{E}3/(1920 \times 9) = 141.3 \text{ N/mm}^2$

From Table 21, with $p_y = 355$, this can be achieved with $a/d = 0.417$, or $a = 801$ mm.

From geometrical considerations use $a = 800$, i.e. 24 intermediate stiffeners. This solution will be economical in terms of tonnage of steel but may be expensive on fabrication due to the large number of stiffeners. As the shear decreases toward the centre of the beam it should be possible to increase the stiffener spacing. The limiting criterion is likely to be the capacity of the web to resist the in-plane effect of the applied load.

Using cl 4.5.2.2 (Section 5.7.5.5)

$$f_{ed} = 1.6 \times 150/t = 240/9 = 26.7 \text{ N/mm}^2$$

Using Equation (5.67(a)) as the compression flange is restrained by stiffeners from rotating relative to the web,

$$p_{ed} = (2.75 + 2/(a/d)^2)E/(d/t)^2$$
$$= (2.75 + 2/(0.417)^2)(205\text{E}3/(213)^2)$$
$$= 64.4 \text{ N/mm}^2$$

For a stiffener spacing of 1100 mm ($a/d = 0.573$), $p_{ed} = 40.0$

For $a/d = 0.573$, $q_{cr} = 85.3$ N/mm^2 (Table 21)

$$V_{cr} = q_{cr}dt$$
$$= 85.3 \times 1920 \times 9/1E3$$
$$= 1474 \text{ kN}.$$

This value of shear occurs at $x = (2442 - 1474)/244.2 = 4.32$ m. This occurs in the sixth panel, thus the stiffener spacing can be increased to 1100 mm in the seventh panel. The resultant stiffener layout is shown in Fig. 5.37.

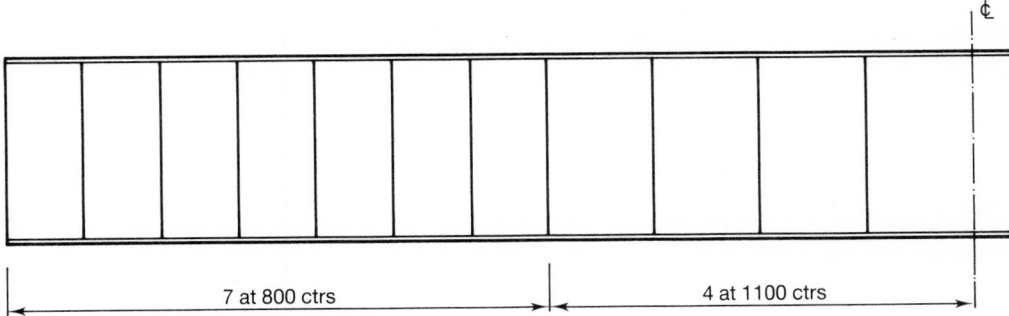

Fig. 5.37   Stiffener spacing for Example 5.7 designed without tension field strength.

Stiffener design

1   Intermediate stiffeners.

For both sizes of panel $a < \sqrt{2}d$, so I_s is given by Equation (5.74).

$$I_s = 1.5d^3t^3/a^2$$

It is normal to overdesign the stiffener by using the actual web thickness rather than the minimum web thickness calculated using tension field action for the given shear and a/d ratio.

So using $t = 9$ mm, and the least value of panel width a of 800 mm,

$$I_s = 1.5 \times (1920 \times 9)^3/800^2$$
$$= 12.1E6 \text{ mm}^4$$

The limiting effective outstand of a stiffener is $13t_s\varepsilon$ (cl 4.5.12, Section 5.7.6.2).

With a yield strength for the stiffener of 355 N/mm^2, $\varepsilon = 0.88$, the limiting outstand $= 13 \times 0.88t_s = 11.44t_s$.

The second moment of area I of the stiffener is given by

$$I = (2 \times 11.44t_s)^3 t_s/12 = 1000t_s^4 \qquad (5.90)$$

Equating I_s and the value of I from Equation (5.90) gives

$$t_s = (12.1E6/1000)^{0.25}$$
$$= 10.5 \text{ mm}$$

Use a 12 × 120 mm flat on both sides of the web.
Actual second moment of area $= (2 \times 120 + 9)^3 \times 12/12 = 15.4E6$ mm^4

This value is slightly in excess of the actual value as the portion of the web between the flats has been included. The effect of this is negligible.

Strictly the stiffeners should be checked for buckling using Equation (5.76), but for equal size panels designed without tension field theory $V_s > V$ and thus F_q is negative. Also in this example although the stiffener spacing has been increased towards the centre of the span, the force F_q in the stiffener between the two panels of unequal length is still negative, and thus no further check need be made.

2 End stiffener.

Try two 16×240 end plates either side of the web.

The outstand b is 240 mm

$$b/T = 245.5/16 = 15.3$$

Max outstand $= 18\varepsilon t_s = 19 \times 0.88 \times 16 = 267.5$, which is satisfactory.
Design core $= 13\varepsilon t_s = 13 \times 0.88 \times 16 = 183$ mm
The details of the stiffener are given in Fig. 5.38.

Check bearing area:

Actual area $= 2 \times 16 \times (240 - 15) = 7200$ mm^2

$$\text{Minimum area, } A_{min} = 0.8 F_x / p_{ys} \tag{5.81}$$

where F_x is the reaction of 2442 kN and p_{ys} is taken as 355 N/mm^2

$$A_{min} = 0.8 \times 2442\text{E3}/355 = 5503 \text{ mm}^2$$

This is satisfactory.

Buckling check:

The nett section used is given in Fig. 5.38 and the section properties are calculated using Equations (5.78) and (5.79).

$$
\begin{aligned}
A_e &= b_{es} t_s + 20 t_w^2 \\
&= 375 \times 16 + 20 \times 9^2 \\
&= 7620 \text{ mm}^2
\end{aligned}
$$

$$
\begin{aligned}
I_e &= b_{es}^3 t_s / 12 + 20 t^4 / 12 \\
&= 16 \times 375^3 / 12 + 20 \times 9^4 / 12 \\
&= 70.3\text{E6 mm}^4
\end{aligned}
$$

$$
\begin{aligned}
r_y &= (I_s / A_s)^{0.5} \\
&= (70.3\text{E6}/7620)^{0.5} \\
&= 96.0
\end{aligned}
$$

$$L_E = 0.7 d = 0.7 \times 1920 = 1344$$

$$\lambda = L_E / r_y = 1344/96 = 14$$

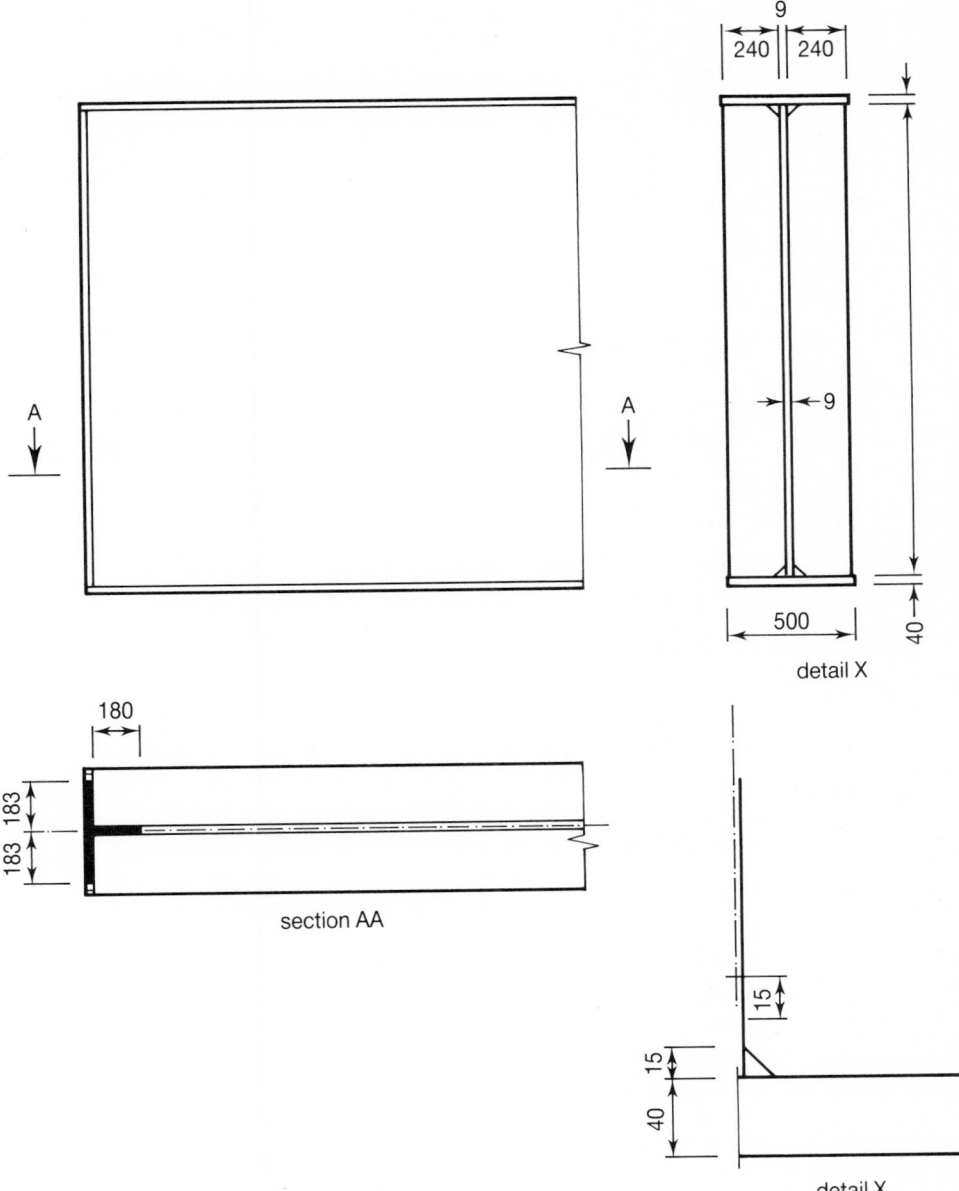

Fig. 5.38 End stiffener details.

Using Table 27(c) with a reduced p_y of $355 - 20 = 335$ (cl 4.7.5, Pt 1), $p_c = 335$ N/mm²

$$P_x = p_c A_e$$
$$= 335 \times 7620/1\text{E}3$$
$$= 2553 \text{ kN} > F_x$$

Design of welds.

1   Intermediate stiffeners:

Force to be taken by weld, s, given by

$$s = t^2/5b_s \qquad\qquad (5.83)$$

$t = 9$ mm, $b_s = 120$ mm,

$$s = 9^2/(5 \times 120)$$
$$= 0.135 \text{ kN/mm}$$

Use a pair of 6 mm fillet welds (0.9 kN/mm)

2   End stiffener:

The weld must be designed to take the reaction in addition to the force from Equation (5.83).

length of weld $= 1920 - 2 \times 15 = 1890$

Total force on each weld $= (2442/2)/1890 + 9^2/(5 \times 245)$
$$= 0.71 \text{ kN/mm}$$

Use a 6 mm fillet weld on both sides of the web.

Flange to web weld:

Force in weld pair q_s is given by

$$q_s = QA_f y_c/I$$

For the derivation of this equation reference should be made to Section 4.3.1

$$I = (500 \times 2000^3 + 491 \times 1920^3)/12$$
$$= 4.37\text{E}10 \text{ mm}^4$$

$$q_s = 2442 \times (500 \times 40) \times (1920/2 - 20)/4.37\text{E}10$$
$$= 1.04 \text{ kN/mm}$$

Use a 6 mm fillet weld on both sides of the web.
Note all welds are to be continuous to avoid problems with corrosion.

(b)   *Web designed using tension field theory except for the end panel*

From the previous design retain $t = 9$ mm and a stiffener spacing for the end panel of 800 mm long.

For the first panel to be designed using tension field theory the design strength should be taken as the basic strength only since if the basic strength q_b exceeds the applied shear strength f_v, the anchor force H_q can be reduced.

There is no need here to check the first panel as the calculations are identical to those for the web designed without tension field theory in the previous section.

For the remainder of the girder, try seventeen panels each of 1082 mm long. A check on the ability of the web to resist the in-plane effect of the applied UDL indicates that this spacing is satisfactory.

Second panel:

$$V = 2442 - 0.8 \times 244.2 = 2247 \text{ kN}$$

Moment at end of second panel, M:

$$M = 2442 \times 1.882 - 244.2 \times 1.882^2/2 = 4163 \text{ kNm}$$

Calculation of q_b:

$$d/t = 1920/9 = 213$$

$$a/d = 1082/1920 = 0.56$$

By interpolation from Table 22 with $p_y = 355$,

$$q_b = 154 \text{ N/mm}^2$$

$$\begin{aligned} V_b &= q_b dt \\ &= 154 \times 1920 \times 9/1\text{E}3 \\ &= 2661 \text{ kN} \end{aligned}$$

This exceeds the applied shear V and thus the contribution from the flange need not be calculated. It is not possible toward the centre of the span to make other than a small adjustment to the stiffener spacing as the critical design condition is that of the ability of the web to support the in-plane effect of the applied loading.

The first panel must be checked under the effect of a moment and shear due to the reaction from the tension field of the adjacent panel.

The anchor force H_q is calculated from Equation (5.66).

$$H_q = 0.75 dt p_{yw}(1 - q_{cr}/0.6p_{yw})^{0.5}$$

q_{cr} is calculated for the first panel under tension field,

$$d/t = 213, \ a/d = 0.56 \text{ and } p_{yw} = 355, \text{ so from Table 21,}$$

$$q_{cr} = 88 \text{ N/mm}^2$$

$$\begin{aligned} H_q &= 0.75 \times 1920 \times 9 \times 355 \times (1 - 88/(0.6 \times 355))^{0.5}/1\text{E}6 \\ &= 3.52 \text{ MN} \end{aligned}$$

Calculate the shear stress f_v in the panel:

$$\begin{aligned} f_v &= V/dt \\ &= 2247\text{E}3/(1920 \times 9) \\ &= 130 \text{ N/mm}^2 \end{aligned}$$

This is less than the basic shear capacity q_b so the above value of H_q may be multiplied by the reduction factor α given by Equation (5.66a),

$$\begin{aligned} \alpha &= (f_v - q_{cr})/(q_b - q_{cr}) \\ &= (130 - 88)/(154 - 88) \\ &= 0.64, \end{aligned}$$

so effective value of H_q is $3.52 \times 0.64 = 2.26$ MN

Using Equation (5.65) to calculate the moment and shear,

$$R_{tf} = H_q/2$$
$$= 2.26/2 = 1.13 \text{ MN}$$

$$M_{tf} = 0.1 \times H_q d$$
$$= 0.1 \times 2.26 \times 1.92$$
$$= 0.43 \text{ MNm}$$

The Code implies that the panel must separately satisfy the design criteria of shear buckling and the ability to act as an anchorage for the reactions from the tension field.

$$\text{Longitudinal shear stress} = R_{tf}/at$$
$$= 1.13\text{E6}/(800 \times 9)$$
$$= 157 \text{ N/mm}^2$$

$$\text{Limiting shear stress} = 0.6 \times p_{yw} = 0.6 \times 355 = 213 \text{ MPa}$$

Panel satisfies in shear.

$$\text{Bending stress} = M_{tf}/(a^2 t/4)$$
$$= 0.43\text{E9}/(800^2 \times 9/4)$$
$$= 299 \text{ N/mm}^2$$

This is less than the panel design strength, and thus the panel passes in bending also. This completes the web design, and so the stiffeners may now be designed.

1   Intermediate stiffeners:

1.1   Stiffeners between panels of the same dimensions:

These need only satisfy the criterion of minimum second moment of area given by Equation (5.74).

Since $a < \sqrt{2}d$,

$$I_s = 1.5(dt)^3/a^2$$
$$= 1.5 \times (1920 \times 9)^3/1082^2$$
$$= 6.61\text{E6 mm}^4$$

From Equation (5.90),

$$I = 1000t_s^4$$

thus,

$$t_s = 9.0 \text{ mm}$$

Use a 100×10 mm double stiffener ($I = 6.85\text{E6 mm}^4$)

1.2   Stiffeners between first and second panel:

The minimum second moment of area should be calculated using the smaller of the two panel lengths, i.e. $a = 800$.
This gives $I_s = 12.1\text{E6 mm}^4$ and $t_s = 10.5$ mm.
Use two flats each 120×12 thick ($I = 15.4\text{E6}$)

The stiffener must also be checked under a force F_q given by Equation (5.76),

$$F_q = V - V_s$$

The shear force at the stiffener V is 2247 kN.
The web capacity V_s is given by

$$V_s = q_{cr}dt$$

with q_{cr} calculated on the larger panel, thus

$$V_s = 88 \times 1920 \times 9/1E3$$
$$= 1521 \text{ kN}$$

and,

$$F_q = 2247 - 1521$$
$$= 726 \text{ kN.}$$

Using Equations (5.78) and (5.79) to calculate the effective properties of the stiffener,

$$A_e = (2 \times 120 + 9) \times 12 + 40 \times 9^2$$
$$= 6228 \text{ mm}^2$$

$$I_e = 249^3 \times 12/12 + 40 \times 9^4$$
$$= 15.5E6 \text{ mm}^4$$

$$r_y = (I_e/A_e)^{0.5}$$
$$= (15.5E6/6228)^{0.5}$$
$$= 50 \text{ mm}$$

$$L_E = 1344 \text{ (as before)}$$
$$\lambda = L_E/r_y$$
$$= 1344/50$$
$$= 27$$

From Table 27(c) with an effective p_y of 355, $p_c = 313$ MPa

$$P_q = A_e p_c$$
$$= 6228 \times 313/1E3$$
$$= 1949 \text{ kN}$$

This is greater than F_q and is therefore satisfactory.

2   End stiffener:

This must be designed for the reaction together with the force induced by the moment M_{tf}.

Force induced by $M_{tf} = M_{tf}/a$
$$= 0.43E3/0.8$$
$$= 538 \text{ kN}$$

Reaction is 2442 kN,

thus total design force $= 2442 + 538$

$$= 2980 \text{ kN}.$$

Try two 245×20 plates.
Maximum outstand $= 19 \varepsilon t_s = 19 \times 0.89 \times 20 = 338$ mm
Actual outstand $= 245$
Design core $= 13 \varepsilon t_s = 13 \times 0.89 \times 20 = 229$ mm

Check bearing area:

Actual bearing area $= 2 \times (245 - 15) \times 20$

$$= 9200 \text{ mm}^2 \text{ (See Fig. 5.39)}$$

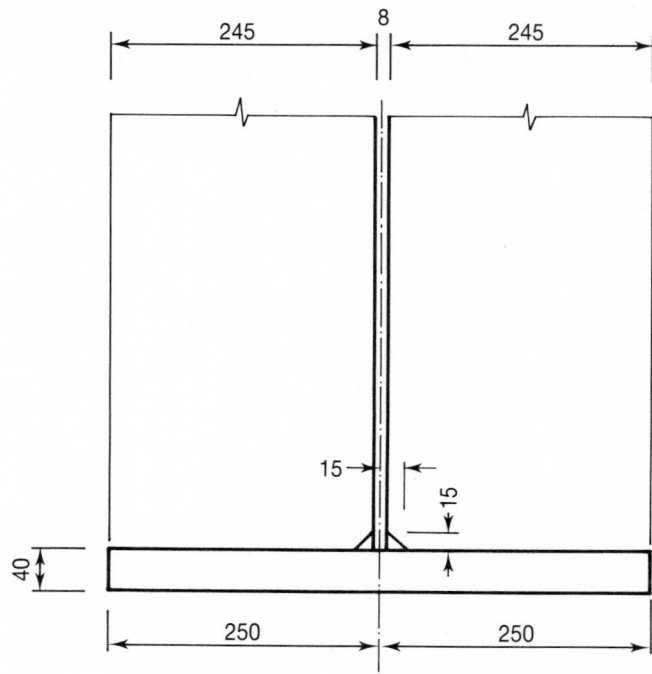

Fig. 5.39 End stiffener details.

Minimum bearing area is calculated from Equation (5.81),

$$A_{\min} = 0.8 F_x / p_{ys}$$
$$= 0.8 \times 2980\text{E}3/345$$
$$= 6910 \text{ mm}^2$$

This check is satisfactory.

Check buckling:

Using Equations (5.78) and (5.79) to calculate stiffener properties $A_e = 10960 \text{ mm}^2$, $I_e = 170\text{E}6 \text{ mm}^4$ and $r_y = 125$ mm

$$\lambda = 1344/125 = 11$$

From Table 27(c) with effective $p_y = 325$, $p_c = 325$ MPa

$$P_q = A_e p_c$$
$$= 10960 \times 325/1E3$$
$$= 3562 \text{ KN}$$

This is greater than the design force and is therefore satisfactory.

The welds are designed exactly the same as in the design of the web without tension field theory, except that the first intermediate stiffener to web weld will need designing to carry the force F_q.

(c)   *All web panels designed using tension field theory*

There are two options here with respect to the design of the end post system at the end of the beam; namely the use of a single or double stiffener system. The web design up to that point is, however, identical.

Again to keep the anchorage force H_q as low as possible, the shear buckling capacity of the first web panel q_{cr} should be kept as high as possible and the actual shear stress f_v kept less than the basic strength q_b i.e. any capacity of the flanges to assist in enhancing the shear capacity should not be used in this first panel.

Try an 8 mm web ($d/t = 240$).

The maximum stiffener spacing will be governed by the ability to support the in-plane effects of the load.

$$f_{ed} = (\text{UDL load})/t$$
$$= 1.6 \times 150/8$$
$$= 30 \text{ MPa}$$

The web capacity, p_{ed}, is given by Equation (5.67a))

$$p_{ed} = (2.75 + 2/(a/d)^2)E/(d/t)^2$$

Equating the values of f_{ed} and p_{ed} gives a maximum panel aspect ratio, a/d, of

$$a/d = 0.6$$

or

$$a = 0.60 \times 1920 = 1152 \text{ mm}$$

For reasons explained above, the end panels will be set at 800 mm with 16 panels at 1150 mm.

The values of bending moment and shear force for the first three panels are given in Fig. 5.40.

Panel 1-2:

$a/d = 800/1920 = 0.42$; $d/t = 240$
From Table 22 with $p_y = 355$, $q_b = 172$ MPa.

$$V_b = q_b dt$$
$$= 172 \times 1920 \times 8/1E3$$
$$= 2642 \text{ kN}$$

This is greater than the applied shear of 2442 kN and is, therefore, satisfactory.

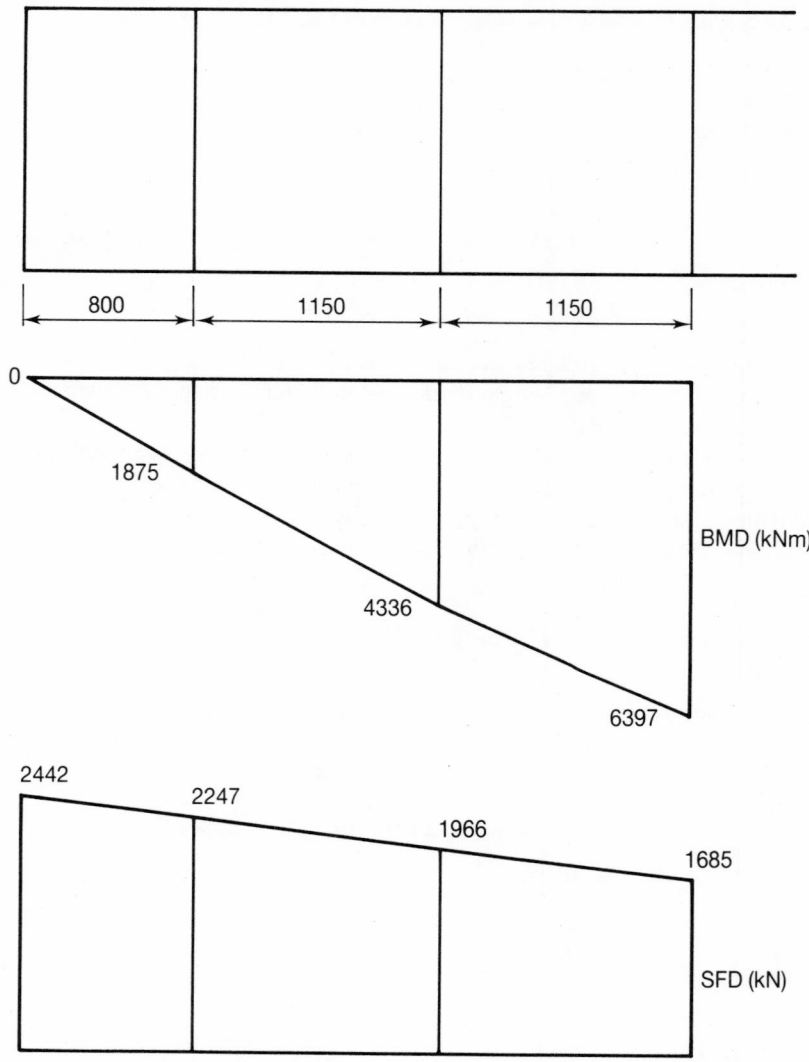

Fig. 5.40   Bending moments and shear forces in the first three panels for the full tension field design.

Panel 2-3:

$a/d = 1150/1920 = 0.60$; $d/t = 240$
From Table 22, $q_b = 137$ MPa

$$V_b = 137 \times 1920 \times 8$$
$$= 2104 \text{ kN.}$$

From Table 23, $q_f = 633$ MPa
The coefficient K_f is calculated from Equation (5.64),

$$K_f = (M_{pf}/4M_{pw})(1 - f/p_{yf})$$

Plastic moment capacity of the flange M_{pf}:

$$M_{pf} = 345 \times (500 \times 40^2/4)/1E6$$
$$= 69 \text{ kNm}$$

Plastic moment capacity of the web, M_{pw}:

$$M_{pw} = 355 \times (1920^2 \times 8/4)/1E9$$
$$= 2.62 \text{ MNm}$$

The stress in the flanges f:

$$f = M/(A_f(D - T))$$
$$= 4298E6/(500 \times 40 \times (2000 - 40))$$
$$= 110 \text{ MPa}$$

$$K_f = ((69/(2.62E3 \times 4))(1 - 110/345)$$
$$= 4.48E - 3$$

Shear capacity due to the flanges V_f:

$$V_f = (q_b\sqrt{K_f})dt$$
$$= (633\sqrt{4.48E\text{-}3}) \times 1920 \times 8/1E3$$
$$= 651 \text{ kN}$$

$$\text{Total shear capacity} = 2104 + 651$$
$$= 2755 \text{ kN.}$$

This is higher than the applied shear.

Panel 3-4:

For this panel, and subsequent panels, the shear strength V_b exceeds the applied shear.

The intermediate stiffeners may now be designed.
    The results for this design will only be summarized as the principles are as in the previous section of the example.

1   between first and second panel:

Minimum stiffness requirement calculated on smaller panel size of $a = 800$ mm.

$$I_s = 8.49E6 \text{ mm}^4$$

Supply two 10×110 flats ($I = 9.88E6$ mm^4)

Check buckling:

$$V = 2247 \text{ kN}, \quad q_{cr} = 61 \text{ MPa}, \quad V_s = 937 \text{ kN}, \quad F_q = 1310 \text{ kN}$$

$A_e = 4840$ mm^2, $I_e = 9.89E6$ mm^4, $r_y = 45.2$ mm, $\lambda = 30$, $p_c = 307$ MPa, $P_x = 1486$ kN, which is satisfactory.

2   Remaining stiffeners:

$a = 1150$ mm, $I_s = 4.11E6$ mm^4. Supply two 8×90 flats ($I = 4.43E6$ mm^4)

Calculation of H_q:

From Table 21, $q_{cr} = 114$ MPa

$$H_q = 0.75 \times 1920 \times 8 \times 355 \times 8(1 - 114/(0.6 \times 355))^{0.5}/1E6$$
$$= 2.79 \text{ MN}$$

$$f_v = 2442E3/(1920 \times 8)$$
$$= 159 \text{ MPa},$$

as $f_v < q_b$ the above value of H_q may be reduced.

$$\alpha = (159 - 114)/(172 - 114)$$
$$= 0.78$$

Effective value of H_q used in design:

$$H_q = 0.78 \times 2.79$$
$$= 2.18 \text{ MN}$$

$$R_{tf} = H_q/2$$
$$= 2.18/2$$
$$= 1.09 \text{ MN}$$

$$M_{tf} = 0.1 \times H_q d$$
$$= 0.1 \times 2.18 \times 1.92$$
$$= 0.42 \text{ MNm}$$

End post design:

The Code allows for two approaches to be made to end stiffener design to carry the anchor force where the entire web has been designed using tension field theory: the first is to use a single stiffener to carry the effect both of the external reaction and the anchor force; the second is to extend the girder beyond the reaction point and use a pair of stiffeners, one to resist the anchor force and the other the external reaction.

1　Single stiffener end post.

The stiffener must be designed to take the external reaction and a force due to the anchor force.

The reaction due to the anchor force:

The reaction due to only 2/3 of the moment M_{tf} need be considered (Section 5.7.5.4)

$$F = (2/3)(M_{tf}/a)$$
$$= (2/3)(0.42E3/0.8)$$
$$= 350 \text{ kN}$$

External reaction $= 2442$ kN

Total force, $F_x = 2442 + 350$
$$= 2792 \text{ kN}$$

Try two 245×20 plates (Fig. 5.41).

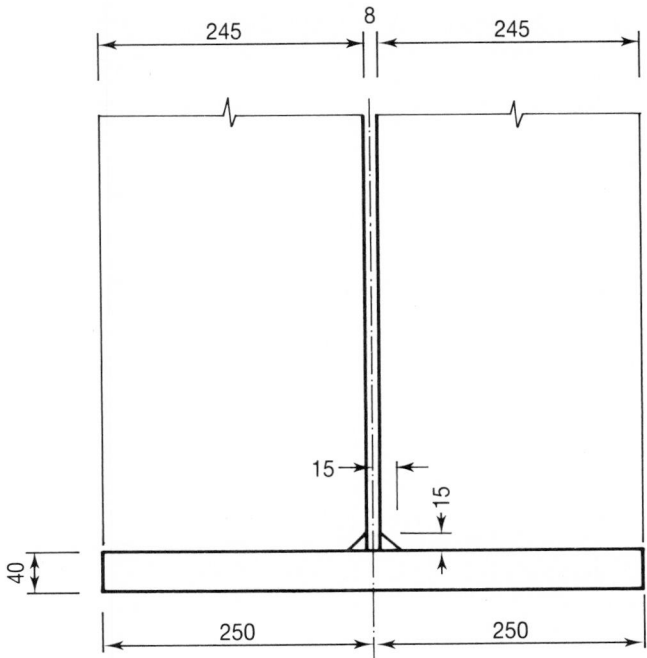

Fig. 5.41   End stiffener details.

Outstand $b = 245$ mm

Maximum outstand $= 19\varepsilon t_s = 19 \times 0.88 \times 20 = 334$

This is satisfactory.

Core section $= 13\varepsilon t_s = 13 \times 0.88 \times 20 = 229$ mm

Actual bearing area $= 2 \times (245 - 15) \times 20$
$$= 9200 \text{ mm}^2$$

Minimum bearing area (Equation (5.81))

$$A_{\min} = 0.8F_x/p_{ys}$$
$$= 0.8 \times 2792E3/345$$
$$= 6474 \text{ mm}^2$$

Bearing area is satisfactory.

Buckling check:

The calculations are only summarized here, as they follow previous calculations of this type.

$A_e = 10600$ mm^2, $I_e = 67.5$E6 mm^4, $r_y = 80$ mm, $\lambda = 17$, $p_c = 321$ MPa, $P_q = 3403$ kN

Thus the stiffener is satisfactory for buckling.

2   Double stiffener (Fig. 5.42):

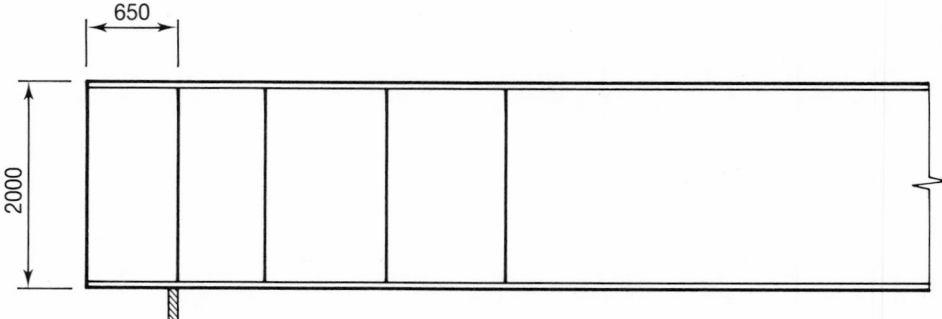

Fig. 5.42   Double stiffener end post.

The extra end panel must be designed to carry the shear due to the horizontal shear force R_{tf}. This is the design criterion for the length of the extra end panel.

Equating shear capacity, assuming the panel can carry the maximum permissible shear stress of $0.6p_{yw}$, and the horizontal shear force R_{tf}, gives

$$e = R_{tf}/(0.6p_{yw}t)$$
$$= 1.09\text{E}6/(0.6 \times 355 \times 8)$$
$$= 640 \text{ mm}$$

Use an end panel of 650 mm.

This value of end panel width is slightly high, but is due to the use of a very thin web. A thicker web might reduce this panel size, and hence the weight of the girder as the flanges must be extended over the whole length of this end panel.

Stiffener design.

1   End stiffener to take reaction from tension field:

$$\text{Force on end stiffener} = M_{tf}/e$$
$$= 0.42\text{E}3/0.65$$
$$= 646 \text{ kN}$$

Taking $p_{ys} = 355$ MPa,

$$\text{Area required} = 646\text{E}3/355$$
$$= 1820 \text{ mm}^2$$

Use a 200×10 flat ($A = 2000$)

2   Load bearing stiffener to take external reaction:

Use a 245×15 mm flat on either side of the web.

Actual outstand $= 245$ mm

Maximum outstand $= 19\varepsilon t_s = 19 \times 0.88 \times 15 = 251$ mm

Core section $= 13\varepsilon t_s = 13 \times 0.88 \times 15 = 172$ mm

Actual bearing area $= (2 \times 245 - 2 \times 15) \times 15 = 6900$ mm^2

$$A_{\min} = 0.8F_x/p_{ys}$$
$$= 0.8 \times 2442\text{E}3/355 \tag{5.81}$$
$$= 5503$$

Stiffener is satisfactory in bearing.

Buckling check:

In this case, unlike all the previously designed stiffeners to take the external reaction, the web extends on both sides of the stiffener, thus enhancing the effective area and second moment of area.

Summarizing the calculations,

$A_e = 7840$ mm^2, $I_e = 54.5$E6 mm^4, $r_y = 83.4$ mm, $\lambda = 16$, $p_c = 333$ MPa, $P_x = 2611$ kN

The stiffener is satisfactory in buckling.

Comparison between the designs:

Figure 5.43 gives the dimensions and weights of the four variant designs in this example.

The first comment is that the actual weights are some 40 to 50 kN above the estimated figure of 60 kN. As pointed out earlier in this example this will increase the maximum design moment by 0.08 MNm, a percentage increase of 0.7% and the end reaction by 35 kN, a percentage increase of 1.4%. Neither of these increases will affect the design as carried out, since in all cases a reasonable margin exists between the actual design forces and the design resistance.

The lightest design is that using tension field theory for the whole web with a single end post stiffener. The heaviest design is that using no tension field theory for the web design. The design using tension field theory with a double end post may well be unacceptable owing to the additional length, in the form of an overhang, of 1.3 m on a span of 20 m.

5.8 Castellated beams

5.8.1 Introduction

Castellated beams are fabricated from rolled sections–beams, columns or joists–by cutting the web and rewelding the cut web sections as shown in Fig. 5.44.

It will thus be observed that the nett effect is to increase the section depth with a series of hexagonal holes in the web. This allows the beams to span further than the parent rolled section, with the advantage that services may be accommodated within the depth of the section. Thus the overall construction may not be unduly increased.

The applied forces are carried out by a combination of pure flexure and Vierendeel action in the webs caused by shear within the discontinuous web panels. It is therefore necessary to consider a series of failure modes.

5.8.2 Failure modes of castellated beams

The possible failure modes are illustrated in Fig. 5.45 (after Kerdal and Nethercot).

5.8.2.1 *Formation of Vierendeel mechanism (Fig. 5.45(a))*

This is caused by excessive shear deformation across one of the voids in the web, and culminates in four hinges forming at the corners of a castellation.

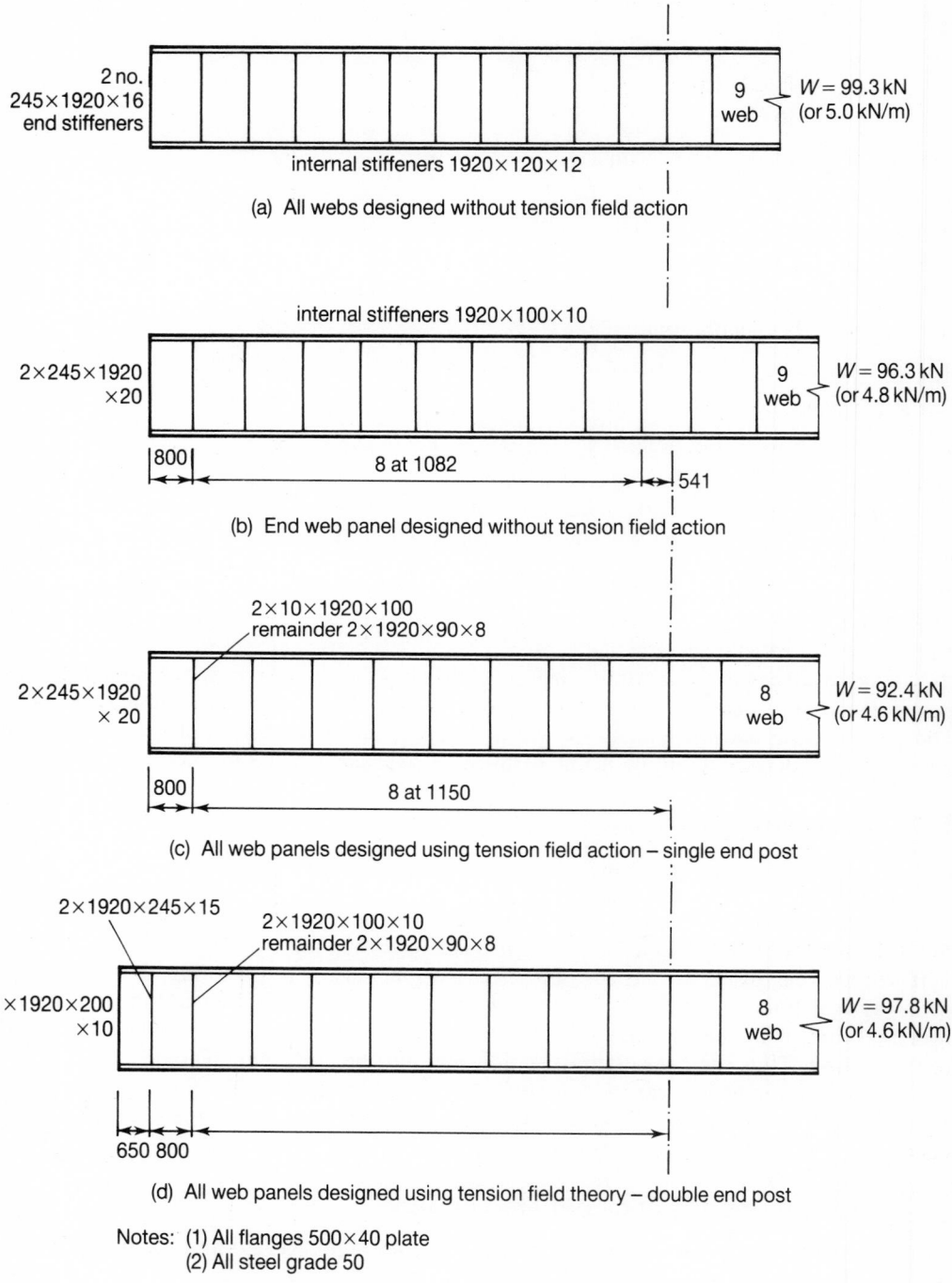

Fig. 5.43 Design summary for Example 5.7.

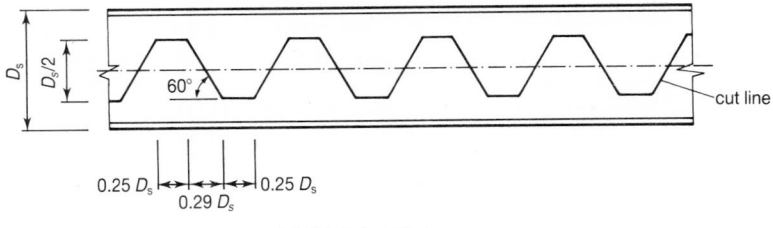

(a) Original section

(b) Fabricated section

Fig. 5.44 Principal dimensions of a castellated section.

5.8.2.2 *Lateral torsional buckling of a web post (Fig. 5.45(b))*

This is caused by excessive shear in the web post. It has, however, been observed that for standard British sections yield at the web weld will occur first (Knowles).

5.8.2.3 *Rupture of the welded joint in the web (Fig. 5.45(c))*

This is caused by excess horizontal shear at the welded joint. Again for British sections this is very rare (Kerdal and Nethercot).

5.8.2.4 *Lateral torsional buckling of the whole span*

Here it has been observed (Nethercot and Kerdal) that there is essentially no difference between the behaviour of welded plate girders and castellated beams under lateral torsional buckling.

5.8.2.5 *Formation of a plastic hinge*

This will occur as for normal beams by the formation of an area of plasticity in the top and bottom flanges due to the spread of yielding caused by a combination of stresses due to primary flexure and Vierendeel action. The hinge will not necessarily form at the point of maximum moment owing to the presence of shear and axial forces. This type of failure only occurs when lateral torsional buckling is prevented.

5.8.2.6 *Web post buckling*

This will only occur under heavy loading. It is more likely to occur on short span heavily loaded beams (Knowles). To avoid the problem at a support it is usual to fill in the first castellation with a plate welded in the hole.

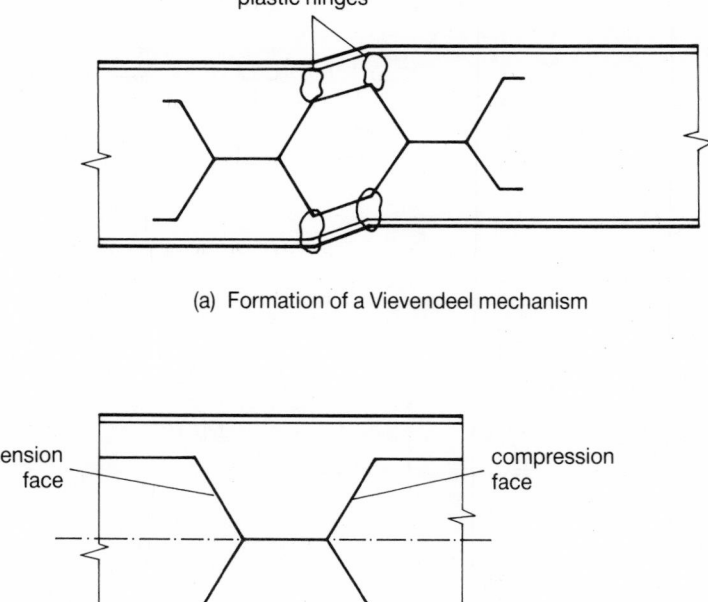

(a) Formation of a Vievendeel mechanism

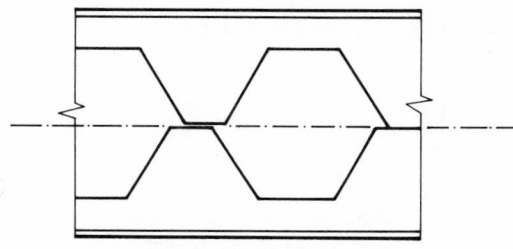

(b) Buckling of web post

(c) Rupture of welded joint

Fig. 5.45   Failure mechanisms for castellated sections.

5.8.3   Design of castellated beams (cl 4.15.3, Pt 1)

5.8.3.1   *Flexure*

Because of the way in which a castellated beam carries the applied loading, the beam can be under high stresses due to primary flexure and additional bending and shear stresses due to Vierendeel action. The effect of the high internal shear and axial forces due to the Vierendeel action may cause a reduction in the plastic moment capacity of the section. This means that the design process becomes complex and also iterative since the critical point for the design of the beam will not necessarily be at the point of maximum moment. In Knowles a series of design interaction graphs are given to facilitate this procedure, the use of which is illustrated in Example 5.8 which follows. For a beam under a UDL (or a loading pattern which is sensibly a UDL such as loading from closely spaced purlins), Knowles suggests that it will be adequate to over-design the beam by around 5% for flexure. A beam carrying heavy point loads should be fully checked for internal forces.

For fully restrained beams, the moment capacity of a castellated beam is given by

$$M = p_y S_x$$

where S_x is calculated at the centre line of a castellation.

For laterally unrestrained beams, the same method should be used as for welded, or plate, girders with the use of Table 12 for determining bending stresses, except that the section properties used should be those on the centre line of the castellation.

5.8.3.2 Shear

At any point the shear stress distribution should be calculated using elastic principles and limited to $0.7p_y$.

5.8.3.3 Local buckling

This check should be carried out according to Table 7.

5.8.3.4 Web buckling

This should be carried out in accordance with the normal procedure except that the contribution of the web in buckling should be limited by taking n_1 equal to the width of the web at the beam mid-height ($0.25D_s$, where D_s is the depth of the parent beam).

5.8.3.5 Deflection

For a castellated beam it is necessary to consider deflection due both to primary flexure and to Vierendeel action.

Following Knowles, the total deflection δ_T may be expressed as the sum of two components, that due to flexure δ_b, and that due to shear δ_s.

(a) Deflection due to flexure:

This is calculated as usual except that the value used for the second moment of area is calculated at the centre of a castellation.

(b) Deflection due to shear:

The shear rigidity is to be taken as GA_{fict}

where G is the shear modulus, is equal to $E/2(1 + v)$, and has a value of 79 GPa, and,

$$1/A_{fict} = 4.3207D_s/(h^2 t) + 235.56\text{E-}6D_s^2/I_T + 116.35\text{E-}3/A_{T,web} \tag{5.91}$$

where D_s is the depth of the parent section, h the distance between the centroids of the T sections at the castellation, t the web thickness, I_T the second moment of area of the T section, and $A_{T,web}$ the area of the web of the T section, respectively.

Note all the parameters in Equation (5.91) MUST be in mm units, and $1/A_{fict}$ has units of mm^{-2}. Values of $1/A_{fict}$ and other extra section properties required for the design of castellated beams are tabulated by Knowles, although it is possible to calculate them from first principles, as will be done in the design example which follows.

EXAMPLE 5.8   Castellated beam design

Design a castellated beam in Grade 50 steel to carry an imposed load of 250 kN over a simply supported span of 15 m. The compression flange is fully restrained.

Assume self weight 20 kN

$$\text{Total load} = 1.6 \times 250 + 1.4 \times 20$$
$$= 428 \text{ kN.}$$

$$M = WL/8$$
$$= 428 \times 20/8$$
$$= 803 \text{ kNm}$$

$$V = W/2$$
$$= 428/2$$
$$= 224 \text{ kN}$$

As the compression flange is restrained and the loading is a UDL, the recommendation by Knowles may be followed and the section overdesigned by some 5%.

$$S_x = M/p_y$$

taking p_y as 345 MPa,

$$S_x = 803\text{E}3/345$$
$$= 2328 \text{ cm}^3$$

Select a 686 × 191 × 74 CUB ($S = 2390$)

Although this is only overdesigned by 3%, the dead load has been overestimated by 45%, giving a required plastic section modulus 3% lower, thus there should be sufficient margin.

Check b/T and d/t ratios for plastic action:

$$\varepsilon = (275/p_y)^{0.5} = (275/345)^{0.5} = 0.89$$

Flange outstand, $b/T = 95.3/14.5 = 6.6$

The limit is that for rolled sections, i.e. $8.5\varepsilon = 8.5 \times 0.89 = 7.6$

Web slenderness, $d/t = 636.5/9.1 = 69.9$

Limiting value $= 79\varepsilon = 79 \times 0.89 = 70.3$

Both ratios satisfy the requirements for plastic action.

A check on the suitability of the chosen section will be made using the interaction charts by Knowles.

The charts prepared by Knowles are for Grade 43 steel, but can be modified for Grade 50 by multiplying the shear force and bending moments by the ratio of the capacities using Grade 50 and Grade 43.

For the shear force axis,

Capacity using Grade 43 = 584.8 kN

Capacity using Grade 50 = 754.9 kN

Ratio = 754.9/584.8 = 1.29

For the bending moment axis,

Capacity using Grade 43 = 656.2 kNm

Capacity using Grade 50 = 847 kNm

Ratio = 847/656.2 = 1.29

Figure 5.46 gives the relevant interaction curve for moment and shear capacity together with the actual loading interaction curve (calculated with a service dead load of 74 kg/m). It will be seen that at all points the loading curve is inside the design interaction curve and is therefore satisfactory. This situation is usual where the loading is uniformly distributed since the positions of maximum shear and maximum moment are distinct.

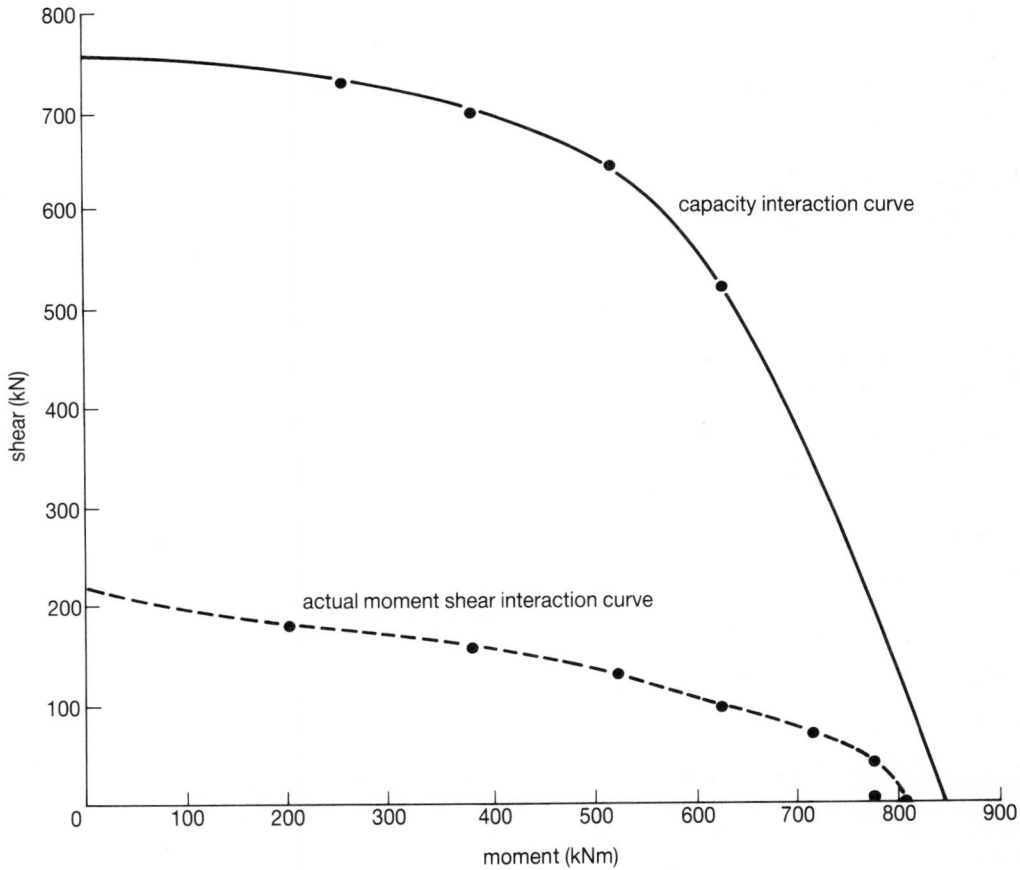

Fig. 5.46 Design interaction diagram.

Shear check.

Check initially for primary shear at a castellation at the end of the beam. The relevant cross section is given in Fig. 5.47.

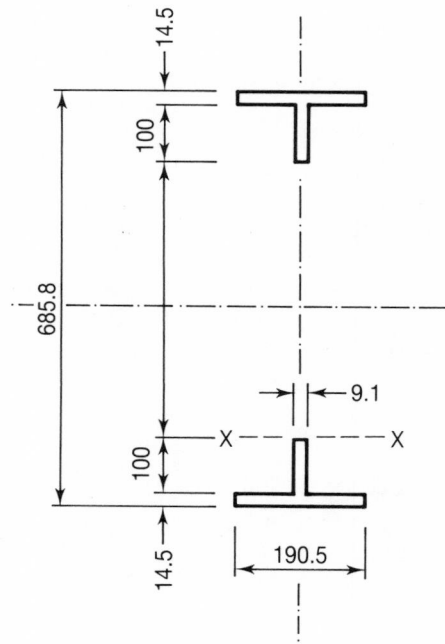

Fig. 5.47   Cross section through web opening.

The elastic shear stress is given by

$$q = QAy_c/tI \tag{4.15}$$

The maximum shear will occur at X–X in Fig. 5.47

$$Ay_c = 190.5 \times 14.5(685.8/2 - 14.5/2) + 9.1 \times 100(456.8/2 + 100/2)$$
$$= 1.18E6 \text{ mm}^3$$

$$Q = 224 \text{ kN}$$

$$t = 9.1 \text{ mm}$$

$$I = 77300 \text{ cm}^4 \text{ (from Section Property Tables)}$$

$$q = 224E3 \times 1.18E6/(9.1 \times 77300E4)$$
$$= 37.5 \text{ MPa}$$

Maximum allowable shear stress $= 0.7p_y = 0.7 \times 345 = 245$ MPa.
   This excess capacity in shear is confirmed by the interaction diagram in Fig. 5.46.

Deflection.

(a) Bending deflection:

$$\delta_b = (5/384)qL^4/EI$$

$$q = 250/15 = 16.67 \text{ kN/m}$$

$$L = 15 \text{ m}$$

$$E = 205 \text{ GPa}$$

$$I = 77300 \text{ cm}^4$$

so,

$$\delta_b = (5/384) \times 16.67 \times 15^4/(205\text{E}6 \times 77300\text{E-}8)$$
$$= 0.069 \text{ m}$$

(b) Shear deflection:

For a UDL,

$$\delta_s = (qL^2/8)/GA$$

where in this case A must be replaced by A_{fict}

Calculation of A_{fict}:

$$1/A_{\text{fict}} = 4.3207 D_s/(h^2 t) + 235.56\text{E-}6 D_s^2/I_\text{T} + 116.35\text{E-}3/A_{\text{T,web}}$$

$$D_s = 457.2 \text{ mm} \tag{5.91}$$

$$t = 9.1 \text{ mm}$$

Calculation of h, $A_{\text{T,web}}$ and I_T:

This may conveniently be done by considering the T section at the centre of the castellation as a standard T stub cut from a UB minus part of the web (Fig. 5.48).

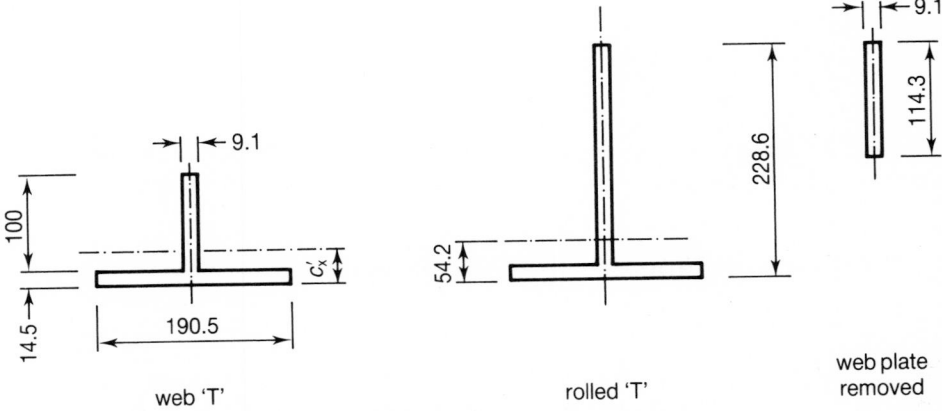

Fig. 5.48 Calculation of web 'T' properties.

Position of centroid of castellated T section, $C_{x'}$:

$$C_{x'} = (47.5E2 \times 54.2 - 114.3 \times 9.1(228.6 - 114.3/2))/(47.5E2 - 114.3 \times 9.1)$$
$$= 21.3 \text{ mm}$$

$$h = D_c - 2C_{x'}$$
$$= 685.8 - 2 \times 21.3$$
$$= 643.2 \text{ mm}$$

$$A_{T,web} = (D_T - T)t$$
$$= (114.3 - 14.5)9.1$$
$$= 908.2 \text{ mm}^2$$

$$A_T = 47.5E2 - 114.3 \times 9.1$$
$$= 3710 \text{ mm}^2$$

Using the parallel axis theorem to calculate I_T:

$$I_T = I_{x,s} - A_T(C_x - C_{x'})^2 - ta^3/12 - ta(1.5a - C_x)^2$$
$$= 2250E4 - 3710(54.2 - 21.3)^2 - 9.1 \times 114.3^3/12 - 9.1 \times 114.3(1.5 \times 114.3 - 54.2)^2$$
$$= 3.053E6 \text{ mm}^4$$

$$1/A_{fict} = 4.3207 \times 457.2/(643.2^2 \times 9.1) + 235.56E\text{-}6 \times 457.2^2/3.5053E6$$
$$+ 116.35E\text{-}3/908.2$$
$$= 6.69E\text{-}4 \text{ mm}^{-2}$$

Knowles gives a value of 669.64E-6.

$$\delta_s = (qL^2/8)/GA_{fict}$$
$$= (16.67 \times 15^2/8) \times 6.69E\text{-}4/79$$
$$= 0.004 \text{ m}$$

Total deflection $= 0.069 + 0.004$
$$= 0.073 \text{ m}$$

Allowable deflection $=$ span/200
$$= 15/200$$
$$= 0.075 \text{ m}$$

The beam is satisfactory in deflection. It should be noted that in this case the shear deflection is 5.5% of the total deflection, which although low is not negligible.

5.9   Tutorial problems

5.9.1   Calculation of moment capacity

Determine from first principles using basic equations the moment capacity of a 305 × 165 × 54 Grade 43 UB loaded in single curvature by moments applied at each end of a simply supported span of 6 m.

Answer

$M_b = 101.8$ kNm ($p_y = 275$ MPa, $\gamma = 0.909$, $M_E = 162.4$ kNm, $M_p = 232.4$ kNm, $\lambda_{LT} = 102.6$, $\lambda_{LO} = 34.3$, $\eta_{LT} = 0.478$)

5.9.2 Rolled section with loading at points of lateral restraint

Prepare a design in Grade 50 steel for the beam shown in Fig. 5.49.

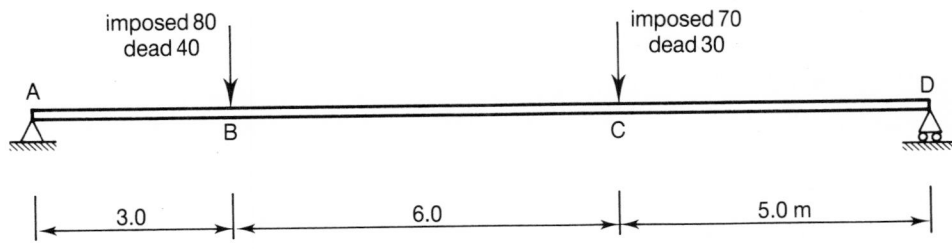

Notes: (1) All loads in kN at SLS
(2) Lateral restraints exist at supports and load points

Fig. 5.49

Answer

Smallest plastic section $686 \times 254 \times 140$ ($M_p = 1573$ kNm). Reactions, $R_A = 199.6$ kN, $R_D = 138.4$ kN. Bending moments, at B 598.8 kNm, at C 692.0 kN. Design criteria: Bending ($n = 1.0$): Section BC, $\beta = 0.865$, $m = 0.93$, $M = 643.6$ kNm, $L_E = 6$ m, $M_b = 739$ kNm. Section CD, $\beta = 0$, $m = 0.57$, $\bar{M} = 394.3$ kNm, $L_E = 5$ m, clearly not critical since both the effective length and design moments are less than for Section BC. Shear, $P_v = 1729$ kN. Web buckling, $\lambda = 124$, $p_c = 99$ MPa, Beam component, $Dtp_c = 839$ kN. Deflection, due to 80 kN load, 1.0 mm; due to 70 kN load, 1.28 mm; total deflection $= 2.28$ mm (span/6140).

Note: all values of permissible stresses from tables have been determined using a design strength of 340 (not 345).

5.9.3 Rolled section with loads applied between points of lateral restraint

Prepare a design in Grade 50 steel for the beam shown in Fig. 5.50.

Answer

Minimum plastic section $610 \times 229 \times 113$ ($M_p = 1134$ kNm) BMD and SFD plotted in Fig. 5.51.

Bending Checks: $m = 1.0$: Span BC, $\beta = 0.6$, $\gamma = -0.53$ ($M_0 = 608$ kNm), $n = 0.89$, $\bar{M} = 353.5$ kNm, $L_E = 8$ m, $\lambda = 108$, $p_b = 114$ MPa, $M_b = 375$ kNm: Span AB, $n = 1.0$, $L_E = 4$ m and is clearly not critical. Web bearing: Beam component, $n_2 = 2.5T = 88$ mm, $t = 11.2$, $p_{yw} = 340$ MPa, $n_2 t p_{yw} = 335$ kN. Max reaction is 400, therefore stiff bearing of 17 mm required to take load of 65 kN. Web buckling: $\lambda = 125$, $p_c = 98$ MPa. Beam component,

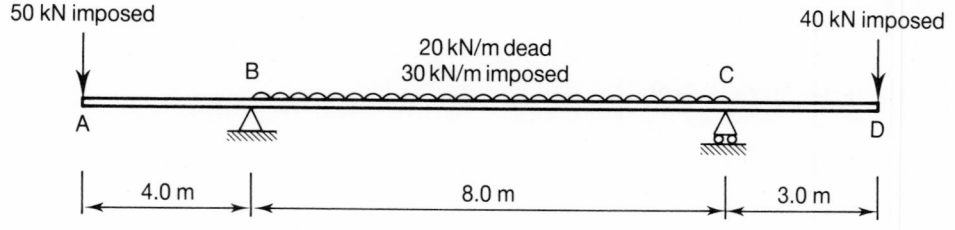

Notes: (1) Lateral torsional restraints exist at C and D
(2) Ends A and D are free
(3) All loading is a single load case

Fig. 5.50

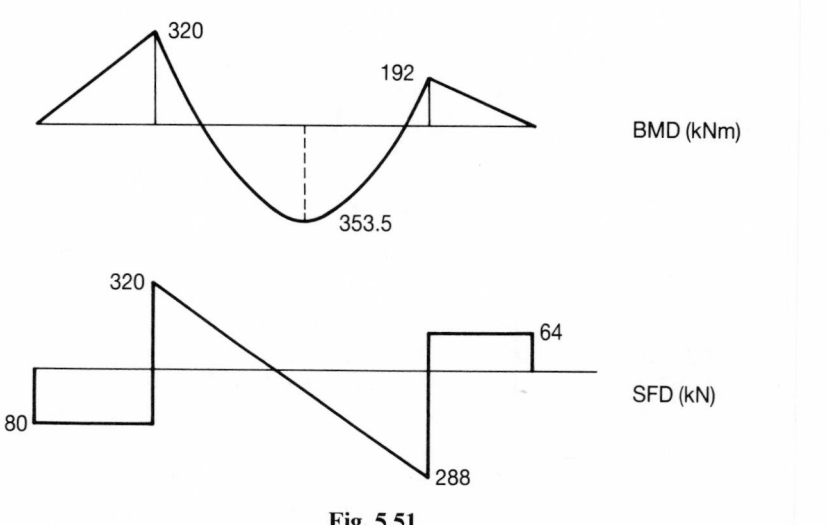

Fig. 5.51

$Dtp_c = 673$ kN. Shear, $P_v = 1388$ kN $(0.6p_v = 833$ kN). Deflection: Loading is given in Fig. 5.52.

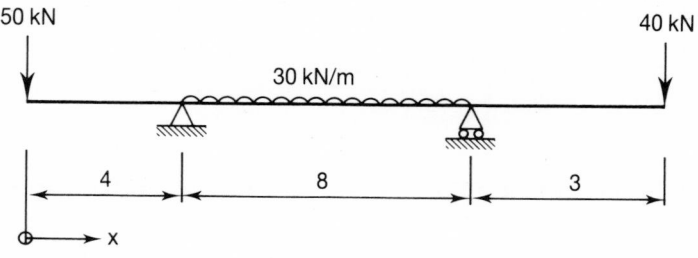

Fig. 5.52

Using Macaulay, and $EI\,\mathrm{d}^2w/\mathrm{d}x^2 = -M_x$

$$M_x = -50x + 180[x-4] + (30/2)[x-4]^2 + 150[x-12] - (30/2)[x-12]^2$$

or

$$EI_w = (50/6)x^3 - (180/6)\,[x-4]^3 + (15/12)[x-4]^4 - (150/6)[x-12]^3$$
$$- (15/12)[x-12]^4 + C_1 x + C_2$$

$w = 0$ at $x = 4$ and 12, giving $C_1 = -453.3$ and $C_2 = 1280$

at $x = 0$, $w = 1280/EI$, at $w = 15$, $w = 200.5/EI$, at $x = 8.33$ (point of zero slope), $w = 324.6/EI$ and at $x = 8$ (centre of span BC), $w = 320.2/EI$. This gives deflections of 22.2 mm at A (span/180) and 40 mm for span BC (span/200).

5.9.4 Compound beam

A $553 \times 210 \times 109$ UB strengthened with a 300×20 plate both in Grade 43 steel has been proposed for a crane beam. The loadings and other relevant design data are given in Fig. 5.53.

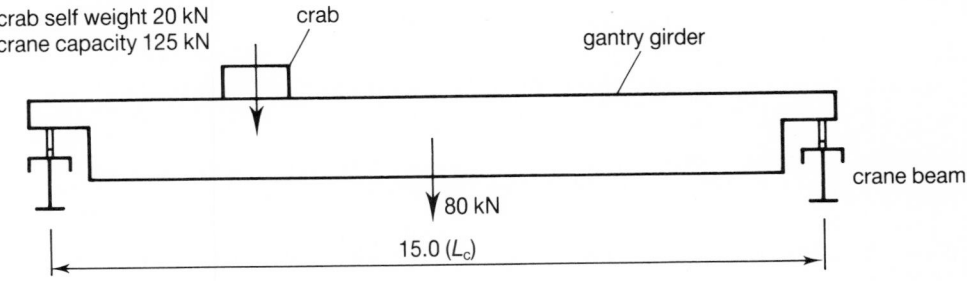

minimum hook approach to c of crane beam 0.75 m

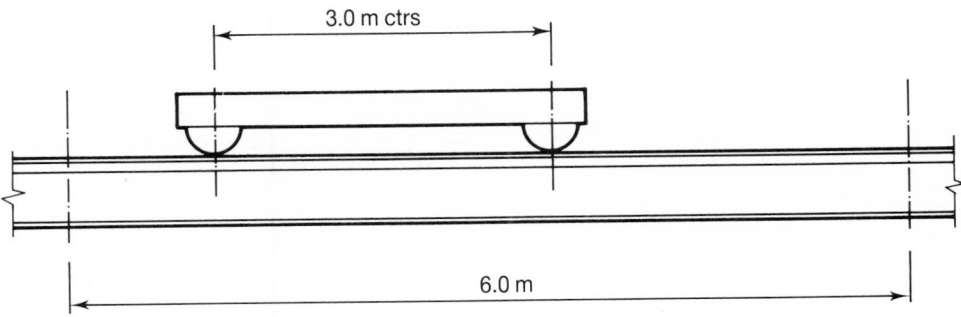

Fig. 5.53

Answer

Loadings: Static wheel load 20 kN self weight, crane load $(125 + 20)(1 - 0.75/15) = 68.9$ kN, total load $= 88.9$ kN or 111.1 with 25% increase, horizontal load/wheel $= 0.1(125 + 20)/4 = 3.6$ kN. Forces (using unfactored loading): Moments, self weight $10 \times 6/8 = 7.5$, live load, $(111.1/(2 \times 6))(6 - 2.5/2)^2 = 208.9$ kNm, surge 6.8 kNm; Shear, self weight, 5 kN, crane load $= 111.1(2 - 2.5/6) = 176.0$ kN, surge 5.7 kN.

Load combinations:

Vertical; Moment 344.7 kNm, Shear, 288.6 kN

Vertical plus surge; Moments, 303 kNm (vertical), 9.5 (horizontal)

Section properties (ignoring fillets) and bending capacity

Total area = 19849 mm^2, Plastic equal area axis is coincident with the underside of the top flange of the UB, $\Sigma A\bar{y} = 210\,000$ mm^3 for the portion above the NA and 3574900 mm^3 below, $S_x = 3.7849E6$ mm^3, $I_y = 29.6E6$ mm^4, $r_y = 38.6$ mm, elastic NA is at 354.3 mm from the soffit of the beam, $I_x = 0.995E9$ mm^4, $\gamma = 0.97$, effective centroid of top flange is at 17.7 mm below the top of the plate, $h_s = 532.4$ mm, effective width of top flange with a thickness of 38.8 mm is 256.7 mm, $J = 5.73E6$ mm^4, $u = 0.84$, $x = 17.74$, $L_E = 1.2 \times 6000 + 2 \times 559.5$, $\lambda = 216.0$, $N = 0.82$, $D_L = 0$, $\psi = 0.483$, $v = 0.546$, $\lambda_{LT} = 99$, $p_b = 121$ MPa, $M_b = 458$ kNm, both b/T and d/t ratios are satisfactory.

Biaxial bending case: Z of top combined flange = 398000 mm^3, $M_{cy} = 105$ kNm, $M_y/M_{cy} + M_x/M_{cx} = 0.09 + 0.3 = 0.39$.

Shear capacity, $P_v = 995$ kN.

Size of weld: strength required = 0.16 kN/mm, minimum weld size 6 mm (0.90 kN/mm).

Web bearing: Beam component = 145 kN, Maximum shear = 289 kN, minimum bearing length = 47 mm.

Web buckling: $\lambda = 103$, $p_c = 118$ MPa, Beam component, calculated on $D/2 = 369$ kN.

Bearing underneath rail: $x_r = 207.6$ mm, Bearing capacity = 683 kN.

Deflections (assuming crane load in centre): Imposed load (both wheels), 3 mm (span/2000), due to horizontal loading = 2.0 mm (span/3000).

5.9.5   Plate girder

It is proposed to use a plate girder fabricated in Grade 50 steel from 600 × 45 plate for the flanges and 2410 × 12.5 plate for the web to carry an imposed load of 175 kN/m run over a span of 25 m. The top flange is fully restrained. Determine suitable panel sizes such that all but the end panels are designed using tension field theory, and further prepare designs for the stiffeners required.

Answer

The critical criteria are that the first panel only carries shear due to basic web buckling and must separately be able to carry shear due to the anchorage force developed by the tension field, and that advantage should be taken of being able to reduce this force by only using basic shear for the remaining panels (i.e. not using the contribution of the flanges). In addition all panels must be capable of carrying the edge load due to the imposed load.

Self weight of beam 6.5 kN/m (excluding stiffeners). Total applied UDL at ULS = 289.1 kN/m.

Maximum reaction 3614 kN.

Stress due to edge load, $f_{ed} = 22.4$ MPa.

Maximum panel aspect ratio to resist this is $a/d = 1.23$ or $a = 2964$ mm.

Shear stress on end panel $= 120$ MPa, $d/t = 192.5$, from Table 21(d) minimum a/d is 0.506 or $a = 1219$ mm. Use $a = 1200$ mm.

For the remaining panels, max. shear $= 3614 - 1.2 \times 289.1 = 3267$ kN, or $v = 108$ MPa. Minimum a/d to satisfy this is 1.0 (from Table 22(d)), or $a = 2410$. This value would give 9.4 panels, so use 10 panels each at 2260 mm (shear capacity for $a/d = 0.94$ is $113 \times 2410 \times 12.5 \times 1\text{E-}3 = 3404$ kN). For such a panel $q_{cr} = 52$ MPa.

Stiffeners:

(a) between the 2260 mm panels.

$a < \sqrt{2} d$, use Equation (5.74) to give $I = 8.03\text{E}6$ mm^4. For a fully effective stiffener this gives $t = 10$ mm, so use two 110×10 mm flats.

Check load carrying capacity (Equation (5.76)). $V = 2614$ kN (at end of second panel), $V_s = 1567$ kN ($q_{cr} = 52$), $F_q = 1047$.

For the stiffener and web, $I_e = 10.6\text{E}6$ mm^4, $A_e = 8575$ mm^2, $r = 35.2$ mm, $\lambda = 0.7 \times 2410/35.1 = 48$, $p_c = 267$ MPa (Table 27(c) with $p_y = 355 - 20$), $P_x = 2290$ kN.

(b) stiffener between the 1200 and 2260 panels (using $a = 1200$)

$I = 28.5\text{E}6$, $t = 13$ mm, use two 150×15 flats. $V = 3267$ kN, $F_q = 1700$ kN. As this stiffener is larger than the previous, its load carrying capacity is adequate.

(c) End post

$H_q = 6.97$ MN (Equation (5.65)), $\alpha = 0.918$, effective $H_q = 6.4$ MN

$R_{tf} = 3.2$ MN (Equation (5.65)), shear capacity of end panel is $0.6 \times 355 \times 1200 \times 12.5 = 3.2$ MN, and therefore satisfactory.

$M_{tf} = 1.54$ MNm (Equation (5.65)), force in end post $= M_{tf}/1.200 = 1.285$ MN.

Design force for end post $= 1.285 + 3.614 = 4.899$ MN, $A_{min} = 0.0115$ m^2 ($p_y = 340$ MPa), i.e. use two 215×30 plates (with a 15 mm fillet to clear web to flange welds) ($A = 0.012$). $I_e = 216.7\text{E}6$ mm^4, $A_e = 16400$ mm^2, $r = 115$ mm, $\lambda = 15$, $p_c = p_y = 340 - 20 = 320$, $P_x = 5.25$ MN, which is satisfactory.

Welds:

(a) Stiffeners between 2260 panels:

Min weld force $= 12.5 \times 12.5/(5 \times 110) = 0.28$ kN/mm (Equation (5.83)), use a 6 mm minimum weld (0.903 kN/mm). This will also cover the stiffener between the 1200 and 2260 panel.

(b) End post:

Force $= 4.899$ MN, length of weld $= 2 \times (2410 - 2 \times 15) = 4760$ mm. Force/unit length $= 1.03$ kN/mm, use an 8 mm weld (1.2 kN/mm).

(c)   Flange to web:

$q_s = QAy_c/I$, $I = 9.6\text{E}10$ mm^4, $A = 600 \times 45$, $y_c = 2500/2 - 45/2$, $Q = 3614$ kN, $q_s = 1.24$ kN/mm, use a 6 mm weld (0.903 per weld).

5.9.6   Castellated beam

Determine the minimum size castellated beam required in Grade 43 steel to carry an imposed load of 280 kN over a span of 17.5 m. The compression flange is fully restrained and it may be assumed that the combination of applied moment and shear is not critical within the beam, and that the self weight is 40 kN.

Answer

$M = 1103$ kNm and $V = 252$ kN. Minimum $S = 4162$ cm^3, increase by 5% to 4370. Use a $915 \times 229 \times 113$ CUB ($S = 4710$, self weight $= 19.4$ kN). Both b/T and d/t checks are satisfactory.

Shear check (Fig. 5.54)

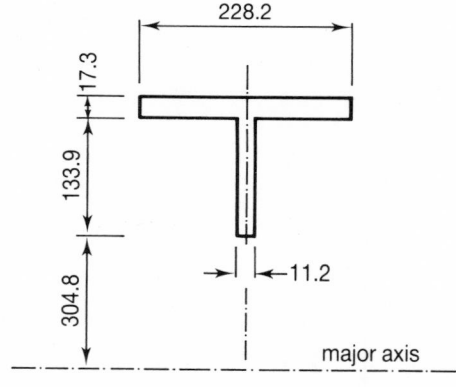

Fig. 5.54

$I = 203\,000$ cm^4, $A\bar{y} = 2.324\text{E}6$ mm^3, $f_v = 252\text{E}3 \times 2.324\text{E}6/(203000\text{E}4 \times 11.2) = 25.8$ MPa, limiting value $= 0.7p_y = 185.5$.

Deflection:

Bending; 47 mm

Shear; $D_s = 607.3$ mm, $t = 11.2$ mm, $A_{\text{T,web}} = 1500$ mm^2, $h = 853.9$ mm   ($C_{x'} = 29.1$), $I_T = 7.614\text{E}6$ mm^4, $1/A_{\text{fict}} = 4.103$ m^{-2}.

Deflection $= 3.2$ mm.

Total deflection $= 50$ mm (span/350).

References

Allen H. G., Bulson P. S. (1980). *Background to buckling*, McGraw Hill.

Anderson J. M., Trahair N. S. (1972). *Stability of mono-symmetric beams and cantilevers*, Journal Structural Division (Proceedings ASCE), **98**, ST1, 269–86.

BS 5400 Part 3: *Steel, concrete and composite bridges – Part 3 – Code of practice for design of steel bridges*. British Standards Institution, London.

Bassler K. (1961). *Strength of plate girders in shear*, Journal Structural Division (Proceedings ASCE), **87**, ST7, 151–80.

Bradford M. A. (1989). *Buckling of beams supported on seats*, Structural Engineer, **69** (23), 411–4.

Bulson P. S. (1970). *The stability of flat plates*, Chatto and Windus.

Coates R. C., Coutie M. G., Kong F. K. (1988). *Structural Analysis* (3rd Edition), Van Nostrand.

Dowling P. J. et al. (1982). *A development in the automated design and fabrication of portal frame industrial buildings*, Structural Engineer, **60A** (10), 311–9.

Ghali A., Neville A. M. (1989). *Structural Analysis*, Chapman and Hall.

Kerdal D., Nethercot D. A. (1984). *Failure modes for castellated beams*, Journal of Constructional Steel Research, **4**, 295–315.

Kirby P. A., Nethercot D. A. (1979). *Design for Structural Stability*, Granada.

Knowles P. R. (1985). *Design of castellated beams*, Steel Construction Institute.

Marshall W. T., Nelson H. M. (1977). *Structures*, Pitman.

Nethercot D. A. (1974a). *Residual stresses and their influence upon the lateral buckling of rolled steel beams*, Structural Engineer, **52**, 3, 89–96.

Nethercot D. A. (1974b). *Buckling of welded beams and girders*, IABSE, **34**, 163–82.

Nethercot D. A. (1975). *Inelastic buckling of steel beams under non-uniform moments*, Structural Engineer, **53** (2), 73–8.

Nethercot D. A., Kerdal D. (1982). *Lateral torsional buckling of castellated beams*, Structural Engineer, **60B** (3), 53–61.

Nethercot D. A., Rockey K. C. (1971). *A unified approach to the elastic lateral buckling of beams*, Structural Engineer, **49** (7), 321–30.

Nethercot D. A., Trahair N. S. (1976). *Lateral buckling approximations for elastic beams*, Structural Engineer, **54** (6), 197–204.

Pillinger A. H. (1988). *Structural steelwork: a flexible approach to the design of joints in simple construction*, Structural Engineer, **66** (19), 316–21.

Porter D. M., Rockey K. C., Evans H. R. (1975). *The collapse of plate girders loaded in shear*, Structural Engineer, **53** (8), 313–25.

Rockey K. C. (1971). *An ultimate load method for the design of plate girders*, IABSE Colloquium – Design of plate and box girders for ultimate strength, London, 253–68.

Rockey K. C., El-gaaly M. A. (1971). Ultimate strength of plates when subjected to in-plane patch loading, *IABSE Colloquium – Design of plate and box girders for ultimate strength*, London, 401–7.

Rockey K. C., Skaloud M. (1971). The ultimate load behaviour of plate girders loaded in shear, *IABSE Colloquium – Design of plate and box girders for ultimate strength*, London, 1–19.

Rockey K. C., El-gaaly, M. A., Bagchi D. K. (1972). *Failure of thin walled members under patch loading*, Journal Structural Division (Proceedings ASCE), **98**, ST12, 2739–52.

Rockey K. C., Evans H. R., Porter D. M. (1978). *A design method for predicting the collapse behaviour of plate girders*, Proc ICE, **65**, 85–112.

Rockey K. C., Valtinat G., Tang K. H. (1981). *The design of transverse stiffeners on webs loaded in shear – an ultimate load approach*, Proc ICE, **71** (2), 1069–99.

Trahair N. S., Bradford M. A. (1988). *The Behaviour and Design of Steel Structures* (2nd edition), Chapman and Hall.

6

Axially Loaded Members

6.1   Introduction and classification of sections

Axially loaded members are either in tension, or in compression, and in practice are generally associated with relatively small bending moments. The members in tension are called 'ties' while members in compression are called 'struts', 'columns' or 'stanchions'.

A tie extends when subjected to an axial load and is deemed to have failed when the yield stress is reached across the full width of the section. The failure load is independent of the length of the member. In contrast, an axially loaded strut reduces in length and is deemed to have failed by buckling when the bending stress reaches yield at the extreme fibres. The failure load is inversely proportional to the square of the length of the member.

A member that is purely in tension does not buckle locally or overall and is therefore not affected by the classification of sections, e.g. compact, semi-compact, as described in Section 1.1.3. The design stress (p_y), as determined from Table 6, Pt 1, is therefore not reduced. However in practice most members in tension are also subjected to a bending moment which produces compression in elements of the section.

6.2   Deductions for holes in tension members (cl 3.4, Pt 1)

Holes are drilled in tension members to accommodate fasteners at connections. This reduces the gross cross-sectional area of a member and has the effect of weakening a tie because the fastener in the hole does not transmit the axial force. A hole in a strut has little effect on the buckling strength of a strut because as the strut compresses the axial force is transmitted by bearing on the shank of the bolt.

When designing a tie the net cross-sectional area is used in calculations to determine the design axial force. The net cross-sectional area is the gross area reduced by the maximum sum of the sectional areas of the holes. These holes may be in line at right angles to the axial stress in the members (see line AA in Fig. 6.1), or staggered (see lines BB and CC in Fig. 6.1).

The diameter to be deducted for a bolt hole is:

Bolt diameter + 2 mm clearance for bolts not exceeding 24 mm in diameter, or bolt diameter + 3 mm clearance for bolts exceeding 24 mm in diameter.

When bolts are staggered the area to be deducted should be the greater of   (cl 3.4.3, Pt 1):

(a)   the deduction for non-staggered holes;

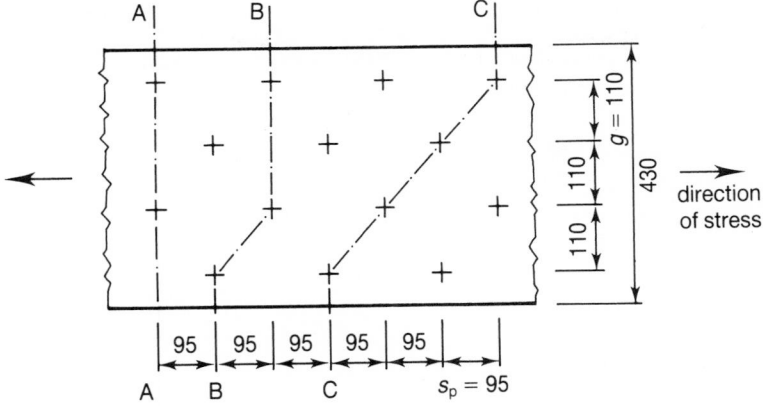

Fig. 6.1   Net area of a plate.

(b)   the sum of the sectional areas of all holes in any zig-zag line extending progressively across the member, or part of the member, less $s_p^2 t/4g$ for each gauge space
in the chain of holes

where

s_p is the staggered pitch (see Fig. 6.1)
g is the gauge (see Fig. 6.1)
t is the thickness of the holed material.

For sections such as angles with holes in both legs the gauge should be taken as the
sum of the back marks to each hole, less the leg thickness.

EXAMPLE 6.1   Net area of a plate with holes

Calculate the net cross-sectional area for the plate shown in Fig. 6.1 which is subjected to
a tensile force. The plate is 20 mm thick and contains four lines of staggered holes drilled
for 24 mm diameter bolts.

From Fig. 6.1, $s_p = 95$ mm, and $g = 110$ mm.
Diameter of hole $d_h = d + 2 = 24 + 2 = 26$ mm.
Gross cross-sectional area perpendicular to the direction of stress

$$= 20 \times 430 = 8600 \text{ mm}^2.$$

Areas to be deducted at possible failure lines are:

$$n_h t d_h - n_{gs} s_p^2 t/(4g)$$

where n_{gs} is the number of gauge spaces in the chain of holes.

Line AA, $2 \times 20 \times 26 = 1040$ mm^2
Line BB, $3 \times 20 \times 26 - 1 \times 95^2 \times 20/(4 \times 110) = 1150$ mm^2
Line CC, $4 \times 20 \times 26 - 3 \times 95^2 \times 20/(4 \times 110) = 849$ mm^2.

Minimum net area for line BB $= 8600 - 1150 = 7450$ mm^2.

6.3   Tension members (cl 4.6, Pt 1)

A member subjected to axial tension extends and tends to remain straight or, if there is a small initial curvature, to straighten out as the load is increased. Ties occur in trusses, bracing and hangers for floor beams. A flat can be used as a tie, but this is generally impractical because it will buckle if it goes into compression. Tie sections are therefore angles and tees for small loads and 'I' sections for larger loads. In practical situations it is not possible to apply the load axially because the connection at the end is eccentric.

Where a tension member is connected to one side of a gusset plate bending moments are introduced in addition to the direct axial force. In a tension member these moments produce lateral deflections which reduce the eccentricity of the load near the middle of the member. Thus under increasing load the bending stresses become concentrated more towards the ends of the member. For angles connected by one leg the principal sectional axes are inclined to the plane containing the bending moment. Secondary deflections therefore occur normal to the plane of bending and, because of the restraints provided by the gusset plates, twisting also takes place.

For angles and tees experiments conducted by Nelson, and by Regan and Salter, demonstrated that the above effects could be compensated for in design by reducing the cross-sectional area of the member. If there are holes, the gross area should be reduced to the net area (cl 3.3.2, Pt 1), but this is reduced further to an effective area.

Angles, channels and T–sections may be treated as axially loaded members provided that the net area is reduced to the effective area (cl 4.6.2, Pt 1).

Tension capacity of a member (cl 4.6.1, Pt 1)

$$P_t = A_e p_y \tag{6.1}$$

where

A_e is the effective area of a member;

p_y is the design strength for steel from Table 6, Pt 1.

Generally the effective area (cl 3.3.3, Pt 1)

$$A_e = K_e \times \text{(net area)}$$

where

$$\begin{aligned} K_e &= 1.2 \text{ for grade 40 or 43 steel;} \\ &= 1.1 \text{ for grade 50 or WR50;} \\ &= 1.0 \text{ for grade 55.} \end{aligned}$$

However for angles and tees the effective area is calculated as follows.

6.3.1   Effective areas of single angles, channels and T-sections (cl 4.6.3.1, Pt 1)

For single angle ties connected through one leg only, single channel sections connected only through the web, and T–sections connected only through the flange the effective area

$$A_e = a_1 + a_2[3a_1/(3a_1 + a_2)] \tag{6.2}$$

where

a_1 is the net sectional area of the connected part(s) less holes;
a_2 is the net sectional area of the unconnected part(s).

For double angle ties, connected to one side of a gusset plate or section, the angles may be designed individually as given above.

6.3.2 Effective area of double angles (cl 4.6.3.2, Pt 1)

For back-to-back double angles connected to one side of a gusset or section which are:

(a) In contact or separated by a distance not exceeding the aggregate thickness of the parts with solid packing pieces;
(b) connected by bolts or welded such that the slenderness of the individual components does not exceed 80;

Effective area

$$A_e = a_1 + a_2[5a_1/(5a_1 + a_2)] \tag{6.3}$$

Note: The area of the leg of an angle should be taken as the product of the thickness by the length from the outer corner minus half the thickness, and the area of the leg of a T–section as the product of the thickness by the depth minus the thickness of the flange.

6.3.3 Effective area of other types of members (cl 4.6.3.3, Pt 1)

The following types of members should be designed using net areas and treated as axially loaded members.

(a) Single angle ties connected through both legs by lug angles or otherwise, single channel sections connected by both flanges, and T–sections connected only through the leg or both the flange and the leg.
(b) Double angle ties connected to both sides of a gusset or section provided that the components are held longitudinally parallel and connected by bolts or welds in at least two places and held apart by solid packing pieces. The outermost of such connections should be at a distance from each end of approximately nine times the smallest leg length. The bolts should be of the same diameter as the end connections.
(c) The internal bays of continuous ties.

EXAMPLE 6.2 Effective area of an angle tie

A $70 \times 70 \times 6$ mm single angle tie is connected through one leg by two 20 mm diameter bolts in line. Determine the effective area as a tension member.

Effective area (cl 4.6.3.1, Pt 1), from Equation (6.2)

$$A_e = a_1 + a_2[3a_1/(3a_1 + a_2)]$$

If A is the leg length of the angle

$$a_1 = t(A - d_h - t/2) = 6(70 - 22 - 6/2) = 270 \text{ mm}^2$$
$$a_2 = t(A - t/2) = 6(70 - 6/2) = 402 \text{ mm}^2$$
$$A_e = 270 + 402 \times 3 \times 270/(3 \times 270 + 402) = 538.7 \text{ mm}^2$$

Compare this value with the gross area $= 813 \text{ mm}^2$.

6.4   Tension members with moments (cl 4.8.2, Pt 1)

Tension members are generally subjected to bending moments from eccentric connections but these moments are not large, and are designed as axially loaded members as described in Section 6.3. In other situations, however, the bending moments from eccentric loads and transverse loads are more likely to be critical, and the stresses from axial load and bending moments about the major and minor axes of bending are combined as an interaction formula of axial forces and bending moments.

Tension members with moments should be checked for resistance to lateral torsional buckling in accordance with 4.3 Pt 1 under moment alone. They should also be checked for capacity under the combined effects of axial load and moment at the points of greatest bending moments and axial loads, usually at the ends.

The following relationship should be satisfied:

$$F/A_e p_y + M_x/M_{cx} + M_y/M_{cy} \leqslant 1 \tag{6.4}$$

where

F is the applied axial force in the member;

A_e is the effective area;

p_y is the design strength;

M_x is the applied moment about the major axis at the critical region;

M_{cx} is the moment capacity about the major axis in the absence of axial load;

M_y is the applied moment about the minor axis at the critical region;

M_{cy} is the moment capacity about the minor axis in the absence of axial load.

$$M_{cx} = p_y S \leqslant 1.2 p_y Z \text{ for plastic and compact sections and}$$
$$= p_y Z \text{ for semi-compact and slender sections with low shear load.}$$

Equation (6.4) is based on the condition that the sum of the elastic axial and bending stresses should not exceed the yield stress, i.e.

$$p_a + p_{bx} + p_{by} \leqslant p_y$$

Alternatively for greater economy in plastic or compact sections the following relationship should be satisfied.

$$(M_x/M_{rx})^{z1} + (M_y/M_{ry})^{z2} \leqslant 1 \tag{6.5}$$

where

M_{rx} is the reduced moment capacity about the major axis in the presence of axial load obtained from published tables;

M_{ry} is the reduced moment capacity about the minor axis in the presence of axial load obtained from published tables;

z_1 is a constant, taken as 2.0 for I and H sections, 2 for solid and hollow circular sections, 5/3 for solid and hollow rectangular sections, 1.0 for all other cases;

z_2 is a constant, taken as 1.0 for I and H sections, 2 for solid and hollow circular sections, 5/3 for solid and hollow rectangular sections, 1.0 for all other cases.

Equation (6.5) is based on full plasticity of the section, not on just reaching yield stress in the extreme fibres as for Equation (6.4) and is therefore less conservative. It should be noted that the axial force is not ignored but reduces the values of M_{rx} and M_{ry}.

6.5   Compression members (cl 4.7, Pt 1)

6.5.1   Introduction

Compression members are located in trusses, bracing and columns, and are generally greater in cross-sectional area than tension members because they fail in buckling. The most efficient cross section for a strut is a tube but this is not always practicable because of the presence of external bending moments and the practical difficulties of making connections. Universal column sections are not as efficient as tubes because they are weak when bending about the y–y axis. However connections are simpler and the thick flanges resist local buckling. Where loads are relatively small angles are used as struts, e.g. roof trusses.

6.5.2   Buckling formulae

Steel members are more susceptible to buckling than concrete members because they are more slender. Short steel members in compression fail by squashing at the yield stress, while long, or more accurately slender, members fail by buckling. Buckling occurs at an axial stress which is less than the yield stress and is related to lack of straightness and non-axial loads. Consequently the load becomes progressively more eccentric to the longitudinal axis of the member and a bending moment is introduced as shown in Fig. 6.2.

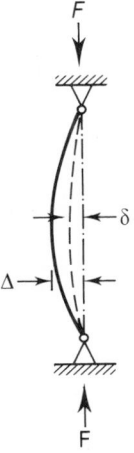

Fig. 6.2   Buckling of a strut.

An ideal pin-ended strut becomes elastically unstable and buckles at the Euler critical stress

$$p_E = \pi^2 E / \lambda^2 \tag{6.6}$$

where

E is Young's elastic modulus;
λ is the slenderness ratio $= l/r$;
l is the length of the strut in this case, but generally the effective length of a member;
r is the radius of gyration.

The ideal conditions assumed in developing the Euler formula, namely homogeneity and isotropy of the material, axial loading, absence of residual stresses, and perfect straightness do not occur in practice and consequently actual failure loads are smaller. Strain hardening, however, increases the buckling load at low slenderness ratios. Small unavoidable eccentricities of loading and lack of initial straightness can be simulated mathematically by assuming an initial curvature which produces a small central deflection δ, empirically defined, as shown in Fig. 6.2. When an axial load F is applied, the member buckles and the deflection increases to Δ, as shown.

The maximum compressive stress p in the extreme fibres on the compressive side of the strut is

$$p = F/A + F\delta/Z_e \tag{6.7}$$

where

A is the cross sectional area;
Z_e is the elastic section modulus.

The critical buckling load is reached when $p = Y_s$, i.e. when yielding commences in the extreme fibres of the strut. By adopting a simple sinusoidal function for the initial curvature it can be shown that

$$p_c = \phi - \sqrt{(\phi^2 - Y_s p_E)} \tag{6.8}$$

where

p_c is the critical average compressive stress, obtained by dividing the critical buckling load by the gross cross-sectional area of the strut;
$\phi = [Y_s - p_E(\eta + 1)]/2$;
Y_s is the specified minimum yield stress;
η is the empirical function representing the initial curvature;
p_E is the Euler critical stress obtained from Equation (6.6) using $E = 205 \, \text{kN/mm}^2$.

Equation (6.8) is known as the Perry formula and its adoption is explained by Dwight (1975). The value of the function η has varied over the years and the value originally obtained experimentally by Perry in 1925 was

$$\eta = 0.003(l/r) \tag{6.9}$$

The value suggested later by Godfrey to give more economical designs, based on experimental work by Duthiel in France, was

$$\eta = 0.3[l/(100r)]^2 \tag{6.10}$$

The value recommended in Appendix C2, Pt 1 is

$$\eta = 0.001a(\lambda - \lambda_0) \text{ but not less than zero} \tag{6.11}$$

where

a varies from 2 to 8 depending on the shape of the section;
$\lambda_0 = 0.2\sqrt{(\pi^2 E/Y_s)}$ and is the limiting slenderness ratio.

For design purposes the specified minimum yield stress Y_s is replaced by the design strength p_y.

As the lowest value of the central deflection δ in Equation (6.7) is related to the initial curvature, bending moments are generated as soon as the axial load is applied and the buckling process starts immediately. Therefore there is no condition of elastic instability as defined by Euler and the average compressive stress may never reach the Euler critical stress for a strut of finite length. Nevertheless the failure of a strut is sudden when compared with the ductile failure of a tension member. For a fuller development of the buckling theory see Trahair and Bradford.

6.5.3   BS 5950 buckling strength (cl 4.7.4, Pt 1)

The compressive strength of a slender member can be expressed as:

$$P_c = A_g p_c \text{ for plastic, compact, or semi compact or}$$
$$P_c = A_g p_{cs} \text{ for slender sections}$$

where

A_g is the gross cross-sectional area.

The previous buckling stress formulae are rather complicated but if included in a computer program present no difficulties. Where hand calculations are required, values of the compressive stress (p_c or p_{cs}) have been calculated for values of the slenderness ratio λ. There are different tables for different shapes of section, residual stress and axis of buckling as shown in Table 25 Pt 1. Buckling stresses are obtained from a selection of Table 27 Pt 1. 'H' sections are assumed to be classified as a UC. These curves are based on experimental results for real columns as described in the ECCS (1976) report, and expressed theoretically as described by Beer and Schultz (1970).

6.5.4   Maximum slenderness ratios (cl 4.7.3.2, Pt 1)

At high values of the slenderness ratio struts become so flexible that deflections under their own weight are sufficient to introduce stresses in excess of the Perry buckling formula. The following empirical limits are therefore imposed by BS 5950.

The value of λ should not exceed the following:

(a) for members resisting loads other than wind loads: 180
(b) for members resisting self weight and wind loads only: 250
(c) for any member normally acting as a tie but subjected to reversal of stress resulting from the action of wind: 350

Members whose slenderness exceeds 180 should be checked for self weight deflections. If this exceeds (length/1000), the effect of the bending should be taken into account in design.

6.5.5   Local buckling (cl 3.5, Pt 1)

Very slender elements of a section, which are primarily in compression, may buckle locally before overall buckling of the member occurs. Local buckling can be checked by calculating the width to thickness ratios of each element of a cross section. Elements are classified as plastic, compact, semi-compact or slender as described in Section 1.1.3 and stresses are adjusted accordingly.

To allow for local buckling BS 5950 reduces the design strength p_y. The strength reduction factor is obtained from Table 8, Pt 1 which is related to the outstand element dimensions (B/T, d/T and d/t) for each type of section. The design stress is reduced further depending on the slenderness ratio λ.

6.5.6   Effective lengths of struts (cl 4.7.2, Pt 1)

The buckling theory, developed in Section 6.5.2, is based on the assumption that the ends of the strut are pinned, i.e. are frictionless joints which prevent linear movement but which can rotate freely about any axis of the section. However pin-ended struts are rare in practice though the pin-ended condition is a useful theoretical concept which can be shown to relate to other end conditions.

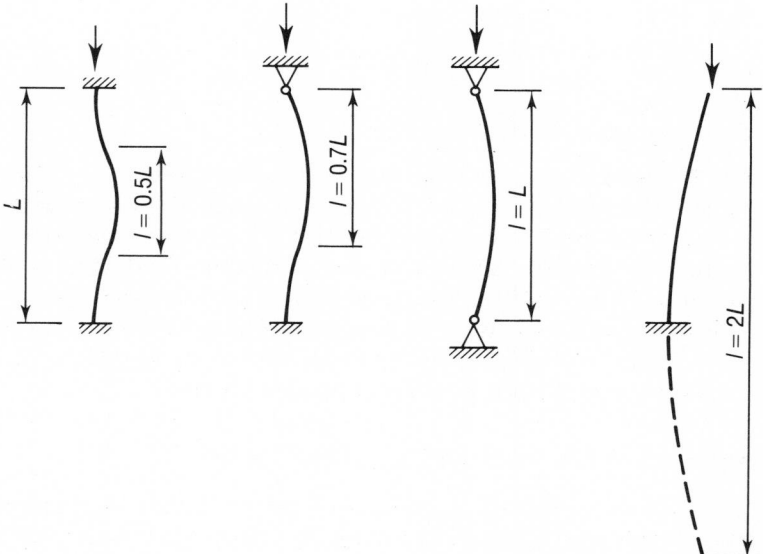

Fig. 6.3   Theoretical effective lengths.

Other end support conditions can be simulated by replacing the actual length of the strut by an effective length. Consider the theoretical end conditions shown in Fig. 6.3 which vary from complete end fixity to complete freedom at the end of a cantilever. The theoretical effective length l is expressed as a proportion of the actual length L. The theoretical effective length is the distance between real or theoretical pins, i.e. points of contraflexure. The effective length is important in design calculations because the buckling load is inversely proportional to the square of the effective length.

In practice the full rigidity of a fixed built-in end is never achieved and some rotation occurs due to the flexibility of the connection or the support. Only a small rotation is necessary to transform a built-in end to a pinned end and thus reduce the buckling resistance of a strut. In comparison small translational movements at support are not so critical and may be limited by supporting members.

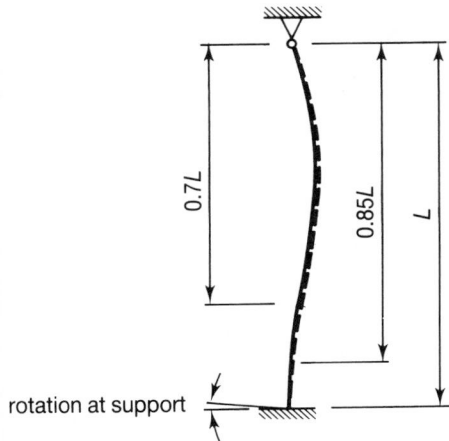

Fig. 6.4   Rotation of the end of a strut.

Practical end conditions therefore allow for some rotation and translation at the ends of a real strut. Consider the strut in Fig. 6.4. The theoretical effective length is 0.7L but because in practice the built-in end can rotate, the real practical effective length is 0.85L, which reduces the axial load at failure. The complete table of nominal effective lengths recommended is in Table 24, Pt 1. The range of practical values is between 0.7L and 2.0L.

When designing a real structure deciding on a value for the effective length in some cases is not easy. Experience is an advantage but if this is lacking some examples are given in Appendix D, Pt 1 for stanchions in single storey buildings of simple construction. Appendix E, Pt 1 gives a method for determining the effective lengths for struts in rigid frames which is based on the relative total column stiffness at a joint to the total stiffness of all members at the joint. For situations where sway of the column does not occur, e.g. where cross bracing is present in simple construction, the effective length is less than the real length (0.75L to 0.85L). Where sway occurs, e.g. in an unpropped portal frame where the effective length is greater than the real length (>1.5L), the effective length method is less accurate and care should be taken and alternative methods of analysis used to check the solution.

For single angles used as struts eccentricity of axial load may generally be ignored. As a rough guide the effective length is approximately 0.85L for two fasteners (or weld) and 1.0L for one fastener. Buckling failure about the minor axis must be considered which for a single equal angle is at 45° to the major axis of bending. For double angle struts the effective lengths are similar to those for a single angle but failure about the 45° may not be possible because of restraints (cl 4.7.10 and Table 28, Pt 1).

Figure 6.5 shows how the effective length of a column varies during the construction of a very simple structure and demonstrates how the erection procedure affects the stability of a structure.

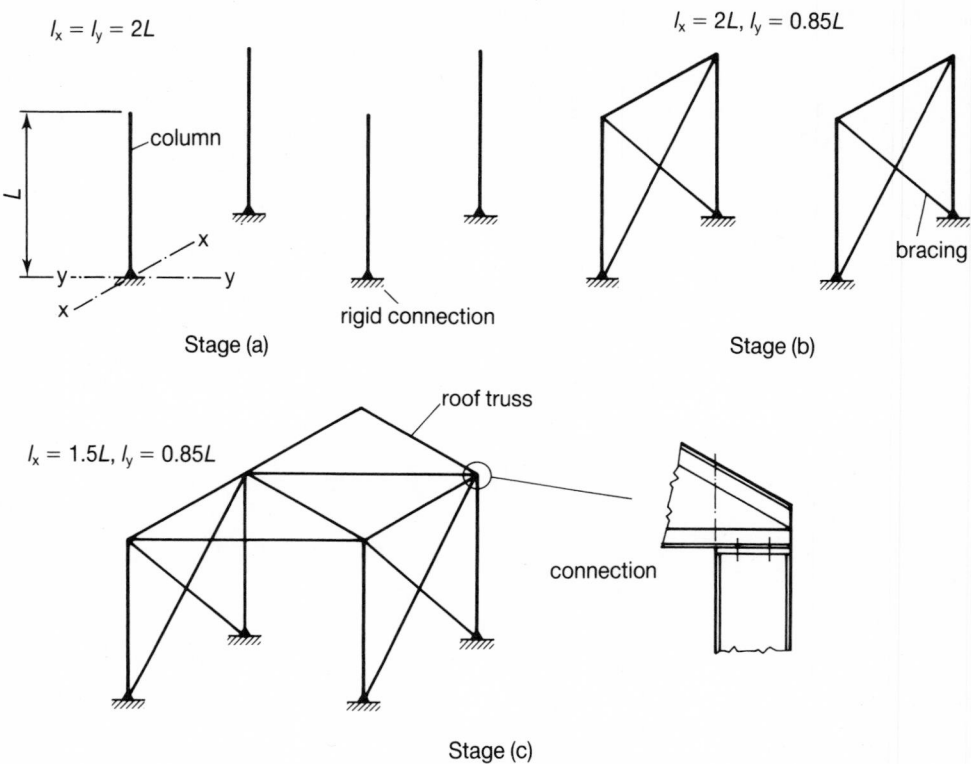

Fig. 6.5   Change in effective lengths during construction.

Stage (a) the columns are cantilevers with a maximum effective length of $l_x = l_y = 2L$ about the major axes of bending.

Stage (b) the bracing prevents the top of the column from movement about the y–y axis of bending so that the effective length is reduced to $l_y = 0.85L$. The bracing connection needs only to be a pin but practically it is capable of resisting a moment. The eaves beam is not essential for stability. There is no change for buckling about the x–x axis.

Stage (c) the roof truss connection gives partial fixity, i.e. prevents some sway and rotation at the top of the column for bending about the x–x axis. This effective length of $l_x = 1.5L$ is an estimate, but it has been used in practice for a number of years and is recommended in Appendix D, Pt 1. In practice there would be cross bracing on the end of the building, and on plan.

EXAMPLE 6.3   Angle strut in a roof truss

A steel roof truss is composed of angles and T-sections and has been analysed assuming pin joints. A particular member is subjected to the following axial forces: dead load (*DL*) -7.2 kN, imposed snow load (*IL*) -10.7 kN, wind load (*WL*) downwards -2.0 kN, and wind load (*WL*) upwards $+19.1$ kN. If an angle, welded at the ends, is chosen to resist the loads, determine the size if the actual length is 2.1 m.

Design load cases:

(a)   no wind

$1.4 \times DL + 1.6 \times IL = 1.4(-7.2) + 1.6(-10.7) = -27.2$ kN (compression).

(b)   wind down

$1.2 \times (DL + IL + WL) = 1.2(-7.2 - 10.7 - 2) = -23.88$ kN (compression).

(c)   wind up

$1 \times DL + 1.4 \times WL = -7.2 + 1.4(+19.1) = +19.54$ kN (tension).

Try $65 \times 50 \times 5$ mm angle grade 43 steel, long leg attached.
From Table 6, Pt 1, design strength for grade 43 steel

$$p_y = 275 \text{ N/mm}^2$$

Designed as a compression member (cl 4.7, Pt 1) and from Table 7, Pt 1.

$$b/t = 65/5 = 13 < 15\sqrt{\varepsilon}, \text{ therefore semi-compact}$$

Axial stress

$$f_c = F_c/A_c = -27.2\text{E}3/554 = -49.1 \text{ N/mm}^2.$$

Slenderness ratios (cl 4.7.10.2, Pt 1)

$$l/r_v = 0.85 \times 2100/10.7 = 166.8$$
$$l/r_{aa} + 30 = 0.7 \times 2100/147 + 30 = 130 < 166.8 < 180, \text{ therefore satisfactory.}$$

From Tables 25 and 27(c), Pt 1, design compressive stress

$$p_c = 57 \text{ N/mm}^2$$

Ratio

$$f_c/p_c = 49.1/57 = 0.86 < 1, \text{ therefore satisfactory.}$$

Designed as a tension member (cl 4.6.1, Pt 1).
Net area of connected leg

$$a_1 = t(A - t/2) = 5(65 - 5/2) = 312.5 \text{ mm}^2$$

Net area of unconnected leg

$$a_2 = t(A - t/2) = 5(50 - 5/2) = 237.5 \text{ mm}^2$$

Effective area

$$A_e = a_1 + a_2[3a_1/(3a_1 + a_2)] = 502 \text{ mm}^2$$

Axial stress

$$f_t = F_t/A_e = 19.54\text{E}3/502 = 38.92 \text{ N/mm}^2$$

Ratio

$$f_t/p_y = 38.92/275 = 0.14 < 1, \text{ therefore satisfactory.}$$

6.5.7   Intermediate restraints (cls 4.7.1.2, Pt 1)

A member that provides intermediate restraint and prevents buckling of a strut reduces
the effective length and increases the strength of a strut. The restraint need not be rigid

and may be elastic provided that its stiffness exceeds a certain value (Trahair and Bradford). According to BS 5950 restraining members associated with built-up members are required to resist not less than 1% of the axial force in the restrained member.

6.6   Compression members with moments

6.6.1   Introduction

In practical situations axial forces in columns are accompanied by bending moments about the major and minor axes of bending. The axial force and bending moment vary along the axis of the member and an exact analysis is complicated, as shown by Culver (1966), and is outside the scope of this book.

Compression members with moments are the most difficult to analyse because they include bending about both axes, local buckling and lateral torsional buckling. Exact theoretical solutions are not available and design equations are based on elastic and plastic limits that are expressed as simply as possible for design purposes. In simple design, where it is assumed that frames can be isolated, the compression members can be isolated and analysed accordingly. In rigid design the compression member is part of the frame and end moments acting on the compression members through the rigid connections increase the end rotations and affect the buckling load.

A steel member subjected to bending moments and axial forces fails when a stress is greater than first yield but less than full plasticity of the section. First yield theories are therefore conservative while full plasticity theories are unsafe. Nevertheless this broad description provides the basis for the equations for slender and semi-compact sections and for compact sections given in BS 5950.

Beam and column structures are often connected together using 'pins' and are braced to prevent sidesway collapse. The 'pin' connections are generally eccentric to the column axis and thus introduce a bending moment to the column. It is obvious from the basic theory that eccentric loads increase the bending moment on a strut and thus reduce the failure load. However small eccentricities are inevitable and are allowed for in design. Situations where the eccentricity may be ignored include beams and roof trusses supported on cap plates; laced and battered struts; and discontinuous angles, channels and T-sections (cl 4.7.6, Pt 1).

In all other cases the load should be taken as acting at a distance from the face of the steel column equal to 100 mm, or at the centre of the stiff bearing whichever gives the greater eccentricity.

Compression members with moments should be checked for two conditions.

(a) Local capacity at the point of greatest bending moment and axial load (usually at the ends of a member). Depending on the section properties this capacity may be limited by;
  (i)   yielding,
  (ii) local buckling.
(b) Overall buckling by either;
  (i)   simplified approach,
  (ii) more exact approach.

6.6.2   Local capacity check (cl 4.8.3.2, Pt 1)

(i) For semi-compact and slender cross sections (and as an alternative simplified approach for plastic or compact cross sections):

$$F/(A_g p_y) + M_x/M_{cx} + M_y/M_{cy} \leqslant 1 \qquad (6.12)$$

where

F is the applied axial force in the member;

A_g is the gross cross-sectional area;

p_y is the design strength;

M_x is the applied moment about the major axis at the critical region;

M_{cx} is the moment capacity about the major axis in the absence of axial load;

M_y is the applied moment about the minor axis at the critical region;

M_{cy} is the moment capacity about the minor axis in the absence of axial load.

(ii) For plastic and compact sections a relationship based on full plasticity of the section is allowed.

$$(M_x/M_{rx})^{z1} + (M_y/M_{ry})^{z2} \leqslant 1 \qquad (6.13)$$

where

M_{rx} is the reduced moment capacity about the major axis in the presence of axial load obtained from published tables;

M_{ry} is the reduced moment capacity about the minor axis in the presence of axial load obtained from published tables;

z_1 is a constant, taken as 2.0 for I and H sections, 2.0 for solid and hollow circular sections, 5/3 for solid and hollow rectangular sections, 1.0 for all other cases;

z_2 is a constant, taken as 1.0 for I and H sections, 2.0 for solid and hollow circular sections, 5/3 for solid and hollow rectangular sections, 1.0 for all other cases.

6.6.3   Overall buckling check (cl 4.8.3.3, Pt 1)

(i) For a simplified approach the following relationship should be satisfied:

$$F/(A_g p_c) + mM_x/M_b + mM_y/p_y Z_y \leqslant 1 \qquad (6.14)$$

where

F is the applied axial load in the member;

A_g is the gross cross-sectional area;

p_c is the compressive strength;

m is the equivalent uniform moment factor obtained from Table 18;

M_x is the applied moment about the major axis at the critical region;

M_b is the buckling resistance moment capacity (about the major axis);

M_y is the applied moment about the minor axis at the critical region;

Z_y is the elastic section modulus about the minor axis;

p_y is the design strength.

When multi-storey buildings are designed using the simple design method then all beams at any one level may be assumed to be fully loaded and the eccentricities from the column produce nominal moments.

In simple design where columns are effectively continuous at their splices the net moment applied at any one level should be divided between the column length above and below that level in proportion to the stiffness (I/L) of each length, except that when the ratio of the stiffnesses does not exceed 1.5 the moment may be divided equally. There are no carry-over moments to other levels (cl 4.7.7, Pt 1). The distance L is between the points where the column axes are restrained.

When nominal moments are used in calculations for the simple design of a steel column, only the overall buckling relationship, Equation (6.14), should be satisfied. In addition, the slenderness correction factor for bending $n = 1$ and the uniform bending factor $m = 1$ (cl 4.7.7, Pt 1) should be used.

When calculating the buckling resistance moment M_b, the equivalent slenderness ratio of the column $\lambda_{LT} = 0.5L/r_y$ provided that the nominal moments are the only applied moments (cl 4.7.7, Pt 1).

(ii)   Alternatively a more exact buckling check is as follows (cl 4.8.3.3.2, Pt 1).

$$mM/M_{ax} + mM_y/M_{ay} \leqslant 1 \tag{6.15}$$

where

M_{ax} is the maximum buckling moment about the major axis in the presence of the axial load, taken to the the lesser of:

$M_{cx}(1 - F/P_{cx})/(1 + 0.5F/P_{cx})$ or

$M_b(1 - F/P_{cy})$ governing lateral/torsional buckling;

M_{ay} is the maximum buckling moment about the minor axis in the presence of the axial load as:

$M_{cy}(1 - F/P_{cy})/(1 + 0.5F/P_{cy})$

where

M_{cx} is the moment capacity about the major axis obtained from 4.2.5 or 4.2.6;

M_{cy} is the moment capacity about the minor axis obtained from 4.2.5 or 4.2.6 but not subject to the restriction $M_c < 1.2p_yZ$;

P_{cx} is the compression resistance about the major axis;

P_{cy} is the compression resistance about the minor axis.

Note. In cases where M_x or M_y approaches zero the more exact approach may be more conservative than the simplified approach. In such situations the values satisfying the simplified approach may be used.

6.6.4   Effect of axial load on the plastic moment of resistance

The local capacity check described in Section 6.6.2 (ii), involves the use of the plastic moment of resistance which is reduced by any axial load. This reduction can be seen from Fig. 6.6 where on the stress diagram the shaded area represents the axial stress produced by the axial load. This means that the reduced area remaining must resist the bending

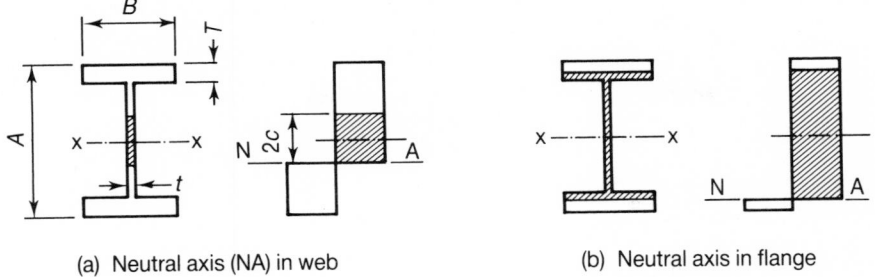

(a) Neutral axis (NA) in web   (b) Neutral axis in flange

Fig. 6.6 Effect of axial force on an 'I' section.

moment. Notice that when an axial load is present the neutral axis is displaced from the equal area position.

A convenient ratio to determine the reduced plastic section modulus is

$$n = p/Y_s$$

where

p is the mean axial stress;

Y_s is the specified minimum yield strength of steel.

If the axial force is small, the neutral axis is in the web. Alternatively if the axial force is large the neutral axis is in the flange as shown in Fig. 6.6. For bending about the x–x axis, the neutral axis moves from the web into the flange when $n > n_c$, where

$$n_c = t(D - 2T)/A$$

where

A is the total area of the section.

Subtracting the plastic modulus of the hatched area from the whole section.

$$S_{xr} = S_x - S_{(\text{shaded area})}$$
$$S_{(\text{shaded area})} = t(2c)^2/4$$

From equilibrium

$$pA = (2c)tY_s$$

and since by definition $n = P/Y_s$

When $n \leqslant n_C$

$$S_{xr} = S_x - an^2$$

When $n > n_c$

$$S_{xr} = b(1 - n)(c + n)$$

where

$$a = A^2/(4t)$$
$$b = A^2/(4B)$$
$$c = 2BD/A - 1$$

For bending about the y–y axis the change point for n is

$$n_c = tD/A$$

where

$$a = A^2/(4D)$$
$$b = A^2/(8T)$$
$$c = 4TB/A - 1$$

The values of a, b, c and n_c for 'I' sections are given in the Section Tables. Values for other shapes of sections are given in the complete version of the Section Tables.

The plastic section modulus determined in this way is applicable to short columns. For longer columns instability effects due to deflection of the column reduce this value.

EXAMPLE 6.4   *Rafter for a roof truss*

A roof truss composed of angles and tees has been analysed assuming pin joints and axial forces calculated. In addition a rafter has been analysed as a continuous member and bending moments determined. The axial forces and bending moments in a rafter member are as follows:

Dead load (*DL*), AF (− 64.5 kN), hogging BM (0.66 kNm), sagging BM (0.47 kNm);
Imposed load (*IL*), AF (− 95.4 kN), hogging BM (1.49 kNm), sagging BM (1.05 kNm);
Wind load (*WL*) down, AF (−12.9 kN), hogging BM (0.28 kNm), sagging BM (0.2 kNm);
Wind load (*WL*) up, AF (+147.1 kN), hogging BM (1.89 kNm), sagging BM (2.69 kNm).

The design forces and bending moments in the rafter member from different load cases are:

(a) no wind
    Axial force = $1.4DL + 1.6IL = 1.4(-64.5) + 1.6(-95.4) = -242.94$ kN
    Hogging bending moment = $1.4 \times 0.66 + 1.6 \times 1.49 = 3.31$ kNm
    Sagging bending moment = $1.4 \times 0.47 + 1.6 \times 1.05 = 2.34$ kNm.
(b) wind down
    Axial force = $1.2(DL + IL + WL) = 1.2(-64.5 - 95.4 - 12.9) = -207.36$ kN
    Hogging bending moment = $1.2(0.66 + 1.49 + 0.28) = 2.92$ kNm
    Sagging bending moment = $1.2(0.47 + 1.05 + 0.2) = 2.06$ kNm.
(c) wind up
    Axial force = $DL + 1.4WL = -64.5 + 1.4(+147.1) = +141.4$ kN
    Hogging bending moment = $0.66 + 1.4 \times 1.89 = 3.31$ kNm
    Sagging bending moment = $0.47 + 1.4 \times 2.69 = 4.24$ kNm.

Try a 165 × 152 × 27 kg structural tee, grade 43 steel, with stem connected.
Class of section (cl 3.5.2, Pt 1), from Fig. 3, Pt 1
$d/t = 155.4/7.7 = 20.2 > 19\sqrt{\varepsilon}$, therefore slender section from Table 7, Pt 1

Reduced design stress for a slender section (cl 3.6.4, Pt 1), from Table 8, Pt 1

$$p_y = 275[14/(d/t\varepsilon - 5)] = 275[14/(20.2 - 5)] = 253 \text{ N/mm}^2.$$

Compression and bending, load case (a) no wind.

Local capacity check (cl 4.8.3.2, Pt 1), Equation (6.12)

$$F/(A_g p_y) + M_x/M_{cx} = 242.94\text{E}3/(3420 \times 253.3) + 3.31\text{E}6/(253 \times 51.5\text{E}3)$$
$$= 0.534 < 1, \text{ therefore satisfactory}$$

Overall buckling check (cl 4.8.3.2, Pt 1), Equation (6.14)
Slenderness ratio for buckling in the plane of the truss

$$l/r_x = 0.7 \times \text{span}/r_x = 0.7 \times 2154/43.1 = 35$$

Slenderness ratio for buckling out of the plane of the truss

$$l/r_y = (\text{purlin spacing})/r_y = 1346/39.4 = 34.2$$

Design stress from Table 27(c), Pt 1

$$p_c = 229 \text{ N/mm}^2 \text{ and}$$

$$\lambda_{LT} = l/r_y = 1346/39.4 = 34.2, \ p_y = 253, \text{ and from Table 11, Pt 1}$$

$$p_b = 253 \text{ N/mm}^2$$

$$F/(A_g p_c) + mM_x/M_b = 242.94\text{E}3/(3420 \times 229) + 1.0 \times 3.31\text{E}6/(253 \times 51.5\text{E}3)$$
$$= 0.563 < 1, \text{ therefore satisfactory}$$

Tension and bending, load case (c) wind up
Effective area of tee section (cl 4.6.3.1, Pt 1)

$$a_1 = t(A - T) = 7.7(155.4 - 13.7) = 1091.1 \text{ mm}^2$$
$$a_2 = TB = 13.7 \times 166.8 = 2285.2 \text{ mm}^2$$
$$A_e = a_1 + a_2[3a_1/(3a_1 + a_2)] = 2436.8 \text{ mm}^2$$

Tension interaction formula (cl 4.8.2, Pt 1), Equation (6.4)

$$F/(A_e p_y) + M_x/M_{cx} = 141.4\text{E}3/(2436.8 \times 253) + 4.24\text{E}6/(253 \times 51.5\text{E}3) = 0.554 < 1,$$
therefore satisfactory.

A lesser weight compact, or semi-compact section would be more economical.

EXAMPLE 6.5   Bending moments in a three storey corner column

The first three storeys of a corner stanchion are shown diagramatically in Fig. 6.7. The beam support reactions are indicated in kN on each beam. Assuming simple design with eccentric pin-joints and cross bracing, determine the bending moments in the columns AB and BC.

Try the following column sizes. From Section Tables:
Columns AB and BC, 203 × 203 × 60 UC

$$D = 209.6 \text{ mm}, \ t = 9.3 \text{ mm}, \ I_x = 6088 \text{ cm}^4, \ I_y = 2041 \text{ cm}^4.$$

Column CD, 203 × 203 × 46 UC

$$D = 203.2 \text{ mm}, \ t = 7.3 \text{ mm}, \ I_x = 4564 \text{ cm}^4, \ I_y = 1539 \text{ cm}^4.$$

This is a simple design (cl 2.1.2.2, Pt 1), i.e. pin connections, and because it is a corner column buckling with bending about the major and minor axes must be considered.

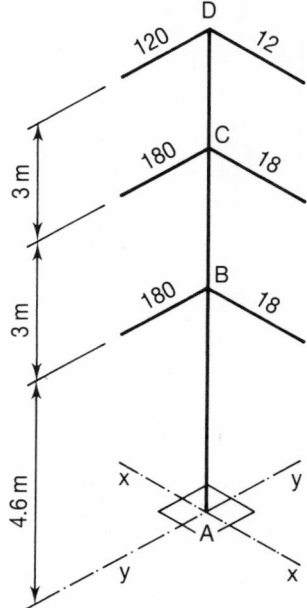

Fig. 6.7 Three storey corner column.

(i)   *Bending about the major x–x axis*

Stiffness of columns (cm units).

>   Column AB, $K_{AB} = I_x/L_{AB} = 6088/460 = 13.23$
>   Column BC, $K_{BC} = I_x/L_{BC} = 6088/300 = 20.29$
>   Column CD, $K_{CD} = I_x/L_{CD} = 4564/300 = 15.21$

(a)   First floor level (at connection B):
Nominal moment from the eccentricity of the connection bending about the x–x axis
(cl 4.7.6, Pt 1)

$$M_{Bx} = (\text{beam reaction}) \times (100 + D/2)/1E3$$
$$= 180 \times (100 + 209.6/2)/1E3 = 36.86 \text{ kNm}.$$

Ratio of stiffnesses of columns BC and AB

$$K_{BC}/K_{AB} = 20.29/13.23 = 1.53.$$

As this exceeds 1.5 the nominal moments are distributed in proportion to the stiffness for
simple multi-storey construction (cl 4.7.7, Pt 1).

$$M_{BA} = M_{Bx}K_{AB}/(K_{AB} + K_{BC}) = 36.86 \times 13.23/(13.23 + 20.29) = 14.55 \text{ kNm}$$

$$M_{BC} = M_{Bx}K_{BC}/(K_{AB} + K_{BC}) = 36.86 \times 20.29/(13.23 + 20.29) = 22.31 \text{ kNm}$$

(b)   Second floor level (at connection C):
Assuming that the splice between columns BC and CD lies above the first floor level, then
the design moment is the same at the first floor, i.e. $M_{Cx} = 36.86$ kNm.

Ratio of stiffnesses between columns BC and CD

$$K_{BC}/K_{CD} = 20.29/15.21 = 1.33.$$

As this does not exceed 1.5 the moments are distributed equally for simple multi-storey construction (cl 4.7.7, Pt 1)

$$M_{CB} = M_{CD} = M_{Cx}/2 = 36.86/2 = 18.43 \text{ kNm}.$$

Nominal moments are assumed to have no effect at levels above and below the level at which they are applied (cl 4.7.7, Pt 1).

(ii)   *Bending about the minor y–y axis at B and C* – see Tutorial problem 6.7.3

EXAMPLE 6.6   Two storey corner column

Determine the size of a universal column section required for a two storey corner column shown in Fig. 6.8. The design loads (in kN) shown are the end reactions from the beams.

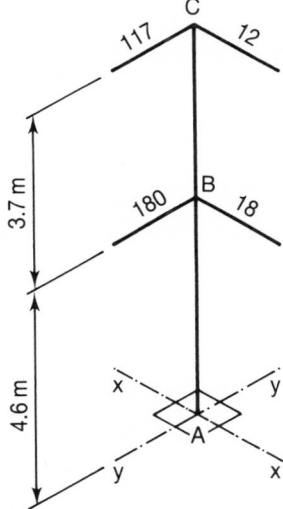

Fig. 6.8   Two storey corner column.

The beams are connected to the column using cleats and bolts and therefore simple design methods are acceptable. The columns and beams are protected from fire by a concrete casing. Assume Grade 43 steel.

From Table 24, Pt 1 the effective length factors for the column are:

Storey	Axis	
	x–x	y–y
AB	0.85	0.85
BC	0.85	1.0

Axial design loads on the columns:

(a) First floor   kN
     Floor beam (self weight included)   117
     Wall beam (self weight included)   12
     Self weight of column BC (estimated)   10
                               Total = 139

(b) Ground floor   kN
     Floor beam (self weight included)   180
     Wall beam (self weight included)   18
     Upper storey   139
     Total (excluding self weight of column AB) $F = 337$

The determination of the optimum section size of column is accomplished by trial and error and a computer program is a useful accessory. However work by hand can be reduced as follows.

The maximum slenderness ratio (cl 4.7.3.2, Pt 1) (see Section 6.5.4) is 180 and the maximum effective length is $0.85 \times 4600 = 3910$ mm. Hence the minimum radius of gyration

$$r_y = l/\lambda = 3910/180 = 21.7 \text{ mm.}$$

All the Universal Column sections satisfy this criterion.

Try $152 \times 152 \times 37$ kg UC from A to C. The relevant section properties are: $D = 161.8$ mm, $B = 154.4$ mm, $t = 8.1$ mm, $T = 11.5$ mm, $A_g = 4740$ mm^2, $r_y = 38.7$ mm, $r_x = 68.4$ mm, $Z_y = 91.78$E3 mm^3, $S_x = 310.1$E3 mm^3.

Class of section (cl 3.5.2, Pt 1) (see Section 4.1).

$b/T = (154.4/2)/11.5 = 6.71 < 8.5\varepsilon$, therefore plastic section from Table 7, Pt 1.

Calculations are now made to determine the values $(F, A_g, p_c, m, M_x, M_b, M_y, p_y, Z_y)$ required for an overall buckling check (cl 4.8.3.3, Pt 1).

For axial load calculation, consider column AB just below B.

$$\lambda_y = l/r_y = 0.85 \times 4600/38.7 = 101$$

From Tables 25 and 27(c), Pt 1 for $p_y = 275$ N/mm^2, $p_c = 124$ N/mm^2.

Nominal bending moments from the eccentricity of the beam reactions R at joint B are (cl 4.7.6, Pt 1):

$$M_{Bx} = R(100 + D/2)/1\text{E}3 = 180(100 + 161.8/2)/1\text{E}3 = 32.56 \text{ kNm}$$
$$M_{By} = R(100 + t/2)/1\text{E}3 = 18(100 + 8.1/2)/1\text{E}3 = 1.87 \text{ kNm}$$

Ratio of upper and lower column stiffnesses

$$K_{BC}/K_{BA} = (I_{BC}/L_{BC})/(I_{BA}/L_{BA}) = (1/3.7)/(1/4.6) = 1.24$$

As this does not exceed 1.5 the moments from the eccentricity of the beam reactions at joint B can be divided equally between the upper and lower columns (cl 4.7.7, Pt 1).

$$M_x = M_{Bx}/2 = 32.56/2 = 16.28 \text{ kNm}$$
$$M_y = M_{By}/2 = 1.87/2 = 0.93 \text{ kNm}$$

Buckling resistance moment capacity (M_b) about the x–x axis (cl 4.3.7.3, Pt 1). The equivalent slenderness of the column (cl 4.7.7, Pt 1) $\lambda_{LT} = 0.5L/r_y = 0.5 \times 4600/38.7 = 59.43$, and from Table 11, Pt 1 for $p_y = 275$ N/mm^2, $p_b = 214$ N/mm^2. Hence for a plastic section

$$M_b = p_b S_x = 214 \times 310.1E3/1E6 = 66.36 \text{ kNm}.$$

For bending resistance about the minor y–y axis $p_y = 275$ N/mm^2 and $Z_y = 91.78E3$ mm^3.

For a steel column using simple design the equivalent uniform moment factor $m = 1$ (cl 4.7.7, Pt 1).

Inserting these values in Equation (6.14) the simplified overall buckling formula (cl 4.8.3.3.1, Pt 1)

$$F/(A_g p_c) + mM_x/M_b + mM_y/(p_y Z_y) = 337E3/(4740 \times 124) + 1.0 \times 16.28E6/66.36E6$$
$$+ 1.0 \times 0.93E6/(275 \times 91.78E3)$$
$$= 0.858 < 1, \text{ therefore satisfactory.}$$

Check the self weight of the column BC. Minimum overall dimensions of cased column

$$D = 161.8 + 100 = 261.8, \text{ say } 270 \text{ mm}$$
$$B = 154.4 + 100 = 254.4, \text{ say } 260 \text{ mm}$$
$$A_c = A_g - A_s = 270 \times 260 - 4740 = 65460 \text{ mm}^2.$$

Total weight = steel column + concrete casing = $Lmg + LA_c mg$

$$= 3.7 \times 37 \times 9.81/1E3 + 3.7 \times 65460 \times 2400 \times 9.81/1E9$$
$$= 7 \text{ kN} < 10 \text{ kN (assumed)}, \quad \text{therefore satisfactory.}$$

These calculations show that the 152 × 152 × 37 kg UC section Grade 43 steel is satisfactory, however a lesser weight of 152 × 152 × 30 kg UC might be suitable (see Tutorial Problem 6.7.5).

6.7 Tutorial problems

6.7.1 Angle strut in a roof truss

A steel roof truss is composed of angles and tee sections. A particular member is subjected to the following axial forces. Dead load +7.1 kN, imposed snow load +10.4 kN, wind downwards +1.9 kN, and wind upwards −18.7 kN. Check if a 50 × 50 × 5 mm angle, Grade 43 steel, welded at the ends, is satisfactory if the actual length of the member is 2.274 m.

Answers

Design load cases.

(a) no wind +26.58 kN
(b) wind down +23.28 kN
(c) wind up −19.08 kN.

As a compression member, $f_c = 39.75$ N/mm^2. $l/r_v = 199.3$, $p_c = 42$ N/mm^2, $f_c/p_c = 0.946 < 1$.

As a tension member, $a_1 = a_2 = 237.5$ mm^2, $A_e = 415.625$ mm^2, $f_t = 63.95$ N/mm^2, $f_t/p_y = 0.233 < 1$.

6.7.2   T-rafter in a roof truss

A steel roof struss is composed of angles and tee sections. The rafter member is subjected to the design forces and bending moments shown in Example 6.4. Check if a $165 \times 152 \times 20$ kg structural tee, Grade 50 steel, stem welded, is satisfactory.

Answers

$d/t = 24.9 > 16.72$, therefore slender.
$p_y = 213.4$ N/mm^2.
As a compression member, $l/r_x = 35$, $p_c = 196$ N/mm^2
In bending, $\lambda_{LT} = l/r_y = 34.96$, $p_b = 213.4$ N/mm^2
Local capacity check, $F/(A_g p_y) + M_x/M_{cx} = 0.836 < 1$, satisfactory.
Overall buckling check, $F/(A_g p_c) + mM_x/M_b = 0.875 < 1$, satisfactory.
As a tension member, $a_1 = 864.4$ mm^2, $a_2 = 1684$ mm^2, $A_e = 1885.4$ mm^2
Tension member check, $F/(A_e p_y) + M_x/M_{cx} = 0.857 < 1$, satisfactory.

6.7.3   Bending moments for a corner column

Determine the distribution of design moments about the minor y–y axis at joints B and C for the three storey corner column in the building in Example 6.5. Assume that the beams are connected by cleats to the web of the columns and that simple design is appropriate.

Answers

$M_{By} = 1.88$ kNm, $M_{BA} = 0.743$ kNm, $M_{BC} = 1.137$ kNm, $M_{Cy} = 1.88$ kNm, $M_{CB} = M_{CD} = 0.94$ kNm.

6.7.4   Two storey corner column

Using simple methods of design carry out the simplified overall buckling check just below C for the two storey corner steel column described in Example 6.6.

Answers

$M_x = 21.17$ kNm, $M_y = 1.25$ kNm, $F = 129$ kN.
$\lambda_y = 95.6$, $p_c = 132$ N/mm^2, $\lambda_{LT} = 47.8$, $p_b = 243$ N/mm^2, $M_b = 75.35$ kNm, $F/(A_g p_c) + mM_x/M_b + mM_x/(p_y z_y) = 0.537 < 1$, satisfactory.

6.7.5   Two storey corner column

Check if a $152 \times 152 \times 30$ kg UC Grade 43 steel section will support the loads described in Example 6.6.

Answers

$\lambda_y = 102.4$, $p_y = 275$, $p_c = 122$, $\lambda_{LT} = 60.21$, $p_b = 213$, $M_b = 52.63$ kNm, $F/(A_g p_c) + mM_x/M_b + mM_x/(p_y z_y) = 1.08 > 1$, not satisfactory.

References

Beer H. and Schultz G. (1070). *Theoretical basis for the European column curves*, Construction Metallique, No. 3.

Culver G. C. (1966). *Exact solution for biaxial bending equations*, American Society of Civil Engineers, Str. Div., **92** (ST2).

Dwight J. B. (1975). *Adaptation of the Perry formula to represent the new European steel column curves*, Steel Construction, AISC, **9** (1).

Dwight J. B. (1978). *Design of axially loaded columns including interactive buckling. The background to British Standards for structural steel work*, Imperial College, London and Constrado.

ECCS (1976). *Manual on stability, Introductory report, Second International Colloquium on Stability*, European Convention for Structural Steelwork, Liege.

Godfrey G. B. (1962). *The allowable stress in axially loaded struts*, Structural Engineer, March.

Nelson H. M. (1953). *Angles in tension*, British Constructional Steelwork Association publication No. 7.

Regan P. E. and Salter P. R. (1984). *Tests on welded angle tension members*, The Structural Engineer, **62B** (2).

Robertson A. (1925). *The strength of struts*, Selected Engineering Paper No. 28, Institution of Civil Engineers.

Trahair N. S. and Bradford M. S. (1988). *The Behaviour and Design of Steel Structures*, Chapman and Hall.

7

Structural Connections

7.1 Introduction (cl 1.6, Pt 1)

Structural connections are required to ensure continuity at the intersection of members and at the foundations. They are also used to form splices and to construct brackets to support loads. Structural connections are composed of components (plates or parts of sections) either bolted and/or welded together. British practice is to weld components together in the controlled conditions of the workshop and to bolt on site. Welding can be carried out on site but it needs to be carefully supervised and therefore is limited because of expense. Generally connections are a combination of plates, shop fillet welds and site bolts.

Structural connections transmit forces which result in linear and rotational movements. The linear movements at a joint are generally small but the rotational movement depends on the stiffness of the type of connection. If the rotational movement is very small the connection is described as 'rigid' (or 'fixed') and if large it is described as a 'pin' or 'simple' connection. If the rotational movement is between these two extremes the connection is described as 'semi-rigid'.

This distinction is important when selecting the method of analysis of the structure and analysing connections (see Sections 5.3.3 and 8.1.3). For example the moment–rotation characteristics of connections are important in the design of frames. For a simple 'pin' connection there should be sufficient free rotation at failure to prevent end moments developing. For a 'rigid' connection the real moment–rotation relationship is required for an accurate analysis of the frame, but further research work is required in this area.

7.2 The ideal structural connection

The types of connection in common use have been developed and modified to suit the manufacturing and assembly processes. Practical connections satisfy some of the ideal requirements which are that the connection should be:

(a) Simple to manufacture and assemble.
(b) Standardized for situations where the dimensions and loads are similar thus avoiding a multiplicity of dimensions, plate thicknesses, weld sizes and bolts.
(c) Manufactured from materials and components that are readily available.
(d) Designed and detailed so that work is from the top of the joint not from below where

the workman's arms will be above his head. There should also be sufficient room to locate a spanner, or space to weld if required.

(e) Designed so that the welding is confined to the workshops to ensure a good quality and to reduce costs.

(f) Detailed to allow sufficient clearance and adjustment to accommodate the lack of accuracy in site dimensions.

(g) Designed to withstand not only the normal working loads but also the erection forces.

(h) Designed to avoid the use of temporary supports to the structure during erection.

(i) Designed to develop the required load-deformation characteristics at the service load and at ultimate load.

(j) Detailed to resist corrosion and to be of reasonable appearance.

(k) Low in cost and cheap to maintain.

7.3 Welding and types of weld

Welding is a method of connecting components by heating the materials to a suitable temperature so that fusion occurs. The most common method for heating steelwork is by means of an electric arc between a coated wire electrode and the materials being joined. The electrical circuit is shown in Fig. 7.1(a). During the process, which is illustrated in Fig. 7.1(b), the coated electrode is consumed, the wire becomes the filler material, and the coating is converted partly into a shielding gas, partly into slag, and some part is absorbed by the weld metal. This method, known as the manual metal arc welding process, is still the most common for structural connections because of low capital cost and flexibility. However for long continuous welds automatic processes are preferred because of greater consistency of quality.

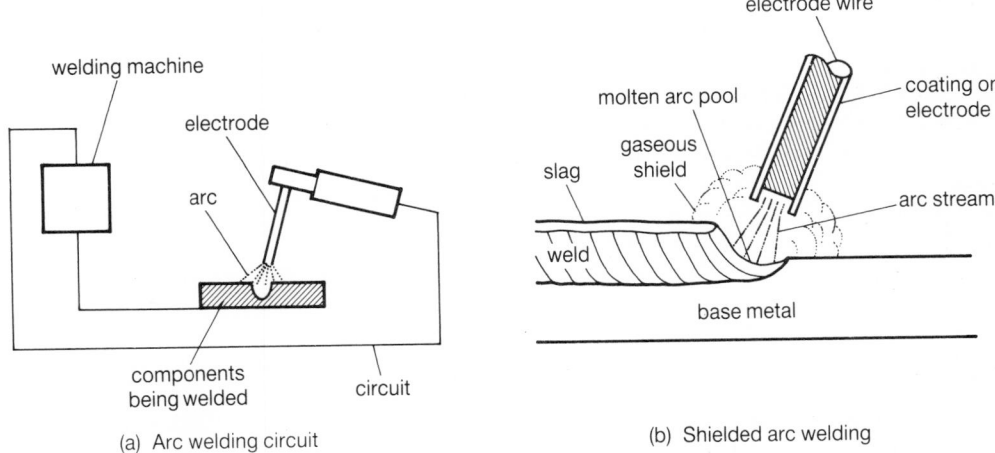

(a) Arc welding circuit   (b) Shielded arc welding

Fig. 7.1 Shielded metal arc welding.

Generally the electrode is stronger than the parent metal. For manual metal arc welding the electrodes should comply with BS 639 where the specified maximum yield stress is 330 N/mm² for type E43 electrode, and 360 N/mm² for type E51 electrode. Information on other mechanical properties is given in BS 639. The main reason for the flux covering to the electrode in the manual metal arc welding process is to provide an inert gas which shields the molten metal from atmospheric contamination. In addition the flux forms a slag to protect the weld until it is cooled to room temperature, when the slag should be easily detachable. Other functions of the flux include arc stabilization, control of surface profile, control of weld metal composition, alloying and deoxidation. However it should

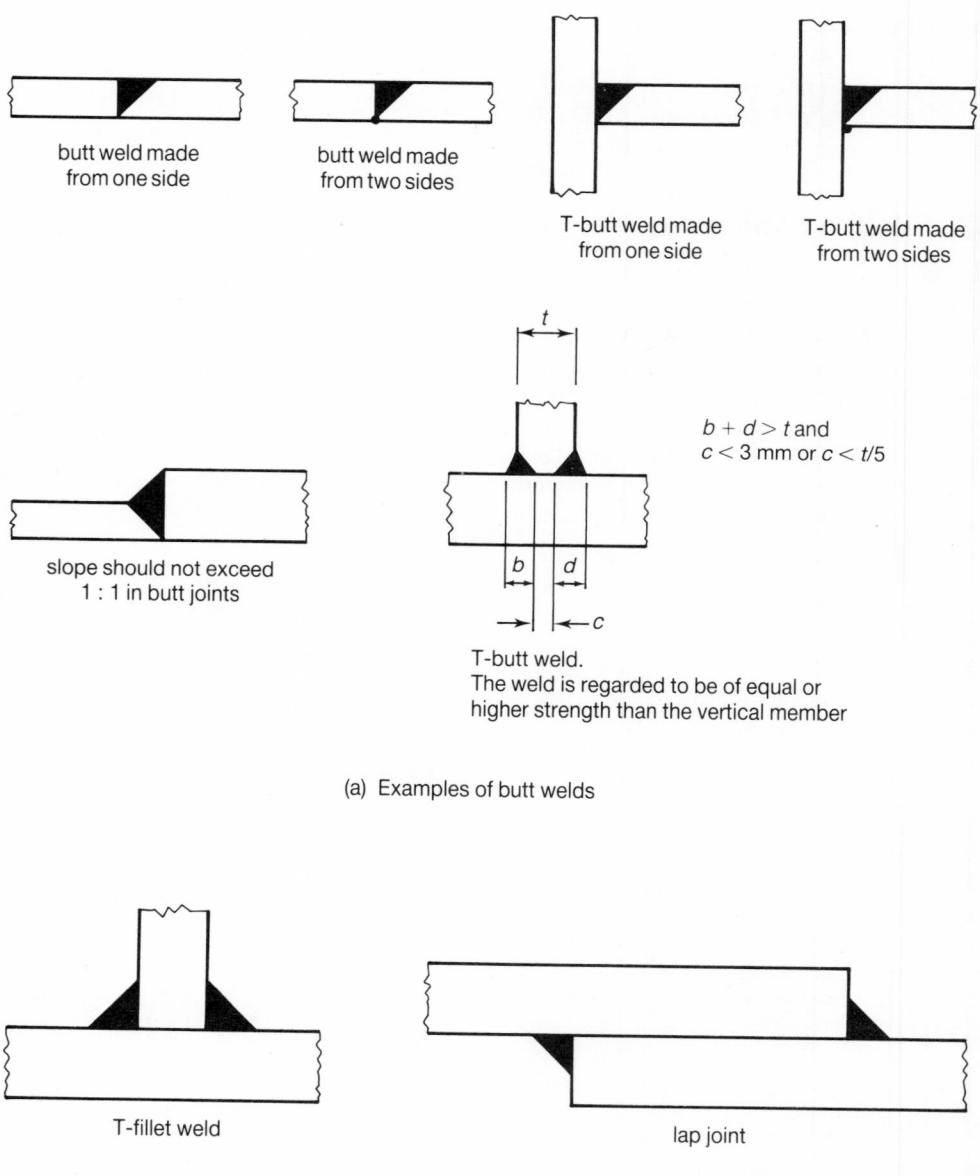

butt weld made
from one side

butt weld made
from two sides

T-butt weld made
from one side

T-butt weld made
from two sides

slope should not exceed
1 : 1 in butt joints

$b + d > t$ and
$c < 3$ mm or $c < t/5$

T-butt weld.
The weld is regarded to be of equal or
higher strength than the vertical member

(a) Examples of butt welds

T-fillet weld

lap joint

(b) Examples of fillet welds

Fig. 7.2   Types of welds in structural connections.

be noted that the flux can be a source of hydrogen contamination from absorbed and chemically combined moisture. The absorbed moisture can be removed by drying. Further information on composition and choice of electrodes is available in Gourd.

The particular advantage of welding is that if forms a rigid connection. However the manufacture of welded connections requires more skill and supervision than bolted connections. Most welded structural connections are effected using the manual metal arc process but long continuous welds, which occur in built-up girders, are laid down with automatic welding equipment. The automatic processes achieve the exclusion of atmospheric pollution by gas shielding, flux core or submerged arc. Further details are given in Gourd.

There are two types of welds, namely butt and fillet as shown in Fig. 7.2. Butt welds, often used to lengthen plates in the end on position, may be considered to be as strong as the parent plate as long as full penetration for the weld is achieved. For thin plates penetration is achieved without preparing the plate, but on thicker plates V or double J preparation is required. Further details are given in BS 5135. Butt welds are also used to connect plates at right angles but the plates require edge preparation. Partial penetration butt welds are not favoured in design and should not be used intermittently or in fatigue situations (cl 6.6.3, Pt 1).

Fillet welds, as shown in Fig. 7.2(b), are most commonly used and are generally formed with equal leg lengths. They do not require special edge preparation of the plates and are therefore cheaper than butt welds. Generally in connections plates intersect at right angles but intersection angles of between 60° and 120° can be used, provided the correct throat size is used in design calculations (cl 6.6.5.4, Pt 1). In order to accommodate lack of fit the minimum size of fillet weld in structural engineering is 5 mm, although 6 mm is often preferred. The maximum size of fillet weld from a single run metal arc process is 8 mm, but 6 mm is preferred to guarantee quality. When larger fillet welds are required they are formed from multiple runs.

The use of intermittent butt and fillet welds is permitted by BS 5950 but should not be used in fatigue situations. Intermittent welds are not favoured in structural engineering because they introduce stress discontinuities, act as stress raisers, may introduce fatigue cracks, may act as corrosion pockets, and are difficult to produce with an automatic welding machine. The longitudinal spacing between effective lengths of weld should not exceed the lesser of 300 mm, or $16t$ for compression elements, or $24t$ for tension elements, where t is the thickness of the thinner part joined   (cl 6.6.2.5, Pt 1).

7.4   Effective length of a weld (cl 6.6.5.2, Pt 1)

The strengths of the start and stop sections of a weld are unreliable and the overall length of a weld is therefore reduced to an effective length for design calculations. The effective length of a fillet weld should be taken as the overall length less one leg length, s, for each end that does not continue round a corner. The minimum length allowed to transmit loading is four times the leg length, and in practice it is generally not less than 40 mm. Fillet welds terminating at the ends, or sides, of parts should be returned continuously round the corners for a distance of not less than twice the leg length, unless this is impracticable. The continuation round the corner is to reduce stress concentrations but its strength is ignored in design calculations.

Although not mentioned in BS 5950, the effective length of a weld is reduced further if a component distorts under load. This can occur in situations similar to that shown in Fig. 7.3, where the deformations in the weld adjacent to the web are greater than those at

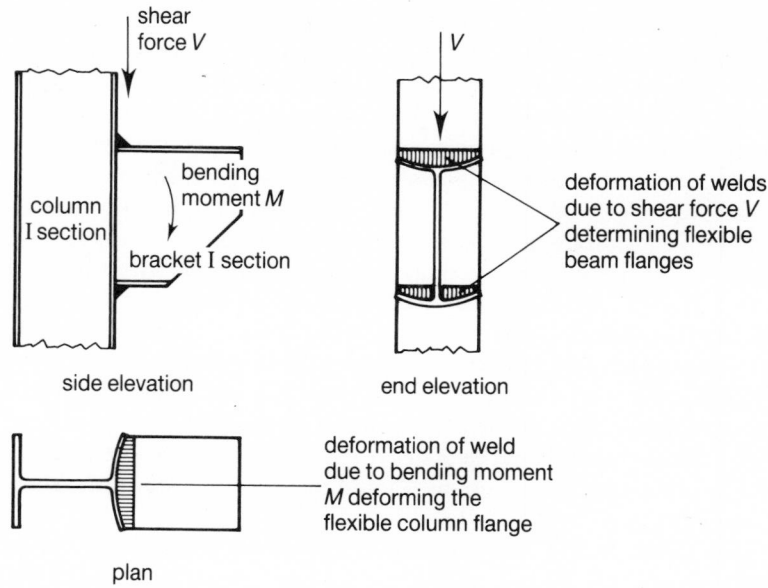

Fig. 7.3   Reduction in strength of welds associated with flange deformations.

the end of the flange. The larger deformations at the web initiate failure in the weld at this point with consequent loss of strength for the total length of the weld. This effect was investigated first by Elzen and later by Rolloos. The International Institute of Welding design rules recommend that the effective length b_{we} of a weld associated with a steel section of web thickness t and flange thickness T is

$$b_{we} = 2t + CT \qquad (7.1)$$

where values of factor C are given in Table 7.1.

Table 7.1   Values of factor C associated with effective weld lengths

Factor C 'I' section	Factor C box section	Position	Type of steel
7	5	tension flange	$F_e 360 (p_y = 235 \text{ N/mm}^2)$
10	7	compression flange	$F_e 360 (p_y = 235 \text{ N/mm}^2)$
5	4	tension flange	$F_e 510 (p_y = 355 \text{ N/mm}^2)$
7	6	compression flange	$F_e 510 (p_y = 355 \text{ N/mm}^2)$

7.5   Failure criteria for fillet welds (cl 6.6.5, Pt 1)

The real external forces acting on a fillet weld are probably those shown in Fig. 7.4(a), according to Clarke (1971). Experiments by Biggs *et al.* on right angled fillet welds of equal length loaded to failure show that the fracture plane varies between 10° and 80° depending on the combination of external forces. The actual distribution of stress on the failure plane is uncertain but a theoretical distribution by Kato and Morita shows peak stresses at the root of the weld which reduce towards the face of the weld. This distribution appears to be confirmed by experimental observations of cracks initiating at the root. The situation is complicated further by residual stresses and variables such as the type of electrode, type of steel, ratio of the size of weld to the plate thickness, the quality of weld, and whether the loading is static or dynamic. If stresses on the failure plane are assumed to be uniform then the relationship between the average shear stress and tensile stress on the failure plane has been shown by Biggs *et al.* to approximate to an ellipse. An ellipse of failure stresses combined with a variable fracture angle can be used theoretically to predict the magnitude of the external forces, but the method is too complicated for general design purposes.

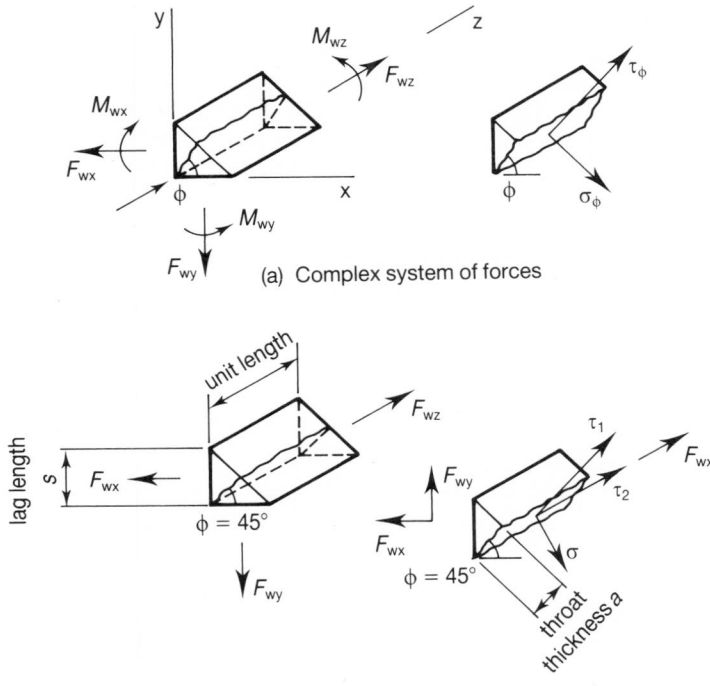

(a) Complex system of forces

(b) Simple system of forces

Fig. 7.4 Forces acting on a fillet weld.

For design purposes a complex system of external forces acting on a fillet weld is reduced to forces acting in three perpendicular directions on a unit length of weld as shown in Fig. 7.4(b). The stresses related to each force are calculated using the throat thickness a. The effective throat size, a, of a fillet weld should be taken as the perpendicular distance from the root of the weld to a straight line joining the fusion faces which lies within the cross section of the weld. It should not, however, be taken as greater than 0.7 times the effective leg length (cl 6.6.5.3, Pt 1).

The vector sum of the design stresses due to all the forces and moments by the weld should not exceed the design strength, p_w (cl 6.6.5.5, Pt 1).

This may be expressed as

$$F_{wx}^2 + F_{wy}^2 + F_{wz}^2 < (ap_w)^2 \tag{7.2}$$

The term ap_w is the design strength of a fillet weld per unit length.

The design strength of a fillet weld has been shown by Ligtenberg to be related to the strength of the parent material. The values given in Table 36, Pt 1 vary between 215 and 275 N/mm² depending on the grade of steel and the electrode strength.

The correct type and strength of electrode must be used for each grade of steel as specified in BS 639. The values of the design shear stresses appear conservative but they have been reduced to allow for inaccuracies in the use of simple theory, lack of knowledge of fatigue and brittle fracture, and the different relative strengths of side and end fillet welds.

An alternative method of design recommended by the European Commission (IIW document XV) also assumes a fixed 45° critical plane but calculates the normal and shear stresses on this plane and combines them using a more accurate failure criterion (see Fig. 7.4(b)). This method is more laborious and introduces the possibility of further errors when resolving the forces onto the 45° plane. The relationship between the forces F_{wx}, F_{wy}, and F_{wz} for this method can be expressed in the same form as Equation (7.2) as shown in Holmes and Martin. The size of the fillet weld obtained using this method is slightly less than using the vector addition method.

7.6   Load-deformation relationships for fillet welds

The strength of the weld in a connection is of primary importance but the load–deformation characteristics of the weld must also be considered. The deformation at the maximum load varies from approximately 0.6 mm to 1.4 mm depending on the orientation of the weld in relation to the applied load as shown by Clarke (1970) in Fig. 7.5. The maximum deformation is for the side fillet weld which is parallel to the applied load. The minimum is for an end fillet weld. Information is not given in BS 5950 on the deformation characteristics of welds but the design stresses in BS 5950 are based on the weaker side fillet welds and as shown in Fig. 7.5 at a design stress of 215 N/mm². At this stress the disparity in deformations for end and side fillet welds is not as marked as at failure.

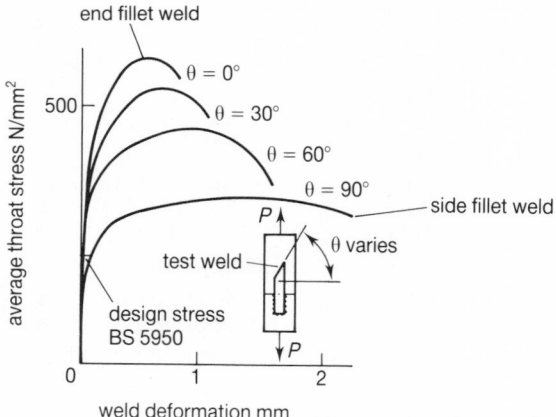

Fig. 7.5   Load-deformation relationships for an 8 mm fillet weld (Clarke 1970).

7.7   Conditions affecting the strength of welded connections

A structural designer should be aware of the following conditions which affect the strength of welded connections.

(a) Use of incorrect steel. The specification for weldable structural steels is given in BS 4360.

(b) Use of an incorrect electrode. Electrodes are related to the parent steel grade as specified in BS 639.

(c) There may be cavities and slag inclusions in the weld. These may be detected by non-destructive testing as described in Section 1.4.4.

(d) Excessive lack of fit between components.

(e) Stress concentrations combined with oscillating loads producing fatigue (see Sections 2.8 and 2.9).

(f) Residual stresses introduced from differential heating during welding (see Section 2.7).

(g) Hydrogen cracks associated with welding occurring when the cooling rate is too rapid (see Fig. 7.6(a)). Excessive hardening occurs in the heat affected zone which cracks under the action of residual stresses if sufficient hydrogen is present in the weld. This defect can be avoided by controlling the cooling and the hydrogen input to the weld as described in BS 5135.

(h) Lamellar tearing may occur when welding plate connections of the type shown in Fig. 7.6(b). Further examples are given in BS 5135. The cracks are produced by a combination of low ductility in the plate in the transverse direction and high joint restraint in the weld which induces tensile forces adjacent to the connection. The low ductility in the plate is produced by inclusions of non-metallic substances formed in the steel making process. When the ingot is rolled to make steel these inclusions form as plates parallel to the direction of rolling. Only a small percentage of plates are susceptible to lamellar tearing, and where it occurs joint details can be changed as recommended in BS 5135 to reduce the chances of it affecting the strength of the connection. Research into lamellar tearing has been carried out by Farrer and Dolby.

(i) Brittle fracture (see Section 2.6).

(j) Corrosion which reduces the size of components or causes pitting which may initiate fatigue cracks (see Section 2.5).

(k) Insufficient penetration of the parent metal which leads to a reduction in strength of the weld. The welder uses a voltage and arc length which produces a stable arc and a satisfactory weld profile. The current then becomes the main factor in controlling penetration. Another important factor in depth of penetration is edge penetration. Plates of 6 mm with square edges can be butt welded from one side, but the edges of thicker material must be bevelled to provide access for the arc.

(l) Lack of side wall fusion occurs if there is poor bond between the parent metal and the weld metal. Good bonding can only occur when the surface of the parent metal has been melted before the weld metal is allowed to flow into the joint.

Further information on faults in welds can be found in Gourd.

7.8 Design table for fillet welds

It is convenient for design purposes to prepare a table of the design strength of fillet welds per unit length for equal leg lengths commonly used in practice. The values given in Appendix A1 are calculated from

$$P_w = (\text{throat thickness}) \times (\text{unit length}) \times (\text{design shear stress})$$
$$= ap_w = 0.7sp_w \tag{7.3}$$

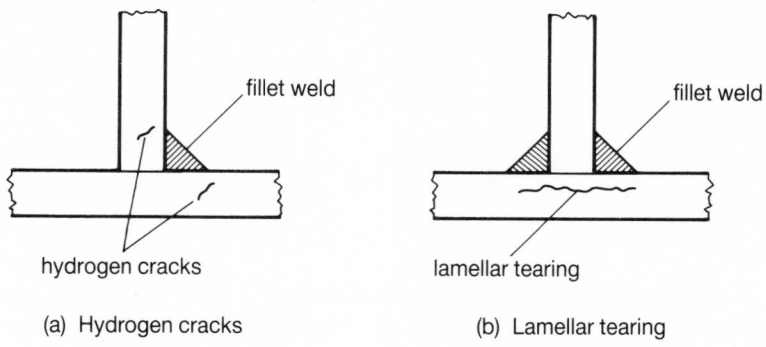

(a) Hydrogen cracks   (b) Lamellar tearing

Fig. 7.6   Faults associated with welding.

It should be noted that although the strength of the weld is calculated using the throat thickness *a* the weld is specified by the leg length *s*.

7.9   Types of bolt

The advantage of bolted connections is that they require less supervision than welding, and therefore are ideal for site conditions. Other advantages are that the connection can be fastened quickly, the connection supports load as soon as the bolts are tightened, and they accommodate minor discrepancies in dimensions.

Disadvantages of bolted connections are that for large forces the space required for the joint is extensive, and the connection is not as rigid as a welded connection even when friction grip bolts are used.

Steel bolts are identified by their gross diameter, strength (grade), and use. The preferred sizes of bolts in general use are 16, 20, 24, 30 and 36 mm diameter. The most common size used in structural connections is 20 mm.

The types of bolt in common use are:

(a)   International Standards Organization (ISO) metric black bolts grade 4.6 to BS 4190. This group also includes foundation bolts (cl 6.7, Pt 1).
(b)   ISO precision bolts, strength grade 8.8 to BS 3692.
(c)   High strength friction grip bolts to BS 4395. There are three grades:
   1   general grade–parallel shank
   2   higher grade–parallel shank, not generally used
   3   higher grade–waisted shank, expensive but may be re-used.

Detailed dimensions of bolts and washers are given in Appendix A. Grade 43 are used most commonly but they are gradually being replaced by grade 8.8. High strength friction grip bolts are used where 'rigid' fatigue resistant joints are required.

The meaning of the bolt grade figures for a grade 4.6 bolt is as follows. The $4 = f_u/10$ and the $6 = (Y_s/f_u) \times 10$. The yield strength of the bolt is obtained by multiplying the grade figures together, i.e. $4 \times 6 = 24$ kgf/mm^2 (approx. 235 N/mm^2).

The grade 4.6 bolt is low in cost, can be installed with the use of simple tools, and requires little supervision during the erection. At fracture the bolt has a relatively large extension of 25%, a property which is preferred in plastic collapse conditions.

Where forces are large, or where space for the connection is limited, or where erection costs can be reduced by using fewer bolts, then the higher grade 8.8 bolts are used. The percentage elongation of 12% at failure is less, but is still acceptable for design purposes.

Where a more rigid connection is required, e.g. in plastic methods of design, high strength friction grip (HSFG) bolts are used. The strength of HSFG bolts is equal to, or greater than that of grade 8.8, but there is an increased cost for the additional site supervision which is necessary to ensure that the bolts are axially pre-loaded in tension to the design values. The object of the preload is to ensure that the friction between the 'faying' surfaces prevents slip when subject to external shear forces. Sometimes HSFG bolts are used untensioned because of their greater strength compared with grade 8.8 bolts.

HSFG bolts are pre-loaded by tightening the nut with a torque wrench which is calibrated in relation to the required axial pre-load. Methods of tightening are specified in BS 4604. A simpler method of measuring the axial force in the bolt is to use load indicating washers under the head of the nut which reduce in thickness to a specified value for a specified pre-load. The washer is less accurate than the torque spanner according to Bahia and Martin (1981). A further alternative method of ensuring that the bolt is pre-loaded is to specify 'turns of the nut'. Investigations by Fisher and Struik showed that in general the pre-load produced by the torque wrench and turn of the nut method on site exceeded the specified value.

Close tolerance turned bolts are used only where accurate alignment of components or structural elements is required. The shank of the bolt is at least 2 mm greater in diameter than the threaded portion of the bolt and the hole is only 0.15 mm greater than the shank diameter. This small tolerance necessitates the use of special methods to ensure that all the holes align correctly.

Foundation bolts, or holding down bolts, are used for connecting structural elements to concrete pads or concrete foundations. Generally the bolts are cast into the concrete before erection of the steelwork and thus require accurate setting out. Where uplift forces occur the bolts must be anchored by a washer plate (cl 6.7, Pt 1). Most bolts used are grade 4.6 but higher strengths are available. Sometimes bolts are grouted in the holes during erection using epoxy resin.

Rivets were used extensively in the past in the fabrication shop and on site. They were difficult and expensive to place but they resulted in a rigid connection because the hot rivet, after driving, expanded to fill the hole. Rivets have now been superseded by welding and bolts.

7.10   Washers

In British practice most bolts have steel washers under the head and under the nut. The washer distributes the bolt force and prevents the nut, or bolt head, from damaging the component or member. However washers are not essential in all cases according to the European Convention for Structural Steelwork (ECCS), and they are now being omitted in British practice.

Washers for grade 4.6 and 8.8 bolts are specified in BS 4320 and an extract table is given in Appendix A2(c). Hardened washers for HSFG bolts are specified in BS 4395, Pt 1 and an extract table is given in Appendix A4(c). The outside diameter of a washer is an important dimension when detailing, e.g. to avoid overlapping an adjacent weld.

7.11   Bolt holes (cl 6.4.6, Pt 1)

Bolt holes are usually drilled, but may be punched full size, or punched undersize and reamed. Holes should never be formed by gas cutting because of the inaccuracy and the effect on the local properties of the steel.

Punched holes are preferred by steel fabricators because punching saves time and reduces cost. However research by Owens *et al.* shows that distortion in the vicinity of a punched hole reduces toughness and ductility and can lead to brittle fracture. Punched holes should not be used in locations where plastic tensile straining can occur.

Bolt holes are made larger than the bolt diameter to facilitate erection and to allow for inaccuracies. The clearance is 2 mm for bolts not exceeding 24 mm diameter and 3 mm for bolts exceeding 24 mm diameter. Oversize and slotted holes are allowable but not often used. Slotted holes are sometimes used for HSFG fasteners to facilitate erection with unusual shaped structures, or alternatively they can be used to accommodate movement in a structure. The clearance for a close tolerance turned bolt is 0.15 mm.

Bolt holes reduce the gross cross-sectional area of a plate to the net cross-sectional area. The net value is used for calculations where the structural element, or parts of an element, are in tension (see Section 6.2). Bolt holes also produce stress concentrations, but is is argued that these are offset by the fact that at yield the highly stressed cross section will work harden before fracture and yield will by then have occurred at adjacent cross sections. The gross cross section of a member is used in compression because at yield the bolt hole deforms and the shank of the bolt resists part of the load in bearing (see Section 6.2).

7.12   Spacing and edge distances for bolt holes (cl 6.2, Pt 1)

The longitudinal spacing between the centre line of bolts in the direction of the axial stress in a member is often called the pitch. The minimum spacing of   2.5 times the nominal diameter of the fastener,   is specified to prevent excessive reduction in the cross-sectional area of a number, to ensure that there is sufficient space to tighten the bolts, and to prevent overlapping the washers.

The maximum pitch is specified in cl 6.2.2, Pt 1 as follows.   The distances between two adjacent fasteners in a line lying in the direction of the stress should not exceed 14*t* where *t* is the thickness of the thinner element. Where the members are exposed to corrosive influences the maximum spacing of fasteners in any direction should not exceed 16*t* or 200 mm, where *t* is the thickness of the thinner outside ply.   These values are specified to prevent buckling of plates in compression between bolts, to ensure that bolts act together as a group to resist forces, and to prevent corrosion.

The minimum edge and end distances are shown in Table 7.2 below.

Table 7.2 Minimum edge and end distances for bolts (Table 31, cl 6.2.3, Pt 1)

Rolled, machine flame cut, sawn or planed edge	1.25 × dia. of hole
Sheared or hand flame cut and any end	1.4 × dia. of hole

The edge distance is from the centre of a hole to the nearest edge measured at right angles to the direction of the stress. The end distance is from the centre of a hole to the adjacent edge in the direction in which the fastener bears. The end distance should also be sufficient for bearing capacity. The edge and end distances are specified to resist corrosion and to provide sufficient space for the bolt head, washer and nut.

The maximum edge distance to the nearest line of fasteners from the edge of any unstiffened part should not exceed 11t. Where the members are exposed to corrosive influences the maximum edge distances should not exceed 40 mm + 4t (cl 6.2.4, Pt 1).

A summary of spacing and edge distances is given in Appendix A6(a). There are also recommended positions, spacing and diameter of holes in standard sections as shown in Appendix A6(b) which are extracts from the Structural Steel Handbook. The distances are based on providing sufficient clearances to the web and adequate edge distances.

7.13 Effective area at connections (cl 3.3.3, Pt 1)

The effective area, A_e, of each element of a member at a connection, where fastener holes occur may be taken as K_e times its nett area, but not more than its gross area, where for steel complying with BS 4360:

$K_e = 1.2$ for grade 40 or 43

$K_e = 1.1$ for grade 50 or WR 50

$K_e = 1.0$ for grade 55

for other steels, $K_e = 0.75 U_s / Y_s$ but < 1.2

where

U_s is the specified minimum ultimate tensile strength;

Y_s is the specified minimum yield strength.

7.14 Design strengths of single ordinary bolts (grades 4.6 and 8.8)

Bolted connections consist of two or more bolts and each bolt may be subject to any combination of tension, shear or bearing forces. The design strength of a single bolt when subject to these forces is now considered in detail.

7.14.1 Design tensile strength of ordinary bolts (cl 6.3.6, Pt 1)

This value is obtained from:

$$P_t = p_t A_t \tag{7.4}$$

where

p_t is the tensile design stress, 195 N/mm^2 for grade 4.6 bolts and 450 N/mm^2 for grade 8.8 bolts (Table 32, Pt 1);

A_t is the reduced area specified in BS 3692 and BS 4190.

Areas of all types of bolts are given in Appendices A2(b) and A4(b). The reduced area $A_t = (\pi/16) \times (\text{effective diameter} + \text{minor diameter})^2$.

A bolt in tension fails at the smallest cross section, i.e. the root of the threads where the net area is approximately 80% of the gross area. The design tensile stress of 195 N/mm^2 for an ordinary grade 4.6 bolt is obtained by dividing the specified minimum yield stress of 235 N/mm^2 (see BS 4190) by a materials factor of 1.2. There is no well-defined yield point for grade 8.8 bolts and therefore the design stress is defined as $0.58 \times$ (specified ultimate tensile stress) $= 0.58 \times 785 = 455 \text{ N/mm}^2$. The vaue given in Table 32, Pt 1 is 450 N/mm². The specified ultimate tensile stress is given in Table 10, BS 3692.

7.14.2   Design shear strength of ordinary bolts (cl 6.3.2, Pt 1)

This value is obtained from:

$$P_s = p_s A_s \qquad (7.5)$$

where

p_s is the design shear stress, 160 N/mm² for grade 4.6 bolts and 375 N/mm² for grade 8.8 bolts (Table 32, Pt 1);

A_s is the reduced area specified in BS 3692 and BS 4190 [see Appendices A2(b) and A4(b)].

Shearing of a bolt can occur on the shank, i.e. the gross area of the bolt, if the thread length on a bolt is carefully specified, but it is safer to assume that it occurs on the reduced area. It also simplifies calculations and avoids confusion.

The design shear strength, P_s is reduced for large grips (see Section 7.16) and for long joints (see Section 7.17).

Where there is a well-defined tensile yield stress, as for grade 4.6 bolts, the design shear stress in BS 5950 is obtained by multiplying by a factor of 0.69. This value is high when compared with experimental evidence and the need to introduce a materials factor. Experiments by Bahia and Martin (1980) and other investigators have found values that vary between 0.62 and 0.71. The value from the Huber-von Mises-Hencky shear distortion strain energy theory is 0.57.

Ordinary bolts deform when subjected to shear stresses but it is important to realize that the shear deformation of the connection is increased by the bearing stresses on the plate. The higher the bearing stresses the greater the deformation as shown in Fig. 7.7.

7.14.3   Design strength of ordinary bolts subjected to shear and tension forces (cl 6.3.6.3, Pt 1)

This value is obtained from the interaction formulae:

$$F_s/P_s + F_t/P_t \leqslant 1.4 \qquad (7.6a)$$

$$F_s/P_s \leqslant 1 \qquad (7.6b)$$

$$F_t/P_t \leqslant 1 \qquad (7.6c)$$

where

F_s is the applied shear;

F_t is the applied tension;

P_s is the shear capacity (see Section 7.14.2);

P_t is the tensile capacity (see Section 7.14.1).

These tri-linear equations are to be compared with experimental results obtained by Chesson *et al.*, based on the nett cross-sectional area of the bolt, and shown in Fig. 7.8. An alternative elliptical relationship is recommended in ECSS and BS 5400, Pt 3.

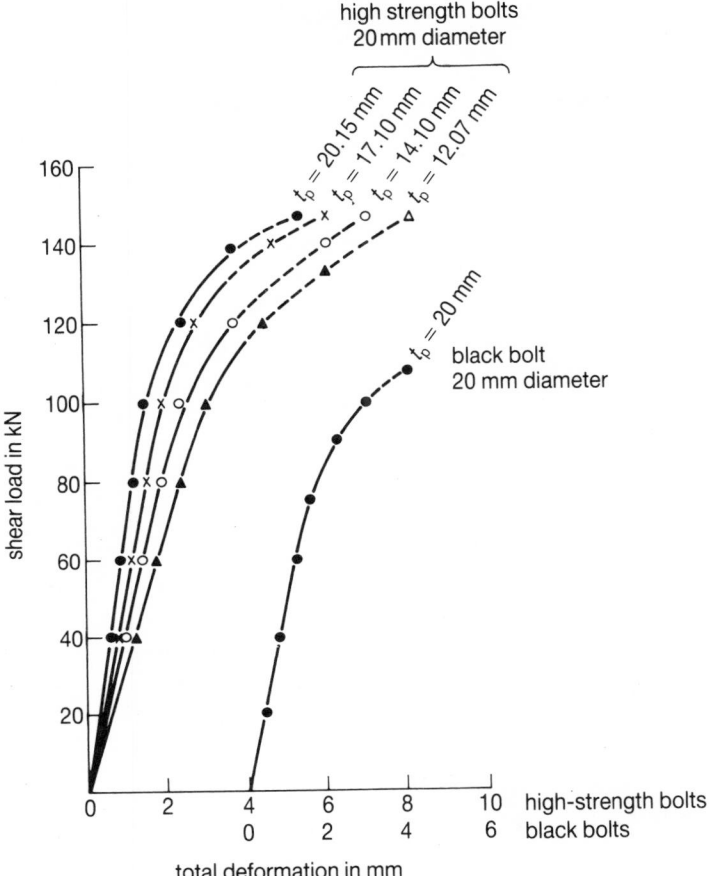

Notes: Permanent deformation of holes shown as a broken line;
bolts failed in single shear across the threads; t_p is
the plate thickness.

Fig. 7.7 Relation between shear load and deformation for single bolt tests (Bahia and Martin 1980).

7.14.4 Design bearing strength for ordinary bolts subjected to shear forces (cl 6.3.3, Pt 1)

An ordinary bolt subjected to shear forces, such as shown in Fig. 7.9, comes in contact with the plate when the shear load is applied and slip occurs. The bearing stresses between the bolt and plate need to be controlled.

The bolt may deform because of high local bearing stresses between the bolt and plate, and consequently it is necessary to calculate the design bearing capacity of the bolt from (cl 6.3.3.2, Pt 1).

$$P_{bb} = dtp_{bb} \tag{7.7}$$

where

d is the nominal diameter of the bolt;

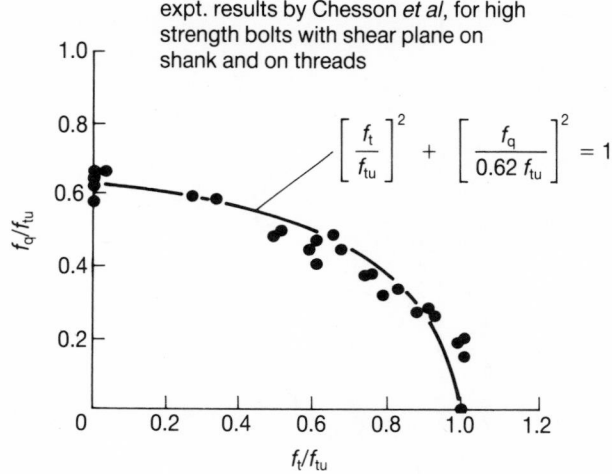

Fig. 7.8   Relationship between shear and tensile stresses for bolts.

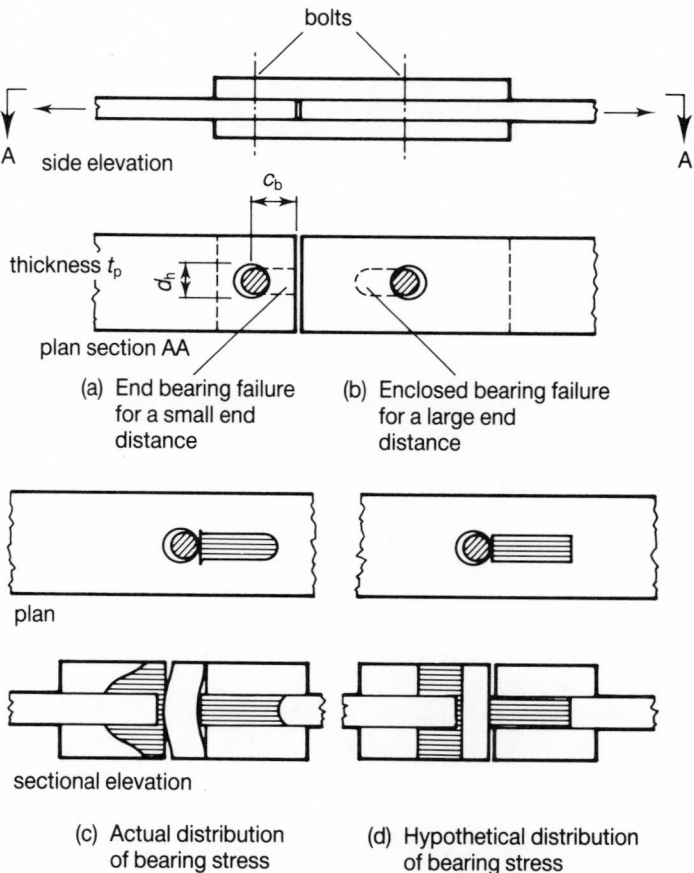

Fig. 7.9   Bearing stresses for ordinary bolts.

t is the thickness of the connected ply, or if the bolts are countersunk, the thickness of the ply minus half the depth of the countersinking;

p_{bb} is the design bearing stress for a bolt obtained from Table 32, Pt 1 (460 N/mm^2 for grade 4.6 bolts and 970 N/mm^2 for grade 8.8 bolts).

Alternatively the small plate may deform because of high local bearing stresses between bolt and plate which elongate the hole. It is therefore necessary to calculate the design bearing capacity of the plate from (cl 6.3.3.3, Pt 1)

$$P_{bs} = dtp_{bs} \tag{7.8}$$

where

d is the nominal diameter of the bolt;

t is the thickness of the connected ply, or if the bolts are countersunk, the thickness of the ply minus the depth of the countersinking;

p_{bs} is the design bearing stress for the connected ply obtained from Table 33, Pt 1 (460 N/mm^2 for grade 43 steel, 550 N/mm^2 for grade 50 steel, and 650 N/mm^2 for grade 55 steel).

A further alternative is that if the bolt hole is close to the end of the plate, the bolt may shear through the end of the plate as shown in Fig. 7.9(a). This can be expressed using the shear strength of the plate as $P_s = 2etp_s$, or if the shear stress is related to the bearing stress ($p_s \approx p_{bs}/4$) then (cl 6.3.3.3, Pt 1)

$$P_{bs} < 0.5etp_{bs} \tag{7.9}$$

where

e is the end distance as defined in cl 6.2.3, Pt 1 (see Appendix A6(a));

t is the thickness of the connected ply, or if the bolts are countersunk, the thickness of the ply minus half the depth of the countersinking;

p_{bs} is the design bearing stress for the connected ply obtained from Table 33, Pt 1 (460 N/mm^2 for grade 43 steel, 550 N/mm^2 for grade 50 steel, and 650 N/mm^2 for grade 55 steel).

Equations (7.7) to (7.9) assume uniform bearing stresses as shown in Fig. 7.9(d) whereas in reality they are closer to those shown in Fig. 7.9(c). It should be noted that the design bearing stresses are high in relation to the yield stress. This is because material subject to bearing stresses is generally confined by other material which restricts deformation. High bearing stresses are not disastrous but lead to excessive deformation of a connection as shown in the experimental results by Bahia and Martin.

7.15 Design tables for ordinary bolts (grade 4.6 and 8.8)

Equations (7.4) to (7.9) express the design tensile, shear and bearing strengths of a bolt and can be presented in the form of tables to avoid repeating these calculations.

The minimum end bearing distance $e = 1.25d$, but this is not often used in practice. If $e = 2d$ then the thickness required for bearing and end bearing is the same. Design values are given in Appendix A2(a) and A3. Note that the plate thickness is determined from the lower bearing stress, i.e. the lower value of the bolt or the plate.

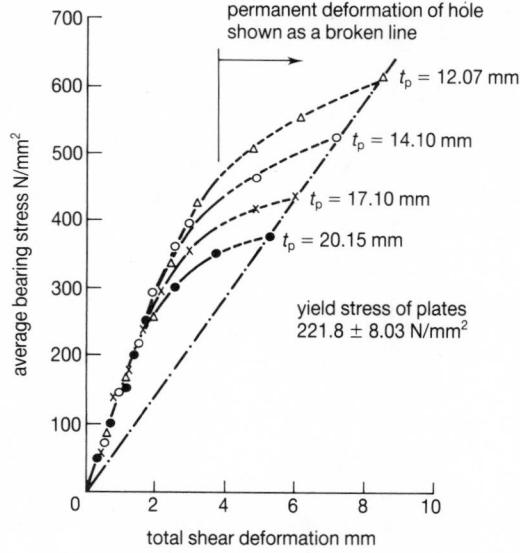

Note: Bolts failed on the threads in single shear at a
shear stress of 600 N/mm²

Fig. 7.10   Relationship between bearing stress and deformation for an M20 high stress single bolt
(Bahia and Martin 1980).

7.16   Large grip lengths for ordinary bolts (cl 6.3.5, Pt 1)

When the total thickness of the connected plies (grip length) for an ordinary bolted joint
exceeds five times the nominal diameter of the bolts the shear capacity is reduced to

$$P_s = p_s A_s 8d/(3d + T_g) \tag{7.10}$$

where

d is the nominal diameter of the bolt;

A_s is the shear area;

p_s is the design shear strength of a bolt;

T_g is the grip length of the plies.

The value for the large grip length calculated using Equation (7.10) should not exceed the
value calculated for long joints (Equation (7.11)), if applicable.

The reason for the reduction in strength is that as the grip length increases the bolt is
subject to greater bending moments from shear forces which move further apart.

7.17   Long joints for ordinary bolts (cl 6.3.4, Pt 1)

When the joint length of a splice or end connection, in tension or compression, exceeds
500 mm the shear capacity is reduced to

$$P_s = p_s A_s (5500 - L_j)/5000 \tag{7.11}$$

where

d is the nominal diameter of the bolt;

A_s is the shear area;

p_s is the design shear strength of a bolt;

L_j is the joint length (see Fig. 7.11).

The value for long joints calculated using Equation (7.11) should not exceed the value calculated for large grips (Equation (7.10)), if applicable

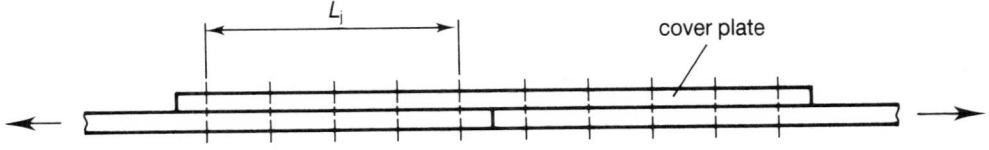

Fig. 7.11   Length (L_j) for long joints.

The reason for the reduction in strength is because the shear force is not evenly distributed to the bolts. The bolts at the end of the joint resist the highest shear forces.

7.18   Prying forces (cl 6.3.6.2, Pt 1)

A prying force is an additional axial tensile force that is induced in a bolt due to the flexing of a component. The conditions necessary for the development of prying forces in a simple tee stub are shown in Fig. 7.12(a). The prying force will develop only when the ends of the flanges are in contact as shown in (ii) and (iii). The plastic hinges do not always form before bolt failure. An example of a prying force in practice is shown in Fig. 7.12(b). The development of a prying force as the external force is applied to a tee stub is shown in Fig. 7.12(c).

Cl 6.3.6.2, Pt 1 states that the prying force need not be taken into account provided that the recommended design strengths of bolts are used. This is because the recommended design strengths are conservative and will accommodate an additional prying force.

If a prying force needs to be calculated then a value can be determined from

$$Q_{be} = [F_{be} - 2EI\delta_b/(a_p b_p^2)]/[2a_p/b_p + (1/3)(a_p/b_p)^2] \tag{7.12}$$

where

$$I = wt^3/12$$

$$\delta_b = (g/A_b)(F_{be} + Q_{be} - P_0)$$

This expression is derived (see Holmes and Martin) from assuming elastic behaviour of the plate and failure of the bolt in tension (see Fig. 7.12(d)). The assumption of elastic behaviour is acceptable because the prying force reduces the bending moment at the plastic hinge at the support and also because the recommended design strengths in BS 5950 for bolts are in the range of elastic behaviour. It should be noted from Equation (7.12) that the prying force is increased by reducing the ratio of (a/b) and by reducing δ which occurs with an HSFG bolt. Equation (7.12) has been shown to agree with the experimental results of Bahia *et al.* (1981).

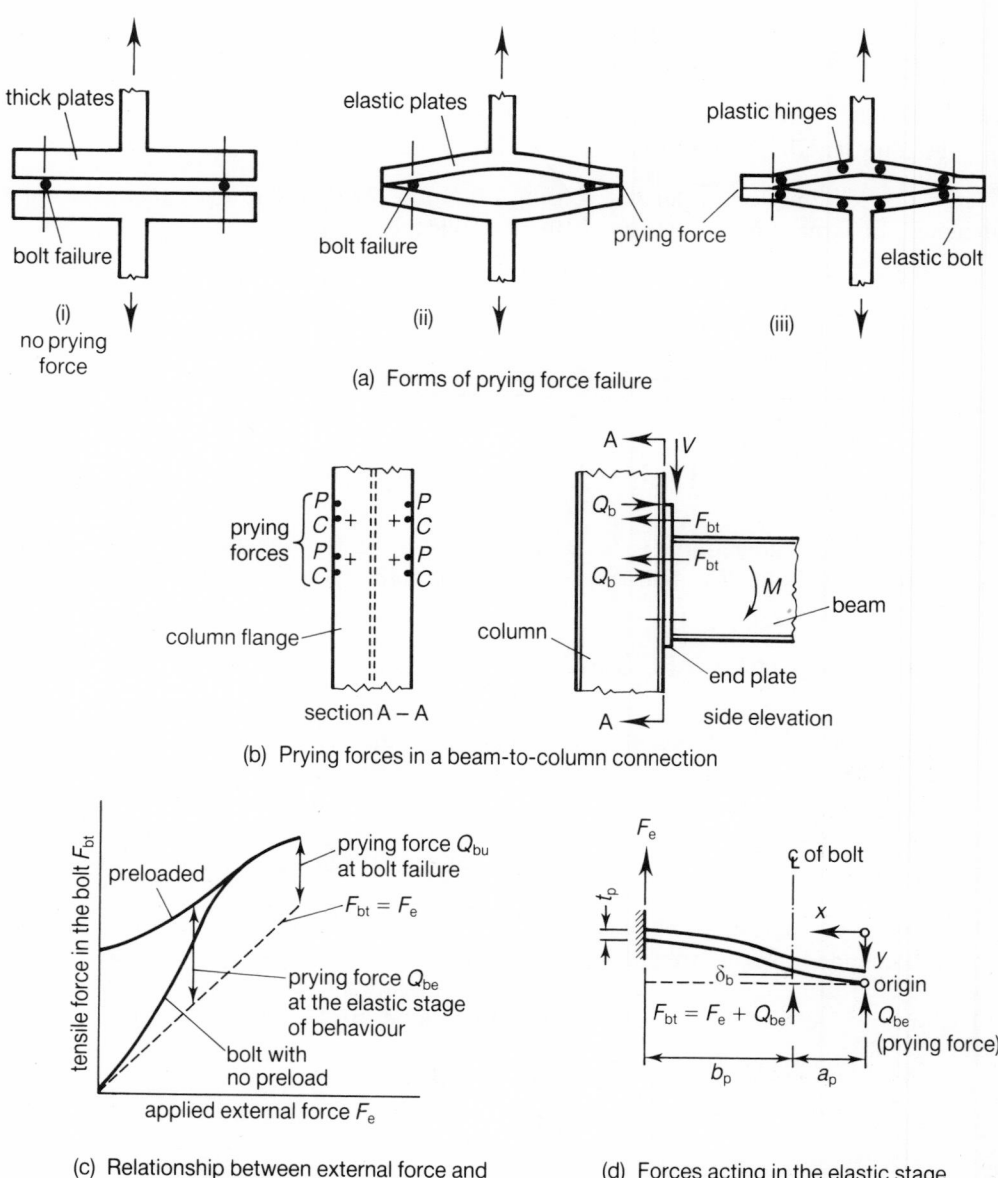

(a) Forms of prying force failure

(b) Prying forces in a beam-to-column connection

(c) Relationship between external force and
     bolt force for a tee stub

(d) Forces acting in the elastic stage
     for prying force theory

Fig. 7.12   Prying forces.

7.19   Design strengths of friction grip bolts (cl 6.4, Pt 1)

Friction grip bolts are subdivided into parallel shank and waisted shank types. A parallel shank bolt, which is the most commonly used, is designed not to slip at serviceability load but slips into bearing at ultimate load. A waisted shank bolt is of greater strength and is designed not to slip at service load and ultimate load. It is therefore more rigid at ultimate load and need not be checked for bearing or long joint capacity (cl 6.4.1, Pt 1).

The design strength of a single friction grip bolt subject to tensile, shear and bearing forces is now considered in detail.

7.19.1   Design tensile strength of parallel and waisted shank bolts (cl 6.4.4.2, Pt 1)

The design tensile strength of parallel shank and waisted shank friction grip bolts is

$$P_t = 0.9P_0 \tag{7.13}$$

where

P_0 is the minimum shank tension (or proof load) as specified in BS 4604 and also given in BS 4395 Pts 1 and 3 (see Appendix A4(a) and A5(a)).

A fastener complying with BS 4395 Pt 2 should not be subjected to an externally applied tension.

Bolt tensions in practice are reduced by movement on the threads, Poisson's ratio effect on the plates, and surface deformation on the plates, but nevertheless remain effective.

7.19.2   Slip resistance of parallel and waisted shank bolts (cls 6.4.2 and 6.4.3, Pt 1)

The design slip resistance of parallel shank and waisted shank friction grip bolts is

$$P_{sL} = kK_s\mu P_0 \tag{7.14}$$

where

P_0 is the minimum shank tension as specified in BS 4604 (see Appendix A4(a));

$k = 1.1$ for a parallel shank bolt and 0.9 for a waisted shank bolt;

$K_s = 1$ for a fastener in a clearance hole;
   $= 0.85$ for a fastener in an oversized and slotted hole, and for a fastener in a long slotted hole loaded perpendicular to the slot;
   $= 0.6$ for a fastener in a long slotted hole loaded parallel to the slot;

μ is the slip factor and is taken as 0.45 for a general grade fastener complying with BS 4395 Pt 1 in connections for untreated surfaces as specified in BS 4604.

Equation (7.14) expresses a serviceability requirement for parallel shank bolts but has been factored using 1.1 so that it may be compared with ultimate limit state requirements. The design slip resistance, P_{sL}, for parallel shank friction grip bolts is reduced for long joints (see Section 7.21) but not for waisted shank friction grip bolts.

Table 7.3   Slip factors for friction grip fasteners (BS 5400)

Faying surface conditions	Slip factor
Weathered and clear or mill scale and loose rust	0.45
Blasted with shot or grit and loose rust removed	0.50
Sprayed with aluminium	0.5
Sprayed with zinc	0.40
Treated with zinc silicate paint	0.35
Treated with etch primer	0.25

The coefficient of friction (or slip factor) is obtained from tests as specified in BS 4604 and is the average for the number of bolts on test. The frictional coefficient μ varies with the condition of the faying surface and Table 7.3 gives values for some surfaces.

7.19.3   Design strength of parallel shank and waisted shank friction grip bolts subject to shear and tension forces (cl 6.4.5, Pt 1)

The design strength of parallel shank and waisted shank friction grip bolts subject to a combination of shear and tension forces is obtained from the interaction formulae:

$$F_s/P_{sL} + 0.8F_t/P_t \leqslant 1 \qquad\qquad (7.15a)$$

$$F_s/P_{sL} \leqslant 1 \qquad\qquad (7.15b)$$

$$F_t/P_t \leqslant 1 \qquad\qquad (7.15c)$$

where

F_s is the applied shear;
F_t is the external applied tension;
P_s is the shear capacity;
P_t is the tensile capacity.

The factor of 0.8 is introduced to allow for the fact that the minimum shank tension may not be achieved.

7.19.4   Design bearing strength of parallel shank friction grip bolts subject to shear forces (cl 6.4.2.2, Pt 1)

The design bearing strength of parallel shank friction grip bolts subjected to shear forces is obtained as follows.

A parallel shank friction grip bolt slips into bearing at the ultimate limit state when subject to shear forces and the bearing stresses between the bolt and plate need to be controlled. At this stage the frictional resistance is low due to plasticity and damage to the frictional surfaces.

The bolt may deform because of high local bearing stresses between bolt and plate, and consequently it is necessary to calculate the design bearing capacity of the plate from

$$P_{bg} = dtp_{bg} \qquad\qquad (7.16)$$

where

d is the nominal diameter of the bolt;
t is the thickness of the connected ply, or if the bolts are countersunk, the thickness of the ply minus half the depth of the countersinking;
p_{bg} is the design bearing stress for the plate obtained from Table 34, Pt 1 (460 N/mm^2 for grade 43 steel, 550 N/mm^2 for grade 50 steel, and 650 N/mm^2 for grade 55 steel).

Note that there is no check on the bearing strength of the bolt because its bearing strength is greater than the plates.

An alternative mode of failure is that if the bolt hole is close to the end of the plate, the bolt may shear through the end of the plate as shown in Fig. 7.9. As in Section 7.14.4 this may be expressed in terms of the bearing stress

$$P_{bg} < 0.333etp_{bg} \qquad (7.17)$$

where

e is the end distance as defined in cl 6.2.3, Pt 1;

t is the thickness of the connected ply;

p_{bg} is the design bearing stress for the connected ply obtained from Table 34, Pt 1 (460 N/ mm^2 for grade 43 steel, 550 N/mm^2 for grade 50 steel, and 650 N/mm^2 for grade 55 steel).

7.20   Design tables for friction grip bolts

Equations (7.13) to (7.17) express the design tensile, shear and bearing strengths of a parallel shank and waisted shank bolts and can be presented in the form of a table (see Appendix A4(a) and A5(a)) to avoid repeating these calculations. The end distance for a parallel shank friction grip bolt can be as small as $1.25d$ (see Appendix A6(a)), but the value adopted to construct the table is $3d$. This means that the plate thickness is the same for bearing and end bearing.

7.21   Long joints for parallel shank friction grip bolts (cl 6.4.2.3, Pt 1)

When the joint length of a splice or end connection, in tension or compression, exceeds 500 mm the shear capacity is reduced to

$$P_{sL} = 0.6P_0(5500 - L_j)/5000 \qquad (7.18)$$

where

P_0 is the minimum shank tension (or proof load) as specified in BS 4604;

L_j is the joint length (see Fig. 7.11);

The value obtained from Equation (7.18) should not exceed the values for slip and bearing.

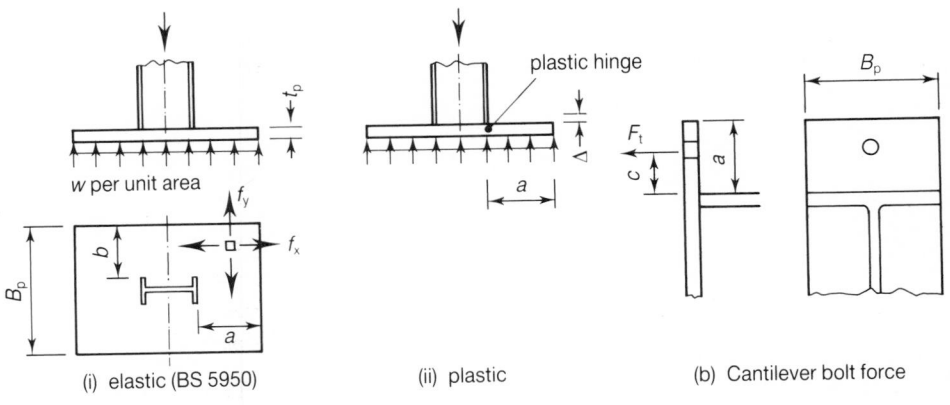

(a) Column base plate

(i) elastic (BS 5950)   (ii) plastic   (b) Cantilever bolt force

Fig. 7.13 Bending strength of plates.

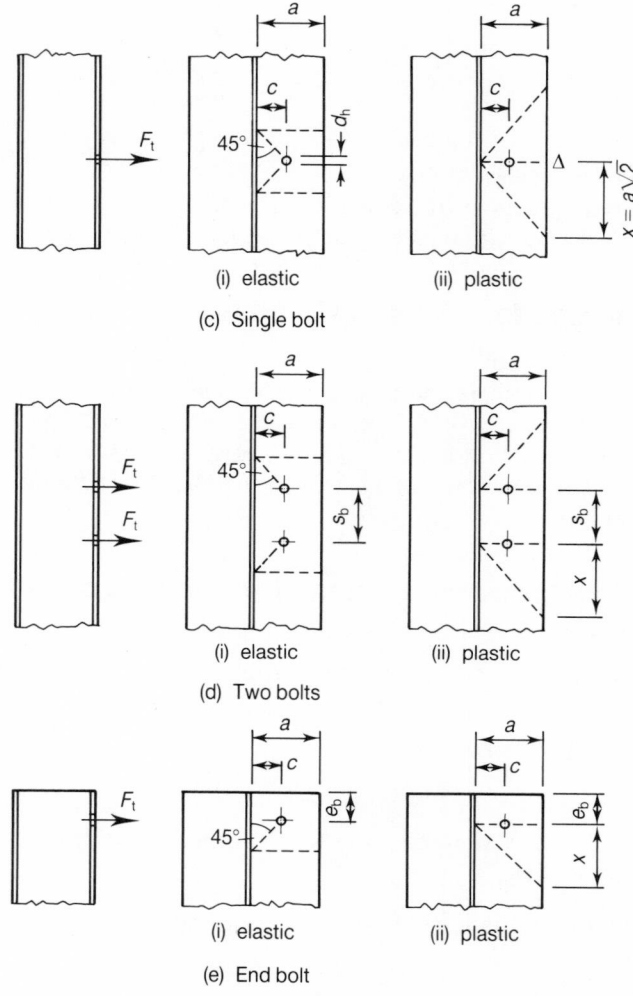

Fig. 7.13 (c–d).

The reason for the reduction in strength for a long joint is because the shear force is not evenly distributed to the bolts, and consequently the bolts at the end of the joint resist the highest shear forces.

7.22   Plate thicknesses for components (cl 4.13, Pt 1)

Plates, or parts of sections acting as plates, often form part of structural connections. The length and breadth of plates are generally determined from the geometry of the connection but the thickness is calculated from the elastic or plastic theory of bending. Examples are given in Figs. 7.13 and 14.

7.22.1 Elastic methods

Elastic analysis using an enhanced design stress of $1.2p_{yp}$ is suggested in BS 5950 for some of the situations shown in Fig. 7.13, and can be applied to other situations. The value of p_{yp} is the design strength from Table 6, Pt 1 but should not be greater than 270 N/mm^2. The use of elastic analysis at ultimate is to ensure that large displacements do not occur and that pressures between components are more evenly distributed.

(a) Plates are used as a base to distribute the load from column sections as shown in Fig. 7.13(a). If the load is axial the pressure beneath the base plate w is uniform. The following theory leads to the equation given in cl 4.13.2.2, Pt 1.

Consider an element of the steel plate adjacent to the edge of the column flange as shown in Fig. 7.13(a). This element is subjected to mutually perpendicular bending stresses f_x and f_y associated with the cantilever lengths a and b. If a is the greater dimension and the maximum strain in the x direction is limited to $1.2p_{yp}/E_s$ then from the assumption of elastic behaviour

$$f_x/E_s - vf_y/E_s = 1.2p_{yp}/E_s \tag{i}$$

where

v is the Poisson's ratio for steel.

From the simple theory of bending for a unit width of steel base plate

$$f_x = M/Z = (wa^2/2)/(t_p^2/6) \tag{ii}$$

and

$$f_y = M/Z = (wb^2/2)/(t_p^2/6) \tag{iii}$$

Combining Equations (i), (ii) and (iii) and rearranging, the thickness of the plate

$$t_p = \sqrt{[2.5w(a^2 - vb^2)/p_{yp}]} \tag{7.19}$$

Poisson's ratio v is given as 0.3 in cl 3.1.2, Pt 1.

Where the base pressure is eccentric to the plate, or for non-rectangular plates, then for a cantilever from the simple theory of bending (cl 4.13.2.3, Pt 1)

$$M = pZ$$

$$wB_p a^2/2 = (1.2p_{yp})B_p t_p^2/6$$

rearranging, the thickness of the plate

$$t_p = a\sqrt{(2.5w/p_{yp})} \tag{7.20}$$

This expression is Equation (7.19) with $b = 0$ and will obviously give a larger value of the plate thickness.

(b) For a single bolt force F_t acting on a cantilever plate of finite width B_p (see Fig. 7.13(b))

$$M = pZ$$

$$F_t c = (1.2p_{yp})B_p t_p^2/6$$

rearranging, the thickness of the plate

$$t_p = \sqrt{[5F_t c/(p_{yp}B_p)]} \tag{7.21}$$

(c)   No recommendations are given in BS 5050 for a column flange subjected to a single bolt force as shown in Fig. 7.13(c). The effective width of the plate is not defined as in the previous case but a solution is to assume a 45° dispersion from the edge of the bolt hole. From the simple theory of bending

$$M = pZ$$

$$F_t c = (1.2 p_{yp})(d_h + 2c)t_p^2/6$$

rearranging, the thickness of the plate

$$t_p = \sqrt{\{5F_t/[p_{yp}(d_h/c + 2)]\}} \qquad (7.22)$$

(d)   Similarly for two bolts where $s_b < d_h + 2c$ as shown in Fig. 7.13(d)

$$M = pZ$$

$$2F_t c = (1.2 p_{yp})(d_h + 2c + s_b)t_p^2/6$$

rearranging, the thickness of the plate

$$t_p = \sqrt{\{10F_t/[p_{yp}(d_h/c + s_b/c + 2)]\}} \qquad (7.23)$$

(e)   Similarly for a single bolt near the end of a member shown in Fig. 7.13(e)

$$M = pZ$$

$$F_t c = (1.2 p_{yp})(e_b + d_h/2 + c)t_p^2/6$$

rearranging, the thickness of the plate

$$t_p = \sqrt{\{5F_t/[p_{yp}(e_b/c + d_h/2c + 1)]\}} \qquad (7.24)$$

(f)   The gusset plate shown in Fig. 7.14(b) is assumed to be subjected to a bending moment which is resisted at section x–x (cl 4.13.2.4, Pt 1). From the simple theory of elastic bending

$$M = pZ$$

$$P_g s_g = (1.2 p_{yp})t_g H_g^2/6$$

rearranging, the thickness of the plate

$$t_g = 5P_g s_g/(p_{yp} H_g^2) \qquad (7.25)$$

This theory can be unsafe especially where H_g/t_g is large. A gusset plate buckles at failure as shown by Martin and therefore the buckling theory in Section 7.22.2(f) is more appropriate.

7.22.2   Plastic methods

Alternatively plastic bending methods of analysis can be used to determine the thickness plates for the situations shown in Fig. 7.13. Plastic methods of analysis are not excluded from BS 5950 (see cls 5.3.7 and 6.1.4, Pt 1) and give similar theoretical expressions and values of plate thickness.

(a)   If a single yield line is assumed for the situation in Fig. 7.13(a) where the width of plate B_p is defined, then by virtual work

$$waB_p\Delta/2 = (B_p t_p^2 p_{yp}/4)\Delta/a$$

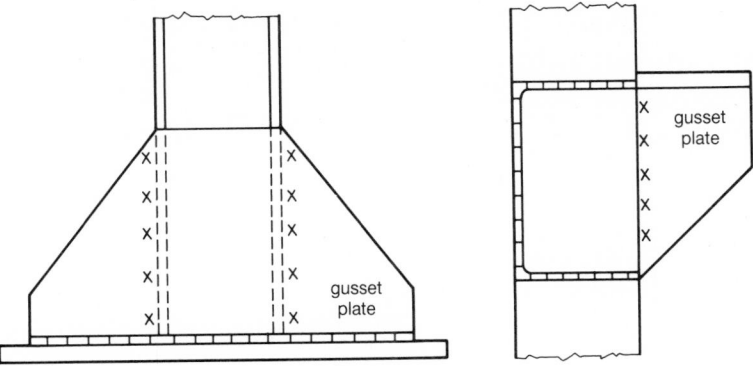

(a) Examples of gusset plates

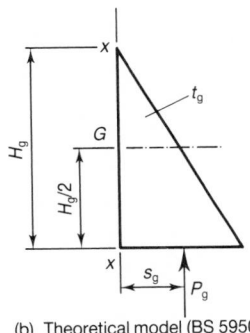

(b) Theoretical model (BS 5950)

Fig. 7.14 Strength of gusset plates.

rearranging, the thickness of the plate

$$t_p = a\sqrt{(2w/p_{yp})} \tag{7.26}$$

(b) If a single yield line is assumed for the situation in Fig. 7.13(b) where the width of the plate is again defined as B_p, then by virtual work

$$F_t c\Delta/a = (B_p t_p^2 p_{yp}/4)\Delta/a$$

rearranging, the thickness of the plate

$$t_p = \sqrt{[4F_t c/(B_p p_{yp})]} \tag{7.27}$$

(c) For a single bolt force assume yield lines as shown in Fig. 7.13(c) then from virtual work

$$F_t c\Delta/a = (t_p^2 p_{yp}/4)\Delta[4a/x + 2x/a] \tag{i}$$

Differentiating with respect to x to determine the value of x for which F_t is a minimum

$$\partial F_t/\partial x = -4a/x^2 + 2/a = 0; \quad \text{hence} \quad x = a\sqrt{2} \tag{ii}$$

Combining Equations (i) and (ii) and rearranging, the thickness of the plate

$$t_p = \sqrt{(F_t c/\sqrt{2}a p_{yp})} \tag{7.28}$$

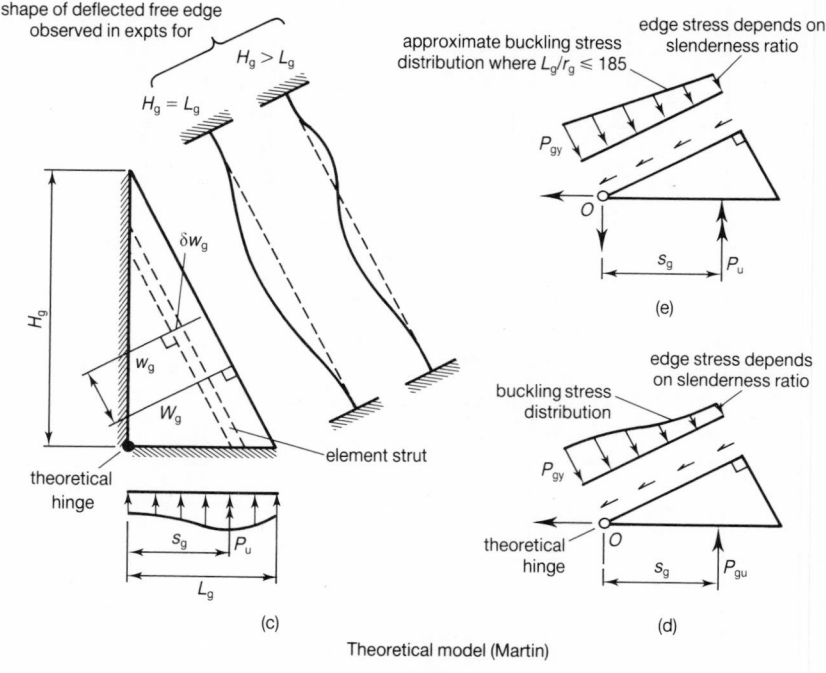

Fig. 7.14   (c–d).

(d)   For the thickness of column flanges when subjected to two bolt forces as shown in Fig. 7.19(d) multiple yield lines are assumed. From virtual work

$$2F_t c\Delta/a = (t_p^2 p_{yp}/4)\Delta[4a/x + (s_b + 2x)/a] \qquad \text{(i)}$$

Differentiating with respect to x to determine the value of x for which F_t is a minimum

$$\partial F_t/\partial x = -4a/x^2 + 2/a = 0; \quad \text{hence} \quad x = a\sqrt{2} \qquad \text{(ii)}$$

Combining Equations (i) and (ii) and rearranging, the thickness of the plate

$$t_p = \sqrt{[(8F_t c/ap_{yp})/(4\sqrt{2} + s_b/a)]} \qquad \text{(7.29a)}$$

The general expression for n bolts in line is

$$t_p = \sqrt{[(4nF_t c/ap_{yp})/(4\sqrt{2} + (n-1)s_b/a)]} \qquad \text{(7.29b)}$$

To reduce t_p, reduce F_t and make s_b/a as large as is practical.

(e)   For a bolt close to the end of a member (see Fig. 7.13(e)) the work equation is

$$F_t c\Delta/a = (t_p^2 p_{yp}/4)\Delta[a/x + (e_b + x)/a] \qquad \text{(i)}$$

Differentiating with respect to x to determine the value of x for which F_t is a minimum

$$\partial F_t/\partial x = -a/x^2 + 1/a = 0; \quad \text{hence} \quad x = a \qquad \text{(ii)}$$

Combining Equations (i) and (ii) and rearranging, the thickness of the plate

$$t_p = \sqrt{[(4F_t c/a p_{yp})/(2 + e_b/a)]} \tag{7.30}$$

The above theory using plastic analysis assumes simple idealized conditions and ignores intersecting plates, welds, washers, limited rotation at the plastic hinge, and strain hardening. However the examples show the basic approach. More complicated expressions related to research can be found in Stark and Bijlaard.

(f) The following theory for the buckling of a gusset plate is based on experimental work by Martin, and Martin and Robinson.

The basic structural unit is a triangular plate with loading applied to one edge as shown in Fig. 7.14(c). For theoretical purposes the plate is assumed to be composed of a series of fixed ended struts parallel to the free edge. The distribution of direct stress across the width, W_g is shown on an element of the gusset plate in Fig. 7.14(d). The buckling stress varies depending on the slenderness ratio of the elemental strut. At the hinge the stress is for a very short strut, i.e. yield. At the free edge the value is for the slenderness ratio of the elemental strut at the free edge.

For simplicity the buckling stress distribution shown in Fig. 7.14(c) can be replaced by a linear distribution as shown in Fig. 7.14(e), provided that the slenderness ratio of the free edge $l_g/r_g < 185$. This restraint is acceptable because slenderness ratios of gusset plates in structural engineering do not often exceed this value.

Taking moments of forces about the theoretical hinge at O (see Fig. 7.14(e)) and ignoring the moment of resistance of the base plate as justified by Martin and Robinson.

$$P_u s_g = \int_0^w w_g f_g t_g \, dw_g \tag{7.31}$$

For each strip the buckling stress f_g is linearly related to the slenderness ratio l_g/r_g. The effective length $l_g = w_g$ when $L_g = H_g$, and from experiments Martin found this to be approximately correct when $L_g \neq H_g$. The buckling stress for each strip can therefore be expressed as

$$f_g = p_{yg}[1 - (w_g/(185 r_g))] \quad \text{provided that} \quad l_g/r_g < 185 \tag{7.32}$$

Combining Equations (7.31) and (7.32), integrating, expressing the radius of gyration as $r_g = t_g/(2\sqrt{3})$, and rearranging

$$t_g = 2P_u s_g/(p_{yg} W_g^2) + W_g/80 \tag{7.33}$$

where from the geometry of the plate

$$W_g = L_g/[(L_g/H_g)^2 + 1]^{1/2} \tag{7.34}$$

The slenderness ratio of the gusset plate may be defined as the slenderness ratio of a strip of unit width parallel to the free edge. From this definition and Equation (7.34)

$$\frac{l_g}{r_g} = \frac{W_g}{t_g/(2\sqrt{3})} = \frac{2\sqrt{3}}{[(L_g/H_g)^2 + 1]^{1/2}}\left(\frac{L_g}{t_g}\right) \tag{7.35}$$

This theory is for non-slender gusset plates, i.e. for $l_g/r_g < 185$. The theory for slender gusset plates is given in Holmes and Martin.

7.23   Connections subject to simple shear forces (cl 6.1.3, Pt 1)

Some simple connections subject to shear forces are shown in Fig. 7.15. The forces in the members are assumed to be axial to act through the centroidal axes of the members. This is true in some situations, e.g. the joint shown in Fig. 7.15(a). It is not correct for the welded lap connection shown in Fig. 7.15(b) because the eccentricity of the force produces a moment which results in distortion at ultimate load. It is not correct for a roof truss connection as shown in Fig. 7.20 because although the centroidal axes intersect and there are axial forces in the members there are also secondary moments.

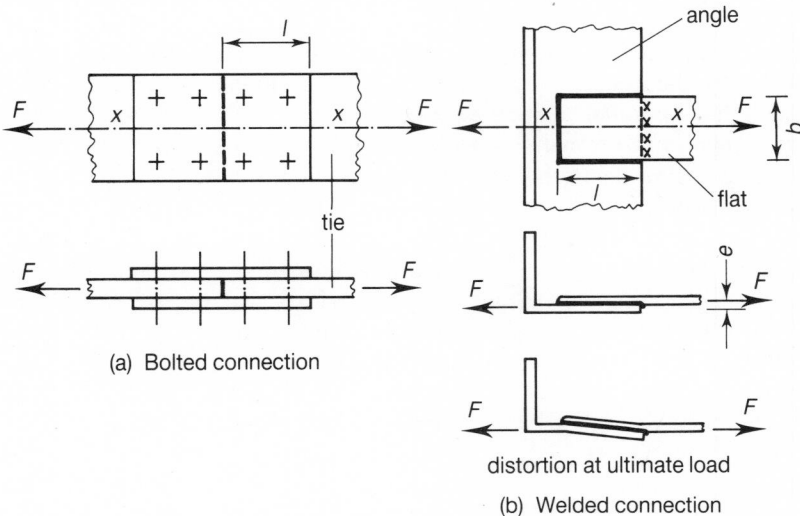

(a) Bolted connection

distortion at ultimate load

(b) Welded connection

Fig. 7.15   Connections subject to shear forces.

A further assumption for simple connections is that the external forces are distributed evenly to the bolts or welds. This is not correct for long bolted joints as explained in Sections 7.17 and 7.21 and allowance must be made for this.

The overlap distance l is important for simple connections. For bolted joints the minimum of two bolts and the required end distances generally ensure that the lap is sufficient. However for welds (see Fig. 7.15) the greater strength may indicate that the lap distance can be small, but it must be appreciated that there must be room for stop and start lengths and that stress concentrations can occur. The minimum lap length should not be less than four times the thickness of the thinner part joined where the weld is continuous. For a connection with side welds only the lap should not be less than the width of the member (cl 6.6.2.2, Pt 1) and there should be end returns of twice the leg length of the weld (cl 6.6.2.1, Pt 1) to reduce stress concentrations.

7.24   Connections subject to eccentric shear forces

Connections, such as shown in Fig. 7.16(a), are subject to eccentric shear forces which tend to rotate the connections. This results in a shear force on a fastener (bolt, or unit length of weld) from the direct shear force and an additional shear force from the moment.

The force acting on a group of fasteners can be idealized as shown in Fig. 7.16(b). The bolt group rotates about the theoretical instantaneous centre of rotation which varies in

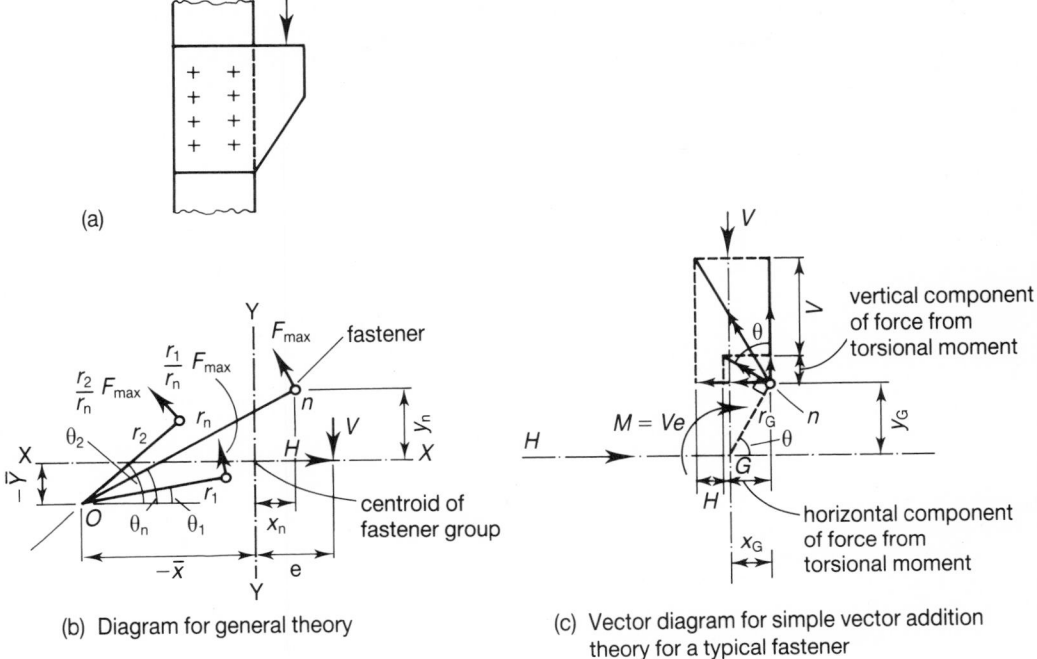

(a)

(b) Diagram for general theory

(c) Vector diagram for simple vector addition theory for a typical fastener

Fig. 7.16   Connections subject to eccentric shear forces.

position depending on the magnitudes of the external forces V and H and the eccentricity e. In the linear elastic stage of behaviour it is reasonable to assume that the force acting on a fastener is proportional to the distance from the centre of rotation. At ultimate load this assumption is not strictly correct but the error involved is not great. A rigorous solution to the theory and accuracy at ultimate load is given by Bahia and Martin.

Although this is the correct approach to the theory there is a simpler more practical method in common use which gives the same values. It is assumed that rotation occurs about the centroid of the fastener group and for convenience the forces acting on a fastener are parallel to the x–x and y–y axes as shown in Fig. 7.16(c). These forces are combined vectorially and the resultant force on a fastener furthest from the centre of rotation is

$$F_R = [F_X^2 + F_Y^2]^{1/2}$$
$$= [(V/n + Mx_G/I_z)^2 + (H/n + My_G/I_z)^2]^{1/2} \tag{7.36}$$

where

n is the number of fasteners in the group;

x_G and y_G are the coordinates of a fastener from the centroid of the fastener group;

$I_z = I_X + I_Y$;

I_X, I_Y and I_z are second moments of area of unit size fasteners about the x–x, y–y and z–z axes.

The method is used in practice for ordinary bolts, high strength friction grip bolts and welds.

7.25   Connections with end bearing

Some connections involve end bearing between components as shown in Fig. 7.17. End bearing can occur in beam-to-column, bracket-to-column, beam-to-beam and column-to-base connections.

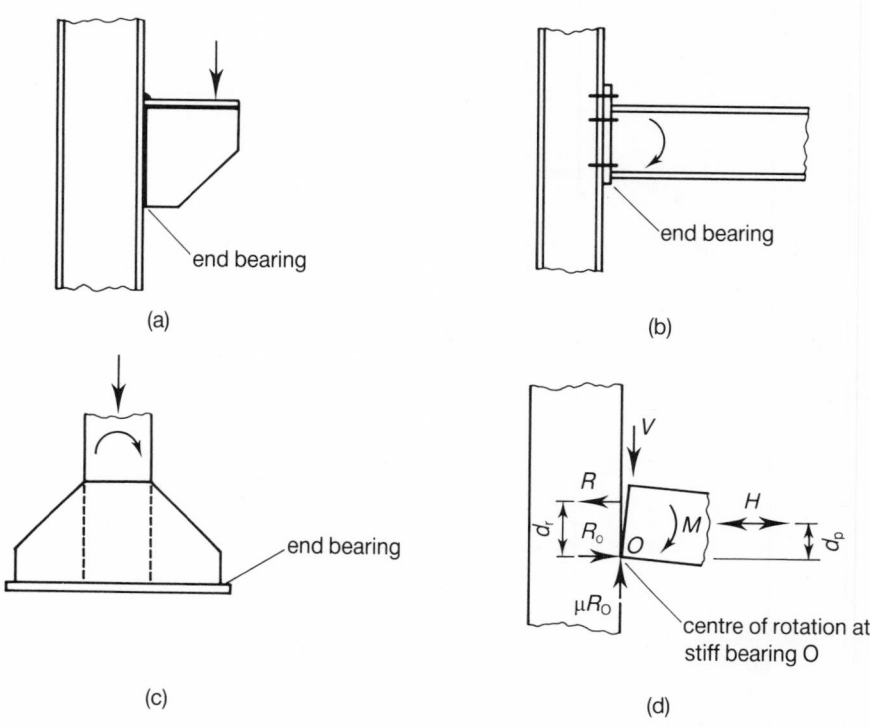

(a)   (b)

(c)   (d)

Fig. 7.17   Connections with end bearing.

Where end bearing occurs rotation takes place about a stiff axis of rotation, axis O–O shown in Fig. 7.17(d). The reaction force R_O is generally large and the bearing may have to be reinforced if it is not to distort under the load. If there is slip at the stiff bearing a frictional force μR_O develops parallel to the stiff bearing surface.

Consider a typical end bearing connection subject to external forces V, H and moment M as shown in Fig. 7.17(d).

If no slip occurs, resolving forces vertically

$$V - \mu R_O < 0 \tag{7.37}$$

Taking moments of forces about force R

$$M \pm H(d_r - d_p) - R_O d_r = 0 \tag{7.38}$$

where R is the resultant force of the fasteners acting at a distance d_r from the axis of rotation O–O.

Combining Equations (7.37) and (7.38) to eliminate R_O, then slip will not occur if

$$\mu[M \pm H(d_r - d_p)]/(V d_r) > 1 \tag{7.39}$$

Most connections with end bearing do not slip and therefore the fasteners are not subject to the external shear force. One exception is a bracket supporting a load with a small eccentricity (see Example 7.6).

In the elastic stage of behaviour it is assumed that the forces acting on a fastener are proportional to the distance from the axis of rotation O–O. If, conservatively, this is also assumed to occur at ultimate load then, taking the moment of forces about axis O–O, the maximum tensile force resisted by a fastener

$$F_{t(max)} = [M \pm Hd_p]y_{max}/I_O + Q \tag{7.40}$$

where

$I = \Sigma\, y^2$ is the second moment of area of unit size fasteners about axis O–O;

Q is the prying force for a bolt, it it exists.

Resolving forces vertically, the shear force on a fastener

$$F_s = (V - \mu R_O)/n = \{V - \mu[M/d_r \pm H(1 - d_p/d_r)]\}/n \tag{7.41}$$

This equation is formed assuming that bolts of the same size and design strength resist equal shear forces and also that slip has occurred.

If the fasteners are welds then $F_{t(max)}$ and F_s are combined vectorially and the size of weld required is obtained from Appendix A1. If the fasteners are ordinary bolts the forces are combined using Equations (7.6) and related to the design strengths given in Appendix A2(a) and A3. If friction grip bolts are used the forces are combined using Equations (7.15) and related to design strengths in Appendix A4(a) and A5(a).

Traditional design methods ignore the existence of the frictional force, which errs on the side of safety. BS 5135 states that it should not be assumed that parts joined are in contact under the joint. However research by Bahia *et al.* has shown that the frictional force does exist. ECCS recommendations allow end bearing to be taken into account.

7.26 'Pinned' connections

Some simple connections, e.g. a tie bar, are connected by real pins as shown in Fig. 7.18(a). Provided that the pins are not corroded, or blocked with debris, they will act as pin joints, i.e. they will resist forces but not moments. Tie bars are rarely used now because of the cost of manufacture, risk of seizure from corrosion or debris, and because safety depends on a single pin.

Other connections shown in Fig. 7.18 are designated as 'pins' because the rotational restraint is small. These joints are also designed assuming that the rotational resistance is zero and the connection resists direct forces only. The general approach to the design of these 'pinned' connections is as follows.

7.26.1 'Pinned' beam-to-column connections (see Figs. 7.18(b), (e) and (f))

In Fig. 7.18(b) it is assumed that the shear force is resisted by the four bolts connecting the bottom cleat to the column flange. During erection the bottom cleat, which is bolted or welded to the column, is used as a marker by the crane operator when the beam is placed. The top cleat is assumed to resist no vertical load but it does provide torsional resistance which is important in lateral stability (see Section 5.2). The top and bottom

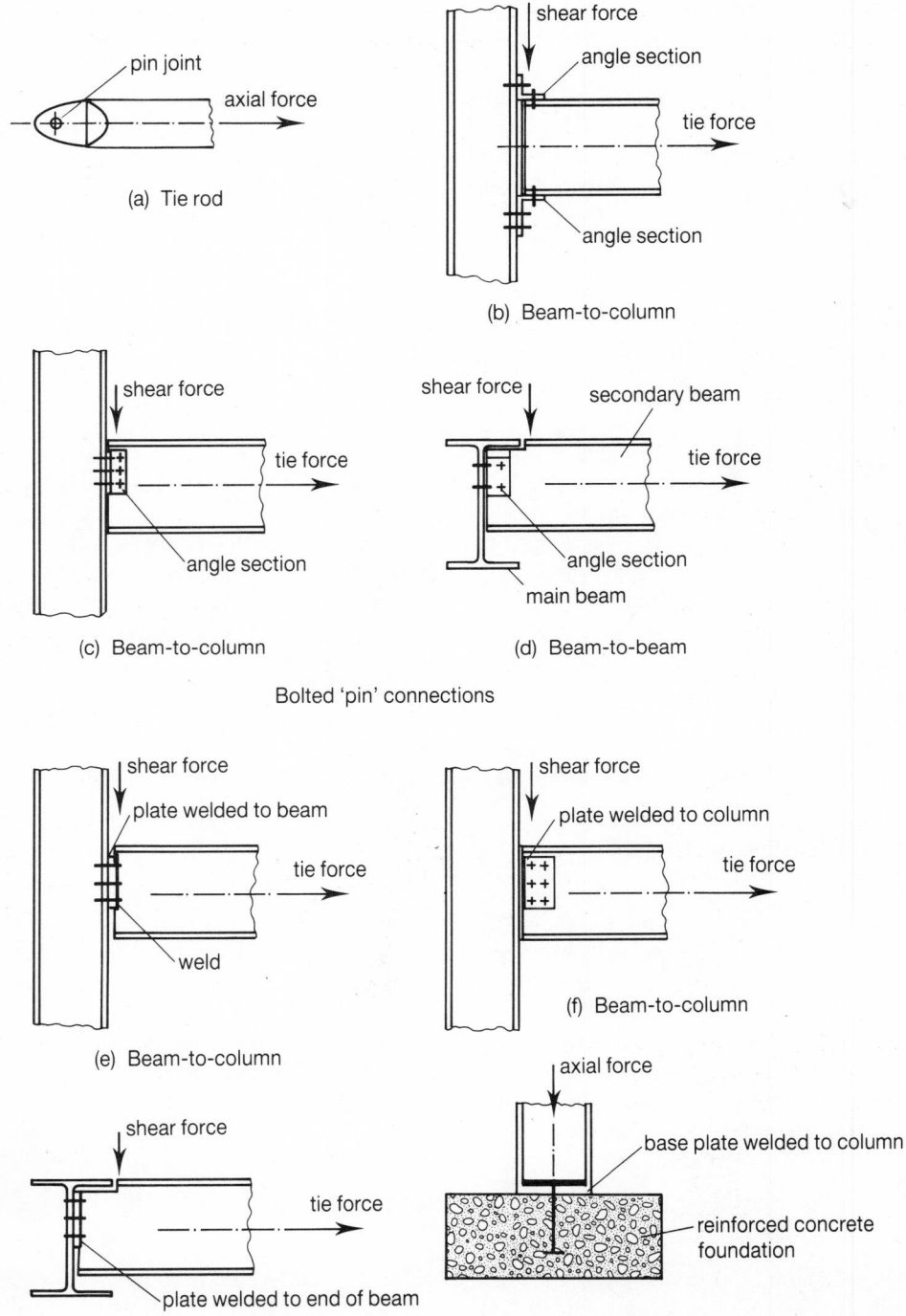

(a) Tie rod

(b) Beam-to-column

(c) Beam-to-column

(d) Beam-to-beam

Bolted 'pin' connections

(e) Beam-to-column

(f) Beam-to-column

(g) Beam-to-beam

(h) Column-to-foundation

Welded-bolted 'pin' connections

Fig. 7.18 Examples of 'pinned' connections.

cleats resist the tie force. The resistance of the web of the beam to shear, bearing and buckling must be checked as described in Section 4.8. At ultimate load the rotation at the end of the beam often introduces end moments. In practice these are assumed to be small and are therefore ignored. A worked design is shown in Example 7.3.

Other types of 'pinned' beam-to-column connections are shown in Figs. 7.18(c), (e) and (f). The end plate depth for Fig. 7.18(e) is kept to a minimum to reduce end moments and the empirical thickness is 8 mm for UB sizes up to 457×191 kg, and 10 mm for sizes greater than 533×210 kg. Examples of 'pin' connections used in practice are given by Pillinger.

7.26.2 'Pinned' beam-to-beam connections (see Fig. 7.18(d) and (g))

The transverse secondary beam is connected to the main beam through angle cleats as shown in Fig. 7.18(d). It is assumed in design that the shear force is transferred to the main beam via the bolts in the web of the main beam. These bolts are therefore designed for single shear and bearing on the web of the main beam and on the angle cleats. It follows therefore that the shear force is eccentric to the bolts in the web of the secondary beam. These bolts are in double shear and bearing on the web of the secondary beam and the angle cleats.

Steel fabricators and erectors often prefer a welded end plate (see Fig. 7.18(g)) as an alternative to the angle cleats shown in Fig. 7.18(d). This results in a more rigid connection and an end moment is introduced to the end of the secondary beam which is dependent on the torsional stiffness of the main beam. If there are secondary beams on both sides of the main beam the secondary moment can be large.

7.26.3 'Pinned' column-to-foundation connections (see Fig. 7.18(h))

The column is fastened to the base plate which is connected to the foundation by foundation bolts. This type of connection is used where the predominant force in the column is axial but there is generally a small shear force. The size of the foundation bolt is based on resisting the forces, but with a minimum size of M 16. The thickness of the base plate is related to the thickness of the web and flange of the column. A worked design is shown in Example 7.5.

7.27 'Rigid' connections

'Rigid' (or 'fixed') connections (see Fig. 7.19) exhibit small rotational displacements in the elastic stage of behaviour. They are useful to limit deflections of members, resist fatigue and resist impact loading. However, generally, the design procedure is more complicated and the components are more highly stressed.

7.27.1 'Rigid' column brackets (see Figs. 7.19(a), (b) and (c))

For the brackets shown in Figs. 7.19(a) and (c) no end bearing is involved and the analysis is carried out as described in Section 7.24. The bracket shown in Fig. 7.19(b) involves end bearing and analysis is carried out as described in Section 7.25. A worked design is shown in Example 7.2.

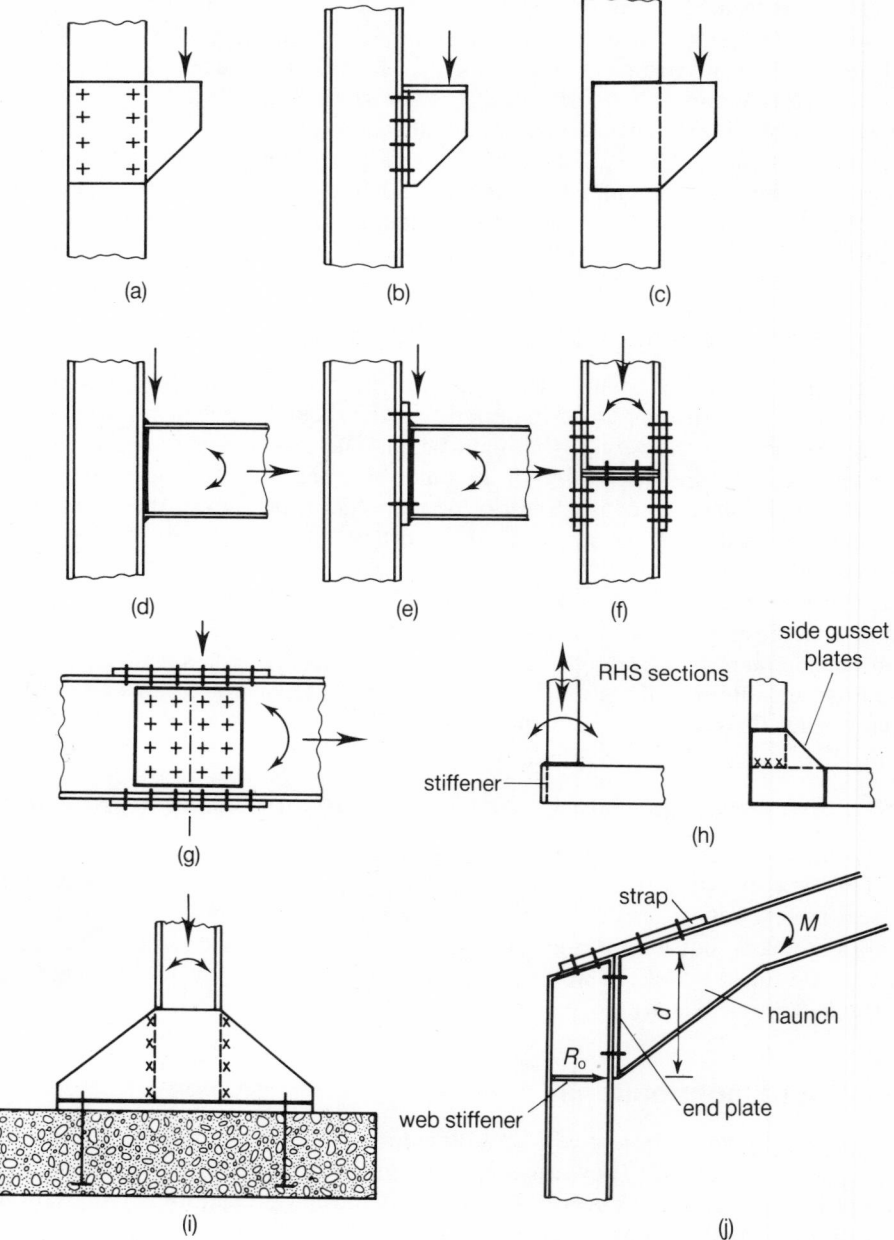

Fig. 7.19 Examples of 'rigid' connections.

7.27.2 'Rigid' beam-to-column connections (see Figs. 7.19(d) and (e))

The most rigid type is where the beam is welded directly to the column on site, as shown in Fig. 7.19(d), but this is expensive and it is difficult to control the quality of the weld. Alternatively stub cantilever beams can be welded to the column in the workshops and a suspended beam site bolted between the ends of the cantilever. The bolted connection is positioned as close to the point of contraflexure as possible. This method is common in the USA for multi-storey buildings but is not used often in Britain.

The method commonly used in Britain is to weld end plates to the beam in the workshops and to bolt these to the columns on site as shown in Fig. 7.19(e). The number of bolts is usually six, as shown, because of the limited depth available. If the moment of resistance needs to be increased then it is necessary to increase the lever arm by haunching the beam at the end. This connection is less rigid than welding the beam directly to the column but it is easier to manufacture and erect. The amount of rotation depends on the thickness of the end plate, thickness of the column flanges and the extensibility of the bolts.

If the connection is close to a plastic hinge then it must be decided whether the plastic hinge should form in the beam, the column or the connection. Recent research favours the formation of the hinge in the beam and therefore the column and the connection must be overdesigned.

End bearing occurs between beam and column and the first step in design is to check whether slip occurs using Equation (7.39). Generally, because the bending moment is large, slip does not occur and the size of the top four bolts required can be determined approximately by taking moments of forces about the compression flange of the beam. The tensile force in a bolt

$$F_t = M/[4(D - T)] \tag{7.42}$$

This ignores the fact that the tensile force in a bolt is increased by a prying force (see Section 7.18) which is condoned by cl 6.3.6.2, Pt 1. Equation (7.42) also assumes that the two bolts close to the compression flange of the beam do not resist any part of the bending moment.

The thickness of the end plate can be determined by assuming the existence of a single plastic yield line at the support for the end plate extension (see Section 7.22).

The thickness of the column flange can be checked (size has been determined from analysis of the frame) by assuming the existence of plastic yield lines adjacent to bolt holes (see Section 7.22).

The force R_O at the axis of rotation may produce failure by bearing or buckling of the web of the column. These situations are checked as shown in Section 4.7. If failure is likely to occur then stiffeners can be welded into the web of the column. However stiffeners are costly and reduce the room available for bolts. The tensile force balancing R_O which acts on the web of the column is often not critical but the strength of the column web must be checked.

The size of the fillet weld connecting the end plate to the beam is determined by assuming the rotation about the axis O–O at the bottom flange of the beam, but an alternative more conservative method is to assume rotation about the centroidal axis of the weld.

If it is required to determine forces more accurately, then the following sequence of calculations can be carried out. The tensile force in the bolt can be determined by taking moments about the stiff bearing axis O–O and including the resistance of the bolts close to the compression flange. In addition the magnitude of the prying force is then determined from Equation (7.12). This force is added to the bolt force and may indicate that a larger size bolt is required.

The thickness of an end plate can be obtained from alternative methods based on the formation of multiple yield lines in the plate. After examining existing literature Mann and Morris recommended the expression

$$\{M/[d_{bf}p_y(4w_p/s_v + d_{bf}/s_h)]\}^{1/2} < t_p < \{Mb_p/(2w_pd_{bf}p_y)\}^{1/2} \tag{7.43}$$

This equation is valid provided that $B_p = 9d_b$, $s_h = 6d_b$, $s_v = 6d_b$, and $e_b > 2.5d_b$.

The thickness of the column flange can also be determined from yield line patterns. A further survey by Mann and Morris recommended:

For unstiffened column flanges the thickness of the flange

$$0.28[M/(d_{bf}p_y)]^{1/2} < T < 0.39[M/(d_{bf}p_y)]^{1/2} \tag{7.44}$$

For stiffened column flanges the thickness of the flange

$$0.23[M/(d_{bf}p_y)]^{1/2} < T < 0.32[M/(d_{bf}p_y)]^{1/2} \tag{7.45}$$

Equations (7.44) and (7.45) are valid provided that $b_c = 2.5d_b$ and $c_c = t_{st} + 5d_b$. Where the column flange thickness is inadequate backing plates can be used. A worked design is shown in Example 7.7.

7.27.3 'Rigid' beam splices (see Fig. 7.19(g))

Beam-to-beam connections (splices) are introduced to extend standard bar lengths, or to facilitate construction and transport. Splices are generally located at sections where the forces are a minimum, and to avoid local geometrical deformations of the structure HSFG bolts are used. The connection is usually made on site and therefore bolts are used.

A simple design method is to assume that the entire shear force is resisted by the web splice and that the flanges resist the entire bending moment. These assumptions are not correct but it simplifies the design and the errors are generally not large. Alternatively part of the moment is assumed to be resisted by the web as shown in Example 7.9. The bolts in the web, in double shear, are subject to an eccentric shear load. The bolts in the flange are also in double shear for two flange plates and the force on a bolt is obtained by a simple moment equation $F = M/(nd)$.

7.27.4 'Rigid' column-to-column connections (see Fig. 7.19(f))

Column splices are used to extend standard bar lengths, to facilitate erection and transport, and for economy by reducing the section size. Where bending moments are large then, as with beam splices, HSFG bolts are used to maintain the axial line of the column. Where sections change size steel packings and an end plate are required to ensure that a good fit is obtained.

Generally the ends of the columns are in contact, or in contact with the end plate, and therefore end bearing occurs. Any shear force will be resisted by the friction at the end bearing and the web plates (or cleats). Where forces are small, sizes of components are decided from experience, practicality and corrosion resistance. A worked design is shown in Example 7.10.

7.27.5 'Rigid' RHS-to-RHS connections (see Fig. 7.19(h))

This is a typical 'L' connection at the support of a Vierendeel girder (see Fig. 1.1(e)) where members of the same width intersect. If forces and moments are not too large then the connection can be made without using stiffeners. However for the connection shown in Fig. 7.19(h) stiffeners are required or larger RHS have to be used.

The weld connecting the two members is continuous and can be either a butt weld throughout, a fillet weld throughout, or part fillet part butt weld (cl 6.6.4, Pt 1). To prevent web buckling of the horizontal RHS for the direction of the moment shown in Fig. 7.19(h) it is necessary to insert a stiffener to resist the end bearing force R_O. An alternative connection is with a plate welded outside on the end of the connection where the connecting welds resist the end bearing force R_O. A worked design is shown in Example 7.12. If the bending moment is in the opposite direction then the connection must be stiffened by using side plates or use larger RHS.

In situations where the width of the vertical RHS is less than the horizontal RHS the strength of the connection is reduced. It is not generally practical to use side gusset plates unless packing is used which entails extra cost. An estimate of the strength of the connection can be obtained by assuming a yield line pattern and method of analysis similar to those shown in Section 7.22.2.

RHS sections are also used in braced triangulated trusses but it is difficult to arrange for the centre lines of the members to intersect at a point. The offsets introduce moments which should be taken into account in the analysis of the structure as shown by Purkiss and Croxton. Further information on analysis and design of the connection is given by Davies.

Circular hollow sections are also used for trusses but the geometry of the connection is more difficult and connections are more expensive to form. A review of methods of analysis of the strength of these connections is given by Stamenkovic and Sparrow.

7.27.6 'Rigid' column-to-foundation connections (see Fig. 7.19(i)) (cl 4.13, Pt 1)

Where bending moments are small a simple slab base is satisfactory but as the bending moment increases, the thickness of the base plate becomes too large and a built-up is used.

For the simple slab base it is assumed that at ultimate load the distribution of stresses beneath the base are as shown in Fig. 7.30. The depth of the compression zone is determined approximately by taking moments of forces about the tensile bolts

$$x = [M/d + N/2]/(Bw) \tag{7.46}$$

From taking moments of forces about the compressive force the tensile force in a bolt is

$$F_t = [M/d - N/2]/n \tag{7.47}$$

The thickness of the base plate is found from the bending of the cantilever length l as described in Section 7.22.

For the built-up base the base plate thickness is determined in the same manner. The gusset plate size is determined by the method shown in Section 7.22. The size of the welds connecting the gusset plate to the base and to the column are determined assuming end bearing. Alternatively end bearing can be ignored and a larger size determined. A worked design is shown in Example 7.11.

7.27.7 'Rigid' knee connection for a portal frame (see Fig. 7.19(j))

This type of connection is similar to the beam-to-column connection shown in Fig. 7.19(e). However because the portal frames are often designed using plastic analysis the magnitude of the reaction $R_O = M/d$ is generally high and consequently the shear stresses in the web are close to the limit. The magnitude of R_O is reduced by haunching the beam and consequently increasing the distance d, but web stiffeners are generally required for the column. To reduce the shear deformation in the column web diagonal stiffeners can be introduced as described and tested by Morris and Newcombe.

The tensile force which balances R_O can be resisted by a strap, or by a group of bolts through the flange of the column. The strap may interfere with the placing of purlins in the roof and the alternative group of bolts may cause distortion of the column flanges if the column flange thickness is small. The distortion can be controlled by the use of flange stiffeners (see Fig. 7.32(c)), or increase the size of the column section, but both increase the cost. A worked design is shown in Example 7.13.

EXAMPLE 7.1   *'Pinned' roof truss joint*

Design the connection for a roof truss shown in Fig. 7.20. The forces, sizes of the angles and tees have been obtained from an elastic analysis made on the assumption of pin joints, and therefore the forces in the members are axial.

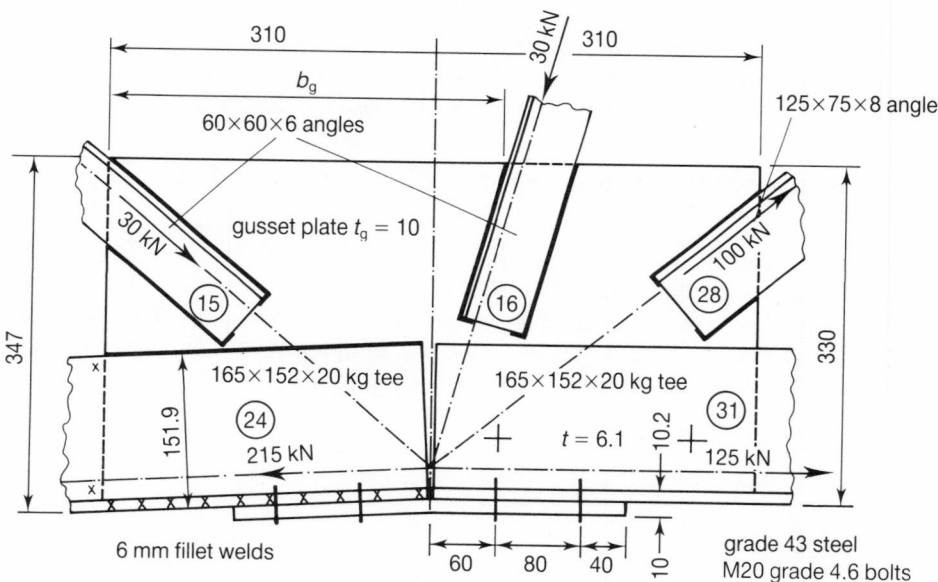

Fig. 7.20 'Pinned' roof truss connection.

The centroidal axes of the members should be arranged to intersect for welded connections, or if bolted the bolt centre lines should intersect. If they do not then the eccentricity should be taken into account in the calculations.

The length and breadth of the gusset plate are obtained from a scale drawing such that the overlaps of the angles and tees are at least equal to the width of the connected leg. The thickness of the gusset plate is at least 6 mm to resist corrosion, and at least equal to the minimum thickness of the angle or tee. Use a 10 mm thick plate grade 43 steel. A rectangular plate is simple to mark and cut and low in fabrication cost. Alternatively a more complicated shape can be used which is aesthetically more acceptable but the fabrication cost is greater.

To avoid distortions in the gusset plate due to forces in the connecting members, according to BS 5400 Pt 3, the maximum unsupported length b_g (shown in Fig. 7.20) is

$$b_g \leqslant 50 t_g (355/p_y)^{1/2}$$

$$b_g \leqslant 50 \times 10 (355/275)^{1/2} = 568.1 \text{ mm}.$$

Actual length 380 mm, therefore satisfactory.

Member 24, structural tee (165 × 152 × 20 kg)

From Section 6.3.1 effective area for a welded connection (cl 4.6.3.1, Pt 1)

$$a_1 = t(A - t) = 6.1(151.9 - 10.2) = 864.4 \text{ mm}^2$$

$$a_2 = TB = 10.2 \times 165.1 = 1684 \text{ mm}^2$$

$$A_e = a_1 + a_2 [3a_1/(3a_1 + a_2)]$$
$$= 864.4 + 1684[3 \times 864.4/(3 \times 864.4 + 1684)] = 1885.4 \text{ mm}^2$$

Design tensile strength of members (cl 4.6.1, Pt 1)

$$P_t = A_e p_y = 1885.4 \times 275/1E3 = 518.5 > 215 \text{ kN, therefore satisfactory.}$$

Assuming a 6 mm fillet weld ($P_w = 0.903$ kN/mm, see Appendix A1) the effective length required to resist the tensile force for member 24

$$T/P_w = 215/0.903 = 238.1 \text{ mm}$$

Two side fillet welds of 150 mm length with end returns would be satisfactory but in practice the lengths would probably be the full overlap. The lengths of weld for members 15, 16 and 28 can be determined using the same method.

This calculation ignores the small eccentricity of the force in the member in relation to the centroid of the weld group. If the eccentricity is taken into account the size of weld is increased slightly.

Member 31, structural tee (165 × 152 × 20 kg)

The connection for member 31 is bolted on site and an arrangement with a strap is shown in Fig. 7.20. The smallest size bolt to be used is M16. From Appendix A2(a) four M20 grade 4.6 bolts in single shear resist a force of

$$\sum P_s = n_b P_s = 4 \times 39.2 = 156.8 > 125 \text{ kN, \quad therefore satisfactory.}$$

The minimum thickness of member for an end bearing distance of $2d$ (see Appendix A2(a)) is 4.3 mm. The tee section and gusset plate thicknesses are greater. The strap is not essential if the four M20 bolts can be fitted in the gusset plate, but the strap increases the out of plane stiffness of the truss.

EXAMPLE 7.2 'Rigid' column bracket

Determine the size of the components required to connect the bracket to the column shown in Fig. 7.21. Use grade 50 steel and design a 'rigid' connection. The forces shown are applied to one gusset plate.

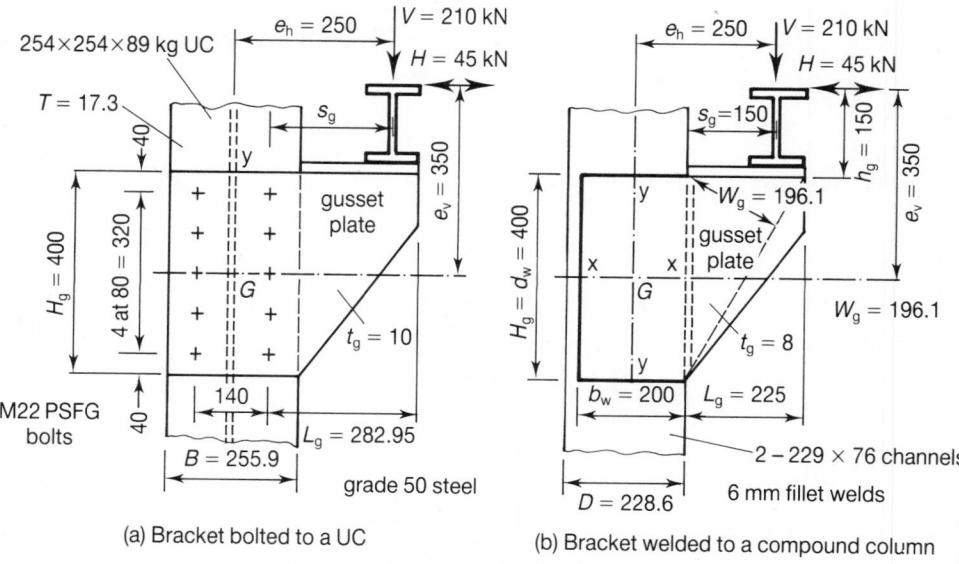

Fig. 7.21 'Rigid' column brackets.

With reference to Appendix B6(a) space the bolts as shown in Fig. 7.21(a). For bolts of unit cross-sectional area the properties of the bolt group are:

Second moment of area of the bolt group about the centroidal x–x axis

$$I_x = \sum (\delta A)y^2 = 4(80^2 + 160^2) = 128\text{E3 mm}^4$$

Second moment of area of the bolt group about the centroidal y–y axis

$$I_y = \sum (\delta A)x^2 = 10(70)^2 = 49\text{E3 mm}^4$$

Second moment of area of the bolt group about the centroidal polar z–z axis

$$I_z = I_x + I_y = (128 + 49)\text{E3} = 177\text{E3 mm}^4$$

From Equation (7.36) the maximum vector force in the direction of the y–y axis on a bolt furthest from the centroid of the bolt group

$$F_y = V/n_b + (Ve_h + He_v)x_n/I_z$$
$$= 210/10 + (210 \times 250 + 45 \times 350)70/177\text{E3} = 48 \text{ kN}$$

The maximum vector force in the direction of the x–x axis on the same bolt

$$F_x = H/n_b + (Ve_h + He_v)y_n/I_z$$
$$= 45/10 + (210 \times 250 + 45 \times 350)160/177\text{E3} = 66.2 \text{ kN}$$

Resultant vector force on this bolt

$$F_r = (F_y^2 + F_x^2)^{1/2} = (48^2 + 66.2^2)^{1/2} = 81.77 \text{ kN}$$

Solution (a) using oridnary grade 4.6 bolts

From Appendix A2(a) an M30 bolt has a design single shear strength of 89.8 kN. However the recommended bolt diameter for a flange width of 254 mm is 24 mm (see Appendix A6(b)), and the solution is not acceptable.

Solution (b) using ordinary grade 8.8 bolts

From Appendix A3 an M20 bolt has a design strength of 91.9 kN. Also from Appendix A3 the required bearing thickness of a grade 50 gusset plate is 8.4 mm, therefore 10 mm is satisfactory. The flange thickness of $T = 17.3$ mm for the column is not critical. The end distance is $2d = 40$ mm. This solution does not produce a fully rigid joint because there is slip from clearance in the bolt holes.

Solution (c) using parallel shank HSFG bolts

From Appendix A4(a) an M22 PSFG bolt has a design single shear strength of 87.6 kN. A parallel shank HSFG slips into bearing at the ultimate limit state and therefore a check on the thickness of the gusset plate is required. From Appendix A4(a) the minimum bearing thickness for grade 50 plate is 3.7 mm for an end distance of $3d = 60$ mm. For an end distance of 40 mm, $t = 3.7 \times 60/40 = 5.55$ mm. The gusset plate must be at least 10 mm thick to avoid crumpling as specified in BS 4604.

Solution (d) using waisted shank HSFG bolts

From Appendix A5(a) an M22 waisted shank HSFG has a design shear strength of 93.4 kN which would be satisfactory. A waisted shank HSFG does not slip into bearing at the ultimate limit state and therefore a check on the thickness of the gusset plate is not required, but the gusset plate must be at least 10 mm thick to avoid crumpling as specified in BS 4604.

The thickness of the gusset plate (see Section 7.22.2) for the bolted connection is determined as follows.

$$L_g = 225 + (255.9 - 140)/2 = 282.95 \text{ mm}$$

$$s_g = 150 + (255.9 - 140)/2 = 207.95 \text{ mm}$$

From Equation (7.34) the width of the gusset plate perpendicular to the free edge

$$W_g = L_g/[(L_g/H_g)^2 + 1]^{1/2}$$
$$= 282.95/[(282.9/400)^2 + 1]^{1/2} = 231.0 \text{ mm}$$

From Equation (7.33), replacing the term $P_u s_g$ by $(Vs_g + Hh_g)$, the thickness of the gusset plate

$$t_g = 2(Vs_g + Hh_g)/(p_{yg}W_g^2) + W_g/80$$
$$= 2(210 \times 207.95 + 45 \times 150)\text{E3}/(355 \times 231^2) + 231/80$$
$$= 8.21 \text{ mm};$$

use a 10 mm thick plate of grade 50 steel.

Check the slenderness ratio of the gusset plate from Equation (7.35)

$$l_g/r_g = 2\sqrt{3}W_g/t_g = 2\sqrt{3} \times 231/10 = 80.02 < 185$$

the limit of the slenderness ratio for the application of this theory, therefore satisfactory.

Alternatively (see Section 7.22.1) the thickness of the gusset plate can be determined using cl 4.13.2.4, Pt 1. From Equation (7.25)

$$t_g = 5P_g s_g/(p_{yg}H_g^2) = 5(Vs_g + He_v)/(p_{yg}H_g^2)$$
$$= 5(210 \times 207.95 + 45 \times 350) \times 1E3/(270 \times 400^2) = 6.88 \text{ mm};$$

use a 10 mm thick plate of grade 50 steel.

Solution (e) using welds

Where it is not possible to bolt to a column, e.g. the compound channel column shown in Fig. 7.21(b), then welds are used. The connection is rigid and for welds of unit size the properties of the weld group are:

Total length of weld

$$L_w = 2(d_w + b_w) = 2(400 + 200) = 1200 \text{ mm}$$

Second moment of area of the weld group about the centroidal x–x axis

$$I_x = \sum (\delta A)y^2 = 2[d_w^3/12 + b_w(d_w/2)^2]$$
$$= 2[400^3/12 + 200(400/2)^2] = 26.67E6 \text{ mm}^4$$

Second moment of area of the weld group about the centroidal y–y axis

$$I_y = \sum (\delta A)x^2 = 2[b_w^3/12 + d_w(b_w/2)^2]$$
$$= 2[200^3/12 + 400(200/2)^2] = 9.33E6 \text{ mm}^4$$

Second moment of area of the weld group about the centroidal polar z–z axis

$$I_z = I_x + I_y = (26.67 + 9.33)E6 = 36E6 \text{ mm}^4$$

Maximum vector force in the direction of the y–y axis on a welded element furthest from the centroid of the weld group from Equation (7.36)

$$F_y = V/L_w + (Ve_h + He_v)x_n/I_z$$
$$= 210/1200 + (210 \times 250 + 45 \times 350)100/36E6 = 0.365 \text{ kN/mm}$$

Maximum vector force in the direction of the x–x axis on the same weld element

$$F_x = H/L_w + (Ve_h + He_v)y_n/I_z$$
$$= 45/1200 + (210 \times 250 + 45 \times 350)200/36E6 = 0.417 \text{ kN/mm}$$

Resultant vector force on this weld element

$$F_r = (F_y^2 + F_x^2)^{1/2} = (0.365^2 + 0.417^2)^{1/2} = 0.554 \text{ kN/mm}$$

From Table A1 (see Appendix) a 6 mm fillet weld (E43 electrode) has a design shear strength of 0.903 kN/mm for grade 50 steel.

The thickness of the gusset plate can be determined from the theory in Section 7.22.2. From Equation (7.34) the width of the gusset plate perpendicular to the free edge

$$W_g = L_g/[(L_g/H_g)^2 + 1]^{1/2}$$

$$= 225/[(225/400)^2 + 1]^{1/2} = 196.1 \text{ mm}$$

From Equation (7.33) replacing the term $P_\mathrm{u} s_\mathrm{g}$ by $V s_\mathrm{g} + H h_\mathrm{g}$

$$t_\mathrm{g} = 2(V s_\mathrm{g} + H h_\mathrm{g})/(p_\mathrm{yg} W_\mathrm{g}^2) + W_\mathrm{g}/80$$
$$= 2(210 \times 150 + 45 \times 150)\mathrm{E}3/(355 \times 196.1^2) + 196.1/80$$
$$= 8.05 \text{ mm};$$

use an 8 mm thick plate of grade 50 steel.

Check the slenderness ratio of the gusset plate from Equation (7.35)

$$l_\mathrm{g}/r_\mathrm{g} = 2\sqrt{3}\, W_\mathrm{g}/t_\mathrm{g} = 2\sqrt{3} \times 196.1/10 = 67.93 < 185$$

the limit of the slenderness ratio for the application of this theory.

Alternatively (see Section 7.22.1) the thickness of the gusset plate can be determined using cl 4.13.2.4, Pt 1. From Equation (7.25)

$$t_\mathrm{g} = 5 P_\mathrm{u} s_\mathrm{g}/(p_\mathrm{yg} H_\mathrm{g}^2) = 5(V s_\mathrm{g} + H e_\mathrm{v})/(p_\mathrm{yg} H_\mathrm{g}^2)$$
$$= 5(210 \times 150 + 45 \times 350) \times 1\mathrm{E}3/(270 \times 400^2) = 5.47 \text{ mm};$$

use an 8 mm thick plate of grade 50 steel.

EXAMPLE 7.3   *'Pinned' beam-to-column connection*

Determine the size of components for the connection shown in Fig. 7.22. The components shown have been obtained by several trials and the following calculations justify the sizes.

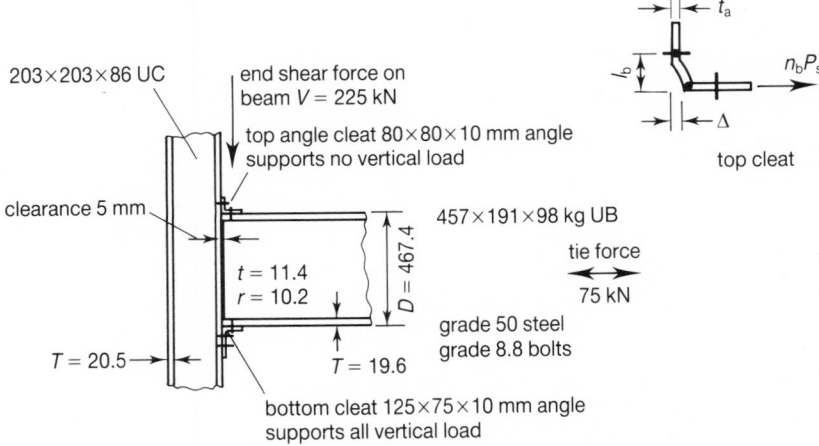

Fig. 7.22   'Pinned' beam-to-column connection.

It is assumed that the shear force of 225 kN is resisted by the bolts in the bottom cleat $(125 \times 75 \times 10 \text{ mm}$ angle). From Appendix A3, using M20 grade 8.8 bolts in single shear the number of bolts required

$$n_\mathrm{b} = V/P_\mathrm{s} = 225/91.9 = 2.45; \quad \text{use 4 bolts.}$$

These bolts are strong enough to resist the additional tie force of 75 kN acting alone (cl 2.4.5.2, Pt 1).

From Appendix A3 the minimum thickness for bearing is 8.4 mm for grade 50 steel, and because this is less than the thickness of the angle (10 mm) and the column flange (20.5 mm) then this is satisfactory. From Appendix A6(b) check the bolt spacing for a $125 \times 75 \times 10$ mm angle and for the $203 \times 203 \times 86$ column.

As an alternative the four bolts in the vertical leg of the bottom cleat could be replaced by fillet welds along the two vertical edges of the cleat with end returns. From Appendix A1 the shear resistance of two 6 mm fillet welds using E51 electrodes, 125 mm in length, is

$$P_w = 2l_w F_w = 2 \times 125 \times 1.071 = 267.8 > 225 \text{ kN}, \quad \text{which is satisfactory.}$$

Check the web bearing, web buckling and web shear strengths.

Dispersion bearing length for the beam

$$n_2 = 2.5(T + r_b) = 2.5(19.6 + 10.2) = 74.5 \text{ mm}$$

Stiff bearing length for the angle

$$b_1 = 2t_a + (2 - \sqrt{2})r_a - \text{clearance}$$
$$= 2 \times 10 + (2 - \sqrt{2})11 - 5 = 21.44 \text{ mm}$$

The design strength for the rolled section, based on the thickness of the flange ($T = 19.6$ mm), is obtained from Table 6, Pt 1, $p_{yw} = 345 \text{ N/mm}^2$.

Web bearing strength (cls 4.5.1.3 and 4.5.3, Pt 1)

$$P_{bg} = (b_1 + n_2)tp_{yw}$$
$$= (21.44 + 74.5)11.4 \times 345/1\text{E}3$$
$$= 377.3 > 225 \text{ kN}, \quad \text{therefore satisfactory.}$$

Width of equivalent strut

$$n_1 = D/2 = 467.4/2 = 233.7 \text{ mm}$$

Slenderness ratio

$$\lambda = 2.5d/t = 2.5 \times 407.9/11.4 = 89.45$$

Buckling stress (cl 4.7.5, Pt 1) for grade 50 steel, $p_y = 345 \text{ N/mm}^2$ (Table 27(c), Pt 1) is $p_c = 160 \text{ N/mm}^2$.

Web buckling design strength (cls 4.5.2.1 and 4.5.1.3, Pt 1)

$$P_w = (b_1 + n_1)tp_c$$
$$= (21.44 + 233.7)11.4 \times 160/1\text{E}3$$
$$= 465.4 > 225 \text{ kN}, \quad \text{therefore satisfactory.}$$

Shear strength of beam (cl 4.2.3, Pt 1)

$$P_v = 0.6p_{yw}Dt$$
$$= 0.6 \times 345 \times 467.4 \times 11.4/1\text{E}3$$
$$= 1103 > 225 \text{ kN}, \quad \text{therefore satisfactory.}$$

The top cleat is used to provide torsional resistance against lateral buckling of the beam and to resist tie and erection forces. The angle must be at least 6 mm thick to resist corrosion and the leg of sufficient length to accommodate M20 bolts. From Section Tables, a $80 \times 80 \times 10$ mm angle is chosen and justified in the following calculations.

Additional calculations

The above calculations are the minimum required in practice but it is instructive to consider calculations based on rotation of the connection. Consider a simply supported beam carrying a uniformly distributed load and developing a plastic hinge at mid-span.

It is assumed that at ultimate load a simply supported beam develops a plastic hinge at mid-span but most of the length of the beam behaves elastically. From the area–moment method (or Fig. 4.11(b)), the rotation of the end of the beam at the support

$$\theta_u = (2/3)(L/2)(M_p/EI) = LM_p/(3EI) \tag{i}$$

If, at service load, the deflection at centre span is Δ, then assuming an average load factor of 1.5

$$\Delta/L = (5/384)(W_u/1.5)(L^2/EI) = (5/48)(M_p/1.5)(L/EI) \tag{ii}$$

This rotation is the minimum that occurs at ultimate load because it is calculated for the onset of a plastic hinge.

Combining Equations (i) and (ii) and eliminating (M_p/EI), the rotation of the end of the beam

$$\theta_u = 4.8(\Delta/L) \tag{iii}$$

If a connection restrains the rotation of the beam and the column does not rotate, then from area–moment (or Fig. 4.11(e)) the moment at the end of the beam

$$M_e = 2EI\theta_e/L = 2EI(\theta_u - \theta_f)/L \tag{iv}$$

where

θ_f is the free angle of movement allowed, e.g. by clearance between the bolts and the holes.

Rotation at ultimate load is sufficient to yield the top angle cleat (see Fig. 7.22) assuming that the bolts do not fail. Equating the work done in shearing the bolts in the top flange to the work done on the yield lines

$$n_b P_s \Delta = 2m_p B\Delta/l_b$$

Rearranging and substituting $m_p = t_a^2 p_{ya}/4$, the thickness of the top angle

$$t_a < \sqrt{[2n_b(l_b/B)P_s/p_{ya}]} \tag{v}$$

End moment developed from rotation which yields the top cleat

$$M_e = n_b P_s D = 0.5(B/l_b)t_a^2 p_{ya} D \tag{vi}$$

This moment must be added to the moment from the eccentricity of the connection to the centre line of the column, to determine the moment transferred to the column.

Continuing the numerical calculations: from (iii) the rotation at the end of the beam at ultimate load assuming $\Delta/L = 1/200$ at service load

$$\theta_u = 4.8(\Delta/L) = 4.8 \times 1/200 = 0.024 \text{ radians.}$$

Movement required at top cleat if no end moment to develop

$$\theta_u D = 0.024 \times 467.4 = 11.2 > 2 \text{ mm (clearance in bolt hole),}$$

therefore end moment develops.

From Equation (v) based on a yield line mechanism the thickness of the top angle

$$t_a < \sqrt{[2n_b(l_b/B)P_s/p_{ya}]} = \sqrt{[2 \times 2(45/192.8)91.9\text{E}3/355]} = 15.55 \text{ mm.}$$

From edge bearing distance the thickness of the top cleat

$$t_a = 2P_s/(ep_{bs}) = 2 \times 91.9\text{E}3/(35 \times 550) = 9.55 \text{ mm}$$

The angle previously selected is 10 mm thick and is therefore satisfactory.

From Equation (vi) the end moment developed from rotation of the beam which produces yield of the top cleat at ultimate load

$$M_e = 0.5(B/l_b)t_a^2 p_{ya} D$$
$$= 0.5(192.8/45) \times 10^2 \times 355 \times 467.4/1\text{E}6 = 35.55 \text{ kNm.}$$

This end moment has a beneficial effect on the beam by reducing the moment at mid-span, but may reduce the strength of the column.

The end moment from yielding of the angle is less than the end moment from full rigidity with allowance for bolt holes given in Equation (vi). Assuming a 10 m span.

$$M_{ef} = 2EI(\theta_u - \theta_f)/L$$
$$= 2 \times 205\text{E}3 \times 457.17\text{E}6(0.024 - 2/467.4)/10\text{E}3 \times 1/1\text{E}6 = 369.7 \text{ kNm.}$$

These figures are only approximations based on the single shear strength of the bolts in the top cleat. More accurate values of moments at the ends of members can be obtained from an analysis of the complete structure using real moment-rotation characteristics of the connections.

EXAMPLE 7.4   *'Pinned' beam-to-beam connection*

Determining the size of the components required for the connection shown in Fig. 7.23. The beam sizes have been determined from the bending calculations.

Assuming that the M20 grade 4.6 bolts through the web of the main beam B are subject to single shear forces

$$n_b = V/P_s = 150/39.2 = 3.82, \quad \text{use four M20 grade 4.6 bolts}$$

The single shear strength P_s of an M20 grade 4.6 bolt is obtained from Appendix A2(a). The minimum thickness for bearing, for an end distance of $2d = 40$ mm in the angle, is 4.3 mm. Choose $70 \times 70 \times 10$ mm angles. The web thickness of beam B (11.9 mm) is not critical in bearing.

Assuming that the bolts connecting the angle cleats to the web of the transverse beam A are in double shear and subjected to an eccentric load.

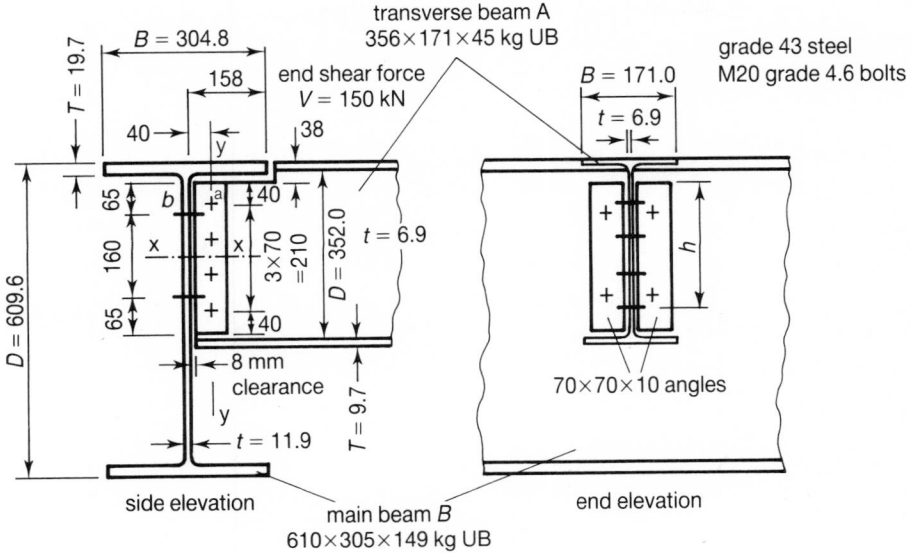

Fig. 7.23 'Pinned' beam-to-beam connection.

Second moments of area of the bolt group about the centroidal axis for bolts of unit area are

$$I_x = \sum (\delta A)y^2 = 2(35^2 + 105^2) = 24.5\text{E}3 \text{ mm}^4$$

$$I_y = 0$$

$$I_z = I_x + I_y = 24.5\text{E}3 \text{ mm}^4$$

Maximum force on a bolt in the x direction from Equation (7.36)

$$F_x = Vey_{max}/I_z = 150 \times 40 \times 105/24.5\text{E}3 = 25.71 \text{ kN}$$

Average force on a bolt in the y direction

$$F_y = V/n_b = 150/4 = 37.5 \text{ kN}$$

Maximum resultant force on a bolt

$$F_r = (F_x^2 + F_y^2)^{1/2} = (25.71^2 + 37.5^2)^{1/2} = 45.47 \text{ kN}$$

Double shear strength of an M20 grade 4.6 bolt from Appendix A2(a)

$$2P_s = 2 \times 39.2 = 78.4 > 45.47 \text{ kN}, \quad \text{therefore satisfactory.}$$

The required bearing thickness for an M20 grade 4.6 bolt in double shear is $2 \times 4.3 = 8.6$ mm. Two 10 mm thick angles are satisfactory, but the web thickness ($t = 6.9$ mm) is not. Revise to $356 \times 171 \times 67$ kg section.

The relatively large end clearance of 8 mm ensures that there is no end bearing and hence no end moment on the transverse beam A.

It is possible for the web of the transverse beam A to fail in shear along a vertical line that passes through the holes. The shear resistance through the holes.

$$V = (h - 3.5d_h)tp_{yw}$$
$$= (250 - 3.5 \times 22)6.9 \times 0.6 \times 275/1E3$$
$$= 197 > 150 \text{ kN}, \quad \text{therefore satisfactory.}$$

EXAMPLE 7.5   *'Pinned' column-to-foundation connection*

Determine the size of the components for the base shown in Fig. 7.24. The concrete cube strength is $f_{cu} = 20 \text{ N/mm}^2$.

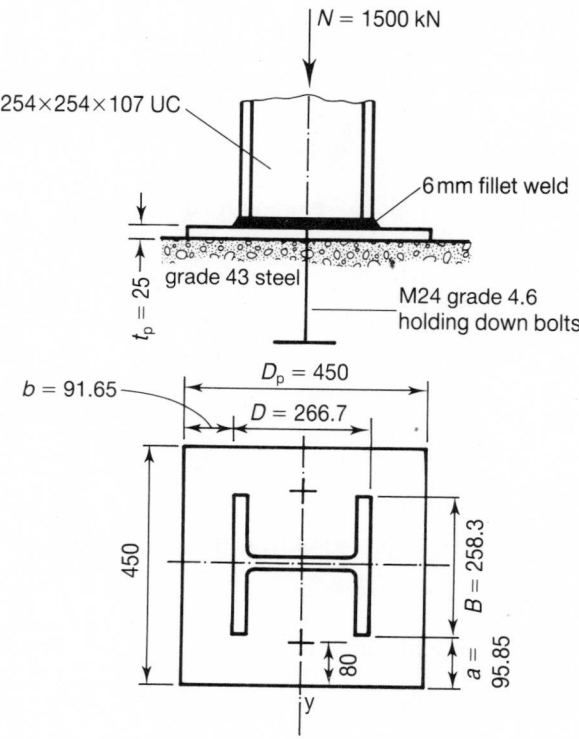

Fig. 7.24   'Pinned' column-to-foundation connection.

Design bearing pressure between the steel base plate and the concrete (cl 4.13.1, Pt 1).

$$w = 0.4f_{cu} = 0.4 \times 20 = 8 \text{ N/mm}^2$$

Minimum area of base plate

$$A_p = N/w = 1500E3/8 = 187.5E3 \text{ mm}^2$$

Length of a side of a square base plate

$$D_p = \sqrt{A_p} = \sqrt{(187.5E3)} = 433 \text{ mm}; \quad \text{use } 450 \times 450 \text{ plate}$$

Maximum base plate extension

$$a = (D_p - B)/2 = (450 - 258.3)/2 = 95.85 \text{ mm}$$

From Equation (7.26) using a plastic method of analysis the base plate thickness for grade 43 steel

$$t_p = a\sqrt{(2w/p_y)} = 95.85\sqrt{(2 \times 8/265)} = 23.55 \text{ mm}$$

Alternatively from Equation (7.19) using an elastic method of analysis the base plate thickness for grade 43 steel

$$a = 95.85$$

$$b = (450 - 266.7)/2 = 91.65 \text{ mm}$$

$$t_p = \sqrt{[2.5w(a^2 - 0.3b^2)/p_{yp}]}$$
$$= \sqrt{[2 \times 8(95.85^2 - 0.3 \times 91.65^2)/265]} = 20.06 \text{ mm.}$$

Use 25 mm thick plate grade 43 steel.

If the end of the column is machined then the load is assumed to be transferred directly to the base plate and a minimum size of fillet weld of 6 mm is used to connect the base plate to the column.

Alternatively if the end of the column is not machined then the force per unit length of weld is approximately

$$F_w = N/(4B + 2D) = 1500/(4 \times 258.3 + 2 \times 266.7) = 0.957 \text{ kN/mm}$$

From Appendix A1 the design strength of an 8 mm weld is 1.204 kN/mm for grade 43 steel.

The base plate is subjected to a compressive force which is not transferred to the holding down bolts. The bolts are therefore subject only to erection forces and if these are not known then experience has shown that a bolt size approximately equal to the plate thickness is suitable. Use two M24 grade 4.6 holding down bolts.

EXAMPLE 7.6 *'Rigid' column bracket*

Determine the size of fillet welds for the bracket shown in Fig. 7.25.

There are two possible solutions based on failure mechanisms assuming (a) rotation about axis G–G (b) rotation about axis O–O. Assuming rotation about axis G–G is the simple traditional method; assuming rotation about O–O is more correct but the calculations are more complicated.

(a) *Rotation about axis G–G*

Second moment of area of the weld group about axis G–G

$$I_G = 2d_w^3/12 + 4B(d_f/2)^2$$
$$= 2(834.9 - 2 \times 18.8)^3/12 + 4 \times 291.6 \times [(834.9 - 18.8)/2]^2$$
$$= (84.7 + 194.21)E6 = 278.68E6 \text{ mm}^4$$

Maximum force per unit length on weld in the x direction (Equation (7.36))

$$F_x = (Ve)(d_f/2)/I_G$$

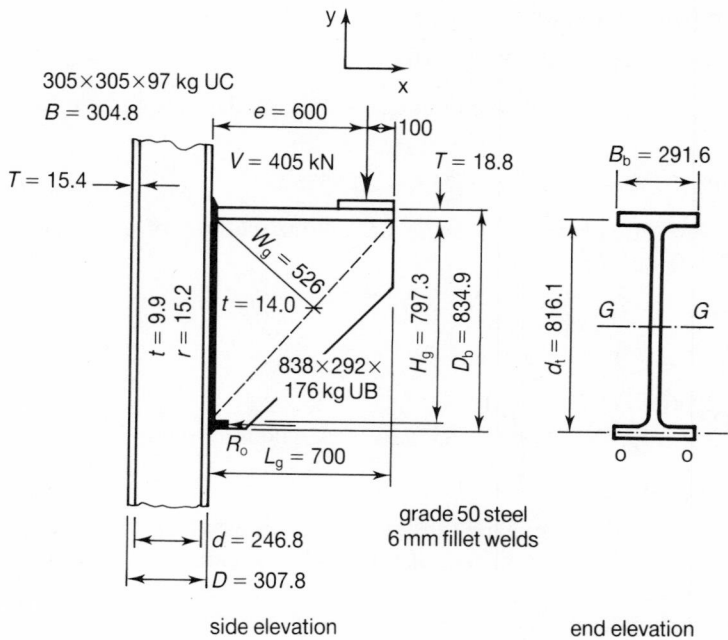

Fig. 7.25 'Rigid' column bracket.

$$= (405 \times 600)[(834.9 - 18.8)/2]/278.68E6 = 0.3558 \text{ kN/mm}$$

Maximum force per unit length of weld in the y direction

$$F_y = V/L_w$$
$$= 405/[4 \times 291.6 + 2(834.9 - 2 \times 18.8)] = 0.1467 \text{ kN/mm}$$

Maximum resultant force per unit length of weld

$$F_r = (F_x^2 + F_y^2)^{1/2}$$
$$= (0.3558^2 + 0.1467^2)^{1/2} = 0.3849 \text{ kN/mm}$$

From Appendix A1 the design strength for a 6 mm fillet weld is 0.903 kN/mm for grade 50 steel and grade E43 electrode.

(b) *Rotation about axis O–O*

The fillet weld is continuous round the bracket section shown in Fig. 7.25. If there are no stiffeners in the web of the column then the strength of the weld around the flanges of the bracket is reduced because of the flexibility of the column and beam flanges (see Section 7.4).

Effective length of the column flange weld from Equation (7.1) with $C = 5$ for Fe510

$$b_{we} = 2t + CT = 2 \times 9.9 + 5 \times 15.4 = 96.8 \text{ mm}$$

This effective length, which is less than the actual length, reduces the second moment of area of the weld group about axis O–O

$$I_{oe} = (2/3)d_w^3 + 2b_{we} d_f^2$$

$$= (2/3)(834.9 - 2 \times 18.8)^3 + 2 \times 96.8 \times (834.9 - 18.8)^2$$
$$= (337.9 + 128.9)E6 = 466.8E6 \text{ mm}^4$$

Maximum force per unit length of weld in the x direction

$$F_x = (Ve)d_f/I_{oe}$$
$$= (405 \times 600) \times (834.9 - 18.8)/466.8E6 = 0.4248 \text{ kN/mm}$$

Assuming that there is vertical slip at the axis O–O then the strength of the weld is affected by the flexibility of the beam flanges. The effective length of the weld (see Equation (7.1))

$$b_e = 2t + CT = 2 \times 14 + 5 \times 18.8 = 122 \text{ mm}$$

This effective length reduces the total length resisting shear

$$L_{we} = 4b_e + 2d_w = 4 \times 122 + 2(834.9 - 2 \times 18.8) = 2082.6 \text{ mm}$$

Distance d_r from the axis O–O to the resultant force in the weld is determined from equating the moments of the forces in the weld group about the axis O–O

moment of the parts = moments of the whole
$$(2b_e + 0.5 \times 2d_f)F_x d_r = F_x I_{oe}/d_f$$

Rearranging and putting $I_{oe} = (2/3)d_f^3 + 2b_e d_f^2$

$$d_r/d_f = (2/3 + 2b_e/d_f)/(1 + 2b_e/d_f)$$
$$= [2/3 + 2 \times 96.8/(834.9 - 18.8)]/[1 + 2 \times 96.8/(834.9 - 18.8)]$$
$$= 0.7306$$

Check whether slip occurs by substituting in Equation (7.39)

$$\mu M/(Vd_r) = \mu_s e/d_r = 0.45 \times 600/[0.7306(834.9 - 18.8)]$$
$$= 0.4528 < 1, \text{ therefore slip occurs.}$$

Maximum force per unit length of weld in the y direction

$$F_y = V/L_{we} - \mu_s R/L_{we}$$
$$= V/L_{we} - \mu_s(Ve/d_r)L_{we}$$
$$= 405/2082.6 - 0.45[405 \times 600/[0.7306(834.9 - 18.8)]]/2082.6$$
$$= 0.1944 - 0.0881 = 0.1063 \text{ kN/mm}$$

Maximum resultant force per unit length of weld

$$F_r = (F_x^2 + F_y^2)^{1/2}$$
$$= (0.4248^2 + 0.1063^2)^{1/2} = 0.4379 \text{ kN/mm}$$

From Appendix A1 the design strength of a 6 mm fillet weld is 0.903 kN/mm for grade 50 steel and grade E43 electrode.

Check the thickness of the web of the 838 × 292 × 176 kg UB acting as a gusset plate. From Equation (7.34)

$$W_g = L_g/[(L_g/H_g)^2 + 1]^{1/2}$$
$$= 700/[(700/797.3)^2 + 1]^{1/2} = 526.0 \text{ mm}$$

Required thickness of the web of the UB acting as a gusset plate from Equation (7.33)

$$t_g = 2P_u s_g/(p_{yg} W_g^2) + W_g/80$$

$$= 2 \times 405\text{E3} \times 600/(355 \times 526^2) + 526/80$$
$$= 11.52 < 14 \text{ mm (thickness of web of the UB)}, \quad \text{therefore satisfactory.}$$

Check the slenderness ratio of the UB acting as a gusset plate from Equation (7.35)

$$l_\text{g}/r_\text{g} = 2\sqrt{3}W_\text{g}/t_\text{g} = 2\sqrt{3} \times 526/14 = 130.2 < 185$$

the limit of application of the theory, therefore acceptable.

Alternatively using Equation (7.25) based on cl 4.13.2.4, Pt 1

$$t_\text{g} = 5P_\text{g}s_\text{g}/(p_{\text{yg}}H_\text{g}^2) = 5 \times 405\text{E3} \times 600/(270 \times 797.3^2)$$
$$= 7.08 < 14 \text{ mm}, \quad \text{therefore satisfactory.}$$

Reaction R_O may buckle or crush the web of the $305 \times 305 \times 97$ kg UC. From Equation (7.38) the reaction

$$R_\text{O} = Ve/d_\text{r} = 405 \times 600/[0.7306(834.9 - 18.8)] = 407.6 \text{ kN}$$

Shear strength of the column web (cl 4.2.3, Pt 1)

$$P_\text{v} = 0.6Dtp_{\text{yw}} = 0.6 \times 307.8 \times 9.9 \times 355/1\text{E3} = 649.1 \text{ kN} > R_\text{O},$$
$$\text{therefore satisfactory.}$$

Check for shear buckling (cl 3.6.2, Pt 1)

$$d/t = 246.5/9.9 = 24.9$$
$$63\varepsilon = 63\sqrt{(275/355)} = 55.4 > 24.9, \quad \text{therefore satisfactory.}$$

Stiff bearing length

$$b_1 = T_\text{b} = 18.8 \text{ mm}$$

Width of equivalent strut

$$n_1 = D_\text{c} = 307.8 \text{ mm}$$

Slenderness ratio

$$\lambda = 2.5d/t = 2.5 \times 246.5/9.9 = 62.25$$

Buckling stress (cl 4.7.5, Pt 1) for grade 50 steel, $p_\text{y} = 355$ N/mm^2 and $T = 15.4$ mm, is $p_\text{c} = 241$ N/mm^2, Table 27(c), Pt 1.

Web buckling design strength of the UC (cl 4.5.2.1, Pt 1), from Equation (4.37)

$$P_\text{w} = (b_1 + n_1)tp_\text{c}$$
$$= (18.8 + 307.8)9.9 \times 241/1\text{E3}$$
$$= 779.2 > 407.6 \text{ kN } (R_\text{O}), \quad \text{therefore satisfactory.}$$

Dispersion bearing length

$$n_2 = 2 \times 2.5(T_\text{c} + r_\text{c}) = 5 \times (15.4 + 15.2) = 153 \text{ mm}$$

Web bearing strength of the UC (cl 4.5.3, Pt 1), from Equation (4.39)

$$P_\text{wbg} = (b_1 + n_2)tp_{\text{yw}}$$
$$= (18.8 + 153)9.9 \times 355/1\text{E3}$$
$$= 603.8 > 407.6 \text{ kN } (R_\text{O}), \quad \text{therefore no stiffeners required.}$$

If stiffeners are required see cl 4.5.4.2, Pt 1.

At axis O–O there is a combination of bearing, shear and direct stresses. No recommendations are given in BS 5950 for this situation but see Section 2.10.

Average shear stress in column web

$$f_q = R_O/(D_c t_c) = 407.6F3/(307.8 \times 9.9)$$
$$= 133.6 \text{ N/mm}^2$$

Bearing stress at the root of the column web

$$f_b = R_O/\{[T_b + 2 \times 2.5(T_c + r_c)]t_c\}$$
$$= 407.6E3/\{[18.8 + 2 \times 2.5(15.4 + 15.2)]9.9\} = 239.6 \text{ N/mm}^2$$

Axial stress in column web

$$f_a = N/A = 405E3/12.33E3 = 32.8 \text{ N/mm}^2$$

Bending stress at the root radius of the column web

$$f_{bc} = V(e + D_c/2)(d_c/2)/1$$
$$= 405E3(600 + 307.8/2)(246.5/2)/222.02E6 = 169.5 \text{ N/mm}^2$$

Inserting these values in the combined stress Equation (2.4)

$$f_e = [(f_{bc} + f_a)^2 + f_b^2 - (f_{bc} + f_a)f_b + 3f_q^2]^{1/2}$$
$$= [(169.5 + 32.8)^2 + 239.6^2 - (169.5 + 32.8)239.6 + 3 \times 133.6^2]^{1/2}$$
$$= 321.6 < 355 \text{ N/mm}^2, \quad \text{therefore satisfactory.}$$

EXAMPLE 7.7 'Rigid' beam-to-column connection

Determine the size of the components for the connection shown in Fig. 7.26.

Check for slip assuming rotation about axis O–O and ignoring friction from HSFG bolts. Inserting values in Equation (7.39)

$$\mu M/[Vd] = 0.45 \times 97.5E6/[60E3 \times (310.9 - 13.7)]$$
$$= 2.46 > 1$$

therefore rotation about the compression flange without slip, and no shear force applied to the bolts.

Assuming rotation about axis O–O the tensile force in one of the top four bolts

$$F_t = M/(4d_f) = M/[4(D - T)] = 97.5E3/[4(310.9 - 13.7)] = 82.02 \text{ kN}$$

From Appendix A4(a) the tensile resistance of an M16 PSFG bolt is 82.9 kN, therefore satisfactory.

Thickness of end plate based on a single plastic yield line in the end plate extension (see Equation (7.27))

$$t_p = \sqrt{[4F_t c/(B_p/p_{yp})]}$$
$$= \sqrt{[4 \times 82.02E3 \times 35/(200 \times 355)]} = 12.72 \text{ mm.}$$

Use 15 mm plate.

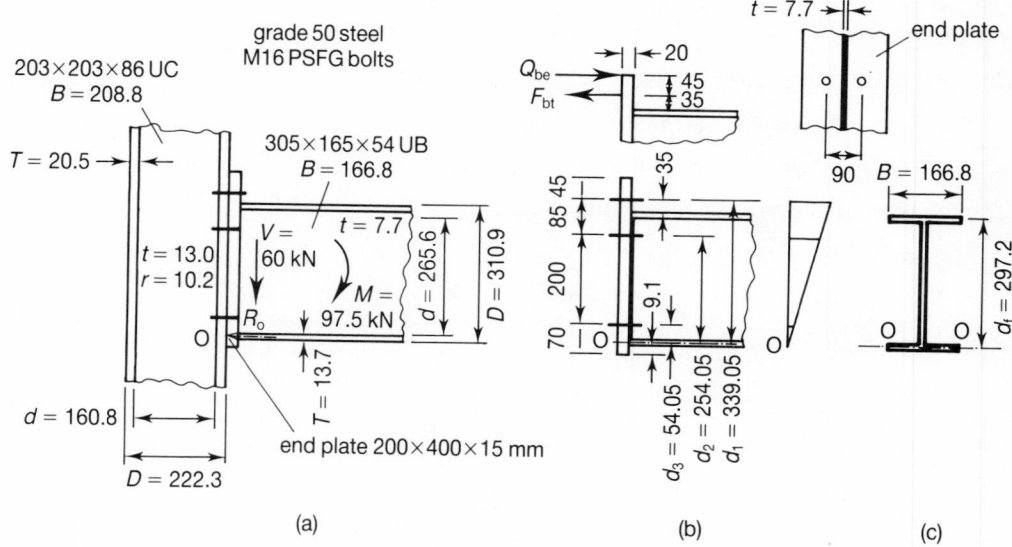

Fig. 7.26 'Rigid' beam-to-column connection.

Check thickness of column flange from yield line theory for two bolts (Equation (7.29a))

$$T = \sqrt{\{[8F_t c/(ap_{yp})]/(4\sqrt{2} + s_b/a)\}}$$
$$= \sqrt{\{[8 \times 82.02E3 \times 38.5/(97.9 \times 345)]/(4\sqrt{2} + 85/97.9)\}}$$
$$= 10.71 < 20.5 \text{ mm,} \quad \text{therefore satisfactory.}$$

Check buckling strength of the web of the column adjacent to axis O–O

Reaction $R_O = 4F_t = 4 \times 82.08 = 328.3$ kN

Shear strength of the column web (cl 4.2.3, Pt 1)

$$P_v = 0.6Dtp_{yw} = 0.6 \times 310.9 \times 7.7 \times 345/1E3 = 495.5 \text{ kN} > R_O,$$
$$\text{therefore satisfactory.}$$

Check for shear buckling (cl 3.6.2, Pt 1)

$$d/t = 160.8/13 = 12.37$$
$$63\varepsilon = 63\sqrt{(275/345)} = 56.25 > 12.37, \quad \text{therefore satisfactory.}$$

Stiff bearing length

$$b_1 = T_b + 2t_p = 13.7 + 2 \times 15 = 43.7 \text{ mm}$$

Width of equivalent strut in column

$$n_1 = D_c = 222.3 \text{ mm}$$

Slenderness ratio

$$\lambda = 2.5d/t = 2.5 \times 160.8/13 = 30.92$$

Buckling stress (cl 4.7.5, Pt 1) for grade 50 steel, $p_y = 345 \text{ N/mm}^2$ and $T = 20.5$ mm, is $p_c = 311 \text{ N/mm}^2$, Table 27(c), Pt 1.

Web buckling design strength of the web of the UC from Table 27(c), Pt 1

$$P_w = (b_1 + n_1)tp_c$$
$$= (43.7 + 222.3)13 \times 311/1\text{E3}$$
$$= 1075.4 \text{ kN} > R_O, \quad \text{therefore satisfactory.}$$

Check bearing strength of the web of the column adjacent to axis O–O.

Dispersion bearing length

$$n_2 = 2 \times 2.5(T_c + r_c) = 5 \times (20.5 + 10.2) = 153.5 \text{ mm}$$

Web bearing design strength of the web of the UC (cl 4.5.3, Pt 1)

$$P_{bg} = (b_1 + n_2)tp_{yw}$$
$$= (43.7 + 153.5)13 \times 345/1\text{E3}$$
$$= 884.4 > 328.1 \text{ kN} (R_O), \quad \text{therefore satisfactory.}$$

For the weld connecting the end plate to the end of the beam the second moment of area of unit size welds about axis O–O

$$I_O = 2(d_f^3/3 + Bd_f^2)$$
$$= 2(297.2^3/3 + 166.8 \times 297.2^3) = 46.97\text{E6 mm}^4$$

Maximum force per unit length of fillet weld

$$F_x = My/I_O = 97.5\text{E3} \times 297.2/46.97\text{E6} = 0.617 \text{ kN/mm}$$

From Appendix A1 the design strength for a 6 mm fillet weld is 0.903 kN/mm for grade 50 steel and E43 electrodes.

Additional calculations

Check the forces on the bolts assuming a linear variation of forces from axis O–O to the bolts furthest from the axis. Second moment of area of unit size connecting bolts about axis O–O

$$I_O = 2(d_3^2 + d_2^2 + d_1^2)$$
$$= 2(54.05^2 + 254.05^2 + 339.05^2) = 364.8\text{E3 mm}^4$$

Maximum tensile force acting on the bolt furthest from axis O–O

$$F_t = My/I_O = 97.5\text{E3} \times 339.05/(364.8\text{E3}) = 90.62 \text{ kN}$$

Prying force for each bolt assuming $\delta = 0$ for a PSFG bolt and elastic behaviour from Equation (7.12)

$$Q_{be} = [F_{be} - 2EI\delta_b/(a_p b_p^2)]/[2a_p/b_p + (1/3)(a_p/b_p)^2]$$
$$= [90.62 - 0]/[2 \times 45/35 + (1/3) \times (45/35)^2] = 29.02 \text{ kN}$$

Maximum total force in a bolt

$$F_{bt} + Q_{be} = 90.62 + 29.02 = 119.6 > 82.9 \text{ kN},$$

therefore increase size of a PSFG bolt from M16 to M20 ($P_t = 129.6$ kN).

Check limits of end plate thickness from Equation (7.43)

$$\{M/[d_{bf}p_y(4w_p/s_v + d_{bf}/s_h)]\}^{1/2} < t_p < \{Mb_p/(2w_p d_{bf} p_y)\}^{1/2}$$
$$\{97.5E6/[297.2 \times 355(4 \times 100/85 + 297.2/90]\}^{1/2} < t_p < \{97.5E6 \times 35/$$
$$(2 \times 100 \times 297.2 \times 355)\}^{1/2}$$
$$10.74 < t_p < 12.7 \text{ mm}$$

The actual plate thickness is 15 mm which is outside the range of values and is safe. However it means that the plate is too thick for the multiple yield line patterns to form before bolt failure.

From Equation (7.44) the limits of the column flange thickness

$$\sqrt{[M/(d_{bf}p_y)]} = \sqrt{[97.5E6/(297.2 \times 345)]} = 30.84 \text{ mm}$$
$$0.28\sqrt{[M/(d_{bf}p_y)]} < T < 0.39\sqrt{[M/(d_{bf}p_y)]}$$
$$0.28 \times 30.84 < T < 0.39 \times 30.84$$
$$8.63 < T < 12.03$$

Actual thickness of column flange is 20.5 mm which is outside the range of values and is safe. However it means that the flange is too thick for the multiple yield line patterns to form before bolt failure.

EXAMPLE 7.8   *'Rigid' beam-to-beam connection*

Determine the size of the components for the connection shown in Fig. 7.27.

Assuming that rotation occurs about axis O–O and the total tension is resisted by the cover plate.

Tensile force in the flange cover plate

$$F_f = M/D = 97.5E6/303.8E3 = 320.9 \text{ kN}$$

Thickness of the flange cover plate assuming $B = 165$ mm and 20 mm diameter bolts

$$t_p = F_f/[(B - 2d_h)p_y] = 320.9E3/[(165 - 2 \times 22)275]$$
$$= 9.64, \quad \text{use 10 mm thick grade 43 steel plate.}$$

Number of M20 PSFG bolts required (see Appendix A4(a)) for the cover plate

$$n_b = F_f/P_{sL} = 320.9/71.3 = 4.50, \quad \text{use six M20 PSFG bolts.}$$

Reaction at the hinge is equal to the force in the flange, $R_O = F_f = 320.9$ kN.

Frictional resistance at the hinge

$$\mu R_O = 0.45 \times 320.9 = 144.4 > 45 \text{ kN (applied shear force), therefore no slip occurs.}$$

Nevertheless check the shear resistance of the bolts through the end plate.

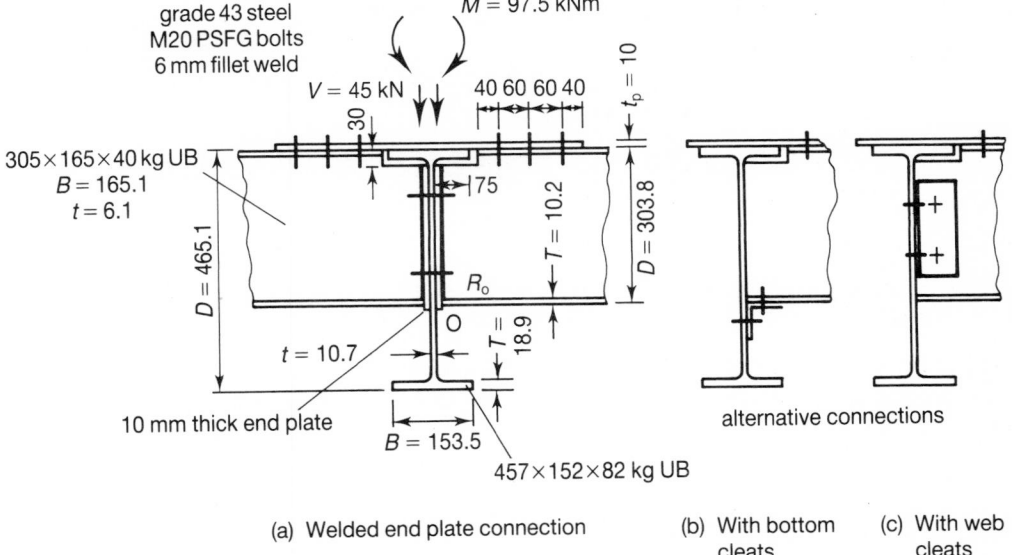

Fig. 7.27 'Rigid' beam-to-beam connections.

Shear resistance to four M20 PSFG bolts (see Appendix A4(a)) through the end plate

$$4P_{sL} = 4 \times 71.3 = 285.2 > 45 \text{ kN (applied shear force)}, \quad \text{therefore satisfactory.}$$

From bearing of the bolt on the plate of the thickness should be greater than 4.3 mm (see Appendix A4(a)). Use a 10 mm thick end plate welded to the end of the $305 \times 165 \times 40$ kg UB. The end plate is bolted to the $457 \times 152 \times 82$ kg UB as shown in Fig. 7.27(a).

Shear resistance of 6 mm fillet welds (see Appendix A1) through the end plate connecting the end plate to $305 \times 165 \times 40$ kg UB

$$2l_w P_w = 2 \times (303.8 - 30 - 10.2)0.903$$
$$= 476.1 > 45 \text{ kN (applied shear force)}, \quad \text{therefore satisfactory.}$$

EXAMPLE 7.9 'Rigid' beam splice

Determine the size of the components for the connection shown in Fig. 7.28.

Check if the beam is in the elastic stage of behaviour

$$f = M/z = 450E6/2076E3 = 216.8 < 355 \text{ N/mm}^2, \quad \text{therefore elastic behaviour}$$
$$I_{web} = t(D - 2T)^3/12 = 10.2(533.1 - 2 \times 15.6)^3/12 = 107.5E6 \text{ mm}^4$$

From Section Tables the gross second moment of area of the beam section

$$I_{gross} = 553.53E6 \text{ mm}^4$$

Proportion of the applied bending moment taken by the web

$$M_{web} = (I_{web}/I_{gross})M = (107.5E6/553.53E6)450 = 87.39 \text{ kNm.}$$

Check the strength of the arrangement of bolts in the web plate (see Fig. 7.28(b)).

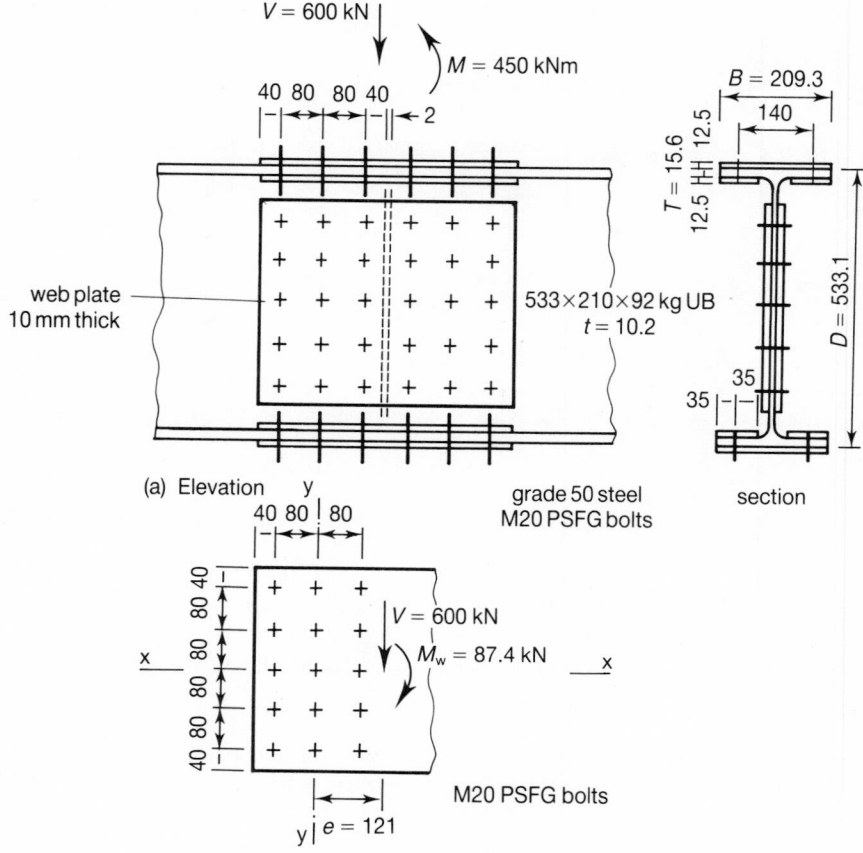

(a) Elevation

grade 50 steel
M20 PSFG bolts

section

(b) Web connection resisting shear force and
web bending moment

Fig. 7.28 'Rigid' beam splice.

Second moment of area of bolts of unit area about the centroidal x–x axis

$$I_x = \sum (\delta A)y^2 = 6(80^2 + 160^2) = 192\text{E3 mm}^4$$

Second moment of area of bolts of unit area about the centroidal y–y axis

$$I_y = \sum (\delta A)x^2 = 10 \times 80^2 = 64\text{E3 mm}^4$$

Second moment of area of bolts of unit area about the centroidal polar z–z axis

$$I_z = I_x + I_y = (192 + 64)\text{E3} = 256\text{E3 mm}^4$$

Eccentricity of the applied shear force to the centroid of half the bolt group is 121 mm (see Fig. 7.28(b)). This eccentricity produces a moment which is increased by the bending moment resisted by the web. The equivalent eccentricity

$$e' = e + M_{\text{web}}/V = 121 + 87.39\text{E3}/600 = 266.65 \text{ mm}$$

Maximum vector shear force in the x–x direction acting on a bolt furthest from the centroid of the bolt group

$$F_x = Ve'y_n/I_z = 600 \times 266.65 \times 160/256\text{E}3 = 99.99 \text{ kN}$$

Maximum vector shear force in the y–y direction acting on the same bolt

$$F_y = V/n + Ve'x_n/I_z = 600/15 + 600 \times 266.65 \times 80/256\text{E}3 = 90.00 \text{ kN}$$

Resultant maximum vector force acting on the same bolt

$$F_r = [F_x^2 + F_y^2]^{1/2} = [99.99^2 + 90.00^2]^{1/2} = 134.53 \text{ kN}$$

Double shear strength of an M20 PSFG from Appendix A4(a) is $2 \times 71.3 = 142.6 > 134.53$ kN, which is satisfactory.

To avoid crumpling and tearing the thickness of the web cover plates should be approximately half the bolt diameter. Use 10 mm thick plates.

Force to be resisted by the flange splice

$$F_f = (M - M_{web})/(D - T) = (450 - 87.39)\text{E}3/(533.1 - 15.6) = 700.7 \text{ kN}$$

Number of M20 PSFG bolts in double shear required

$$n_b = F_f/(2P_{sL}) = 700.7/(2 \times 71.3) = 4.91, \quad \text{use 6 bolts.}$$

Thickness of the flange plates

$$t_p = F_f/(B - 2d_h + 2w_p - 2d_h)p_y$$
$$= 700.7\text{E}3/(209.3 - 2 \times 22 + 2 \times 70 - 2 \times 22)355 = 7.55 \text{ mm.}$$

Use 10 mm plates as shown in Fig. 7.28(a). Length of the joint does not reduce the strength.

EXAMPLE 7.10 'Rigid' column splice

Determine the size of the components for the connection shown in Fig. 7.29.

Where column sections are of the same serial size it is possible to connect them directly with web and flange plates.

The ends of the column are machined and will be in contact. Rotation will take place about an axis near the outer edge of the flange of the upper column.

Thickness of the flange plate, from moments of forces about the axis of rotation

$$t_p = [M - N(D_u/2)]/[(B_p - 2d_h)p_y D_u]$$
$$= [480\text{E}6 - 712.5\text{E}3(355.6/2)]/[(365 - 2 \times 24)355 \times 355.6]$$
$$= 8.83 \text{ mm,} \quad \text{use 10 mm thick plate.}$$

Number of M22 PSFG bolts in single shear in the flange plates (see Appendix A4(a))

$$n_b = [M - N(D_u/2)]/(P_{sL}D_u)$$
$$= [480\text{E}6 - 712.5\text{E}3(355.6/2)]/(87.6\text{E}3 \times 355.6)$$
$$= 11.34, \quad \text{use 12 bolts.}$$

Length of the lap is not sufficient to reduce the shear strength of the connection (see Section 7.21).

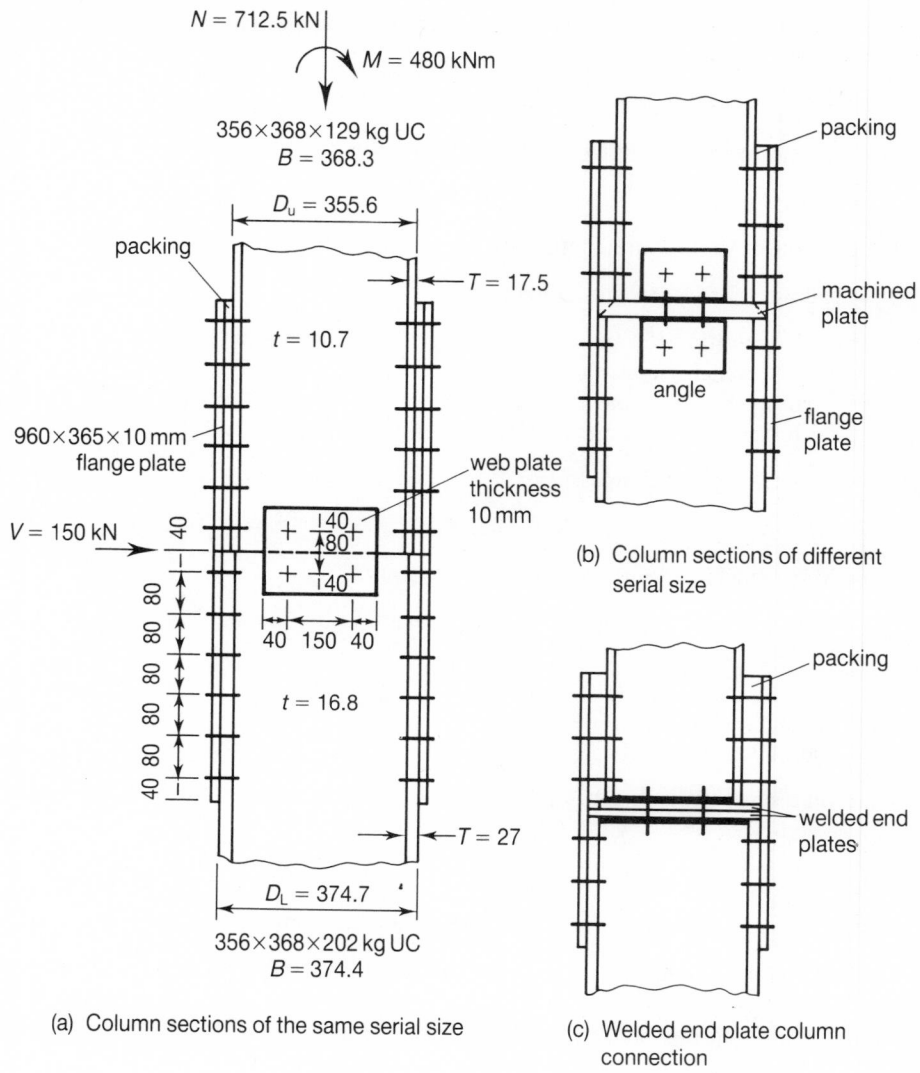

$N = 712.5$ kN

$M = 480$ kNm

356×368×129 kg UC
$B = 368.3$

$D_u = 355.6$

packing

$T = 17.5$

$t = 10.7$

960×365×10 mm
flange plate

$V = 150$ kN

40

80 80

80 80

80 80

40 80

40

|40|
|80|

+

+ |40|

40 150 40

$t = 16.8$

web plate
thickness
10 mm

$T = 27$

$D_L = 374.7$

356×368×202 kg UC
$B = 374.4$

(a) Column sections of the same serial size

packing

machined
plate

angle

flange
plate

(b) Column sections of different
serial size

packing

welded end
plates

(c) Welded end plate column
connection

Fig. 7.29 'Rigid' splices in steel columns.

Where the ends of the column are machined and in contact the horizontal shear force on the column is resisted by the friction force, in part of whole, at the point of contact, i.e. at the axis of rotation.

From Section 7.25 assuming that for machined surfaces $\mu = 0.15$ the frictional resistance

$$= \mu(M/D_u + N/2)$$
$$= 0.15(480\text{E}3/355.6 + 712.5/2) = 255.9 > 150 \text{ kN (applied shear force)},$$

therefore theoretically no shear connection is required. However in practice a web plate is generally provided.

If the frictional resistance is ignored then the splice is designed to resist the entire shear force as follows.

Second moments of area of two bolts of unit area on one side of the web connection about the centroidal axes are

$$I_x = 0$$
$$I_y = 2 \times 75^2 = 11.25E3 \text{ mm}^4$$
$$I_z = I_x + I_y = 11.25E3 \text{ mm}^4$$

Vector force on a bolt in the x–x direction

$$F_x = V/n_b = 150/2 = 75 \text{ kN}$$

Vector force on a bolt furthest from the centre of rotation in the y–y direction

$$F_y = (Ve)x_n/I_z = 150 \times 40 \times 75/11.25E3 = 40 \text{ kN}$$

Maximum vector shear force on the same bolt

$$F_r = (F_x^2 + F_y^2)^{1/2} = (75^2 + 40^2)^{1/2} = 85 \text{ kN}$$

Double shear resistance of an M22 PSFG from Appendix A4(a) is $2P_{sL} = 2 \times 87.6 = 175.2 > 85$ kN, therefore satisfactory.

EXAMPLE 7.11 'Rigid' built-up column base connection

Determine the size of the components for the connection shown in Fig. 7.30. The concrete cube strength is $f_{cu} = 20 \text{ N/mm}^2$.

Distance between holding down bolts (see Equation 7.47), assuming M36 grade 4.6 (see Appendix A2(a) for tensile strength)

$$d_p = M/[(N/2) + nP_t] = 264E3/(198/2 + 2 \times 159.3) = 632.2 \text{ mm}$$

Assume a bolt edge distance of approximately $2d = 2 \times 36 = 72$ mm, use 70 mm.

Total length of base plate

$$D_p = d_p + 2e_b = 632.2 + 2 \times 70 = 772.2 \text{ mm}, \quad \text{use length of 800 mm.}$$

Minimum width of base plate

$$B_p = B_c + 2t_g + 2s + d_w + 2e_b$$
$$= 208.8 + 2 \times 12.5 + 2 \times 10 + 66 + 70 = 459.8 \text{ mm. Use width of 460 mm.}$$

The thickness of gusset plate t_g and size of weld(s) are assumed. The diameter of the washer d_w for an M36 holding down bolt is obtained from Appendix A2(c).

Cantilever length (see Fig. 7.30(b))

$$a = 0.5(B_p - B - 2t_g) = 0.5(460 - 208.8 - 2 \times 12.5) = 113.1 \text{ mm}$$

Design bearing strength for the concrete (cl 4.13.1, Pt 1)

$$w = 0.4f_{cu} = 0.4 \times 20 = 8 \text{ N/mm}^2$$

Length of the concrete compression zone (see Equation (7.46)) assuming lever arm $l_a = 660$ mm

$$x_p = (M/l_a + N/2)/(B_p w) = (264E6/660 + 198E3/2)/(460 \times 8) = 135.6 \text{ mm}$$

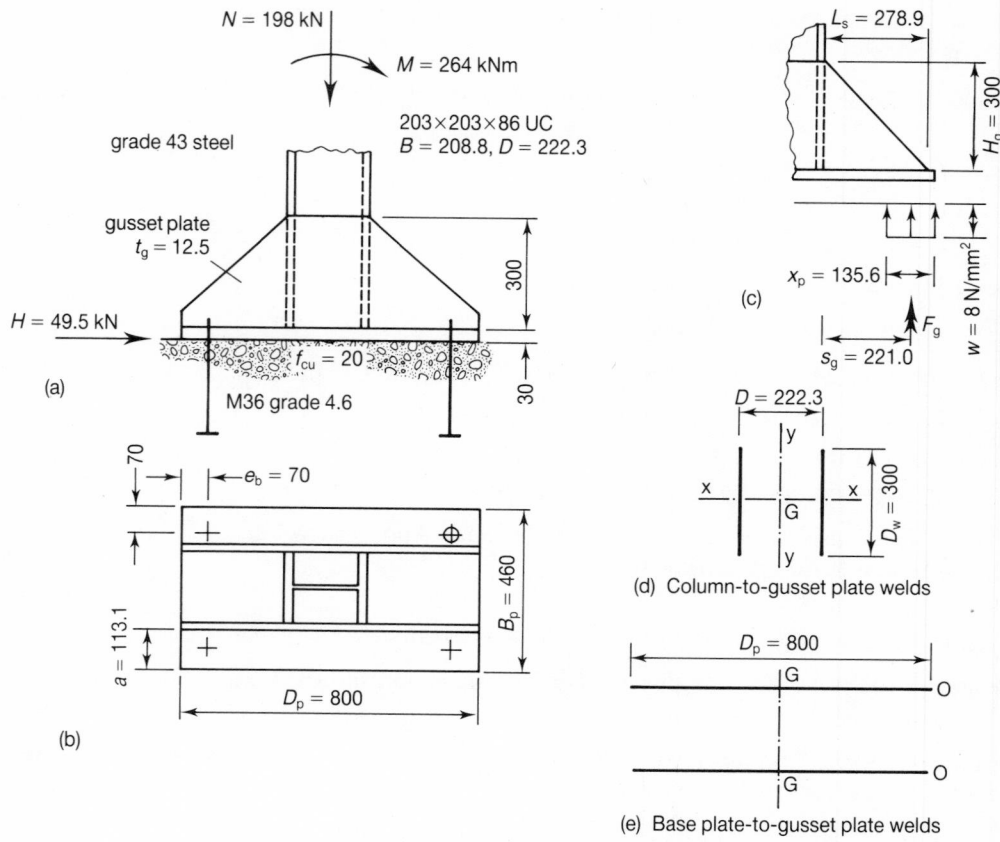

Fig. 7.30 'Rigid' column-to-foundation connection.

Check lever arm for resistance of concrete in bending at ultimate limit state

$$l_a = D_p - e_b - x_p/2 = 800 - 70 - 135.6/2 = 662.2 \text{ mm}$$

Tensile force in a holding down bolt (see Equation (7.47))

$$F_{bt} = [M - N(D_p/2 - x_p/2)]/(n_b l_a)$$
$$= [264E3 - 198 \times (800/2 - 135.6/2)]/(2 \times 662.2) = 149.7 \text{ kN}$$

From Appendix A2(a) the tensile design strength of an M36 grade 4.6 bolt is 159.3 kN.

Thickness of base plate based on plastic analysis and on the cantilever length (see Equation 7.26)

$$t_p = a\sqrt{(2w/p_y)} = 113.1 \times \sqrt{(2 \times 8/265)} = 27.79 \text{ mm}$$

The design strength p_y for the steel plate is obtained from Table 6, Pt 1.

Thickness of base plate based on plastic analysis and the tensile bolt force (see Equation (7.30))

$$t_p = \sqrt{\{4F_{bt}c/[ap_y(2 + e_b/a)]\}}$$

$$= \sqrt{\{(4 \times 149.7\text{E}3 \times 43.1/[113.1 \times 265(2 + 70/113.1)]\}}$$
$$= 18.13 \text{ mm}$$

Use $800 \times 460 \times 30$ mm base plate grade 43 steel.

Force applied to each gusset plate (see Fig. 7.30(c))

$$F_g = w(B_p/2)x_p = 8 \times (460/2) \times 135.6/1\text{E}3 = 249.5 \text{ kN}$$

Length of gusset plate allowing for 10 mm for weld

$$L_g = (D_p - 2 \times 10 - D_c)/2 = (800 - 20 - 222.3)/2 = 278.85 \text{ mm}$$

Let height of gusset plate $H_g = 300$ mm.

Eccentricity of force F_g in relation to the inner corner of the gusset plate (see Fig. 7.30(c))

$$s_g = (D_p - D_c)/2 - x_p/2 = 288.85 - 135.6/2 = 221.05 \text{ mm}$$

Width of gusset plate (see Equation (7.34))

$$W_g = L_g/[1 + (L_g/H_g)^2]^{1/2} = 278.85/[1 + (278.85/300)^2]^{1/2} = 204.2 \text{ mm}$$

Thickness of gusset plate grade 43 steel (see Equation (7.33))

$$t_g = 2F_g s_g/(p_{yg} W_g^2) + W_g/80$$
$$= 2 \times 249.5\text{E}3 \times 221.05/(275 \times 204.2^2) + 204.2/80 = 12.17 \text{ mm}$$

Use 12.5 mm thick grade 43 steel gusset plate.

Check slenderness ratio of gusset plate (see Equation (7.35))

$$l_g/r_g = 2\sqrt{3}W_g/t_g = 2\sqrt{3} \times 204.2/12.5$$
$$= 56.59 < 185, \quad \text{therefore satisfactory.}$$

Alternatively based on elastic bending, Equation (7.25) (cl 4.13.2.4, Pt 1)

$$t_g = 5P_g s_g/(p_{yg} H^2) = 5 \times 249.5\text{E}3 \times 221.05/(270 \times 300^2) = 11.3 \text{ mm}.$$

Minimum length of foundation bolt (see Holmes and Martin)

$$L_b = \sqrt{[F_{bt}/(\pi p_{tc})]} = \sqrt{[149.6\text{E}3/(\pi \times 0.36\sqrt{20})]} = 172.0 \text{ mm}$$

Use four M36 grade 4.6 holding down bolts, 300 mm long anchored by a washer plate (cl 6.7, Pt 1).

If the surfaces between the gusset plate and the base plate are not machined and no contact is assumed (cl 4.13.3, Pt 1), then the size of the fillet weld is obtained as follows.

Length of weld (see Fig. 7.30(e))

$$L_w = 2D_p = 2 \times 800 = 1.6\text{E}3 \text{ mm}$$

Second moment of area about centroid of the weld group for unit size weld

$$I_{wG} = 2D_p^3/12 = 2 \times 800^3/12 = 85.33\text{E}6 \text{ mm}^4$$

Force per unit length of weld in the y direction

$$F_{wy} = N/L_w + M(D_p/2)/I_{wG}$$
$$= 198/1.6\text{E}3 + 264\text{E}3 \times (800/2)/85.33\text{E}6 = 1.3613 \text{ kN/mm}$$

Force per unit length of weld in the z direction

$$F_{wz} = H/L_w = 49.5/1.6E3 = 0.0309 \text{ kN/mm}$$

Resultant vector force per unit length of weld

$$F_{wr} = (F_{wy}^2 + F_{wz}^2)^{1/2} = (1.3613^2 + 0.0309^2)^{1/2} = 1.362 \text{ kN/mm}$$

Use 10 mm fillet weld which from Appendix A1 has design strength of 1.505 kN/mm for grade 43 steel.

Alternatively if the surfaces between the base plate and the edge of the gusset plate are machined and bearing is assumed at the axis of rotation O–O (see Fig. 7.30(e)).

From Equation (7.38)

$$\begin{aligned} R_O &= (ND_p/6 + M)/(2D_p/3) \\ &= (198 \times 800/6 + 264E3)/(2 \times 800/3) = 544.5 \text{ kN} \end{aligned}$$

Frictional resistance at R_O

$$= \mu R_O = 0.15 \times 544.5 = 81.65 > 49.5 \text{ kN } (H), \quad \text{therefore satisfactory.}$$

Second moment of area about axis O–O for unit size weld (see Fig. 7.30(e))

$$I_{wO} = 2D_p^3/3 = 2 \times 800^3/3 = 341.3E6 \text{ mm}^4$$

Force per unit length of weld in the y direction

$$\begin{aligned} F_{wy} &= (M - ND_w/2)/I_{wO} \\ &= (264E3 - 198 \times 800/2)/341.3E3 = 0.5415 \text{ kN/mm} \end{aligned}$$

Use 6 mm fillet weld which from Appendix A1 has design strength of 0.903 kN/mm for grade 43 steel.

If the end of the column is not machined then rotation is assumed to be about axis G–G and the size of the weld connecting the column to the gusset plate is obtained as follows.

Length of weld (see Fig. 7.30(d))

$$L_w = 4D_w = 4 \times 300 = 1.2E3 \text{ mm}$$

Second moments of area about the centroid of the weld group (axis G–G) for unit size welds

$$\begin{aligned} I_{wGx} &= 4D_w^3/12 = 4 \times 300^3/12 = 9E6 \text{ mm}^4 \\ I_{wGy} &= 4D_w(D/2)^2 = 4 \times 300(222.3/2)^2 = 14.83E6 \text{ mm}^4 \end{aligned}$$

Polar second moment of area

$$I_{wGz} = I_{wGx} + I_{wGy} = (9.0 + 14.83)E6 = 23.83E6 \text{ mm}^4$$

Force per unit length of weld in the y direction on an element furthest from the axis of rotation

$$F_{wy} = N/L_w + (M - HD_w/2)(D_c/2)I_{wGz}$$
$$= 198/1.2E3 + (264E3 - 49.5 \times 300/2) \times (222.3/2)/23.83E6$$
$$= 1.362 \text{ kN/mm}$$

Force per unit length of weld in the x direction on the same element

$$F_{wx} = H/L_w + [M - HD_w/2](D_w/2)/I_{wGz}$$
$$= 49.5/1.2E3 + (264E3 - 49.5 \times 300/2) \times (300/2)/23.83E6$$
$$= 1.656 \text{ kN/mm}$$

Resultant vector force per unit length of weld

$$F_{wr} = (F_{wx}^2 + F_{wy}^2)^{1/2} = (1.656^2 + 1.362^2)^{1/2} = 2.144 \text{ kN/mm}$$

Use 15 mm fillet weld which from Appendix A1 has design strength of 2.257 kN/mm for grade 43 steel.

Alternatively if the contact between the base plate and the end of the column is machined then the axis of rotation is O–O (see Fig. 7.30(d)).

Second moments of area about centroid of the weld group for unit size weld about axis O–O

$$I_{wOx} = 4D_w^3/3 = 4 \times 300^3/3 = 36E6 \text{ mm}^4$$
$$I_{wby} = 2D_w D^2 = 2 \times 300 \times 222.3^2 = 29.65E6 \text{ mm}^4$$

Polar second moment of area

$$I_{wOz} = I_{wOx} + I_{wOy} = (36 + 29.65)E6 = 65.65E6 \text{ mm}^4$$

Force per unit length of weld in the y direction on an element furthest from the axis of rotation

$$F_{wy} = (M - ND_c/2)D_c/I_{wOz}$$
$$= (264E3 - 198 \times 222.3/2) \times 222.3/65.65E6 = 0.819 \text{ kN/mm}$$

Force per unit length of weld in the x direction on the same element

$$F_{wx} = [M - ND_c/2]D_w/I_{wOz}$$
$$= (264E3 - 198 \times 222.3/2) \times 300/65.65E6 = 1.106 \text{ kN/mm}$$

Resultant vector force per unit length of weld

$$F_{wr} = (F_{wx}^2 + F_{wy}^2)^{1/2} = (1.106^2 + 0.819^2)^{1/2} = 1.376 \text{ kN/mm}$$

Use 10 mm fillet weld which from Appendix A1 has design strength of 1.505 kN/mm for grade 43 steel.

EXAMPLE 7.12   *'Rigid' RHS connection*

Determine the size of the components for the connection shown in Fig. 7.31.

The connection between the vertical and horizontal member is a butt weld which necessitates a 45° chamfer on the end of the vertical member. Failure of the weld is assumed to

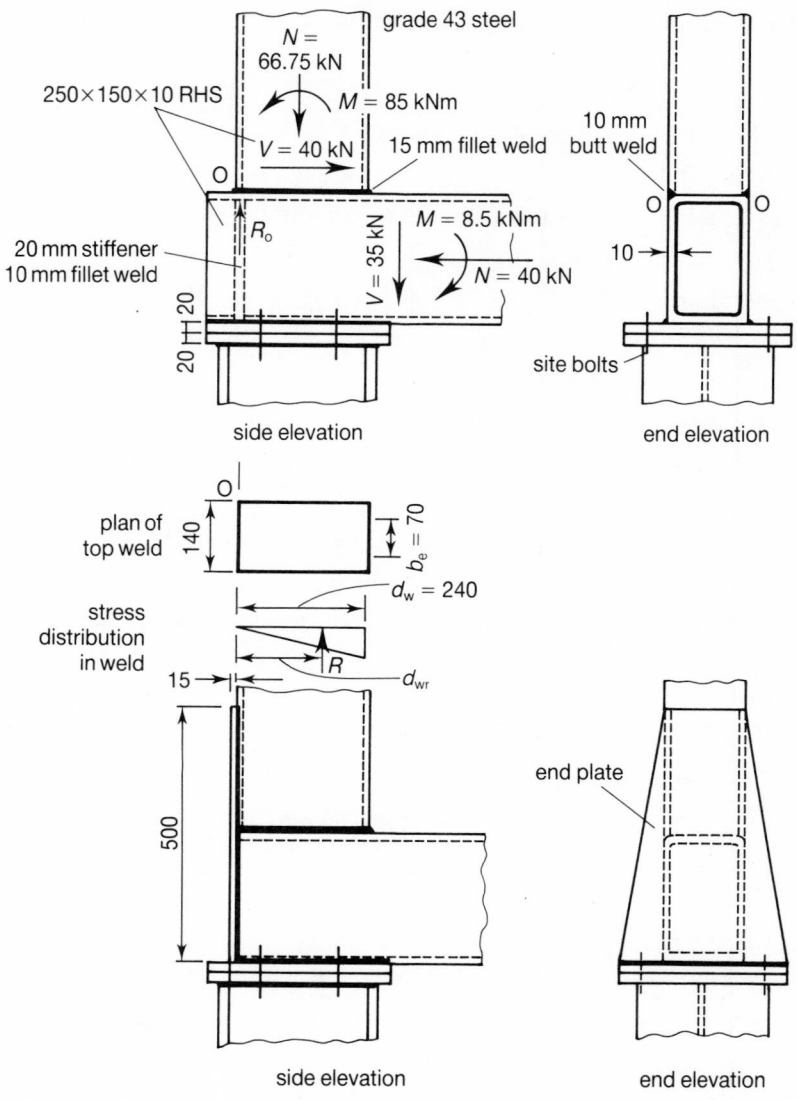

Fig. 7.31 'Rigid' rectangular hollow steel (RHS) connection.

take place by rotation about the axis O–O (see Fig. 7.31). The axis O–O is assumed to be a stiff bearing because the intersecting members are of the same width.

From Table 7.1 the effective length of the weld furthest from the axis O–O

$$b_{we} = 2t + CT = 2 \times 10 + 5 \times 10 = 70 \text{ mm}$$

Distance d_r from the axis O–O from the resultant force in the weld is obtained as follows.

moment of the parts = moment of the whole

$$(b_{we} + 1/2 \times 2d_w)F_x d_r = F_x I_{oe}/d_r$$

rearranging and substituting $I_{oe} = 2d_w^3/3 + b_{we}d_w^2$

$$d_r/d_w = (2/3 + b_{we}/d_w)/(1 + b_{we}/d_w)$$
$$= (2/3 + 70/240)/(1 + 70/240) = 0.742$$

Resultant reaction from Equation (7.38)

$$R_O = [M + N(d_r - d_w/2)]/d_r$$
$$= [85E3 + 66.75(0.742 \times 240 - 240/2)]/(0.742 \times 240) = 499.1 \text{ kN}$$

Frictional force at the stiff bearing if the end of the vertical RHS is machined

$$\mu R_O = 0.15 \times 499.1 = 74.87 > 40 \text{ kN (applied shear force)}.$$

The weld group is subjected to the actions from N and M only.

Second moment of area of the weld group (see Fig. 7.31) for unit size welds

$$I_O = 2d_w^3/3 + b_{we}d^2$$
$$= 2 \times 240^3/3 + 70 \times 240^2 = 13.25E6 \text{ mm}^4$$

Moment applied about axis O–O

$$M' = M - Nd_w/2 = 85 - 66.75 \times 240/(2 \times 1E3) = 76.99 \text{ kNm}$$

Force per unit length of weld furthest from the axis of rotation

$$F_{wy} = M'd/I_O = 76.99E3 \times 240/13.25E6 = 1.395 \text{ kN/mm}$$

From Appendix A1, $P_w = 1.505$ kN/mm for a 10 mm weld for grade 43 steel.

The web buckling and web bearing clauses were developed for 'I' beams but are assumed to apply to this situation.

Slenderness ratio

$$\lambda = 2.5d/t = 2.5 \times (250 - 2 \times 10)/10 = 57.5$$

Buckling stress (cl 4.7.5, Pt 1) for grade 43 steel, $p_y = 275$ N/mm^2 and $t = 10$ mm, is $p_c = 205$ N/mm^2, Table 27(c), Pt 1.

Web buckling design strength of the RHS (cl 4.5.2.1, Pt 1), from Equation (4.37)

$$P_w = (b_1 + n_1)tp_c$$
$$= (10 + 250/2)10 \times 205/1E3$$
$$= 553.5 > 499.1 \text{ kN } (R_O), \quad \text{therefore satisfactory.}$$

Web bearing design strength of the RHS (cl 4.5.3, Pt 1), from Equation (4.39)

$$P_{wbg} = (b_1 + n_2)tp_{yw}$$
$$= (10 + 2.5 \times 10)10 \times 275/1E3$$
$$= 192.5 < 499.1 \text{ kN } (R_O), \quad \text{therefore stiffeners required.}$$

The force R_O may crush the webs of the horizontal RHS section and a stiffener is welded in the end of the horizontal section (see Fig. 7.31).

Crushing strength of a 20 mm thick stiffener from cl 4.5.4.2, Pt 1

$$P_s = 0.8b_s t_s p_y = 0.8 \times (150 - 20) \times 20 \times 275/1E3 = 572 > 499.1 \text{ kN } (R_O)$$

A stiffener inside the horizontal tube is difficult to place and weld. An alternative to prevent web crushing is to weld a plate outside the end of the joint as shown in Fig. 7.31. The vertical welds must be capable of transferring the force R_O to the horizontal RHS. If 10 mm fillet welds are used then the effective length required

$$L_w = R_O/P_w = 499.1/1.505 = 331.6 \text{ mm}$$

Use two 10 mm fillet welds of total length 500 mm as shown and a plate thickness of 10 mm to match the thickness of the RHS.

EXAMPLE 7.13   *'Rigid' knee connection for a portal frame*

Determine the size of the components for the connection shown in Fig. 7.32.

Design moment about axis O–O

$$M_O = 550.2 - 116.8 \times 0.5379 = 487.4 \text{ kNm}$$

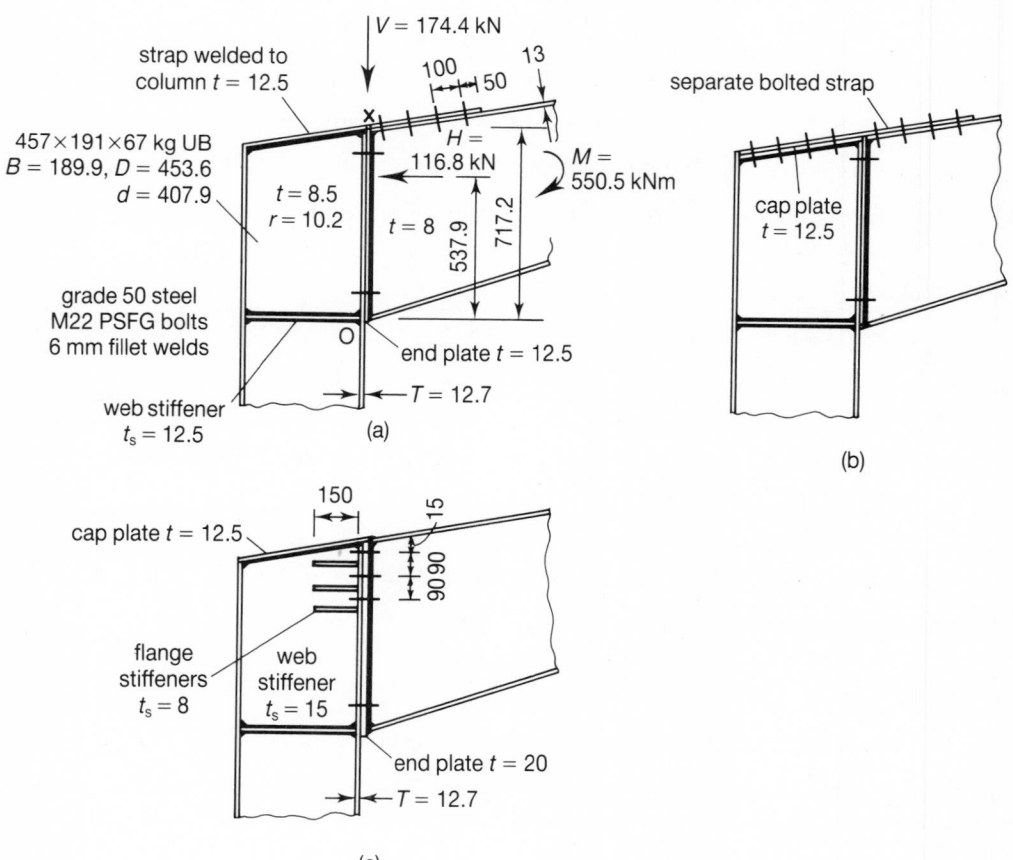

Fig. 7.32 'Rigid' knee connection for a portal frame.

Solution (*a*) using a strap welded to the top of the column and bolted to the beam as shown in Fig. 7.32(a).

Number of M22 PSFG bolts in single shear in the strap

$$n_b = M_O/(l_a P_{sL}) = 487.4E3/(717.2 \times 87.6) = 7.76, \quad \text{use 8 bolts.}$$

Thickness of strap assuming 200 mm wide

$$
\begin{aligned}
t_{st} &= M_O/(l_a b_{st} p_{yst}) \\
&= 487.4E6/[717.2 \times (200 - 2 \times 24) \times 355] \\
&= 12.6 \text{ mm.} \quad \text{Use a 12.5 mm thick strap.}
\end{aligned}
$$

Force per unit length of weld connecting strap to head of column.

$$F = M_O/(l_a n_w l_w) = 487.4E3/[717.2 \times 2 \times 407.9] = 0.833 \text{ kN/mm.}$$

Use a 6 mm fillet weld with E51 electrode, $P_w = 1.071$ kN/mm (see Appendix A1). This calculation assumes conservatively that the force is resisted by the welds along the web, but the strap is also welded to the flanges of the column.

Moments of forces about X to determine the reaction R_O about axis O–O

$$-717.2R_O + 550.2E3 + 116.8 \times (717.2 - 537.9) = 0$$

hence

$$R_O = 796.35 \text{ kN}$$

Vertical frictional force

$$= \mu R_O = 0.45 \times 796.35 = 358.4 > V = 174.4 \text{ kN,} \quad \text{therefore no slip.}$$

Shear strength of the web of the column (cl 4.2.3, Pt 1)

$$P_v = 0.6Dtp_{yw} = 0.6 \times 453.6 \times 8.5 \times 355/1E3 = 821.2 \text{ kN} > R_O,$$
$$\text{therefore satisfactory.}$$

Check for shear buckling of the web of the column (cl 4.2.3, Pt 1)

$$d/t = 407.9/8.5 = 48$$

$$63\sqrt{(275/p_{yw})} = 63 \times \sqrt{(275/355)} = 55.4 > 48, \quad \text{therefore satisfactory.}$$

Web bearing strength of the column (cl 4.5.3, Pt 1) from Equation (4.39)

$$
\begin{aligned}
P_{wbg} &= [b_1 + n_2]tp_{yw} \\
&= [13 + 2.5 \times (12.7 + 10.2)] \times 8.5 \times 355/1E3 \\
&= 212 \text{ kN} < R_O, \quad \text{therefore not satisfactory.}
\end{aligned}
$$

Web buckling strength of the column (cl 4.5.2.1, Pt 1) from Equation (4.37)

$$\lambda = 2.5d/t = 2.5 \times 407.9/8.5 = 120, \quad p_c = 106 \text{ N/mm}^2$$

$$
\begin{aligned}
P_w &= (b_1 + n_1)tp_c \\
&= (13 + 453.6) \times 8.5 \times 106/1E3 \\
&= 420.4 \text{ kN} < R_O \quad \text{therefore not satisfactory and stiffeners required.}
\end{aligned}
$$

Thickness of a load bearing web stiffener (cl 4.5.4.2, Pt 1)

$$t_s = 0.8R_O/[(B_c - t_c - 2r_c)p_{yw}]$$
$$= 0.8 \times 796.35E3/[(189.9 - 8.5 - 2 \times 10.2) \times 355]$$
$$= 11.15 \text{ mm}. \quad \text{Use 12.5 mm thick stiffener.}$$

Force per unit length of weld connecting the web stiffener to the column web

$$F = 0.8R_O/(n_w l_w) = 796.35/[2 \times 407.9] = 0.976 \text{ kN/mm}.$$

Use a 6 mm fillet weld with E51 electrode, $P_w = 1.071$ kN/mm (see Appendix A1). It is assumed conservatively that the force is resisted by the welds along the web but the stiffener is also welded to the flanges.

Solution (b) using a separate strap as shown in Fig. 7.32(b).

The design is similar to solution (a). The strap in solution (a) is welded to the head of the column in the workshop and may be distorted in transportation. A further alternative is to provide a strap that is separate and bolt it to the column and beam on the site as shown in Fig. 7.32(b). If the strap is long it may be bent and continued vertically along the flange of the column.

Solution (c) using bolts and welds only and no strap as shown in Fig. 7.32(c).

Try eight M22 PSFG bolts arranged as shown in Fig. 7.32(c).

Second moment of area of the upper six M22 PSFG bolts about the axis O–O

$$I_O = \sum ny^2$$
$$= 2 \times [(717.2 - 45 - 6.5)^2 + (717.2 - 135 - 6.5)^2 + (717.2 - 225 - 6.5)^2]$$
$$= 2.021E6 \text{ mm}^2$$

Moment of resistance of bolts about axis O–O

$$M_{rO} = P_t I_O/y_{max}$$
$$= 159.3 \times 2.021E6/[(717.2 - 45 - 6.5) \times 1E3]$$
$$= 483.6 \text{ kNm}, \quad \text{approximately equal to 487.4 kNm, therefore accepted.}$$

Distances (relating to the beam)

$$c = (s_b - t_b)/2 = (90 - 8)/2 = 41 \text{ mm}$$
$$a = (B_b - t_b)/2 = (172.1 - 8)/2 = 82.05 \text{ mm}$$

From Equation (7.29(b)) the thickness of the end plate

$$t_p = \sqrt{\{4nF_t c/(ap_{yp})/[4\sqrt{2} + (n-1)s_b/a]\}}$$
$$= \sqrt{\{4 \times 3 \times 159.3E3 \times 41/(82.05 \times 345)/[4\sqrt{2} + 2 \times 90/82.05]\}}$$
$$= 18.78 \text{ mm}, \quad \text{therefore use a 20 mm thick end plate.}$$

Column flange thickness $T_c = 12.7 < 18.78$ mm, therefore flange stiffeners are required on both sides of the column web as shown in Fig. 7.32(c). If a flange stiffener is required to resist the tensile bolt force then the thickness of the stiffener

$$t_s = F_t/(b_s p_{ys}) = 159.3E3/[(189.9 - 8.5 - 2 \times 10.2) \times 355]$$
$$= 2.79, \quad \text{use 8 mm.}$$

Force per unit length of weld connecting the stiffener to the column flange

$$F = P_t/(n_w l_w) = 159.3/[2 \times (189.9 - 8.5 - 2 \times 10.2)] = 0.495 \text{ kN/mm}.$$

Use a 6 mm fillet weld with E51 electrode, $P_w = 1.071$ kN/mm (see Appendix A1).

Force per unit length of weld connecting the stiffener to the column web

$$F = P_t/(n_w l_w) = 159.3/[2 \times 150] = 0.531 \text{ kN/mm}.$$

Use a 6 mm fillet weld with E51 electrode, $P_w = 1.071$ kN/mm (see Appendix A1).

The clear distance between stiffeners should be large enough to accommodate a bolt washer (see Appendix A4(c)) plus two fillet welds $= 50 + 2 \times 6 = 62 < 82$ mm, therefore satisfactory.

Distances (relating to the column)

$$c = (s_b - t_c)/2 = (90 - 8.5)/2 = 40.75 \text{ mm}$$
$$a = (B_c - t_c)/2 = (189.9 - 8.5)/2 = 90.7 \text{ mm}$$

The yield lines are restricted by the clear distance between the stiffeners $2d_s$. From Equation (7.28) where $x = d_s$ for a single bolt the thickness of the end plate

$$t_p = \sqrt{\{4F_t c/(a p_{yp})/[4a/d_s + 2d_s/a]\}}$$
$$= \sqrt{\{4 \times 159.3\text{E}3 \times 40.75/(90.7 \times 355)/[4 \times 90.7/(82/2) + 82/90.7]\}}$$
$$= 9.09 \text{ mm},$$

therefore thickness of the column flange $T_c = 12.7$ mm is satisfactory.

Size of the weld connecting the beam to the end plate, welded all round the beam profile, is obtained by assuming rotation about the axis O–O.

Second moment of area of the weld group about axis O–O

$$I_O = 2(717.2^3/3 + 172.1 \times 717.2^2) = 423\text{E}6 \text{ mm}^4$$

Maximum force per unit length of weld

$$F_x = M_O y/I_O = 487.4\text{E}3 \times 717.2/423\text{E}6 = 0.826 \text{ kN/mm}.$$

Use a 6 mm fillet weld with E51 electrode, $P_w = 1.071$ kN/mm (see Appendix A1).

The position of the reactive force R in the bolt group (see Section 7.25) is determined from

moment of parts = moment of whole

$$F_t I_O/y_{max} = nF_t d_r$$

hence the distance

$$d_r = I_O/(n y_{max}) = 2.021\text{E}6/[6 \times (717.2 - 45 - 6.5)] = 506 \text{ mm from axis O–O}$$

Taking moment of forces about R to determine the magnitude of R_O

$$-d_r R_O - (d_H - d_r)H + M = 0$$
$$-506 R_O - (537.9 - 506) \times 116.8 + 550.23\text{E}3 = 0; \quad \text{hence} \quad R_O = 1080 \text{ kN}$$

Shear strength of the web of the column (cl 4.2.3, Pt 1)

$$P_v = 0.6Dtp_y = 0.6 \times 453.6 \times 8.5 \times 355/1E3 = 821.2 \text{ kN} < R_0,$$

therefore not satisfactory and the web area of the column must be increased.

A 533 × 210 × 82 kg UB ($D = 528.3$ and $t = 9.6$ mm) is one solution but the weight of the section is increased.

Thickness of a load bearing stiffener (cl 4.5.4.2, Pt 1)

$$t_s = 0.8R_0/[(B_c - t_c - 2r_c)p_{yw}]$$
$$= 0.8 \times 1080E3/[(208.7 - 9.6 - 2 \times 12.7) \times 355]$$
$$= 13.0 \text{ mm.} \quad \text{Use 15 mm thick stiffener.}$$

A further alternative is to use a 457 × 191 × 98 kg UB for the column which is strong enough to resist R_0, and has a flange thickness of 19.6 mm which avoids the use of flange stiffeners.

Solutions (a) and (b) require different depth purlins over the end strip. Solution (c) is more expensive due to the increased amount of steel, or the increased work in welding flange stiffeners.

7.29   Tutorial problems

7.29.1   Bolted tie

A connection for a tie is of the type shown in Fig. 7.11. The size of the plate and the number of M24 grade 8.8 bolts on one side of the joint is shown in Fig. 6.1. Determine the tensile capacity of the tie if the plate is grade 43 steel.

Answer

T(plate) = 2279, T(bolts) = 1853.6, T(long joint) = 1827.65 kN.

7.29.2   'Pinned' roof truss joint

Check the strength of the connection of member 28 to the gusset plate as shown in Fig. 7.20. The member, which is in tension, is connected by the 75 mm leg. Use grade 43 steel.

Answer

$a_1 = 568$, $a_2 = 968$, $A_e = 1185.3$ mm^2, $P = 326 > 100$ kN, $L_w = 110.7 < 200$ mm.

7.29.3   'Pinned' beam-to-column connection

Check a connection of the type shown in Fig. 7.22 but with a 610 × 229 × 101 kg UB. The applied shear force $V = 300$ kN, the top angle is 100 × 100 × 10 mm and the bottom angle is 200 × 150 × 18 mm with the longer leg vertical. Use M20 grade 8.8 bolts, 3 mm clearance and grade 50 steel.

Answer

P(bolts) = 367.6, P_v = 1359.6, P_w = 341.66, P_{wbg} = 415.96 kN, 7.43 < t_a < 15.82 mm (for top angle), M_e = 44.23 kNm.

7.29.4 'Pinned' beam-to-beam connection

Replace the two angle cleats for the connection shown in Fig. 7.23 by a grade 43 plate welded to the end of the transverse beam A.

Answer

Assume the plate subjected to a shear force with no eccentricity. Minimum $D_p = V/(0.6p_y t)$ = 131.75, $D_p = 2e + s$ = 130, $D_p = V/(2P_w)$ = 83.1, $B_p = 2e + s$ = 130, t_p > 4.3 mm. Use 200 × 150 × 8 mm grade 43 plate with two 6 mm fillet welds.

7.29.5 Flange-web connection for a plate web girder

Determine the size of the fillet welds connecting the flange to the web for a plate web girder. The section is symmetrical about the principle axes with B = 540, T = 20, D = 1840, and t = 10 mm. The maximum shear force is 2000 kN and the steel is grade 43.

Answer

I_x = 22.75E9 mm^4, $VA\bar{y}/I_x$ = 0.864 kN/mm, 2–6 mm fillet welds = 1.806 kN/mm.

7.29.6 'Rigid' bracket

A 'rigid' bracket is welded to a column as shown in Fig. 7.25. The bracket is formed from a 406 × 140 × 39 kg UB and the column is a 152 × 152 × 23 kg UC. V = 70 kN, H = 0, e = 250 mm and L_g = 300 mm. Check the strength of the connection assuming 6 mm welds, grade 43 steel and rotation about the axis O–O.

Answer

H_g = 380.1, t_g = 5.24 < 6.3, l_g/r_g = 129.5 < 185, I_{woe} = 54.68E6 mm^4, d_r/d_f = 0.745, F_R = 0.131 kN/mm, R_O = 60.43, P_{wbg} = 135.2, P_w = 215.1 kN.

References

Astill A. W., Holmes M., and Martin L. H. (1980). *Web buckling of steel I beams*, CIRIA report, technical note 102.

Bahia C. S., Graham J., and Martin L. H. (1981). *Experiments on rigid beam-to-column connections subject to shear and bending forces*, Conference Proc., Joints in Structural Steelwork, Teeside Polytechnic.

Bahia C. S. and Martin L. H. (1980). *Bolt groups subject to torsion and shear*, Proc. I.C.E. Pt 2, V69.

Bahia C. S. and Martin L. H. (1981). *Experiments on stressed and unstressed bolt groups subject to torsion and shear*, Conference Proc., Joints in Structural Steelwork, Teeside Polytechnic.

Biggs M. S. A. B., Crofts M. R., Higgs J. D., Martin L. H., and Tzogius A. (1981). *Failure of fillet welded connections subject to static loading.* Conference Proc., Joints in Steelwork, Teeside Polytechnic.

BS 639 *Covered electrodes for the manual metal arc welding of carbon and carbon manganese steels*, British Standards Institution, London.

BS 3643 *ISO Metric screw threads*, British Standards Institution, London.

BS 3692 *ISO Metric precision hexagon bolts, screws and nuts*, British Standards Institution, London.

BS 4190 *ISO Metric black hexagon bolts, screws and nuts*, British Standards Institution, London.

BS 4320 *Metal washers for general engineering purposes*, British Standards Institution, London.

BS 4360 *Specification for weldable structural steels*, British Standards Institution, London.

BS 4395 *High strength friction grip bolts and associated nuts and washers for structural engineering: Pt 1 General grade, Pt 2 Higher grade bolts and nuts and general grade washers, Pt 3 Higher grade bolts (waisted shank), nuts and general grade washers*, British Standards Institution, London.

BS 4604 Pt 1 *The use of high strength friction grip bolts in structural steel work, metric series: Pt 1 General grade, Pt 2 Higher grade .(parallel shank), Pt 3 Higher grade (waisted shank)*, British Standards Institution, London.

BS 5135 *Specification for the process of arc welding of carbon and carbon manganese steels*, British Standards Institution, London.

BS 5400 *Steel concrete and composite bridges, Pt 3 Code of practice for the design of steel bridges*, British Standards Institution, London.

Butler L. J., Pal S., and Kulak G. L. (1972). *Eccentrically loaded welded connections*, Proc A.S.C.E. (Struct. Div.), V98.

Chesson E. Jr., Faustino N. L., and Munse W. H. (1965). *High strength bolts subject to torsion and shear*, A.S.C.E. (Struct. Div.), V91, ST5.

Clarke A. (1970). *The strength of fillet welded connections*, MSc thesis, Imperial College, University of London.

Clarke P. J. (1971). *Basis for the design of fillet welded joints under static loading*, Conf. Proc., Welding Institution, Improving Welding Design Paper 10, V1.

Crawford S. F. and Kulak C. L. (1971). *Eccentrically loaded bolted connections*, A.S.C.E. (Struct. Div.), V97, ST3.

Davies G. (1981). *Estimating the strength of some welded lap joints formed from rectangular hollow section members*, Conference Proc., Joints in Structural Steelwork, Teeside Polytechnic.

Douty R. T. and McGuire W. (1965). *High strength bolted moment connections*, A.S.C.E. (Struct. Div.), V91, ST2.

Elzen L. W. A. (1966). *Welding beams in beam-to-column connections without the use of stiffening plates*, Report 6-66-2, I.I.W. document XV-213-66.

European Convention for Structural Steelwork (1981). *European recommendations for steel construction*, Construction Press.

Farror J. C. M. and Dolby R. E. (1972). *Lamellar tearing in welded steel fabrication*, Welding Institute, Cambridge, England.

Fisher J. W. and Struik J. H. A. (1974). *Guide to Design Criteria for Bolted and Riveted Joints*, John Wiley and Sons.

Gourd L. M. (1980). *Principles of Welding Technology*, Edward Arnold.

Holmes M. and Martin L. H. (1983). *Analysis and Design of Structural Connections*, Ellis Horwood Ltd.

International Institute of Welding (1964). *Calculation formula for welded connections subjected to static loads*, Welding in the World, V2.

International Institute of Welding (1964). *Design rules for arc welded connections in steel subject to static loads*, Welding in the World, V14.

Johnson L. G. (1959). *Tests on welded connections between 'I' sections*, British Welding Journal, V6.

Kato B. and Morris K. (1974). *Strength of transverse fillet welded joints*, Welding Research.

Ligtenberg F. K. (1968). *International test series, final report*, Stevin Laboratory, Technological University of Delft, Doc XV-242-68.

Mann A. P. and Morris L. J. (1979). *Limit state design of extended plate connections*, A.S.C.E. (Struct. Div.), V105, ST3.

Martin L. H. (1979). *Methods for limit state design of triangular steel gusset plates*, Building and Environment.

Martin L. H. and Robinson S. (1981). *Experiments to investigate parameters associated with the failure of gusset plates*, Conference Proc., Joints in Structural Steelwork, Teeside Polytechnic.

Morris, L. J. (1988). *Design rules for beam-to-column connections in Europe*, Steel Beam-to Column Building Connections, Elsevier Applied Science.

Morris L. J. and Newsome C. P. (1981). *Bolted corner connection subjected to an out of balance moment – the behaviour of the web panel*, Conference Proc., Joints in Structural Steelwork, Teeside Polytechnic.

Owens G. W., Driver P. J., and Kriege G. J. (1981). *Punched holes in structural steelwork*, Jnl. Constructional Steel Research, **1**, No. 3.

Packer J. A. and Morris L. J. (1977). *A limit state design method for the tension region of bolted beam-to-column connections*, Struct. Eng., V55, 10.

Pillinger A. H. (1988). *Structural steelwork: a flexible approach to the design of simple construction*, Struct. Eng., V66, 19/4, October.

Purkiss J. A. and Croxton P. C. L. (1981). *Design of eccentric welded connections in rolled hollow sections*, Conference Proc., Joints in Structural Steelwork, Teeside Polytechnic.

Rolloos A. (1969). *The effective weld length of beam-to-column connections with stiffening plates, final report*, I.I.W. document XV-276-69.

Salmon C. G., Buettener D. D., and O'Sheridan T. C. (1964). *Laboratory investigation of unstiffened triangular bracket plates*, Proc. A.S.C.E. (Struct. Div.), V90.

Stamenkovic A. and Sparrow K. D. (1981). *A review of existing methods for the determination of the static axial strength of welded T, Y, N, K, and X joints in circular hollow steel sections*, Conference Proc., Joints in Structural Steelwork, Teeside Polytechnic.

Stark J. W. B. and Bijlaard F. S. K. (1988). *Design rules for beam-to-column connections in Europe*, Steel Beam-to-Column Building Connections, Elsevier Applied Science.

Sections Book (1988). British Steel Corporation Sections in association with British Constructional Steel work Association Ltd.

Timoshenko S. (1959). *Theory of Plates and Shells*, McGraw Hill.

8

Frames and Framing

8.1 Introduction

The previous chapters have dealt with element design; beams, columns and connections. This chapter, and the next, deals with the way these individual components are assembled together and also with design problems associated with the whole structure, and is illustrated with a series of examples.

Initially, it is worth considering how a choice is made of the structural form that will be used to carry the imposed loading, and it is convenient to divide the survey into two categories: (a) single storey structures, and (b) multi-storey structures.

8.1.1 Single storey structures

Typical examples of single storey structures are factory units, sports complexes or assembly buildings, including schools. In general, unless architectural considerations prevail, the most economical solutions will be obtained using one-way spanning structures rather than space frame structures. It should be noted that roof systems which appear to be two-way spanning are often only one-way spanning. Roof systems can be conveniently divided into flat and pitched roof systems.

8.1.1.1 *Flat roof systems*

For spans up to 15 m rolled sections form the most economic solution. It should however be noted that the usual maximum length of rolled section available from the rolling mills is 12 m. The upper limit of 15 m may require the use of special runs, and should only be contemplated if a large number of such beams is required. For spans between 14 and 20 m, castellated beams become economic. Although fabrication costs are high with castellated beams, and the overall depths are increased, the hexagonal holes in the webs allow services to be contained within the depth of the beam instead of requiring additional construction depth between the soffit of the beam and the ceiling. For spans over 20 m it is usual to use parallel chord lattice trusses fabricated either from rolled hollow sections or from lightweight cold formed sections for top and bottom chords and bar members in the web.

The roof decking can take a variety of forms ranging from woodwool slabs, through timber decking with asphaltic waterproofing to precast prestressed concrete slab units. One problem that must be given due consideration with flat roofs is drainage and any ponding effects on the roofing system. It is usual to camber the roof to provide adequate run-off.

8.1.1.2 *Pitched roof systems*

These are usually economic above spans of 20 m, and may take the form of the conventional large pitch triangulated roof truss, low pitch triangulated lattice trusses or, more commonly, pitched roof portal frames.

The conventional large pitch (around 20°) roof truss fabricated from 'T' sections and angle sections with bolted or welded joints has largely become obsolete, except where extensions to existing structures using such frames are concerned, partly due to the wasted roof space requiring heating and partly to the development of roofing materials capable of remaining watertight at low pitches (below 10°).

The low pitch lattice truss fabricated from rolled hollow sections, generally in Grade 50 steel, has taken over as the usual form of truss and overcomes the drawbacks of the older forms of truss.

Pitched roof portal frames have probably become the most common form of single storey construction system due mainly to ease of fabrication and erection. Frame spacing is generally 7.5 m or 9.0 m on very high spans: 6.0 m may however be used on low spans of below 20 m (Horridge).

The end bay of a portal frame structure is usually designed as a much lighter frame with intermediate gable end columns, partly to ease the problem of supporting the sheeting rails and partly because the frame used in the remaining bays would be over-strong. Only where there is likely to be an extension of the structure should the end frame be identical to the interior frame. In this latter case the sheeting rails are supported on temporary gable end stanchions designed to be easily removable.

The roofing systems for all these types of structure will be proprietry sheeting systems with insulation supported on cold formed section purlins.

All the methods covered for pitched roof construction are capable of being used with spans up to 60 m, although above about 30 m, it is worthwhile considering the use of multi-bay structures. The unit weight of steel work will be less with multi-bay structures, but this will need to be offset by the reduced flexibility in working space caused by intermediate stanchions, although with truss systems this can be mitigated by the use of internal lattice girder support systems, and the need to supply valley guttering and drainage, together with the additional problems of the build up in snow loading.

8.1.2 Multi-storey construction

8.1.2.1 *Multi-storey steel skeleton*

The major variations here are not with the steel skeleton, which is usually designed only to resist vertical loading–a concrete or masonry core being used to resist wind forces–but with the flooring systems used.

8.1.2.2 *Flooring systems*

The traditional in situ concrete flooring cast using normal-weight concrete on traditional shuttering systems, either timber or proprietary, needing propping often over two lower storeys, has been superseded either by (i) the use of precast prestressed normal-weight concrete floor units, either supported on the top flange of the floor beams or supported on shelf angles welded to the web of the floor beam at such a height that the upper surface of the precast units are level with the top flange of the beam, or (ii) by the use of steel decking acting compositely with the concrete deck, usually light-weight and with the support beam system. There is some advantage in the use of shelf angle floor systems in that the overall construction depth is decreased and the fire performance much improved often to the extent of needing no additional fire protection (Section 11.9). A possible disadvantage of this system however is that the precast units have to be slotted in between the shelf angle and the top flange of the beam.

The advent of the use of composite slabs with profile sheet steel decking has brought a number of advantages: the steel decking gives a clear and clean working environment on lower storeys since no propping is required, the requirements for specialist and expensive trades such as carpenters are much reduced, erection time is short, and hook time on cranage is much reduced as the concrete is usually pumped. A further reason why the use of traditional in situ concrete work has declined is the emergence of methods of fire protection such as gypsum plaster boarding or lightweight vermiculite sprays rather than concrete encasement which is both expensive as shuttering needs to be erected, and is very heavy, increasing substantially the loading on the foundations and the individual members. Although concrete encasement gives some increase in the load carrying capacity of the encased members, this enhancement was never fully realized by design codes, and it should be noted that profiled steel decking, provided it is fully coupled to the support beam system using in situ through deck stud welding, will give full restraint to the beam system allowing full design stresses to be used. Where very high fire resistance periods are required, most notably in basements used as underground car parks where a fire resistance period of 4 hrs may be required, concrete encasement may still be necessary and consideration should be given to casting the concrete encasement around the steel members off site. It will then only be necessary to encase the connections on site thereby obviating the need to erect substantial formwork and reducing potential delays whilst the concrete cures.

8.1.3   Influence of connection design and detailing

In the previous chapter dealing with connection design it would have been noticed that connections between beams and columns could either be designed to resist the effect of the vertical reaction of the beam in the form of shear, together with a nominal design moment to allow for eccentricity at the connection, or to resist the vertical reaction and the moment transfer between the beam and column. The first type is generally described as a 'pin' joint in which no moment transfer between the beam and column is allowed for and therefore there is no resultant compatibility between the resultant rotations in the beam and the column. The second is generally described as a 'fixed' joint in which rotational compatibility is enforced between the beam and column, and thus moment transfer will occur. Clearly the type of connection will have a marked effect on the behaviour of the structure when loaded, and it is this effect which must now be discussed. It should also be noted that the connection detail will have an influence on the flexural capacity of the beams (Section 5.3.3). However it will be convenient to consider the type of loading a structure will carry and how this loading is distributed.

8.2   Loading

8.2.1   Structural loading

This can conveniently be divided into two categories: (a) gravity loading, and (b) non-gravity loading.

8.2.1.1   *Gravity loading*

This covers dead load, superimposed dead load (e.g. finishes) and imposed load. This will not be discussed further as it has been covered in Chapter 3 of this text except to note that it may not be sufficient to consider only uniform snow loading (see Section 8.10.2).

8.2.1.2   *Non-gravity loading*

This may for convenience be divided into a series of sub-headings:

1   Wind Loading.   This is covered by CP 3 Ch V Pt 2 (due to be revised and issued as BS 6399, Pt 2). It is recommended that the full text be available when working through the design examples in this chapter and the next.
2   Inertia and Impact Loads.   These have to be considered where dynamic loading is involved, for example cranes. For crane loading, the static loading is increased by a percentage to allow for non-static effects (BS 2573 and BS 6399, Pt 1). Impact loading needs considering where parts of the structure, e.g. columns in a multi-storey car park or in a storage warehouse where fork lift trucks may be used, can be subject to vehicular impact, and is generally treated by the imposition of an equivalent static force (BS 6399, Pt 1, cl 10).
3   Seismic Loading.   Whilst this may not be critical or entirely relevant to the UK, it must be remembered that in certain parts of the world seismic loading is important and cannot be omitted.
4   Accidental Loads.   These comprise loadings due to explosions or fire. To avoid progressive collapse following an explosion, BS 5950, Pt 1 specifies the need for tying forces (this is dealt with in Section 8.5). The design of structures and structural elements to withstand the effects of fire is considered in Chapter 11.

It is now necessary to consider how structures resist the forces due to applied load. The next section deals with single storey structures.

8.3   Single storey structures under horizontal loading

Consider first a simple encastré base flat roof portal frame loaded solely with vertical loading (Fig. 8.1). As far as vertical load carrying is concerned it does not matter whether the connections at B and C are rigid or pinned. In either case the structure can be analysed and the members designed. If the joints at B and C are pinned the stanchions will be lighter (the vertical reaction from the beam is the same in either case), the connection detail will be simpler, with lower fabrication costs, but the rafter will be heavier as the bending moment will be larger.

If the use of non-moment resisting connections at B and C is adopted and the more usual practice of pinned feet is also adopted, i.e. all joints are now effectively pinned, the frame will now be unstable if any horizontal forces are applied. Clearly this is not

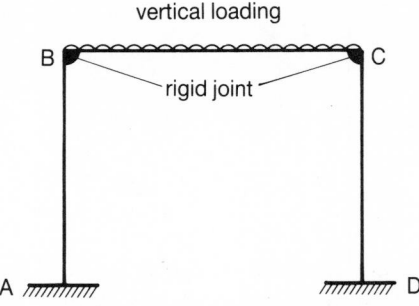

(a) Basic frame with rigid joints at B and C

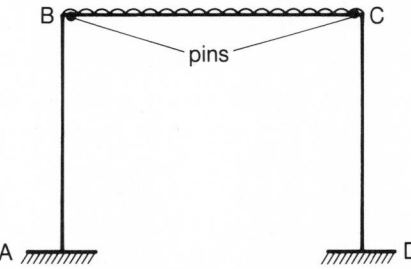

(b) Basic frame with pinned joints at B and C

Fig. 8.1   Simple single bay flat roof portal.

satisfactory, and a method must be found to give stability and resist the application of horizontal forces.

There are two alternative solutions, assuming, for reasons that become apparent in Section 8.8.1, that a return to encastré bases cannot be made: the first is to restore the rigid connections at B and C, the second is to keep all joints pinned and find an alternative method of resisting the horizontal loads. The first alternative has drawbacks in that any connection will have to deal with wind reversals (Fig. 8.2) causing moment reversal in the connection and the attendant problems of detailing. The second alternative is to supply diagonal bracing.

For the wind blowing in the direction of *P* (Fig. 8.3(a)), a tie is provided to resist the wind force and has the effect of triangulating the structure. If, however, the wind reverses in the direction of *P'* (Fig. 8.3(b)) this tie, member AC, becomes a strut and hence is far weaker owing to buckling. It is therefore necessary to provide a further tie, member BD. Thus the frame becomes cross-braced. Such a cross-bracing system is conventionally designed by ignoring the strength capacity of any compression members. No structure, however, exists as an isolated single frame, and therefore methods which can be applied to multi-bay frames must be sought.

Consider a structure single bay in one direction, multi-bay in the other (Fig. 8.4(a)). If it were to be braced in each frame as described above to resist the wind forces *P*, the structure would be unusable owing to the cross-bracing occurring in every bay of the frame. This problem is overcome by removing the bracing in internal frames and connecting the frames together at rafter level in the plane of the rafter by cross-bracing (Fig. 8.4(b)).

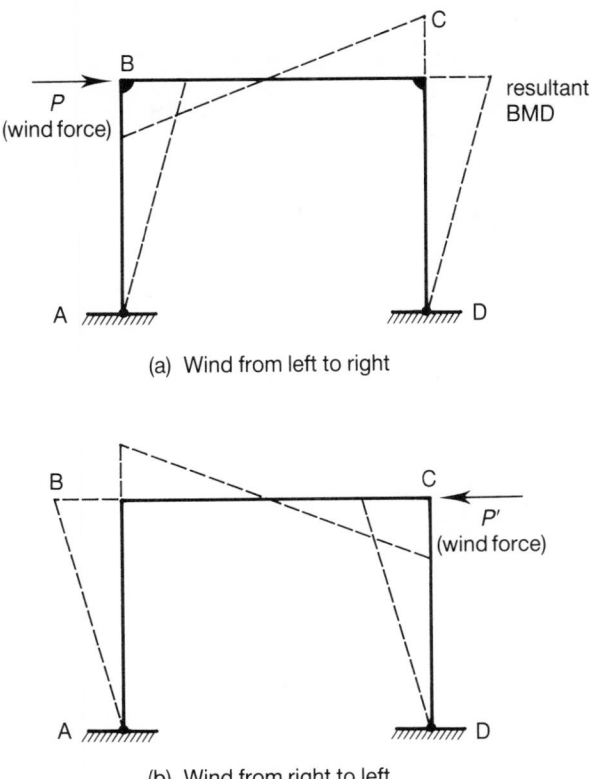

(a) Wind from left to right

(b) Wind from right to left

Fig. 8.2 Effect of wind reversal.

Such a bracing system is known as a wind girder, and has the effect of transmitting the wind forces at eaves level to the individual frames and of equalizing the resultant deflection at eaves level, since the end frames are often designed as lighter structures as they carry only half the loading of an internal frame. Deflections also need to be equalized to reduce shear stresses on the roof cladding which is normally incapable of resisting such forces. This is not true if stressed skin construction is used. This will not be covered here, and reference should be made to Davies and Bryan. To resist the wind forces normal to the end frames vertical bracing will also be needed in the end bays only, as the wind girder will transmit forces through the structure.

For a multi-bay structure the principles described above must be applied to each face of the structure (Fig. 8.5).

For pitched roof portal frames which will have rigid connections at the eaves, wind bracing is still necessary to distribute the wind forces since the end frame is usually designed as a rafter spanning over gable end columns and is therefore substantially more flexible.

Under certain circumstances bracing can be replaced by alternatives. Vertical bracing can be replaced by shear wall construction or masonry cladding. Shear wall construction is more often met with in multi-storey construction (Section 8.4). Masonry cladding may be used to resist the horizontal forces: for the design of masonry cladding and shear walls

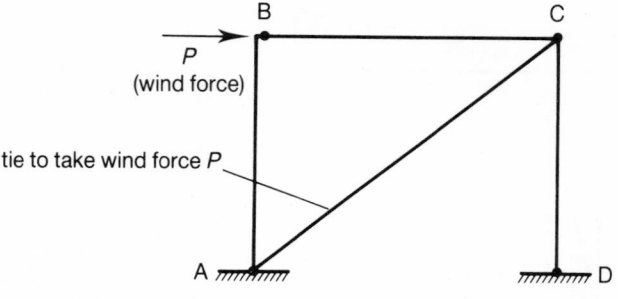

(a) Triangulation to take effect of wind force *P*

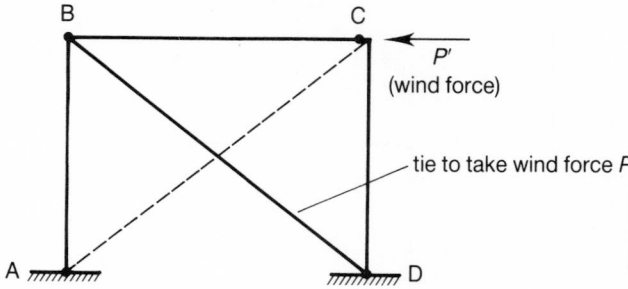

Note: AC now acts as a strut, and for calculating the
       force in BD is assumed to have zero strength

(b) Triangulation to take effect of wind force *P′*

Fig. 8.3   Elementary wind bracing.

reference should be made to Hendry or Curtin *et al.* However two points should be noted: first, the masonry must be adequately tied to the steel skeleton using, for example, wall ties spot welded to the stanchions, and second, that consideration must be given to the situation when the frames are erected and the roof sheeting placed. This latter situation can often be critical for the structure as the pattern of wind loading will be different to that assumed in the design, and therefore temporary bracing will need to be specified by the design engineer (Section 8.7.3).

The roof wind girder may be omitted when an in situ roof slab is adopted, as the roof will resist any shear stress and is stiff enough to distribute the wind forces. However, the critical design case for the frame may be whilst the concrete is being placed, since the shuttering system may offer no lateral restraint to the compression flange of the rafter. With profile sheet steel decking this situation does not occur when the ribs of the profile sheet steel decking run perpendicular to the beam provided, since as is usual, the decking is adequately fastened by through welded shear studs. Any beams running parallel to the ribs of the sheet steel decking cannot be assumed to be restrained and must be checked for lateral torsional buckling during the construction stage (see Section 10.4.3.1).

8.4   Multi-storey structures under horizontal loading

It is convenient to consider bracing in the vertical and horizontal planes separately.

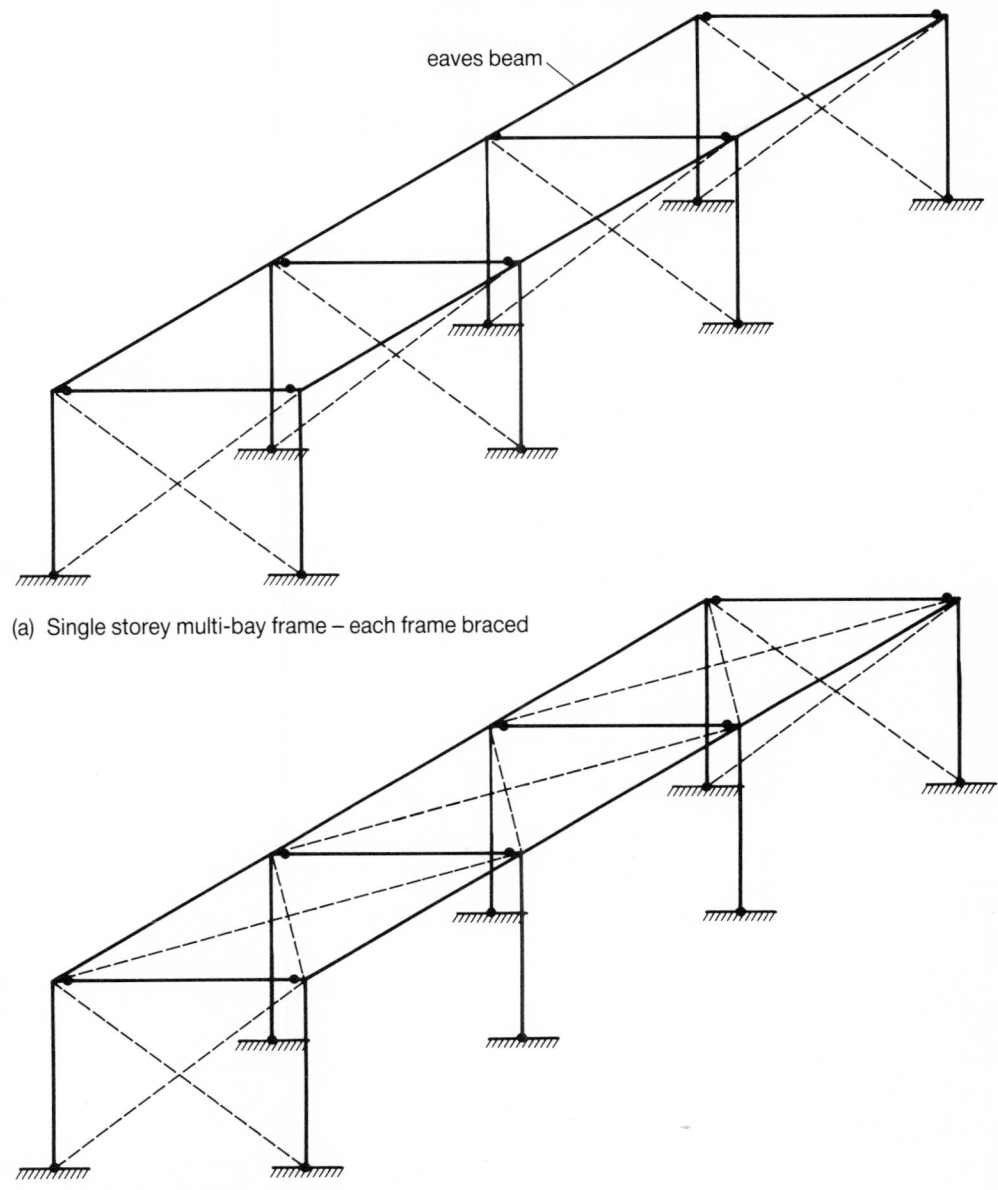

(a) Single storey multi-bay frame – each frame braced

(b) Single storey multi-bay frame with wind girder

Fig. 8.4 Wind girder bracing.

8.4.1 Vertical plane bracing

There are four alternatives to be considered here.

The first is to design all the connections between the floor beams and the stanchions to resist the wind moments. This is likely to produce large, deep connections with their inherent detailing problems. This solution is not common. Note, the frame is then described as unbraced, and will act as a sway frame (see Section 8.10.2).

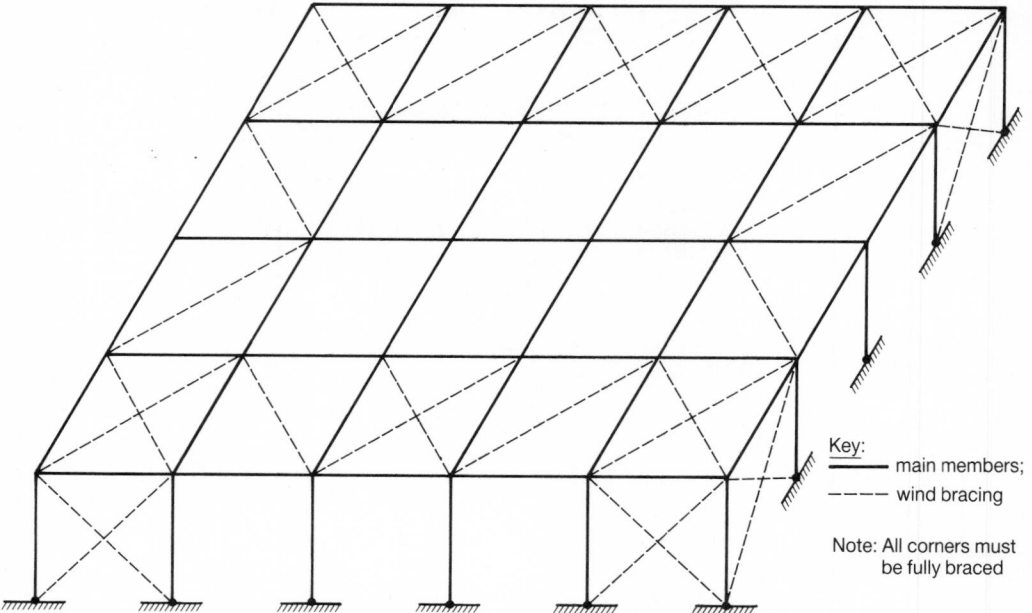

Fig. 8.5 Schematic layout of wind bracing for a multi-bay structure.

The second is to use diagonal bracing up the whole of the structure, often making an architectural feature of such bracing. The bracing is either set in each corner of the structure or in a central bay with the forces transmitted internally by horizontal bracing at each floor level.

The third is to use masonry shear walls to resist the horizontal forces. The problem here is that the wall must be virtually unpierced, i.e. free from significant or dominant openings. This does impose severe restrictions on the architecture, but the shear wall will also act as structural cladding.

The fourth, and probably the most common, is to allow the wind forces to be resisted by the stairwells and lift shafts. Since these are protected members in that they either provide access for firemen or are used as fire escapes, they are relatively unpierced and are therefore inherently stiff and capable of resisting such forces. They may be of masonry construction, which is usually restricted to stairwells only in relatively low rise construction, though for lift shafts and stairwells in medium or high rise construction, in situ concrete is much more likely. There are problems in marrying up steelwork and concrete construction, most notably that the two techniques demand differing constructional tolerances (those for concrete work being much greater) and that the long term deformations in the concrete due to creep and shrinkage must be allowed for. These problems can usually be overcome. For the analysis and design of cores reference should be made to CIRIA Report No. 112.

8.4.2   Horizontal plane bracing

This can be considered under three cases.

The first is where profiled steel decking is used. Here no bracing is required as the steel decking, provided it is properly fixed, will give adequate restraint during the pouring of the concrete slab to beams spanning perpendicular to the ribs, and, after curing, the slab itself will be adequate. It is unlikely that the beams running parallel with the decking will need bracing, as intermediate beams will give lateral torsional restraint to the main beams. Only if there are no intermediate beams will temporary bracing be necessary.

The second is where conventional shuttering is used and bracing will be required to restrain the compression flange during pouring.

The third involving any other type of flooring including precast concrete units will need full bracing to transmit the horizontal forces through to the vertical restraint system.

8.5 Tying forces (cl 2.4.5.1 to 5, Pt 1)

All building frames should be effectively tied together at all levels. Columns should be tied at each end in two directions. These ties should wherever possible form a continuous whole at each floor level. Ties may take the form of either steel members (either members of the main structural frame or if necessary additional members) or reinforced concrete and masonry provided each are adequately coupled to the steel frame by suitable fixings. Normal members may be used provided the connections are capable of taking a factored tensile force of not less than 75 kN at floors or 40 kN at roof level. This loading in the member is not additional to the normal loads carried as part of the frame. Ties are not required at roof level where the cladding has a load intensity of less than 0.7 kN/m^2 and carries roof loading only.

For large multi-storey structures the tying force requirements are enhanced to $0.5w_f s_t L_a$
for an internal tie or half this value for an external tie (w_f is the total factored dead plus imposed load at the level being considered, s_t is the mean spacing of the transverse ties and L_a is the greatest distance, in the direction of the tie, between adjacent lines of columns or supports). In no case should these values be less than 75 kN at a floor or 40 kN at the roof. At the periphery, ties anchoring columns should be checked for the force given above, which should be taken as not less than 1% of the factored vertical load in the column.

Column splices should be checked for a tensile force equal to two thirds of the factored vertical loading. Where required by a Regulatory Body e.g. Local Authority, the structure may need checking for progressive collapse. If the removal of any member should cause collapse of the structure, it is designated a key element and the structure checked under the loading specified in cl 2.4.5.4. It is recognized that where collapse does not ensue following the removal of a member, substantial deflection and permanent deformation is acceptable.

8.6 Transmission of loading

8.6.1 Transmission of loading from floor systems

Loading, whether dead, imposed or wind, is usually quoted as a load per unit area. For one way spanning systems, including in situ concrete floor slabs with composite trough decking, loads are transmitted as in Fig. 8.6(a).

Figure 8.6(b) indicates the transfer of loading from a UDL on roof sheeting through line reactions on purlins through point reactions on the rafter supports. It is sufficient to assume that the sheeting consists of a series of simply supported spans over the purlins, and that the purlins are themselves simply supported over the rafters. For the interior span of a large multi-span system this assumption will be correct, but will be less accurate as the end span is reached, even though the error will be acceptable. Precast concrete decking and in situ concrete decking with composite action trough decking can be treated similarly.

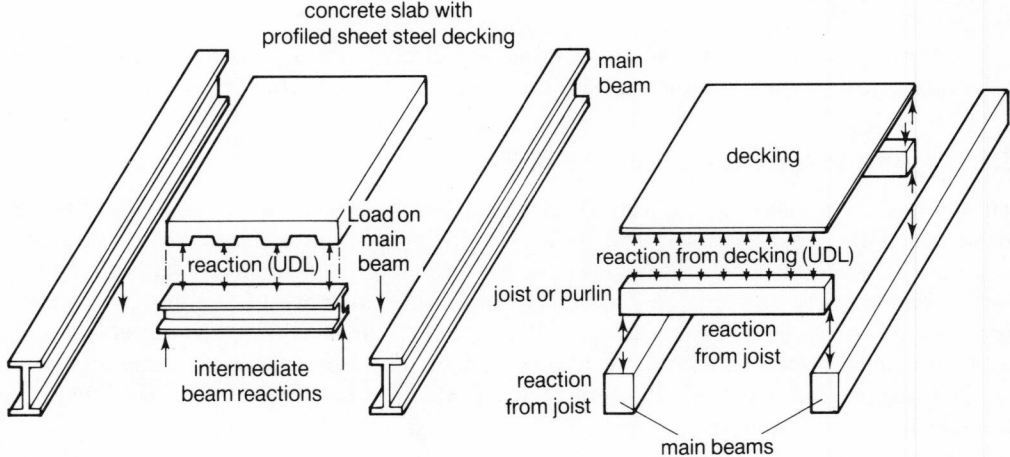

Note: The loading on the main beam will be applied through web cleats (not shown)

(a) Transfer of load from one way spanning concrete deck

(b) Load transfer from decking through a joist or purlin system

Fig. 8.6 Load transfer for one way spanning system.

However, this is not true for in situ concrete slabs cast with traditional formwork, designed to be two way spanning. For such slabs designed by the Hillerborg strip method the reactions on the support beams are found directly. For slabs designed using the BS 8110 approximate methods Table 3.16 of BS 8110, Pt 1 may be used to determine the reactions from the slabs onto the support beams. Where yield line methods have been used, the solution is more complex and reference should be made to a textbook such as Jones and Wood, or Moy. It should be noted that for the last two cases it may well be accurate enough to use a 45° distribution of loading (Fig. 8.7).

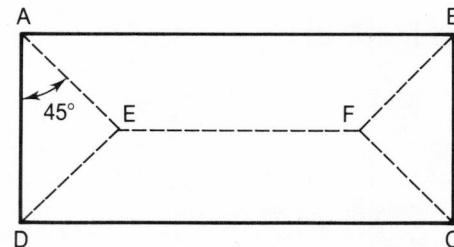

Note: (1) Load on area AED is supported by beam AD
(2) Load on area AEFB is supported by beam AB

Fig. 8.7 Approximate load transfer from two way spanning slabs.

8.6.2   Loading on lintels

This is covered by BS 5977, Pt 1, and it is only intended to provide a brief resumé in this text. The basic configuration to be considered is shown in Fig. 8.8.

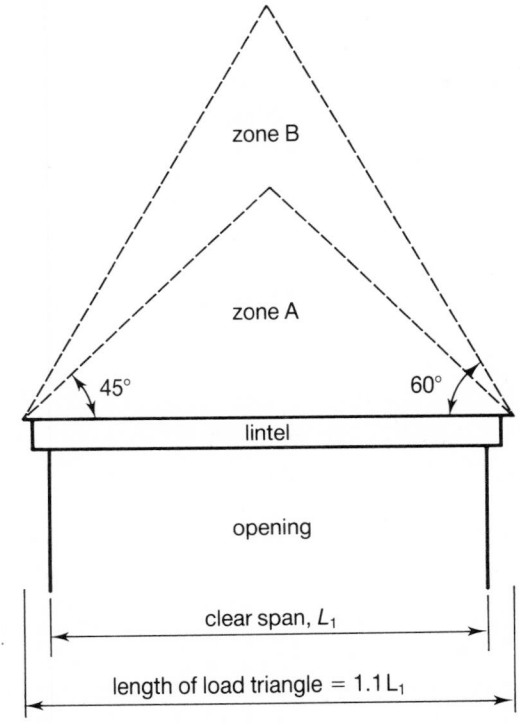

Fig. 8.8   Loading on a lintel.

In zone A, all the loading including any imposed loading from floors and the self weight of the masonry is taken by the lintel. In zone B, the interaction zone, 50% of any imposed load is taken by the lintel, after dispersion through the masonry at an angle of 45°, except that the self weight of the masonry is assumed to be carried by arching action and is not included. Outside both zones A and B, no loading is carried by the lintel, all the loading is carried by arching action. The presence of openings complicates the above, and for such cases reference should be made to the Code. Note, the maximum clear spans to which the above rules can be applied are 4.5 m in single storey construction and 3.6 m for two or three storey construction, and the masonry must extend at least 0.6 times the clear span above the lintel, and in addition these provisions only apply to domestic type structures. It is suggested that the above rules may be used as a guide in alternative situations, but should be interpreted with care.

8.7   Design of bracing

8.7.1   Permanent bracing

The forces to be applied when designing bracing are the maximum of

(a) the factored wind forces at that level including where appropriate wind drag on the roof, or

(b) the maximum of   1.0% of the factored dead load from that level or 0.5% of the factored dead and imposed load acting at that level   (cl 2.4.2.3, Pt 1).

In addition, in case (a), the relevant dead and imposed load with appropriate load factors must be considered, and in case (b)   1.4 × (unfactored dead load) + 1.3 × (imposed load).   The application of the load combination 1.4 × (unfactored dead load) + 1.3 × (imposed load) only applies to the second section of case (b).

It is generally assumed in triangular bracing systems that the analysis and design may be undertaken assuming pinned connections and that diagonal members, in compression, are ignored. Roof bracing should have sufficient flexural stiffness to avoid undue sag under its self weight. Cl 4.7.3.2, Pt 1 states that for members whose slenderness exceeds 180, a check should be made for self weight deflection. Should this deflection exceed one hundredth of the length then bending effects must also be taken into account in the design of the member.

8.7.2   Restraint bracing to compression flanges

Bracing designed to give lateral restraint to compression flanges of beams, whether permanent or temporary, should be designed to resist a force equal to   2.5% of the maximum factored force in the compression flange of chord, divided equally between the points of restraint   (cl 4.3.2, Pt 1). Where a wind girder also supplies restraint to a compression flange it must also be checked for this 2.5% force. This check is in addition to that indicated in Section 8.7.1.

Where a series of parallel flanges require restraint, it is not sufficient to tie the flanges as in Fig. 8.9(a). The restraining members must be cross-braced as Fig. 8.9(b), or alternatively anchored to a part of the structure sufficiently robust to resist the restraint forces, e.g. a lift shaft or stairwell (Fig. 8.9(c)). Any such restraining member should have sufficient flexural stiffness to avoid undue sag under the action of self weight (see Section 8.7.1).

8.7.3   Temporary or erection bracing

It is the design engineer's responsibility to ensure that the structure has adequate stability during construction and to ensure that the erection contractor adopts a safe method of working. It is not merely sufficient to ensure adequate stability in the completed state. The need to specify, as part of the original design, temporary bracing during construction cannot be over-emphasized. A large number of accidents on site occur owing to lack of such bracing. The classic case is where only the end bays of a single storey structure are braced against wind, but in order to reduce the effect of fabrication tolerances, erection starts from a centre bay and proceeds towards the end bays, with the resultant effect that there is no bracing in the vertical plane until half the structure has been erected! Another reason for the need of temporary bracing is that whilst imposed loading may be lower, wind loads may not be, and furthermore elements of the structure may have different restraint or support conditions from those envisaged in the completed structure.

A further problem may be caused where additional loading is imposed on a partially completed structure by equipment, such as cranes used to erect the remainder of the structure or props supporting conventional formwork whilst concrete is being poured. It is also necessary, as has already been pointed out, to consider cases where incomplete

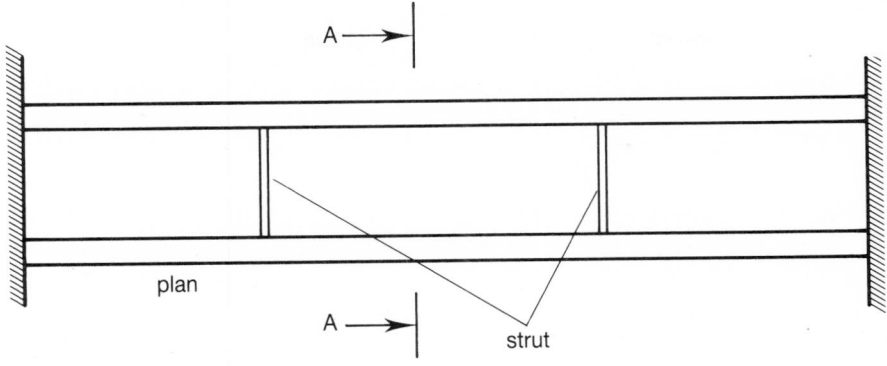

(a) Inadequate lateral restraint

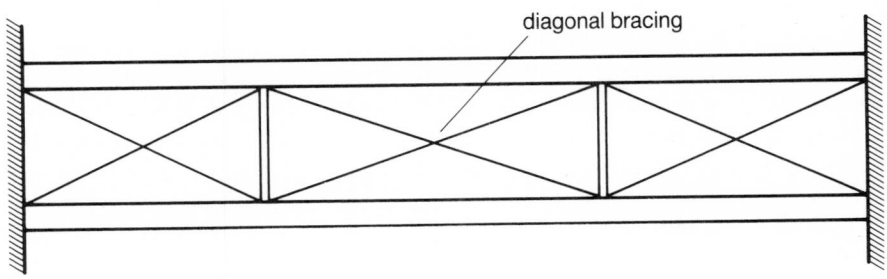

(b) Adequate lateral restraint using bracing

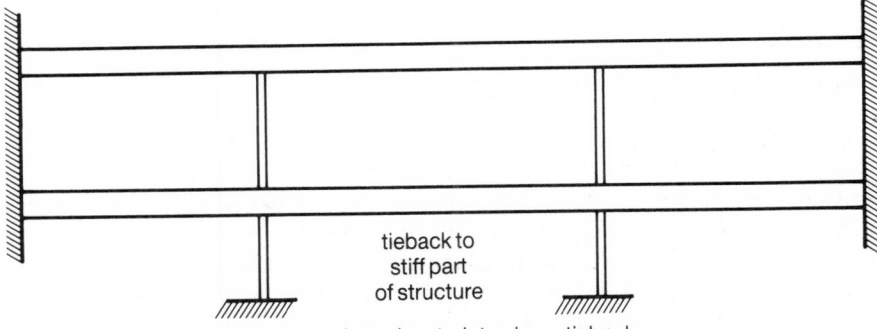

(c) Adequate lateral restraint using a tieback

Fig. 8.9 Lateral restraint to beams.

restraint to a compression flange may occur during construction. These may either be due to the absence, say, of the in situ concrete slab or due to different support conditions. This latter most often occurs when the completed sections of a structure are being lifted into their final position by cranes. The top chord during lifting has no restraints to out of plane buckling except the lifting points which are usually toward the ends of the truss. This condition will give rise to an unacceptably high slenderness ratio and will lead to premature buckling. This can be avoided by lifting the trusses as a pair of trusses spaced apart at the correct centres and adequately braced to keep the resultant slenderness ratio within allowable limits. This bracing should be designed and detailed by the design engineer, and should form part of the erection specification.

It must be reiterated that it is the design engineer's responsibility to ensure the adequacy of any structure during erection; though this does not remove the contractor's contractural obligation to employ safe methods of working. For further information on this topic the reader's attention is drawn to Allen and Lovejoy, BS 5531 and the British Constructional Steelwork Association's (BCSA) National Specification, although it is this author's belief (J. A. Purkiss) that this last document is flawed in that it considers that the design engineer should only specify the forces to be carried by temporary bracing and that the bracing should then be detailed and designed by the steelwork fabricator or erector. Given the number of collapses caused by the lack of erection bracing, it is important that all bracing is designed and properly detailed.

8.7.4   Restraint bracing at plastic hinges

A torsional restraint must be provided at each hinge position when continuous construction is used (cl 5.3.5, Pt 1). This restraint should preferably be placed at the hinge itself but under no circumstances must it be placed further than $D/2$ away from the hinge position, where D is the depth of the member being restrained. Morris and Randall recommend that the plastic moment capacity of any such restraining members should equal the plastic moment capacity about the minor axis of the member being restrained. Note that for portal frames, a restraint need not be provided at the last hinge to form provided that hinge can be clearly identified (cl 5.5.3.1, Pt 1). This will not be possible if simple plastic collapse analysis is used. The order of hinge formation may only be determined if an incremental elasto-plastic analysis is used.

It is therefore recommended that since only a very small extra cost is incurred that restraints are provided at all hinge positions as often the last hinges to form do so at loads very close to the final collapse load.

8.8   Additional design constraints

8.8.1   Ground conditions and foundations

This section is neither intended to be a comprehensive survey of foundation design or construction, nor of problems associated with ground conditions. It is merely intended to bring to the attention of the reader some of the implications of such problems in the design of the structure. For full consideration, reference should be made to a text such as Henry.

Foundation selection is generally based on the bearing capacity of the ground and its susceptibility to absolute or differential settlement. The cheapest type of foundation is the pad foundation. This however requires a reasonably high bearing capacity, otherwise it becomes unduly large. Where a large number of heavy loads exist, any set of pad foundations will start to overlap and become a large base slab or combined foundation. A combined foundation may also be necessary when only moderate bearing capacities are available. For cases where very low bearing capacities exist, the alternatives of ground stabilization, replacement by imported fill or piling should be considered. In all cases the advice should be sought of a geotechnical engineer.

Pad or raft foundations should be designed to have uniform bearing pressure to avoid rotation of the foundations. If this is not possible the smaller bearing pressure should be limited to about three quarters of the maximum under dead and imposed loading at serviceability limit state. A slightly lower ratio can be allowed with wind loading which is of short duration. Under no circumstances should uplift on the foundation be comtemplated. This means that where structures only impose low vertical loading on the foundations, e.g. single storey portal frames, pinned feet should be used at the base of the stanchion. It will prove totally uneconomic to attempt to use an encastré base to the stanchion as the foundation will be impossibly large. For multi-storey structures this becomes less important as the axial load will be much higher and the resultant eccentricity of loading lower.

A pad or raft foundation should be sized using serviceability loading, and bearing pressures are usually quoted at working loads. This will entail a serviceability analysis of the structure. It is not sufficient, unless the structure is isostatic, to divide the reactions calculated at ultimate limit state by the relevant partial safety factors, since there will have been a redistribution of forces between the service and ultimate limit states. The reinforcement in the foundation is then calculated using the application of ultimate loading. Similar remarks apply to the selection of the number and size of piles.

The transmission of horizontal reactions caused by the imposed vertical loading can be avoided by tying the bases of the frame together at ground level. The floor slab or ground beams can be used for this purpose. With wind loading, since the horizontal reactions will not be equal, the nett out of balance horizontal force occurring will still need resisting by the foundations. Where piles are used it may be necessary to provide raked piles to resist this force.

For the construction of large multi-storey buildings it is usual to employ tower cranes to facilitate erection, and these will need substantial foundations.

Differential settlement may either be caused by non-uniform loading imposed by a structure on its foundations or by variable ground conditions beneath the foundations to the structure. The effects of non-uniform loading may be reduced by tying all the foundation systems together so that they are restrained to settle as a single unit. This may be achieved, for example, by supporting the ground beams carrying masonry cladding and the floor slab on the pad foundations to the stanchions. The alternative is to allow differential settlement to take place and to ensure that the structure is designed and detailed to ensure that, or full movement joints between the masonry cladding, the floor slab and the stanchions and their bases are provided. This does however mean that the cladding cannot be relied upon to provide the equivalent of wind bracing, and that a full bracing system must be designed independent of the cladding and floor or roof system.

The effect of variable ground conditions may be reduced by using a single foundation system, such as a raft, beneath the whole structure giving uniform bearing pressure distribution and allowing the whole structure to settle as a unit thereby inducing no extra forces. An alternative is to articulate the structure such that this movement can be accommodated by the structural frame without inducing any extra forces. A typical form of articulation to overcome differential settlement is the cantilever–suspended span arrangement (Fig. 8.10), whereby hinges, or pins, are introduced at the points of contraflexure, turning a hyperstatic structure into an equivalent isostatic structure.

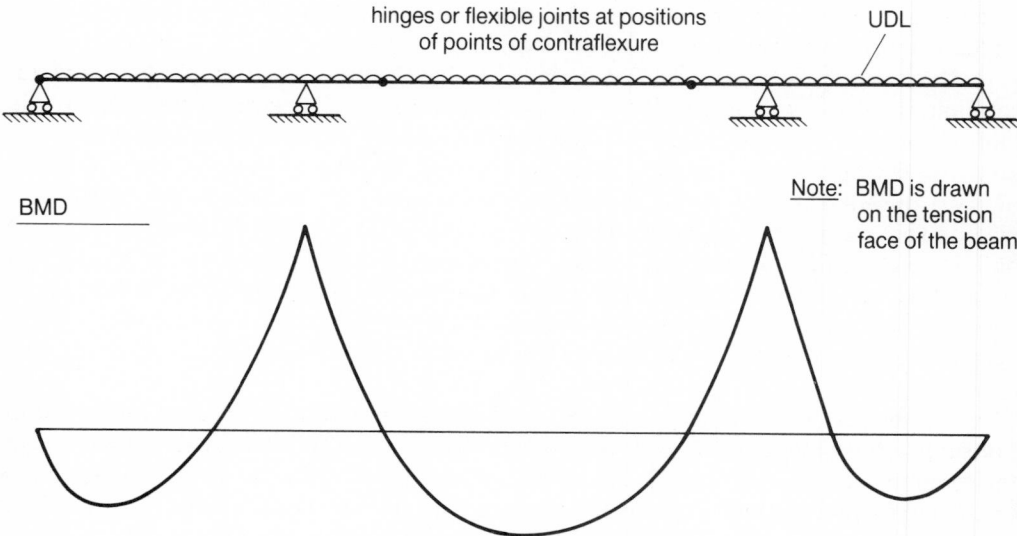

Fig. 8.10   Cantilever-suspended span structure.

This type of arrangement is often found in bridge structures. An alternative approach is to design and fabricate the structure as a braced pin jointed frame with lightweight non-structural cladding and design the bracing system to be adjustable to take up any resultant movement. Special care must be taken in areas of differential settlement where dimensional stability is required for, say, crane rails in factory units. It is not only necessary to ensure that a uniform settlement pattern occurs within the structure, but provision will need to be made for adjustments both in line and level of sensitive parts of the structure. This also means that a reliable estimate will be needed of such movements.

Mining subsidence can cause similar problems to differential settlement, except that any slab or raft foundation will need to be designed to span across any areas likely to subside. It may in certain cases be possible to backfill the mined area to give stability. Further alternatives when there are variable ground conditions, are either to consider the use of replacement fill or soil stabilization techniques.

8.8.2   Expansion and contraction joints

It can be a matter of some debate as to how often these should be provided. Certainly concrete floor slabs will need expansion/contraction joints at around 20 m to 30 m centres in both directions and masonry requires such joints at 7 m centres (BS 5628, Pt 3). It is recommended that steel frames should have movement joints also at around 20 to 30 m centres. This means that with a temperature range of -5 to $+35°C$ (cl 2.3, Pt 1), a coefficient of thermal expansion of 12 microstrain/deg C (cl 3.1.2, Pt 1), a movement of a length of 30 m of $0.6 \times (35 - (-5)) \times 12E\text{-}6 \times 30E3 = 8.6$ mm can be expected. The 0.6 factor is introduced in recognition of the fact that erection (or the datum from which movements are measured) does not take place at either of the extremes of temperature. If this movement were to be completely restrained a stress of $(8.6/30\text{E}3)\text{E} = (8.6/30\text{E}3) \times 205\text{E}3 = 60 \text{ N/mm}^2$ would be induced. Even in rolled members this stress would be significant. Thus the provision of movement joints is essential.

Where movement joints are provided, both parts of the structure either side of the joint must be independently braced against wind. Such a joint will also remove the continuity in any other form of bracing, such as that to compression flanges of beams. Not only must the movement joint allow movement in the main frame, but the purlins, sheeting rails and cladding must also be designed to take up this movement. For further information on movement joints Alexander and Lawson should be consulted.

8.8.3 Stability

This exists at three levels; element stability, member stability and frame stability.

8.8.3.1 *Element stability*

This requirement is necessary to ensure that the member behaves in accordance with the assumptions made in the design. The principal requirement is that the member should have sufficient capacity for rotation when a plastic hinge has formed without either local flange or web buckling taking place. This requirement can be satisfied by placing limits on the flange and web slenderness (see Section 1.1.3).

8.8.3.2 *Member stability*

This is achieved by ensuring that the member does not suffer elastic buckling causing premature collapse at a load lower than that which could be carried by the frame at the formation of the requisite plastic collapse mechanism. For individual elements in simple construction a slightly different approach can be adopted, i.e. the loading can be limited to that just causing elastic instability. This option cannot be used for rigid frames designed plastically.

Premature collapse by elastic instability is prevented in rigid jointed frames by the provision, if required, of bracing to the compression flange of the relevant member.

8.8.3.3 *Frame stability*

This requirement takes two forms: for single bay or multi-bay pitched roof portal frames there is the need to ensure that deflections are not excessive and for multi-bay frames only that 'snap through' of the rafters cannot occur.

The deflection check is to ensure that no significant geometry changes can occur in the frame at collapse. This check is needed so that normal plastic collapse theory can be used in which second order effects are neglected. The check is either made using a calculation of the eaves deflection using notional horizontal loads or by limiting the span/depth ratio of the rafter. BS 5950 does not specify any deflection limits for pitched roof portal frames for either horizontal deflections at the eaves or vertical deflections at the ridge. It is usual in a design to calculate such deflections, but it should be realized that the deflection occurring in the clad frame will be substantially less than that calculated on the bare frame without the stiffness of the cladding being taken into account (see Section 8.10.1).

8.9   Design philosophies (cl 2.1.2)

BS 5950, Pt 1 lays down guidance on three possible methods that may be used in design. They are discussed in turn below.

8.9.1   Simple design

This in effect assumes that the connections are incapable of transmitting other than nominal moments. It therefore follows that methods of analysis which allow the assumption of pin-joints in triangulated frame structures or simply supported beams in frame structures are permissible. This design method is applicable either to triangulated trusses, in which any secondary moments caused by practical joint rigidity may be disregarded (see Example 9.3), or to frame structures in which any sway due to the imposition of horizontal forces is resisted either by bracing or similar features, e.g. lift shafts or stairwells (Section 8.4). It also follows that only elementary analysis techniques are required as the structure can be considered isostatic.

8.9.2   Rigid design

Here the connections are considered capable of full moment transfer between members. Analysis may be either by elastic or plastic methods. For analysis techniques applicable to rigid jointed structures it is recommended that the reader consults standard textbooks such as Marshall and Nelson, Neville and Ghali, or Coates *et al.* on elastic techniques or Horne and Morris, Neale or Moy on plastic techniques.

8.9.3   Semi-rigid design

Here only partial moment transfer is permitted between members, i.e. some relative rotation is allowed between members. Semi-rigid design may either be carried out using moment–rotation characteristics derived experimentally for the actual connection being used (refer to Roberts and Taylor for a more detailed discussion of this approach) or by allowing a transfer limited to 10% of the mid-span moment to the supports. Certain design constraints are also laid down in BS 5950, Pt 1, and it is likely that any savings in the basic quantity of steelwork will be outweighed by the increase in design time over the simple design method.

   A recent development is to design low-rise multi-storey frames using the 'wind connection method', in which connections are assumed to be pinned for the purpose of calculating the forces in the frame for dead and imposed loading (including the nominal column restraint moment assuming the reaction acts at 100 mm from the column face) but that the connections are designed to resist the sway forces induced by the wind (Cunningham (1990)). The column will almost certainly need designing under factored dead and imposed load with the column moments based on the sum of the moments due to the nominal eccentricity plus 10% of the free bending moment in the beam plus the moment due to the nominal horizontal stability force of 0.5% of the applied factored vertical loading defined in cl 5.1.2.3, Pt 1. The columns should also be checked under the appropriate factored combinations involving wind loads. In these combinations the nominal stability moments are not included. Cunningham also gives methods of calculating horizontal deflections due to wind, as well as data required to permit semi-rigid analysis using modified beam stiffnesses.

To illustrate the principles involved a series of design examples will be presented preceded by some brief notes on various aspects of the design method.

8.10 Design techniques

8.10.1 Single bay pitched roof portal frames

To illustrate the principles involved both an internal frame with rigid connections and a gable end (Section 8.10.1.2) will be designed.

8.10.1.1 *Notes on the design of a pitched roof portal (Example 8.1)*

1 A heavier section is usually used for the stanchion than the rafter. This will, however, mean that a haunch will be needed both at the eaves and the ridge in order to accomodate the connections at these points. The resultant frame is still more economic than one with a uniform section.

2 In order to avoid problems with instability it is essential that stresses at the rafter end of the haunch remain below yield. The usual economic length for the haunch is frame span/10. The haunch is usually taken as having twice the depth of the basic rafter section. This is economic since the haunch may be fabricated by welding on a cut rafter section.

3 Plastic hinges will be assumed to occur at the base of the haunch, i.e. 1.5 times the depth of the rafter below the point of intersection of the rafter and stanchion centre lines. It will be necessary to check the moment at the first purlin point afterwards.

4 It is usual to design the frame under factored dead and imposed loads and then check the frame as an analysis problem, under the effects of the appropriate wind loading combination.

5 Frame stability is checked using an empirical equation (cl 5.5.3.2, Pt 1) which gives a maximum allowable span/depth ratio for the rafter. An alternative approach is to calculate the deflection δ at the top of the stanchion due to two horizontal loads applied in the same sense at the top of the stanchion, each equal to 0.5% of the factored dead and imposed load on the frame. The resultant deflection δ should be less than stanchion height/1000. This latter method is seldom necessary for single storey portal frames, except where there are unusual features such as fixed feet, since the empirical equation was derived for the symmetric condition with pinned feet.

6 It is necessary to check the elastic stability of the stanchion, in that it must be capable of generating the plastic moment at the base of the haunch before elastic buckling takes place. The reason for this is that the only restraint along the length of the stanchion is offered by the sheeting rails which are on the tension flange. The Code allows two methods of achieving this; an empirical approach in cl 5.5.3.4, Pt 1 or a more exact approach in Appendix G, Pt 1. Both are illustrated, for comparison, in the example which follows.

7 It is necessary to check the rafter stability at the ridge owing to the presence of the plastic hinge at either the first or second purlin point before the ridge.

8 The haunch must also be checked for stability since the only restraints to the compression flange will be at the haunch connection and at the bracing usually supplied at the first purlin point beyond the end of the haunch in the rafter. It is usual that the haunch will not satisfy the stability criterion and will need additional restraints in the form of K bracing from the compression flange to the purlin. The bracing should be designed to resist at least 1% of the force in the compression flange.

9   The remaining stability check where the rafter compression flange is unrestrained is between the purlin beyond the end of the haunch and the first purlin beyond the point of contraflexure in the rafter.

10   In the example which follows neither the design of the connections at the ridge nor the eaves is covered, neither is the baseplate detail and design at the foot of the stanchion. For the methods of designing such connections reference should be made to Chapter 7.

11   BS 5950 gives no guidance on deflections for portal frames either for vertical deflections at the ridge or lateral deflections at the eaves. This omission is in part due to the fact that measured deflections will be substantially less than those calculated using the section properties of the bare steel frame as no account is taken of the extra stiffness of the cladding.

In Australia, Woolcock and Kitipornchai carried out a survey on deflection criteria on portal frames and from their findings have recommended limits of $h/150$ for horizontal deflections at the eaves (h is the height to eaves) and $L/360$ for the vertical deflection at the ridge for frames with a roof pitch of greater than $3°$ (L is the span of the portal). Both deflection limits are under service imposed load only.

12   Note that Davies (1990) has suggested that single bay portal frames should be treated as sway frames (see Section 8.10.3.2) and a plastic load factor determined that should satisfy Equation (8.13). The relevant equations to carry out this check are given at the end of the following example.

For the theoretical background to the methods employed to check facets such as rafter or haunch stability and as a more general introduction reference should be made to Morris and Randall, Horne and Morris, or Morris.

EXAMPLE 8.1   Design of a single bay pitched roof portal frame

Prepare a design in Grade 50 steel for an interior bay of the frame whose general arrangement is given in Fig. 8.11.

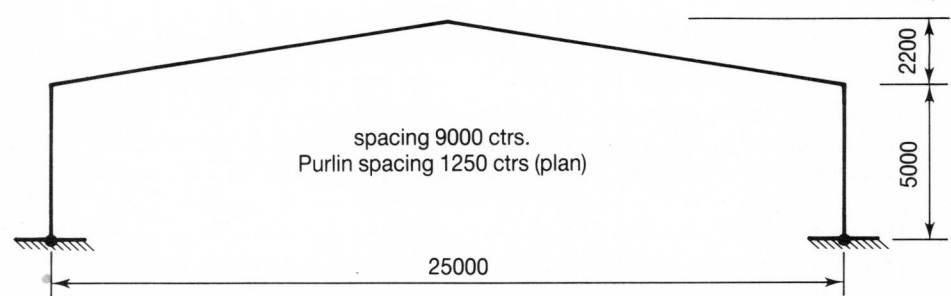

spacing 9000 ctrs.
Purlin spacing 1250 ctrs (plan)

2200

5000

25000

Fig. 8.11   Basic frame geometry for Example 8.1.

Dead load: Sheeting 0.18 kN/m²
        Purlins 0.06
        Frame 0.18
        Total 0.42

UDL due to dead load, $g_k = 0.42 \times 9.0 = 3.78$ kN/m

Imposed load (BS 6399, Pt 3, cl 4.3.1)

Snow load $= 0.60 \text{ kN/m}^2$

UDL due to imposed load, $q_k = 0.60 \times 9.0 = 5.40 \text{ kN/m}$

Wind load:

This will be dealt with later.

The frame will be designed under dead and imposed load and then checked under wind loading.

Design load, $1.4g_k + 1.6q_k = 1.4 \times 3.78 + 1.6 \times 5.40$
$$= 13.95 \text{ kN/m.}$$

It will be assumed that the loading can be treated as a UDL, whereas the loading is actually applied at discrete points through the purlins. Any error caused by this assumption will be small since the purlin spacing of 1.25 m is small compared with the span of 25 m.

The geometry used in the next section is defined in Fig. 8.12.

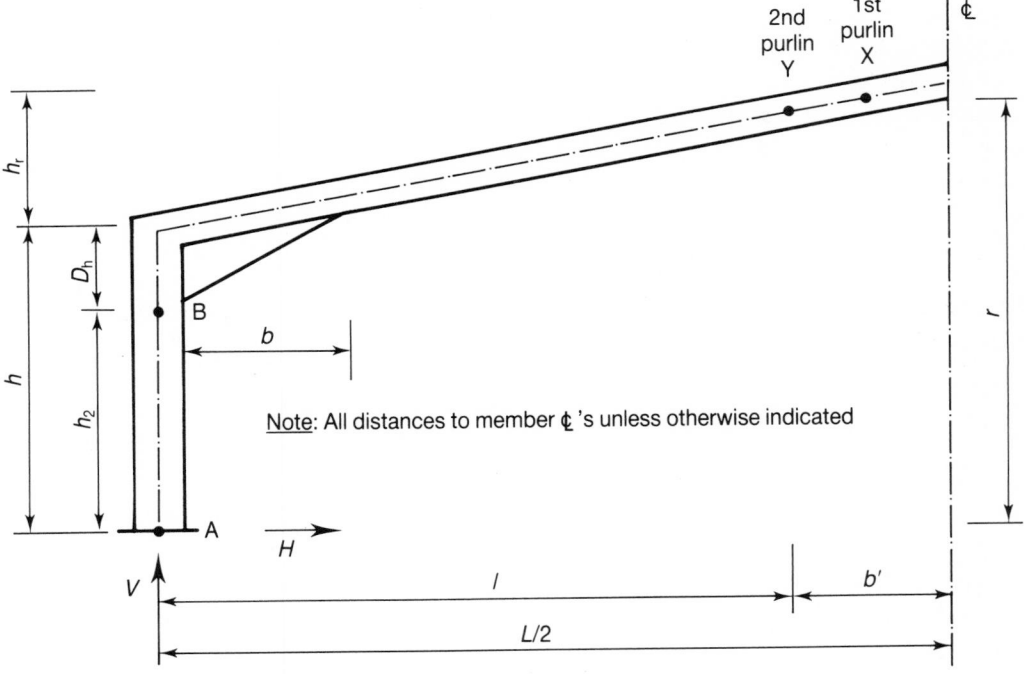

Fig. 8.12   Detail frame geometry.

Preliminary design:

Assume $D_h = 0.65 \text{ m}$

$$h_2 = h - D_h = 5.00 - 0.65 = 4.35 \text{ m}$$

Distance from apex to second purlin point, $b' = 2.5 \text{ m}$

so,

$$l = L/2 - b' = 12.5 - 2.5 = 10.00 \text{ m}$$

Gradient of rafter, $s_r = 2.2/12.5 = 0.176$

Height to second purlin point, r:

$$r = h + ls_r = 5.00 + 0.176 \times 10.00 = 6.76 \text{ m}$$

Moment at base of haunch, $M_{B1} = Hh_2$

$$= 4.35 H \qquad (8.1)$$

Moment at second purlin point, $M_Y = Vl - Hr - ql^2/2$
By symmetry, $V = qL/2 = 13.95 \times 25/2 = 174.4 \text{ kN}$.

So,

$$M_Y = 174.4 \times 10.00 - 6.76H - 13.95 \times 10^2/2$$

$$= 1047 - 6.76H \qquad (8.2)$$

Equations (8.1) and (8.2) are plotted in Fig. 8.13.

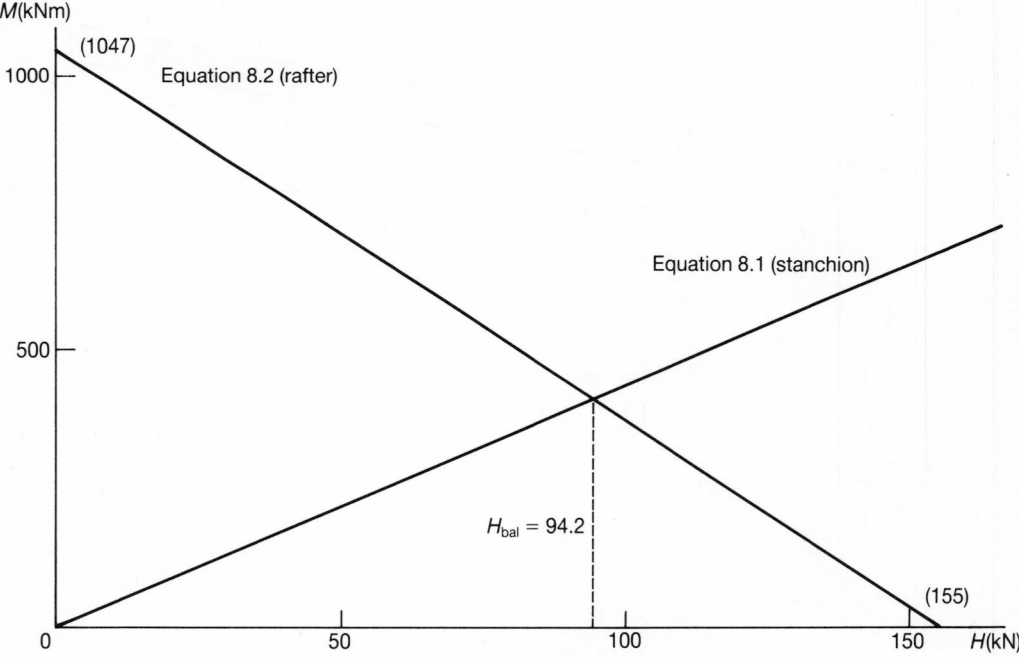

Fig. 8.13 Graphical solution for H_{bal}.

The intersection point of the two lines is at $H_{bal} = 94.2 \text{ kN}$ and $M = 410 \text{ kNm}$. This would give the solution for a uniform section. It is however more economic to use differing sections for the rafter and stanchion. A practical solution is obtained by increasing the value of H_{bal} obtained above by around 15–20%, so increase H to 110 kN (an increase of 17%).

Thus $(M_p)_{stanchion} = 4.35 \times 110 = 478.5$ kNm

and $(M_p)_{rafter} = 1047 - 6.76 \times 110 = 303.4$ kNm

From Table 6, assuming $T > 16$ mm, $p_y = 345$ N/mm^2

for the stanchion, $S = 475.8\text{E}3/345 = 1379$ cm^3

select a $457 \times 191 \times 67$ UB $(S = 1470)$

for the rafter, $S = 303.4\text{E}3/345 = 879$ cm^3

select a $356 \times 171 \times 57$ UB $(S = 1010)$

Since both flange thicknesses are less than or equal to 16, the full design strength of 355 N/mm^2 could have been used, with a consequent slight reduction in section sizes. No change will be made until after the sections have been checked for plastic action, although the maximum design strength will be used.

Design Checks:

Check for suitability of plastic action (Table 7):

$$\varepsilon = (275/p_y)^{0.5} = (275/355)^{0.5} = 0.88$$

Flange outstand (b/T):

Actual values: stanchion 7.48
             rafter 6.62

Maximum allowable value is $8.5\varepsilon = 8.5 \times 0.88 = 7.5$

Web (d/t):

Actual values: stanchion 48.0
             rafter 39.0

Maximum allowable value is $79\varepsilon = 79 \times 0.88 = 69.5$.

Both sections satisfy.

Moment capacity check (cl 4.2.5):

In general for single storey frames neither the shear nor the axial load is high enough to cause reductions in the moment capacity of the section.

However, for plastic sections, $M_c = S_x p_y < 1.2 Z_x p_y$

For the stanchion $S_x/Z_x = 1470/1300 = 1.13$, thus the moment capacity is given by

$$M_c = S_x p_y = 1470 \times 355/1\text{E}3 = 521.9 \text{ kNm}$$

For the rafter, $S_x/Z_x = 895/796 = 1.12$, so the moment capacity of rafter,

$$M_c = S_x p_y = 895 \times 355/1\text{E}3 = 317.7 \text{ kNm}.$$

Check haunch depth:

Depth of rafter $(D) = 355.6$, so depth to base of haunch, $h_2 = 1.5 \times 355.6 = 533$ mm.

Thus hinge at B_1 occurs at $5.00 - 0.533 = 4.467$ m,

so,

$$H = (M_c)_{\text{stanchion}}/h_2$$
$$= 521.9/4.467$$
$$= 116.8 \text{ kN}$$

Check moment capacity of frame using actual moment capacities of the stanchion and rafter.

It can be shown from an elastic distribution of forces that the first hinge to form will be at the eaves. Since the horizontal reaction has been calculated assuming that this hinge occurs, then it remains only to calculate the load (q_1) to cause the formation of the hinge at the second purlin point.

$$(M_p)_{\text{rafter}} = (25q_1/2) \times 10 - 116.8 \times 6.76 - q_1 \times 10^2/2$$

$$(M_p)_{\text{rafter}} = 317.7 \text{ kNm giving } q_1 = 14.8 \text{ kN/m}.$$

This is greater than the design load of 13.95 kN/m and is therefore satisfactory.

Check the moment at the first purlin point, X:

$$l = 11.25 \text{ m}, r = 5.00 + 0.176 \times 11.25 = 6.98 \text{ m}$$

$$M_X = Vl - H_r - ql^2/2$$
$$= 174.4 \times 11.25 - 116.8 \times 6.98 - 13.95 \times 11.25^2/2$$
$$= 264.0 \text{ kNm}$$

This is less than the plastic moment in the rafter and is therefore satisfactory.

Check moment in the rafter at the end of the haunch, M_H:

$$b = 2.5 + 0.5 \times D_{\text{stanchion}}$$
$$= 2.5 + 0.5 \times 453.6 = 2.727 \text{ m}$$

This figure will be rounded to 2.73 m for convenience, and will in fact slightly overestimate the moment.

$$M_H = Vb - H \times (5.00 + 0.176 \times b) - qb^2/2$$
$$= 174.4 \times 2.73 - 116.8 \times (5.00 + 0.176 \times 2.73) - 13.95 \times 2.73^2/2$$
$$= -216.0 \text{ kNm}$$

The negative sign indicates that the moment is causing compression on the bottom flange.

It is recommended (Morris and Randall) that the numerical value of this moment be limited to $0.87(M_p)_{\text{rafter}}$

$$0.87(M_p)_{\text{rafter}} = 0.87 \times 317.7 = 276.4 \text{ kNm}.$$

This condition is satisfied, and thus the haunch length is suitable.

Frame stability (cl 5.5.3.2, Pt 1).

The effective span to depth ratio for the rafter, $(L - L_h)/D$, where L_h is the average length of the haunches, should be less than the value from the following equation,

$$((L - L_h)/D)_{\text{max}} = (44/\Omega)(L/h)(\rho/(4 + \rho L_r/L))(275/p_{\text{yr}}) \tag{8.3}$$

Stiffness ratio, ρ:

For a single bay portal frame, $\rho = (2I_c/I_r)(L/h)$

where I_c and I_r are the second moments of area of the stanchion and the basic rafter section.

So, $\rho = (2 \times 29400/16100)(25/5)$
$\qquad = 18.3$

Note it is sufficient to use consistent units for each of the quantities in the relevant ratios, thus the second moments of area can be left in cm^4 and the frame dimensions in m.

Ratio L_r/L:

L_r is the total developed, or slope, length of the rafter, thus $L_r/2 = (L/2)/\cos \theta$, owing to symmetry

where θ is the roof slope.

Thus $L_r/L = 1/\cos 10 = 1.105$.

Arching ratio, Ω:

This is the ratio of the applied ultimate load (W) to the ultimate load (W_0) that could be carried by the rafter if it acted as an encastré beam of the same total span of the frame. This ratio is a measure of the increased capacity of the member when acting as part of a frame.

$$\Omega = W/W_0 = q/q_0$$

where q and q_0 are the respective UDLs.

$$q_0 = 16(M_p)_{rafter}/L^2$$
$$\quad = 16 \times 317.7/25^2$$
$$\quad = 8.13 \text{ kN/m}$$

Thus, $\Omega = 13.95/8.13 = 1.72$.

Strength ratio, $275/p_{yr}$:

$$275/p_{yr} = 275/355 = 0.775$$

Span to height ratio (L/h):

$$L/h = 25/5 = 5$$

So,

$$((L - L_h)/D)_{max} = (44/1.72)(5)(18.3/(18.3 \times 1.105 + 4))(0.775)$$
$$\qquad = 74.9$$

Actual value of $(L - L_h)/D = (25000 - 2730)/358.6 = 52.1$. This is less than the limiting value and is therefore satisfactory.

An alternative approach to frame stability is given in cl 5.5.3.2, Pt 1, whereby horizontal loads of 0.5% of the total factored dead and imposed loading are applied in the same

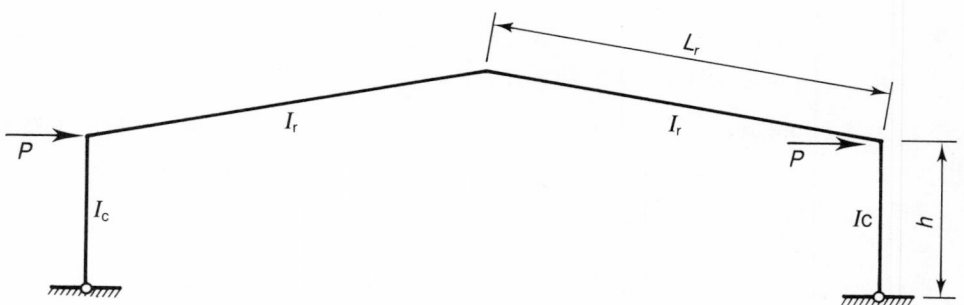

Fig. 8.14   Idealized frame for stanchion deflection.

sense at the top of the stanchions and the elastic horizontal deflection calculated. The limiting value for this deflection is stanchion height/1000.

For an unhaunched pitched roof portal frame, whose geometry is given in Fig. 8.14, the horizontal deflection, δ_h is given by,

$$\delta_h = (Ph^3/3EI_c)(1 + L_r I_c/hI_r)(L_r/(h + L_r)) \tag{8.4}$$

$$
\begin{aligned}
L_r &= (L/2)/\cos\theta \\
&= 12.5/\cos 10 \\
&= 12.693
\end{aligned}
$$

$$
\begin{aligned}
L_r/(h + L_r) &= 12.693/(5 + 12.693) \\
&= 0.72
\end{aligned}
$$

$$
\begin{aligned}
L_r I_c/hI_r &= 12.693 \times 29400/(5 \times 16100) \\
&= 4.64
\end{aligned}
$$

$$
\begin{aligned}
P &= 0.5\% \text{ factored total load} \\
&= 0.5 \times 13.95 \times 25/100 \\
&= 1.74 \text{ kN}
\end{aligned}
$$

$$
\begin{aligned}
\delta_h &= (1.74 \times 5^3/(3 \times 205E6 \times 29400E\text{-}8) \times (1 + 4.64) \times 0.72 \\
&= 0.0049 \text{ m}
\end{aligned}
$$

$$
\begin{aligned}
\text{Allowable deflection} &= h/1000 \\
&= 5/1000 = 0.005
\end{aligned}
$$

The frame therefore just satisfies the deflection criterion. It should be pointed out that if the effect of the haunch were to be taken into account when performing the deflection calculations, the deflection would be lower than that predicted by Equation (8.4).

Column Stability (cl 5.3.5, Pt 1):

The Code gives two alternative approaches to this problem; an approximate solution which is likely to be conservative and a more exact solution. The approximate solution is likely to be faster on design time but may be more restrictive. In this example both methods will be illustrated, although in practice, obviously, only one would be used.

(a)   Approximate approach (cl 5.3.5, Pt 1)

The clear distance from a restraint to the nearest hinge L_m is given by

$$L_m = 38r_y/(f_c/130 + (p_y/275)^2(x/36)^2)^{0.5} \qquad (4.36)$$

where f_c is the actual stress in the member and all other symbols have their usual meaning.

$$f_c/130 = (174.4E3/85.4E2)/130 = 0.16$$

$$r_y = 41.2 \text{ mm}, \quad x = 37.9$$

$$(p_y/275)^2(x/36)^2 = ((355/275)(37.9/36))^2 = 1.85$$

so, $L_m = 38 \times 41.2/(0.16 + 1.85)^{0.5}$

$$= 1104 \text{ mm}.$$

Thus a restraint must be placed not less than 1104 mm below the bottom of the haunch. This is in addition to the restraint required at the hinge itself (cl 5.3.5, Pt 1). It is still necessary to check the column below the restraint for buckling. The moments in the column are given in Fig. 8.15.

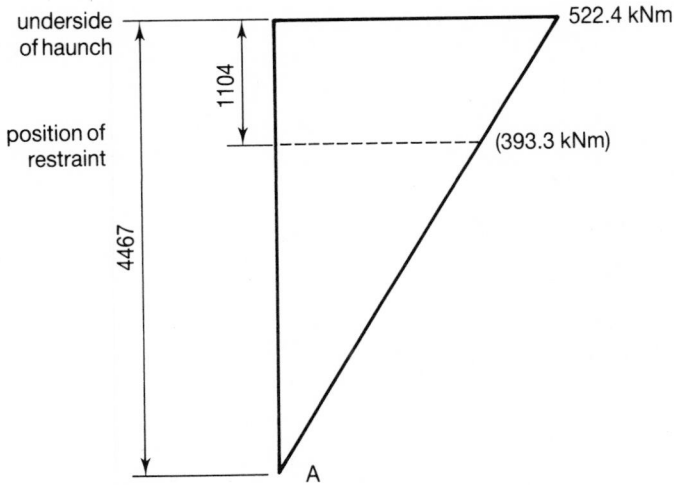

Fig. 8.15   BMD in stanchion AB.

Using the simplified approach (cl 4.8.3.3.1, Pt 1), it is necessary to check,

$$F/A_gp_c + mM_x/M_b + mM_y/p_yZ_y \leqslant 1.0 \qquad (6.14)$$

$M_y = 0$.

Calculation of F/A_gp_c:

$$F = 174.4 \text{ kN}, \quad A_g = 85.4 \text{ cm}^2$$

Calculation of p_c:

For the x–x axis the effective length, L_E is taken as the distance between the base of the haunch and the bottom of the stanchion, since adequate restraints occur at each of these points, so

$$\lambda_x = 4400/r_x = 4467/185 = 24.1$$

Table 25 indicates that for x–x buckling of a UB, Table 27(a) is to be used, thus $p_c = 348$ N/mm^2.

For the y–y axis, the effective length may be taken as the distance between the bottom of the stanchion and the additional restraint, so

$$\lambda_y = (4467 - 1104)/41.2 = 81.7.$$

Using Table 27(b), as directed by Table 25, $p_c = 207$ N/mm^2

Using the lesser of these two values,

$$F/A_g p_c = 174.4E3/(85.4E2 \times 207) = 0.10$$

Calculation of mM_x/M_b:

Equivalent moment factor, m (Equation (5.30)):

$$m = 0.57 + 0.33\beta + 0.1\beta^2$$
$$\beta = 0, \quad \text{so} \quad m = 0.57$$

Calculation of M_b (cl 4.3.7.3 et seq.):

$$\lambda_{LT} = nuv\lambda$$

$$n = 1.0 \text{ (Table 13)}$$

$$u = 0.873$$

$$\lambda = 81.7 \text{ (as for y–y axis strut buckling)}$$

$$\lambda/x = 81.7/37.9 = 2.16$$

From Table 14, with $N = 0.5$, $v = 0.95$

$$\lambda_{LT} = 1.0 \times 0.873 \times 0.95 \times 81.7$$
$$= 67.8$$

From Table 11, $p_b = 229$ N/mm^2

$$M_b = S_x p_b$$
$$= 1470 \times 229/1E3$$
$$= 336.6 \text{ kNm}$$

$$mM_x/M_b = 0.57 \times 393.3/336.6$$
$$= 0.67$$

$$F/A_g p_c + mM_x/M_b + mM_y/Z_y p_y = 0.10 + 0.67 + 0$$
$$= 0.77 < 1.0$$

The remaining length of the column is therefore satisfactory.

(b)   Exact approach (Appendix G, Pt 1):

Since the member is in compression on the flange unrestrained by the sheeting rails and has no plastic hinges between the points of torsional restraint (the underside of the haunch and the stanchion base), the member, which is of uniform section, need only be checked for elastic stability, i.e. Condition G.2 (a) (1).

This takes the form,

$$F/P_c + \bar{M}/M_b \leqslant 1.0$$

where F is the applied axial load, P_c is the compression resistance determined using the usual methods except that a modified slenderness ratio λ_{TC} is to be used, $\bar{M}$ is a moment defined as $m_t M_A$ (m_t is a modified uniform moment factor and M_A is the maximum applied moment), and M_b is the bending resistance calculated using a modified slenderness ratio λ_{TB}.

$$F = 174.4 \text{ kN}.$$

Calculation of λ_{TC}:

$$\lambda_{TC} = y\lambda$$

$$y = ((1 + (2a/h_s)^2)/(1 + (2a/h_s)^2 + (\lambda/x)^2/20))^{0.5}$$

$$\lambda = L/r_y$$

$a = $ distance from the member centre line to any lateral restraint (e.g. sheeting rails)

$h_s = $ distance between the shear centres of the flanges ($= D - T$).

Suitable sheeting rails for this frame are 202 mm deep,

so,

$$a = 453.6/2 + 202 = 428.8 \text{ mm}$$

$$h_s = 453.6 - 12.7 = 440.9$$

$$\lambda = 4467/41.2 = 108.4$$

$$(2a/h_s)^2 = (2 \times 428.8/440.9)^2 = 3.78$$

$$(\lambda/x)^2/20 = (108.4/37.9)^2/20 = 0.41$$

$$y = ((1 + 3.78)/(1 + 3.78 + 0.41))^{0.5} = 0.93$$

$$\lambda_{TC} = 108.4 \times 0.93 = 100.8$$

From Table 27(b), $p_c = 153 \text{ N/mm}^2$

$$F/P_c = 174.4\text{E}3/(153 \times 85.4\text{E}2) = 0.13$$

Calculation of $\bar{M}$:

$$M_A = 521.9 \text{ kNm}$$

Calculation of m_t (cl G3.4):

There are no intermediate loads so use Table 39;

$$\beta_t = 0, \quad \text{and} \quad y = 0.93, \quad \text{so} \quad m_t = 0.52$$

$$\bar{M} = m_t M_A = 0.52 \times 521.9 = 271.4 \text{ kNm}$$

Calculation of M_b:

Calculation of λ_{TB} (cl G3.3)

$$\lambda_{TB} = n_t u v_t c \lambda$$

$n_t = 1.0$ (no intermediate loads)

$c = 1.0$ (uniform section)

$$v_t = ((4a/h_s)/(1 + (2a/h_s)^2 + (\lambda/x)^2/20))^{0.5}$$

$4a/h_s = 4 \times 428.8/440.9 = 3.89$

$$v_t = (3.89/(1 + 3.78 + 0.41))^{0.5}$$
$$= 0.84$$

$u = 0.873$

$$\lambda_{TB} = 1.0 \times 0.873 \times 0.84 \times 1.0 \times 108.4$$
$$= 79.5$$

From Table 11, $p_b = 189$ N/mm^2.

$M_b = S_x p_b = 1470 \times 189/1E3 = 277.8$ kNm.

$M/M_b = 271.4/277.8 = 0.98$

$F/P_c + \bar{M}/M_b = 0.13 + 0.98 = 1.11 > 1.0$

Thus a restraint will be required.

　　Note this exact method may in some cases indicate that the column will be stable from elastic buckling and that no restraint will then be needed between the bottom of the haunch and the base.

　　Since the value obtained in the interaction equation is only slightly greater than unity, it may be worthwhile, in practice, to consider the use of a heavier section for the stanchion to avoid the extra costs of supplying and erecting additional bracing.

Check rafter stability at ridge (cl 5.3.5, Pt 1):

　　Any increase in depth caused by a haunch in order to allow an efficient connection to be made will be slight and can be ignored in stability calculations.

　　It is first necessary to calculate the axial force in the rafter (Fig. 8.16).

At the eaves, $F = H \times \cos 10 + V \times \sin 10 = 116.8 \times \cos 10 + 174.4 \times \sin 10$
$$= 145.3 \text{ kN}$$

At the ridge, $F = H \times \cos 10 = 116.8 \times \cos 10 = 115.0$ kN.

Since the variation in axial force is small, the maximum value of 145.3 kN will be used in all calculations.

　　The maximum distance to a restraint L_m is given by,

$$L_m = 38r_y/(f_c/130 + (p_y/275)^2 (x/36)^2)^{0.5} \tag{4.36}$$

$f_c/130 = (145.3E3/72.2E2)/130 = 0.15$

$x = 28.9$, $r_y = 39.2$ mm

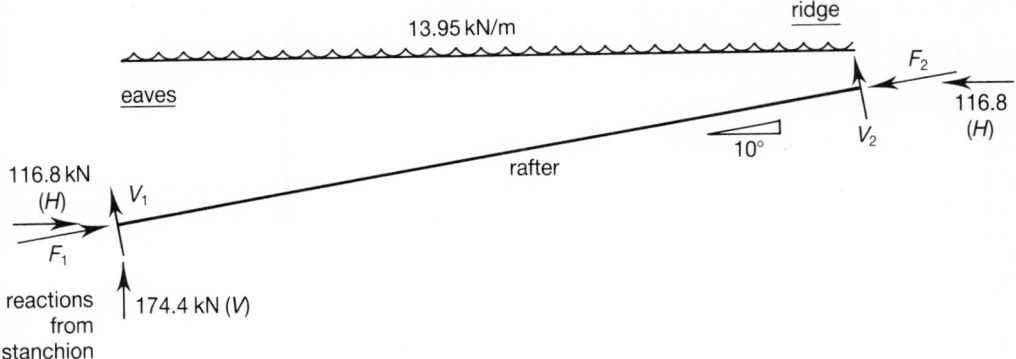

Fig. 8.16 Axial and shear forces in the rafters.

$$((p_y/275)(x/36))^2 = ((355/275)(28.9/36))^2 = 1.07$$
$$L_m = 38 \times 39.2/(0.15 + 1.07)^{0.5}$$
$$= 1.350 \text{ m}$$

This is greater than the purlin spacing and thus no additional bracing need be supplied. The restraint at the first purlin should take the form of a K brace to the compression flange of the rafter.

Check haunch stability:
  Relevant dimensions are given in Fig. 8.17.

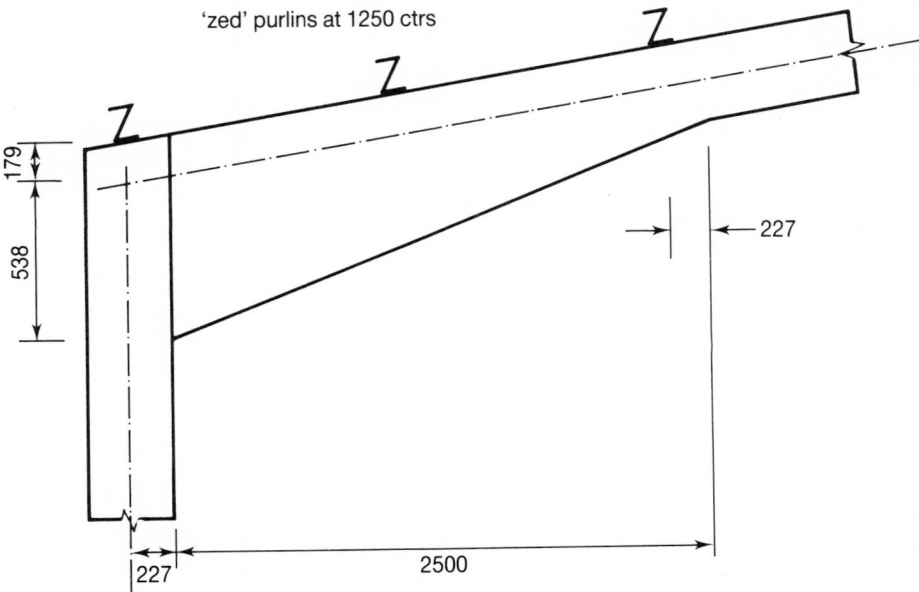

Fig. 8.17 Haunch geometry.

At any point the moment acting on the haunch is given by

$$M = 174.4s - H(5 + 0.176s) - (13.95/2)s^2$$

where s is the distance measured from the stanchion centre line.

The end of the haunch adjacent to the stanchion will be checked first. The compression flange of the haunch is restrained by K bracing up to the first purlin on the rafter. It will be conservative to take no account of the moment gradient along the member, i.e. $m = 1$, so it is necessary to check $M_A < M_b$.

Calculate M_A:

$$s = 0.20 \text{ m}, \quad \text{so} \quad M_A = -574.7 \text{ kNm}$$

giving $M = 574.7 \text{ kNm}$.

Calculate M_b:

$$\lambda_{LT} = nuv\lambda$$

$n = 1.0$ (cl B3, $R_f = 1.0$).

Effective length (L_E) is taken as the distance between restraints, i.e. $(1250 - 227)/\cos 10 = 1039$ mm.

Calculate section properties (Fig. 8.18):

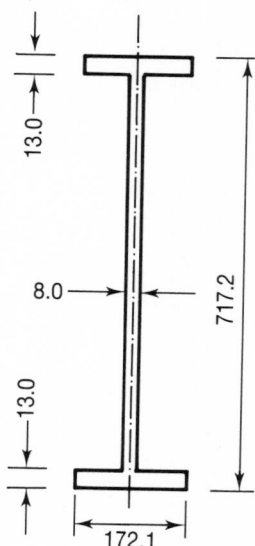

Note: Centre flange is ignored

Fig. 8.18   Section through haunch at stanchion face.

$A_f = 172.1 \times 13.0 = 2237 \text{ mm}^2$

$A_w = (717.2 - 2 \times 13) \times 8 = 5530 \text{ mm}^2$

$A = 2 \times A_f + A_w = 10004 \text{ mm}^2$

$I_x = (172.1 \times 717.2^3 - (172.1 - 8) \times (717.2 - 2 \times 13)^3)/12$

$\quad = 7.749E8 \text{ mm}^4$

$I_y = (2 \times 13 \times 172.1^3 + (717.2 - 2 \times 13) \times 8^3)/12$

$I_y = 11.07E6 \text{ mm}$

$S_x = (D - T) \times A_f + A_w d/4$

$\quad = 2237 \times (717.2 - 13) + 5530 \times (717.2 - 2 \times 13)/4$

$\quad = 2.531E6 \text{ mm}^3$

$r_y = (I_y/A)^{0.5} = (11.07E6/10004)^{0.5}$

$\quad = 33.3 \text{ mm.}$

Calculate u (cl B.2.5.1) (Equation (5.11))

$$u = (4S_x^2 \gamma/(A \times h_s)^2)^{0.25}$$

$$\gamma = 1 - I_y/I_x = 1 - 11.07E6/7.749E8 = 0.986$$

Distance between shear centres of the flanges (h_s)

$$h_s = D - T = 717.2 - 13 = 704.2 \text{ mm}$$

$$u = (4 \times 2.531E6^2 \times 0.986/(10004 \times 704.2)^2)^{0.25}$$

$$\quad = 0.845$$

Calculate x (cl B2.5.1) (Equation (5.11)):

$$x = 0.556h_s(A/J)^{0.5}$$

Torsional second moment of area J

$$J = \left(\sum bt^3 \right)/3$$

$$\quad = (2 \times 171.2 \times 13^3 + 691.2 \times 8^3)/3$$

$$\quad = 0.370E6 \text{ mm}^4$$

$$x = 0.556 \times 704.2 \times (10004/0.370E6)^{0.5}$$

$$\quad = 64.4$$

$$\lambda = L_E/r_y = 1039/33.3 = 31.2$$

$$\lambda/x = 31.2/64.4 = 0.48.$$

From Table 14, with $N = 0.5$, $v = 1.0$

$$\lambda_{LT} = 1.0 \times 0.845 \times 1.00 \times 31.2 = 26.4$$

From Table 11, $p_b = 355 \text{ N/mm}^2$

$$M_b = S_x p_b = 2.531E6 \times 355/1E6 = 898.5 \text{ kNm.}$$

This is in excess of the applied moment, and is therefore satisfactory.
   Check elastic stability of haunch (cl G.2(a).2, Pt 1).
   At any point in the haunch, $F/A + M/S_x = p_b$.

In order to apply certain of the equations needed, the total length of the haunch must be divided into four equal lengths, i.e. $s = 227, 852, 1477, 2102$ and 2727 mm.

Calculation of section properties (Fig. 8.19).

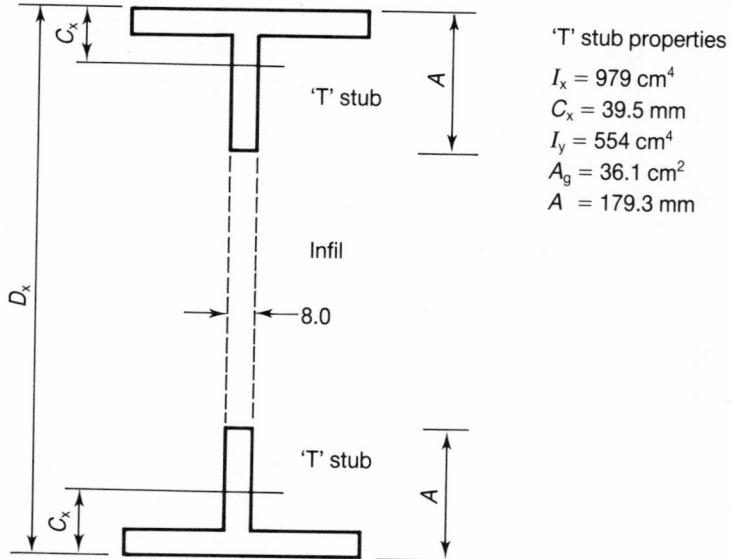

'T' stub properties

$I_x = 979$ cm^4
$C_x = 39.5$ mm
$I_y = 554$ cm^4
$A_g = 36.1$ cm^2
$A = 179.3$ mm

Fig. 8.19   Calculation of haunch properties.

Note, all calculations are in mm units.
It is easier to consider the haunch made up of two structural T-sections cut from a $356 \times 171 \times 57$ UB with the intervening web space filled with 8.0 mm plate. The properties required for the T-sections are also given in Fig. 8.19.
The depth D_s at any point s is given by

$$D_s = 717.2 - 358.6 \times (s - 227)/2500$$

Second moment of area about the y–y axis I_{yT}:

$$I_{yT} = 2I_y + t^3(D_s - 2A)/12$$
$$= 1108\text{E}4 + 42.67(D_s - 358.6)$$

Total Area A_T:

$$A_T = 2A_g + t(D_s - A)$$
$$= 7220 + 8.0(D_s - 358.6)$$

Plastic section modulus, S_x:

$$S_x = 2A_g(D_s/2 - C_x) + t(D_s - 2A)^2/4$$
$$= 3610(D_s - 79) + 2.0(D_s - 358.6)^2$$

These values are to be found in Table 8.1, together with calculations of F/A, M/S_x and $F/A + M/S$.

Table 8.1   Calculations for haunch stability

Distance s	227	852	1477	2102	2727	mm
Depth D_s	717.2	627.6	537.9	448.3	358.6	mm
I_{yT}	11.10	11.11	11.09	11.08	11.08	E6 mm^4
A_T	10089	9372	8654	7938	7200	mm^2
S_x	2.561	2.125	1.721	1.349	1.009	E6 mm^3
M	550.5	458.0	372.0	291.4	216.3	kNm
M/S_x	215.0	215.5	216.2	216.0	214.4	N/mm^2
F	145.3	145.3	145.3	145.3	145.3	kN
F/A	14.4	15.5	16.8	18.3	20.2	N/mm^2
$M/S_x + F/A$	229.4	231.0	233.0	234.3	234.6	N/mm^2

Calculation of p_b (cl G.3.3, Pt 1)

$$\lambda_{TB} = n_t u v_t c \lambda$$

$u = 1.0$

v_t:

$$v_t = ((4a/h_s)/(1 + (2a/h_s)^2 + (\lambda/x)^2/20)^{0.5}$$

This must be calculated on the smallest section.

A suitable size purlin depth to support the roof loads is 232 mm (i.e. $a = 232$ mm)

$a = 232 + D/2 = 232 + 358.6/2 = 411.2$ mm

$h_s = D - T = 358.6 - 13.0 = 345.6$ mm

L is taken as the distance from the start of the haunch adjacent to the stanchion to the first purlin point, which will require K bracing to the bottom flange to give full torsional restraint, beyond the end of the haunch, i.e.

$L = (3 \times 1250 - 227)/\cos 10 = 3577$ mm

$r_y = 39.2$ mm

$\lambda = L/r_y = 3577/39.2 = 91.3$

$x = 28.9$

$\lambda/x = 91.3/28.9 = 3.16$

$4a/h_s = 4 \times 411.3/345.6 = 4.76$

$(2a/h_s)^2 = (2 \times 411.3/345.6)^2 = 5.67$

$(\lambda/x)^2/20 = (3.16)^2/20 = 0.50$

$v_t = (4.76/(1 + 5.67 + 0.50))^{0.5}$

$\quad = 0.81$

Calculation of c

$c = 1 + (3/(x - 9))(R - 1)^{2/3} q^{0.5}$

q is defined as the ratio of the tapered length (2500 mm) to the length between torsional restraints (3523 mm),

$q = 2500/3523 = 0.71$

R is the ratio of the depths at either end of the haunch,

$R = 2$

$c = 1 + (3/(28.9 - 9))(2 - 1)^{2/3}0.71^{0.5}$
$\quad = 1.13$

Calculation of n_t (cl G3.6, Pt 1):

$$n_t = ((N_1/M_1 + 3N_2/M_2 + 4N_3/M_3 + 3N_4/M_4 + N_5/M_5 + 2(N_S/M_S - N_E/M_E))/12)^{0.5}$$

The N values are the applied moments at points 1 to 5, and the M values are the moment capacities at these points assuming the attainment of full stresses.

The values of actual moment should be increased by aF (a is defined as the distance to the restraint axis and F is the axial force N). In this example, and indeed in most single bay frames, the axial stresses are low and may be ignored in the calculations of N. Thus N becomes equal to the applied moment. The sign convention adopted for N is that N is positive when causing compression on the unrestrained flange, and that if N is negative (compression on restrained flange) it is set equal to zero.

N_E/M_E is the maximum of N_1/M_1 and N_5/M_5

N_S/M_S is the maximum of N_2/M_2 to N_4/M_4

$N_1/M_1 = 550.5/2.561 \times 355 = 0.61$

$N_2/M_2 = 458.0/2.125 \times 355 = 0.61$

$N_3/M_3 = 372.0/1.721 \times 355 = 0.61$

$N_4/M_4 = 291.4/1.349 \times 355 = 0.61$

$N_5/M_5 = 216.3/1.009 \times 355 = 0.60$

$N_S/M_S - N_E/M_E = 0.61 - 0.61 = 0.$

This latter value is zero in the calculation of n_t

$n_t = ((0.71 + 3 \times 0.61 + 4 \times 0.61 + 3 \times 0.61 + 0.60)/12)^{0.5}$
$\quad = 0.78$

$\lambda_{TB} = 0.78 \times 1.0 \times 0.81 \times 1.13 \times 91.3$
$\quad = 65$

From Table 11, $p_b = 239$ N/mm^2

Since the values of $F/A + M/S$ in Table 8.1 do not exceed the allowable value, no further restraints are required.

Check elastic stability between the end of the haunch and the purlin beyond the point of contraflexure in the rafter.

The point of contraflexure occurs at $s = 4.87$ m from the stanchion. The check, therefore, needs carrying out between $s = 2.727$ and $s = 5.00$ m. This distance is divided into four sections. The procedure is similar to that carried out for the haunch, and is carried out in Table 8.2.

Table 8.2   Rafter stability

Distance s	2727	3295	3863	4431	5000	mm
M	−216.3	−152.7	−99.2	−39.3	10.8	kNm
M/S	214.2	151.2	98.2	38.9	–	N/mm²
F	145.3	145.3	145.3	145.3	145.3	kN
F/A	20.1	20.1	20.1	20.1	20.1	N/mm²
$M/S + F/A$	234.3	171.3	118.8	59.0	–	N/mm²

Note

1 Moments have been given their conventional signs.
2 Calculations have not been made of total stress where the moment induces tension on the bottom flange.

$$\lambda = L/r_y = (1250/\cos 10)/39.2 = 32.4$$

$v_t = 0.82$ (as for the haunch)

$c = 1.0$ (cl G3.3)

$u = 0.84$

Calculation of n_t:
All moment capacities are 358.6 kNm (full plastic capacity of rafter section)

$N_1/M_1 = 0.60$

$N_2/M_2 = 0.43$

$N_3/M_3 = 0.28$

$N_4/M_4 = 0.11$

$N_5/M_5 = 0$

$N_S/M_S - N_E/M_E = 0.43 - 0.60 = -0.17$, which is set equal to zero.

$n_t = ((0.60 + 3 \times 0.43 + 4 \times 0.28 + 3 \times 0.11)/12)^{0.5}$
$\quad = 0.53$

$$\lambda_{TB} = 0.53 \times 0.83 \times 0.82 \times 1.0 \times 32.4$$
$$= 11.8$$

From Table 11, $p_b = 355$ N/mm².
   This is nowhere exceeded, and thus there are no additional bracing requirements between the end of the haunch to beyond the point of contraflexure.
   This completes the checks required on the structure under dead and live load (except deflection which will be calculated at the end), it remains to check that wind loading is not critical.

Wind loading check:

Calculation of wind loads:

Note all Code references in this section are to CP 3 Ch V Pt 2.

Location of structure–Birmingham.

Basic wind speed, 42 m/s (Fig. 1)

$S_1 = 1.0$ (cl 5.4)

$S_3 = 1.0$ (design life 50 yr, cl 5.6)

Conservatively set $S_4 = 1.0$

S_2:

Structure is city centre (Category 4), length greater than 50 m (Class C), height above ground level 7.2 m, thus

$S_2 = 0.54$ (Table 3)

Calculation of characteristic wind speed, V_s (cl 5.1)

$$V_s = S_1 S_2 S_3 S_4 V$$
$$= 1.0 \times 0.54 \times 1.0 \times 1.0 \times 42 = 22.7 \text{ m/s}$$

Calculation of dynamic pressure, q (cl 6):

$$q = 0.613 V_s^2$$
$$= 0.613 \times 22.7^2$$
$$= 315 \text{ N/m}^2$$

Calculation of global C_{pe} values:

Walls (long sides only) (Table 7)

height to width ratio $(h/w) = 5/25 = 0.2 < 0.5$

length of width ratio (L/w): assume $2 < L/w < 4$

Wind normal to long side $(\alpha = 0)$: $C_{pe} = 0.7$ and -0.25

wind normal to end $(\alpha = 90°)$: $C_{pe} = -0.5$ and -0.5

Roof (Table 8)

$h/w < 0.5$, roof slope is 10°

for $\alpha = 0$, $C_{pe} = -1.2$ and -0.4

for $\alpha = 90°$, $C_{pe} = -0.8$ and -0.6

Calculation of C_{pi} (Appendix E)

Because the proportion of any dominant openings in relation to the total wall area of any face is not known at this stage, C_{pi} should be taken as 75% of the value of the external coefficient (C_{pe}) outside the opening (i.e. from values relating to the walls).

Thus, for $\alpha = 0$, $C_{p1} = 0.75 \times (-0.25) = -0.19$ (Case 1) or $0.75 \times (0.7) = 0.52$ (Case 2)

for $\alpha = 90°$, $C_{pi} = 0.75 \times (-0.5) = -0.38$

Total pressure coefficients $(C_{pe} - C_{pi})$ for the roof:

$$\alpha = 0: \text{Case 1: } C_{pe} - C_{pi} = -1.2 - (-0.19) = -1.01$$
$$\text{and} \quad -0.40 - (-0.19) = -0.21$$
$$\text{Case 2: } C_{pe} - C_{pi} = -1.2 - 0.52 = -1.72$$
$$\text{and} \quad -0.40 - 0.52 = -0.92$$
$$\alpha = 90: C_{pe} - C_{pi} = -0.80 - (-0.38) = -0.42 \quad \text{and} \quad -0.60 - (-0.38) = -0.22$$

For the walls, the corresponding values are:

$\alpha = 0$: Case 1: 0.51 and -0.06:   Case 2: 0.18 and -0.75
$\alpha = 90$: both walls, -0.12.

None of these wind load cases produce downthrust, thus the load case $1.2(g_k + q_k + w_k)$ does not need considering. The only wind load case which will be examined is that producing maximum suction ($\alpha = 0$, Case 2).

The easiest way of dealing with this is to load the structure with $1.4w_k + 1.0g_k$ and demonstrate that the structure has a factor of safety γ greater than unity.

Dead load: 3.78 kN/m
Wind Load:

Walls: LH: $0.18 \times (1.4 \times 0.315) \times 9.0 = 0.71$ kN/m
    RH: $-0.75 \times (1.4 \times 0.315) \times 9.0 = -2.98$ kN/m

Roof: LH: $-1.72 \times (1.4 \times 0.315) \times 9.0 = -6.83$ kN/m
    RH: $-0.92 \times (1.4 \times 0.315) \times 9.0 = -3.65$ kN/m

The loading and salient dimensions are given in Fig. 8.20.

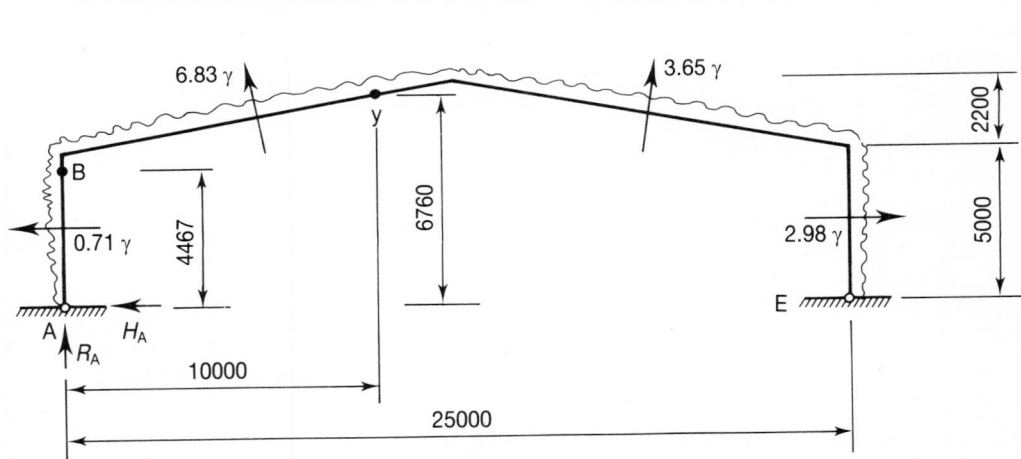

Notes: (1) All values of loads are in kN/m run
    (2) Point B is the base of the haunch and Y is
        the second purlin point from the ridge

Fig. 8.20 Frame data for wind loading case.

Taking moments about E:

$$25R_A + 0.71\gamma \times 5^2/2 + 6.83\gamma \times 12.5 \times (12.5/2 + 12.5) - 6.83\gamma \times 2.2$$
$$\times (5 + 2.2/2) + 3.65\gamma \times 12.5^2/2 + 3.65\gamma \times 2.2 \times (5 + 2.2/2) + 2.98\gamma$$
$$\times 5^2/2 - 3.78\gamma \times 25^2/2 = 0$$

or,

$$R_A = -28.32\gamma$$

Moment at B,

$$M_B = 4.467H_A + 0.71\gamma \times 4.67^2/2$$

or,

$$H_A = 0.224M_E - 1.73\gamma$$

Moment at Y,

$$M_Y = 10R_A + 6.76H_A - 0.71\gamma \times 5 \times (6.76 - 5/2) + 6.83\gamma \times 1.76^2/2$$
$$+ 6.83\gamma \times 10^2/2 - 3.78\gamma 10^2/2$$
$$= 10R_A + 6.76H_A + 148.0\gamma$$

Substituting for R_A and H_A

$$M_Y = 1.51M_B - 146.9\gamma$$

or,

$$\gamma = (M_Y - 1.51M_B)/(-146.9)$$

M_B is the moment capacity of the stanchion, 521.9 kNm and M_y the moment capacity of the rafter, 317.7 kNm.

Hence $\gamma = 7.53$. Thus an additional load factor greater than unity exists on this load case, indicating that wind uplift is not a problem.

Deflection under imposed load:

For a single bay symmetric unhaunched frame the vertical deflection, δ_v, at the ridge is given by

$$\delta_v = (qL^4 \cos \theta/(768EI_r))(10 - (8 + 5h_1/h)(3 + 2h_1/h)/ \atop ((h_1/h)^2 + 3h_1/h + 3 + I_r h/(I_c L_r))) \tag{8.5}$$

where q is the imposed load, L the span of the portal frame, θ the pitch of the rafter, I_r and L_r the second moment of area and the slope length of the rafter respectively, I_c the second moment of area of the stanchion, h the height to eaves and h_1 the height of the ridge above the eaves.

The horizontal deflection, δ_H, at the eaves is given by

$$\delta_H = \delta v \tan \theta \tag{8.6}$$

$q = 5.40$ kN/m
$E = 205$ GPa
$h = 5.0$ m
$h_1 = 22.2$ m
$L = 25$ m
$\theta = 10°$
$I_c = 29400$ cm^4
$I_r = 16100$ cm^4

$\cos \theta = 0.985$

$L_r = (L/2)/\cos \theta = 12.5/0.985 = 12.692$ m

$h_1/h = 2.2/5.0 = 0.44$

$I_r h/I_c L_r = 16100 \times 5/(29400 \times 12.692) = 0.216$

$$\delta_v = (5.4 \times 25^4 \times 0.985/(768 \times 205E6 \times 16100E\text{-}8))(10 - (8 + 5 \times 0.44)$$
$$(3 + 2 \times 0.44)/(0.44^2 + 3 \times 0.44 + 3 + 0.216))$$
$$= 0.134 \text{ m}$$

This is equivalent to span/188, which is not excessive although double that of the Australian recommendations, remembering that no allowance has been made either for the stiffening effects of the haunches or the cladding.

$$\delta_H = \delta v \tan \theta$$
$$= 0.133 \times \tan 10$$
$$= 0.023 \text{ m}$$

This is equivalent to height/217, which again is acceptable, and is less than Australian $h/150$ recommendation.

It is instructive to compare the final results obtained in this design example with a solution obtained using the design charts in the IStructE/ICE Steelwork Design Manual.

For this portal frame, calculate:

1 Span to height to eaves ratio (L/h)

$$L/h = 25/5 = 5,$$

2 Rise to span ratio (r/L)

$$r/L = 2.2/25 = 0.088$$

Horizontal reaction (H)
Using Fig. 13 of the manual, determine the non-dimensional horizontal coefficient, H_{FR}. By interpolation, $H_{FR} = 0.331$,

$$H = H_{FR} WL$$

where W is the factored dead and imposed load per unit run

$$L = 25 \text{ m}, \quad W = 13.95 \text{ kN/m}, \quad \text{thus}$$
$$H = 0.331 \times 25 \times 13.95 = 115.4 \text{ kN}$$

The value used in the main design was 110 kN
Plastic Moment in the rafter (M_{pr})
From Fig. 14 in the manual, $M_p = 0.0321$

$$M_{pr} = M_p WL^2$$
$$= 0.0321 \times 13.95 \times 25^2$$
$$= 279.9 \text{ kNm}$$

The value used in the actual design was 303.4 kNm
Plastic Moment in the stanchion (M_{ps})

From Fig. 15 in the manual, $M_p = 0.0574$

$$M_{ps} = M_p WL^2$$
$$= 0.0574 \times 13.95 \times 25^2$$
$$= 500.8 \text{ kNm}$$

The value used in the actual design was 478.5 kNm.

It is noted that the two sets of values are close, thus justifying the method used to determine the horizontal reaction in this example.

Deflection check at the eaves (d_e)

Using Fig. 16 of the manual,

roof slope, θ, $= 10°$

Span to height to eaves ratio, $L/h = 5$,

So, from Fig. 16, $D = 0.49$

$$d_e = D(hL/d_r)(p_y/\gamma_p) \times \text{E-6}$$

where h_r is the depth of the unhaunched rafter, p_y the design strength, and γ_p the load factor at collapse, which may be taken with little loss of accuracy as 1.5.

$$d_e = 0.49(5000 \times 25000/356)(355/1.5) \times \text{E-6}$$
$$= 41 \text{ mm}$$

This is almost double that calculated using a full frame analysis, even neglecting the haunches.

It should be noted that Fig. 16 is from the 'Plastic Design Supplement' issued originally by Constrado to accompany Morris and Randall, and the method used therein is known to be conservative since it almost certainly makes the assumption of equal sized rafter and stanchion.

Consider Davies' hypothesis that single storey frames should be considered as sway frames. The following equations may be used (from Davies):

$$\lambda_{cr} = 3EI_r/L_r(hP_c + 0.3L_r P_r)$$

where

$P_r = qL^2(3 + 5m) \cos \phi/(16Nh) + qL \sin \phi/4$
$P_c = qL/2$
$m = 1 + h_1/h$
$N = 2(1 + m + m^2 + I_r h/L_r I_c)$
$L = 25 \text{ m}$, $\quad h = 5.0 \text{ m}$, $\quad h_i = 2.2 \text{ m}$, $\quad L_r = 12.692 \text{ m}$, $\quad I_c = 29400 \text{ cm}^4$, $\quad I_r = 16100 \text{ cm}^4$, $q = 13.95 \text{ kN/m}$, $\phi = 10°$, giving
$m = 1.44$, $N = 9.459$, $P_r = 130.8 \text{ kN}$, $P_c = 174.4 \text{ kN}$ and $\lambda_{cr} = 5.69$

From Equation (8.13), since $4.6 \leqslant \lambda_{cr} \leqslant 10.0$,

$$\lambda_p = 0.9\lambda_{cr}/(\lambda_{cr} - 1) = 1.09$$

Actual load carrying capacity is 14.8 kN/m, so actual plastic load factor is $14.8/13.95 = 1.06$.

This is slightly below the value calculated using the theory postulated by Davies, but it should be pointed out that Davies' theory is derived excluding the effect of haunches. It also should be pointed out that a slightly low value of λ_p could have been expected since the frame only just satisfies the deflection limit when the two 0.5% horizontal loads are applied to the eaves.

8.10.1.2   *Notes on the design of a gable end frame (Example 8.2)*

1   The rafter is usually designed as a continuous beam spanning over the gable end columns, with simple, non-moment transferring connections to the frame. This will mean that if the spans are equal the first bay will be critical since it will only be continuous at one end.
2   The centre lines of the gable end columns should be coincident with the roof purlins in order to resist the reaction from the wind loading on the gable end columns without subjecting the rafters to bi-axial bending. It is also preferable if the spacing of the gable end columns can be similar to the frame spacing in order to keep the same section for the sheeting rails, although this is not essential.
3   Often the loading on the gable end will require the use of only very small sections, but detailing requirements and compatibility with the rest of the frame may make it necessary to use larger sections.

EXAMPLE 8.2   Gable end frame design

Design a gable end frame in Grade 50 steel for the pitched roof portal of Example 8.1. The gable end columns will be spaced at 6.25 m centres, giving columns at the quarter and mid-span points. The resultant frame geometry is given in Fig. 8.21.

   All the loading data are given in Example 8.1 and will therefore not be derived here. The reader is referred to that Example.

Rafter design:

The rafter will be designed as continuous over the internal columns and simply supported at the external columns.

   The UDL is half that for the internal frame designed in Example 8.1, i.e. $13.95/2 = 7.0$ kN/m.

   Using the reactant bending moment diagram approach (Fig. 8.22(a)).

Span AB

It is sufficiently accurate to assume the span hinge occurs at mid-span. The error in so doing will be around 2%.

$$M_p + 0.5 \times M_p = 34.2$$

or,

$$M_p = 22.8 \text{ kNm.}$$

Span BC:

$$2M_p = 34.2$$

or,

$$M_p = 17.1 \text{ kNm}$$

Using the larger value of M_p for design

$$S_x = M_p/p_y = 22.8\text{E}3/355$$
$$= 64 \text{ cm}^3$$

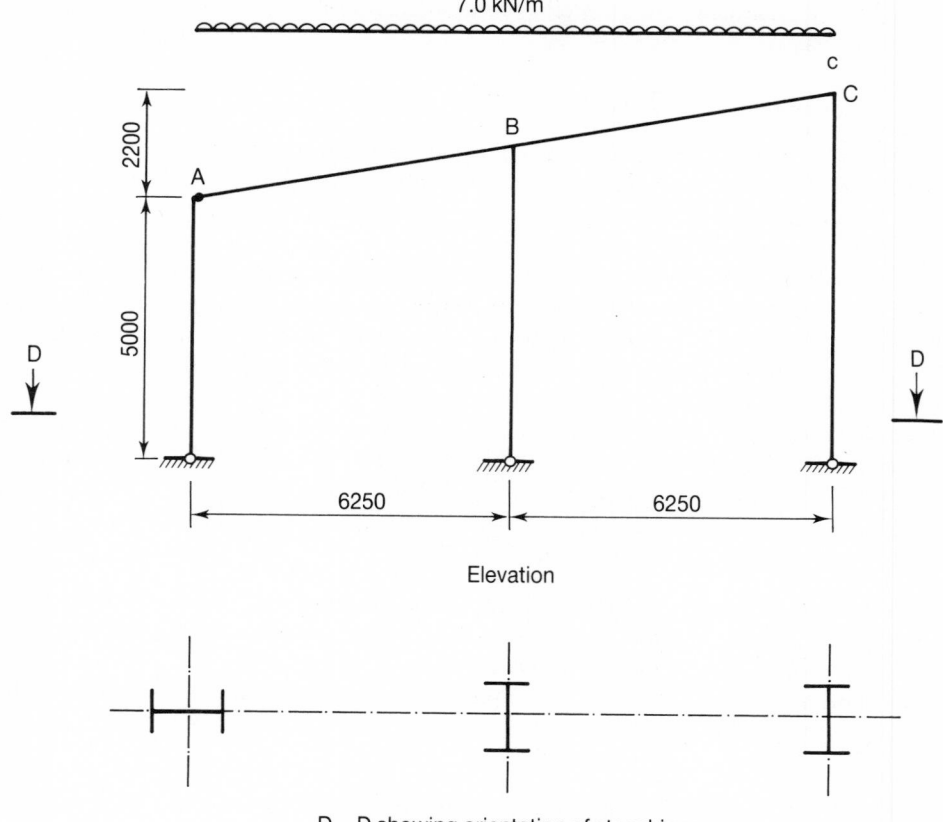

Fig. 8.21   Elevation of half gable end.

Note that span BC remains partly elastic. The final bending moment diagram is plotted in Fig. 8.22(b).

Select a $254 \times 102 \times 25$ UB ($S = 306$).

Check suitability for plastic action (Table 7):

$$\varepsilon = (275/p_y)^{0.5} = (275/355)^{0.5} = 0.88$$

Flange outstand (b/T):
Actual value $= 6.07$
Allowable value $= 7.5\varepsilon = 7.5 \times 0.88 = 6.6$

Web slenderness (d/t):
Actual value $= 36.9$
Allowable value $= 79\varepsilon = 79 \times 0.88 = 69.5$
Both criteria are satisfied.

Given the large amount of overdesign, even for the smallest UB section, full plastic action may not be strictly necessary, i.e. a compact section would have sufficed.

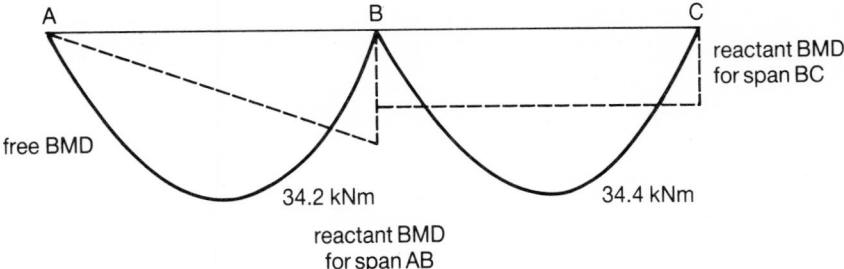

(a) Plastic collapse diagrams

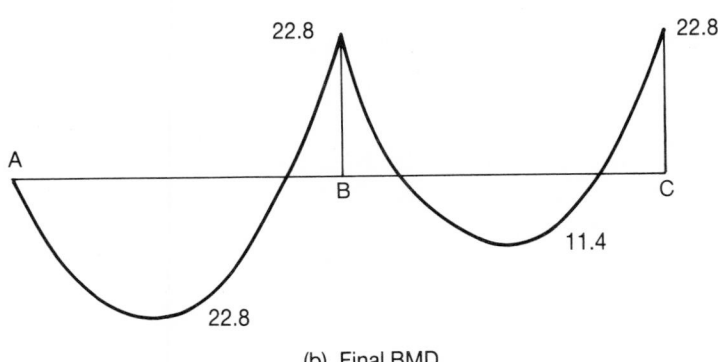

(b) Final BMD

Fig. 8.22   Moments in gable rafter.

Moment capacity check (cl 4.2.5, Pt 1):

$$M_c = S_x p_y < 1.2 Z_x$$

$$S_x / Z_x = 306/265 = 1.15$$

$$M_c = S_x p_y = 306 \times 355/1E3 = 108.6 \text{ kNm}.$$

Shear check (cl 4.2.3 et seq., Pt 1):

$$R_{B1} = 2 \times 17.4 + 108.6/6.25 = 52.2 \text{ kN}.$$

Shear capacity, P_v

$$P_v = 0.6 p_y D t$$
$$= 0.6 \times 355 \times 257.0 \times 8.4/1E3$$
$$= 460 \text{ kN}.$$

No reduction in moment capacity is needed if the shear is less than 0.6×460. This condition is satisfied.

Since there is a point of contraflexure within the span, cl G2 (Pt 1) can be used to check stability. A check must be made to see whether the restraint length is less than the critical value L_T given by

$$L_T = L_k / c n_t$$

Calculation of L_k (cl G3.5, Pt 1):

$$L_k = (5.4 + 6000p_y/E)r_y x/(5.4(p_y/E)x^2 - 1)^{0.5}$$

$r_y = 21.4$ mm, $x = 31.4$

$E = 205$ kN/mm^2 (cl 3.1.2, Pt 1)

$p_y/E = 355/205E3 = 1.73E\text{-}3$

$L_k = (5.4 + 6000 \times 1.73E\text{-}3) \times 21.4 \times 31.4/(5.4 \times 1.73E\text{-}3 \times 31.4^2 - 1)^{0.5}$
$\quad = 3700$ mm

$c = 1.0$ for uniform members (cl G3.5, Pt 1)

Calculation of n_t (cl G3.6, Pt 1):

$$n_t = ((N_1/M_1 + 3N_2/M_2 + 4N_3/M_3 + 3N_4/M_4 + N_5/M_5 + 2(N_S/M_S - N_E/M_E))/12)^{0.5}$$

For notes on the application of this equation the reader is referred to Example 8.1.

Values of N are as follows:

$N_1 = 0$, $N_2 = 21.7$, $N_3 = 11.0$, $N_4 = -32.6$ and $N_5 = -108.6$ kNm respectively.

At all points, the moment capacity M is 108.6 kNm, thus

$N_1/M_1 = 0$, $N_2/M_2 = 0.20$, $N_3/M_3 = 0.10$, and N_4/M_4 and N_5/M_5 are both set to zero, as they are negative.

$N_S/M_S = 0.20$ and $N_E/M_E = 0$, giving
$N_S/M_S - N_E/M_E = 0.20$,

thus,

$n_t = ((3 \times 0.20 + 4 \times 0.10 + 2 \times 0.2)/12)^{0.5}$
$\quad = 0.34$

and,

$L_T = 3700/(1.0 \times 0.34)$
$\quad = 10880$ mm.

This is greater than the span of the rafter between stanchions, and thus no additional restraints are required.

Reactions from purlin loads in the rafter spans are:
At A, 17.4 kN, at B, 102.6 kN and at C, 19.2 kN

The axial loads on the columns become:

at A: $17.4 + 17.4/2 = 26.1$ kN
at B: $102.6 + 17.4 = 120$ kN
at C: $2 \times 19.2 + 17.4 = 55.8$ kN

Stanchion Design:
Lateral loading due to wind:
C_{pe} values (CP 3 Ch V Pt 2, Table 7):

$\alpha = 0$: -0.6 for both ends
$\alpha = 90$: 0.7 and -0.1

From Example 8.1, C_{pi} values are for $\alpha = 0$, -0.19 and for $\alpha = 90$, -0.38
Thus total pressure coefficients, $C_{pe} - C_{pi}$ are

for $\alpha = 0$: -0.41 on both ends
for $\alpha = 90$: 1.08 and 0.28

The worst value is 1.08 (causing compression on the unrestrained face). Assume the bottom 0.5 m of the structure to be clad in masonry (giving no restraint to the columns, but taking the wind loads below 0.5 m above floor level). The sheeting rails are at 1.5 m centres. The resultant arrangement is given in Fig. 8.23.

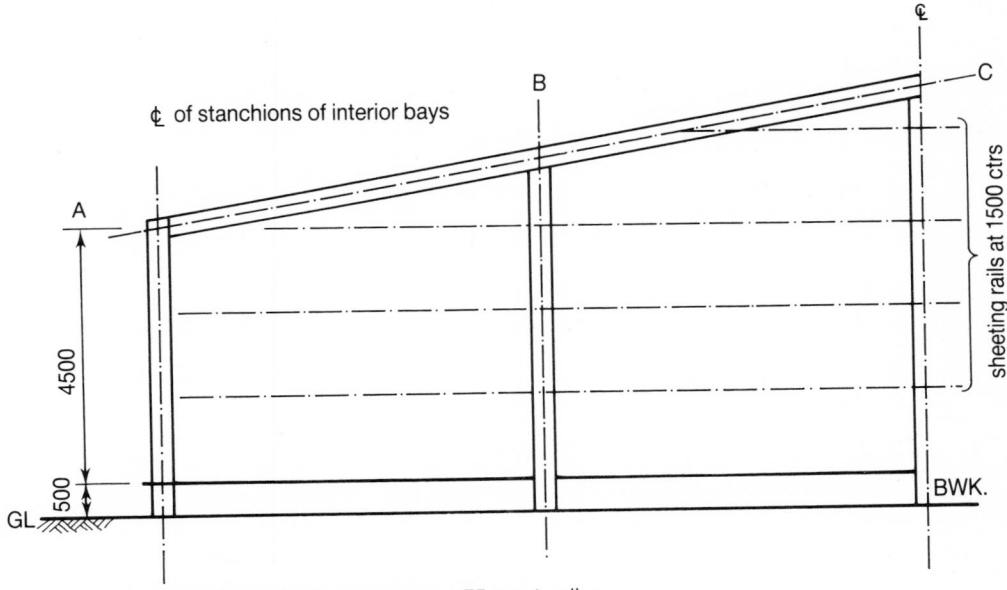

Notes: (1) Stanchion at A offset by approx. 75 mm to align
outside faces for sheeting rails
(2) All steelwork 254×102×25 UB grade 50

Fig. 8.23   Elevation of gable end.

The column at B carries the highest axial load but is shorter than that at C, and carries lower bending moments than that at C. Design that at C and check those at A and B.

Column at C:
Assume the same size as the rafter (254 × 102 × 25 UB). This has already been checked for plastic action.
    The bending moment diagram is given in Fig. 8.24
Maximum moment is 18.7 kNm.
Moment capacity of section is 108.6 kNm.

Local capacity check (cl 4.8.3.2 (b), Pt 1):
Axial stress is $55.8E3/32.2E2 = 17$ N/mm^2

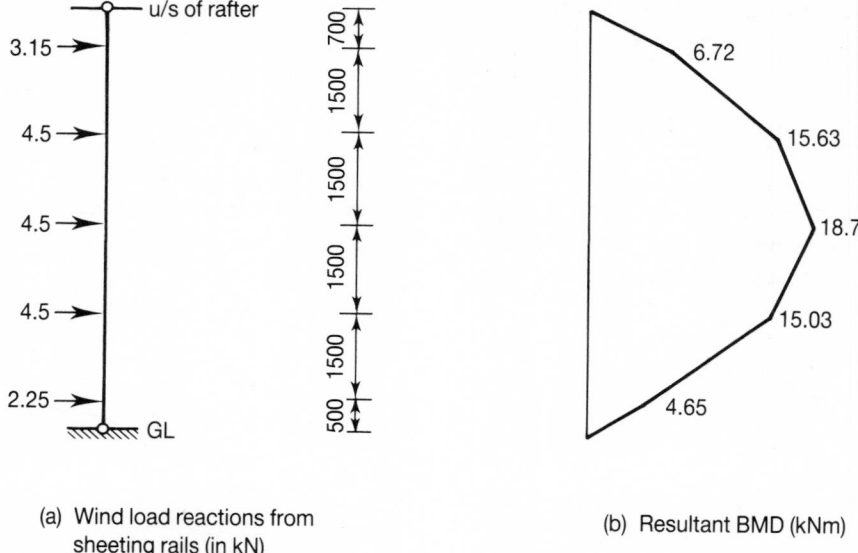

(a) Wind load reactions from sheeting rails (in kN)

(b) Resultant BMD (kNm)

Fig. 8.24   Wind loading on stanchion at C.

This is low and will cause only about 1 to 2% reduction in moment capacity.
Since $M_y = 0$, the local capacity check reduces to

$$M_x > M_{rx}$$

where M_x is the applied moment and M_{rx} is the reduced moment capacity allowing for axial stress.
Clearly this is satisfied.

Overall buckling check (cl 4.8.3.3.1, Pt 1):

$$F/A_g p_c + mM_x/M_b \leq 1.0 \quad (M_y = 0)$$

Calculation of $F/A_g p_c$:
calculation of p_c:
for the x–x axis: $L_{Ex} = 7200$ mm, $r_x = 103$ mm,

$$\lambda = 7200/103 = 70$$

using Table 27(a), $p_c = 270$ N/mm^2
for the y–y axis: $L_{Ey} = 1500$, $r_y = 21.4$ mm,

$$\lambda = 1500/21.4 = 70$$

using Table 27(b), $p_c = 242$ N/mm^2
Thus, $F/A_g p_c = 55.8\text{E}3/242 \times 32.3$

$$= 0.07$$

Calculation of mM_x/M_b:
calculation of m (Equation (5.30)):

$$m = 0.57 + 0.33\beta + 0.1\beta^2$$

maximum value of $\beta = 15.83/18.7 = 0.87$,

so $m = 0.57 + 0.33 \times 0.87 + 0.1 \times 0.87^2$

$\quad = 0.93$

Calculation of M_b:

calculation of p_b:

Since no loads are applied between restraints, $n = 1.0$ (*Fig. 5.6*)

$\qquad u = 0.864$

$\qquad \lambda = 1500/21.4 = 70$

$\qquad \lambda/x = 70/31.4 = 2.22$

from Table 14, with $N = 0.5$, $v = 0.95$

$\qquad \lambda_{LT} = nuv\lambda$

$\qquad\qquad = 1.0 \times 0.864 \times 0.95 \times 70$

$\qquad\qquad = 57.5$

From Table 11, $p_b = 263 \ \text{N/mm}^2$

$\qquad mM_x/M_b = 0.93 \times 18.7\text{E}3/(306 \times 263)$

$\qquad\qquad\quad = 0.22$

Thus $F/A_g p_c + mM_x/M_b = 0.07 + 0.22 = 0.29 < 1.0$

Note, this stanchion has been assessed under a load combination which strictly does not exist ($1.4g_k + 1.6q_k$ on the rafter giving maximum axial load and the wind load $1.4w_k$ on the stanchion giving maximum bending moments). It is however conservative.

The same section can safely be used at A and B. At B, even though the axial load is double, the bending moments are less, and the value of the design interaction equation so much below unity that this change will not be significant. At A, even though the stanchion is turned through 90° (Fig. 8.21), and the column is in biaxial bending with a nominal moment applied due to the rafter connection, the applied forces are less than half those on the stanchion at C.

8.10.2   Multi-bay portal frames

It is not intended to present a design example of such a frame since the principles and methods are similar to those for single bay portals. It is intended, rather, to point out the differences and the additional calculations required.

1   The critical bay, assuming, as is normal practice, that all the spans are equal, is the end bay. However it must be noted that it is no longer possible to assume a constant uniform snow loading. The case of non-uniform snow loading is to be treated as an accidental load and reduced load factors used to calculate the forces in the structure. The calculations required are carried out using BS 6399, Pt 3.

EXAMPLE 8.3   Snow drift loading

Determine the snow loading in the valley between two bays of pitched roof portal frame of spans 25 m and ridge height (above eaves) of 2.2 m.

All references are to BS 6399, Pt 3.

(i)   Determine the maximum likely snow intensity exceeded once in 50 years, s_b (Fig. 1) e.g. Birmingham 0.53 kN/m² (by interpolation).

(ii) Correct this value for height above sea level to obtain design value, s_0,

$$s_0 = s_b + s_{alt} \times (A - 100)/100$$

where s_{alt} is the altitude correction factor taken from Table 1 and A is the altitude in metres.

For Birmingham, $s_{alt} = 0.14$ kN/m² and $A = 110$ m

so, $s_0 = 0.53 + 0.14 \times (110 - 100)/100$
$\qquad = 0.54$ kN/m²

(iii) Calculate the design load, s_d:

This is given by $s_d = \mu_i s_0$

where μ_i is a shape coefficient from Fig. 5 of BS 6399, Pt 3. The particular case for multi-bay pitched roofs is reproduced in Fig. 8.25.

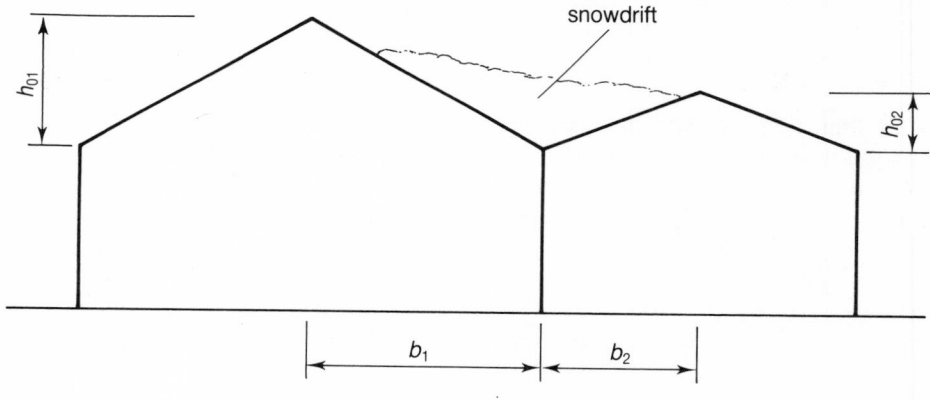

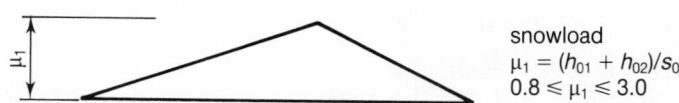

snowload
$\mu_1 = (h_{01} + h_{02})/s_0$
$0.8 \leqslant \mu_1 \leqslant 3.0$

Note: Both b_1 and b_2 are to be taken $\leqslant 15$ m

Fig. 8.25   Calculation of snow drift loads.

For b_i, where b_i is the half span of either of the two portal frames, less than 15 m, ls_1 and ls_2, the extent of the drift on either frame, are taken as the half spans, i.e. 12.5 m and h_{01} and h_{02}, the height of the ridge above the eaves for either portal as

2.2 m, calculate value of $(h_{01} + h_{02})/s_0$:

$$(h_{01} + h_{02})/s_0 = (2.2 + 2.2)/0.54 = 8.15$$

Value calculated is greater than 5, but the height difference between the ridges is less than 1 m, so

$$\mu_i = 3.0$$

so,

$$s_d = 0.54 \times 1.0 \times 3 = 1.62 \text{ kN/m}^2$$

Note, the extreme rafters in the end bays should still be loaded with an uniform loading of 0.60 kN/m², assuming no access.

2 Although with balanced loading the interior stanchions carry no moment, and the rafters have a higher reserve of strength as both ends are continuous, it is normal practice to detail all inner bays identical to the end bays. This is partially to gain economies in fabrication by repetition, and partly to avoid deflection and 'snap-through' modes of failure.

3 The suitability of sway behaviour may be checked either by using the 0.5% load approach applied to all columns or by using the empirical approach illustrated in Example 8.1 but with the value of ρ calculated as $(I_c/I_r) \times (L/h)$.

4 In frames with more than two bays it is possible for 'snap-through' failure to occur. This is where instability occurs owing to the spread of the tops of the stanchions and inversion of the rafter then follows (Fig. 8.26).

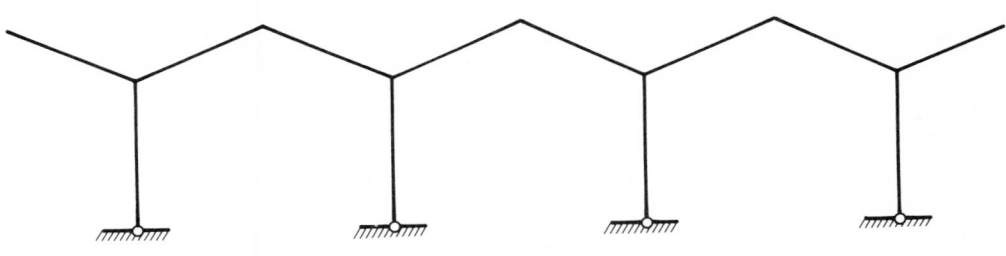

(a) Interior bays before snap-through

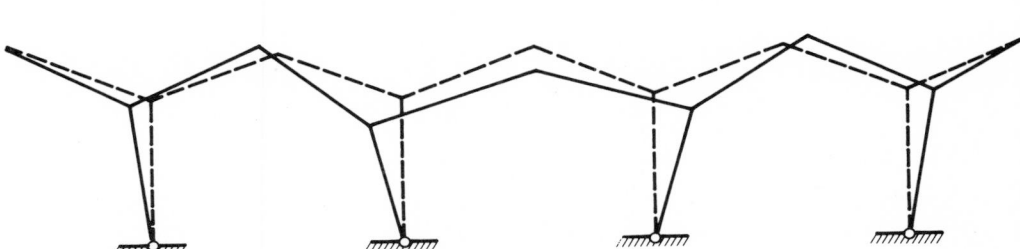

(b) Exterior bays after snap-through

Fig. 8.26 Snap-through failure of multi-bay portal frames.

This is prevented by limiting the span/depth ratio $((L - L_h)/D)$ (cl 5.5.3.3, Pt 1) of the rafter using the expression

$$(L - L_h)/D < ((22(4 + L/h)/\Omega(\Omega - 1)) \times (1 + I_c/I_r) \times (275/p_{yr}) \times \tan(2\theta_r)) \qquad (8.7)$$

Symbols are those defined for calculating the permissible value of $(L - L_h)/D$ for frame deflection (Example 8.1), except

$$\theta_r = \arctan(2h_1/L) \qquad (8.8)$$

where h_1 is the height of the ridge above the eaves.

No limit is placed on $(L - L_h)/D$ if $\Omega < 1.0$.

8.10.3 Multi-storey structures

A distinction must be drawn between simple construction frames in which the joints between the beams and the stanchions are designed to carry the end shears from the beams only with all horizontal forces taken by bracing or stairwells and lift shafts, and continuous construction in which all beam–to–stanchion connections are capable of resisting moments and the horizontal forces may either be taken by bracing or by the connections. Each will be dealt with in turn.

8.10.3.1 *Simple construction frames*

The beams supporting the flooring units which will be, in general, either proprietary precast prestressed concrete with a non-structural topping or profiled steel decking with in situ lightweight concrete slabs, in which case the supporting beams are designed as composite, and in either case are designed as simply supported. The stanchions may be continuous over more than one floor and are to be designed in accordance with cl 4.7.7. This effectively states that:

Nominal moments are to be applied at connections calculated on the assumption that the shear from the beams act 100 mm from the face of the column. These moments may if the column is continuous above and below the connection be distributed above and below the connection according to the stiffness ratio of the column above and below, except that if the stiffness ratio is less than 1.5 the moments may be divided equally.

With only nominal moments due to the beam end shears, the stanchions should satisfy the design relationship of cl 4.7.7, Pt 1, i.e.

$$F_c/A_g p_c + M_x/M_{bs} + M_y/p_y Z_y \leqslant 1 \qquad (6.14)$$

Note, for the purpose of calculating M_{bs}, $\lambda_{LT} = 0.5 \times (L/r_y)$, where L is the distance between levels at which both axes are restrained.

8.10.3.2 *Continuous construction (Section 5, Pt 1)*

With continuous construction, where there is moment transfer betweem the beams and the stanchions owing to the use of rigid joints, it is necessary to distinguish between sway and non-sway frames. In the former there is lateral deflection between the ends of a stanchion between two floors which will be sufficient to give possible problems with instability caused by the additional moments generated in the columns due to the $P - \delta$

effect from the axial loads in the stanchion and its lateral deflection. In non-sway frames this effect is deemed to be negligible and thus it may be ignored in the design procedure. Sway may be reduced by the existence of wind bracing, shear walls or the inherent stiffness of the frame itself. One method of taking account of any possible geometry changes is to analyse the structure using an incremental elasto-plastic analysis which takes account of second order effects (the $P - \delta$ effect). This, however, except for the very simple frames requires the use of computer techniques and is therefore very expensive and time consuming.

Because of the possible severity of the $P - \delta$ effect which can reduce the collapse load of a structure significantly due to geometry changes which are neglected in normal analysis techniques, it is necessary to provide a design criterion to separate sway and non-sway frames. It is insufficient to rely on pure judgement as the bracing in what appears to be a non-sway frame may possess little stiffness and therefore not prevent lateral deflection between floor levels to a sufficient degree to prevent geometry changes.

To determine whether a frame is to be classified as sway or non-sway an elastic analysis is carried out on the frame with notional horizontal loads equal to 0.5% of the factored dead and imposed load at any floor or roof level applied at that level. The resultant horizontal deflections determined from the elastic analysis are then used to determine whether the frame is to be classified as sway or non-sway. If the ratio of the deflection to the storey height is less than 1/2000 for clad frames where the cladding is not taken into account in the calculation or 1/4000 if the cladding is included, then the frame is classified as non-sway. If these deflection limits are exceeded then the frame is classified as sway, even though the frame may be braced.

The loading system used to determine whether the structure is a sway or non-sway frame is also that required to calculate the elastic critical load factor, λ_{cr}, following a method first suggested by Horne (who in his original analysis used a horizontal load of 1.0% of the unfactored total imposed load).

Having determined whether a particular frame has been classified as sway or non-sway, the methods used to analyse and design such frames may be considered. It is possible to use either elastic design methods or plastic design methods on either type of frame. In either case the analysis is carried out using factored loading. In an elastic analysis the use of subframes as in reinforced concrete frames may be used to determine the effect of vertical loading. For further information on the use of subframes in analysis reference should be made to Martin *et al.* The effect of horizontal loading must be determined by an analysis on the whole frame. Redistribution of up to 10% of the peak elastic moment provides that after redistribution the forces and moments in the frame remain in equilibrium, all members in which moments are reduced are classified as compact and that no moments are reduced about the minor axis of stanchions.

1   Non-sway frames

(i)   Elastic design

Following the linear elastic analysis, members are then designed as 'beam–columns' using Section 4 of the Code. The effective length factors for the columns are to be taken from Appendix E of the Code.

The graphs in Appendix E which determine the effective lengths of columns in both non-sway, sway and partially restrained sway frames are obtained from analyses carried out by Wood which take account of the elastic critical loads determined by using stability functions. A separate determination of these critical loads that may be sustained by columns in either a sway or non-sway subframe must be carried out (Fig. 8.27).

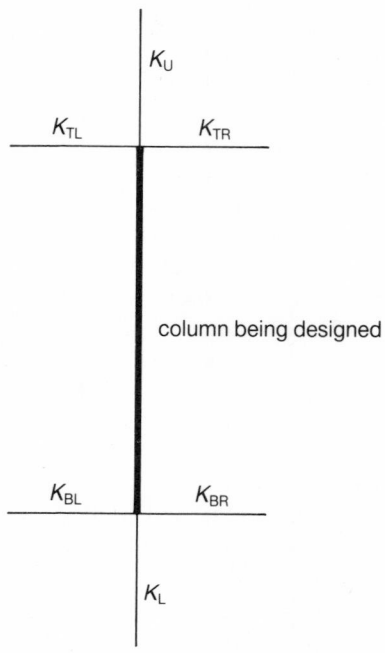

Note: The K values are the stiffnesses of the upper and lower beams, and the upper and lower columns in CONSISTENT units

Fig. 8.27 Data for determination of column effective lengths.

The effective length factors are determined in relation to the relative stiffness of joints at either end of the column whose effective length is being determined. For non-sway frames Fig. 23 of Part 1 is to be used.

(ii) Plastic design

For plastic design methods to be used the frame must be adequately braced and must not rely on its inherent stiffness to satisfy the deflection limits under notional horizontal loading. Members should then be checked for overall buckling under the forces obtained from a suitable plastic analysis using cl 4.8.3.3, Pt 1 to carry out the buckling check.

2   Sway frames

(i)   Elastic design

Initially the frame should be designed as a broad non-sway frame with the effect length factors determined from Appendix E of Part 1 assuming the frame to be braced.

The effect of sway should then be considered under all combinations of loading. This means that where vertical loading only is applied the notional horizontal forces must also be included. Either of the following two methods may then be used to check the frame:

(a)   The frame may be checked as if it were a simple design but with the effective length factors taken from Appendix E allowing for sidesway (Fig. 8.24). Note that the effective length factors obtained from Fig. 8.24 may well exceed unity because of the reduced elastic critical loads that can occur with sidesway.

Williams and Sharp (1990) have derived equations which may be used in place of the Code graphs for determining effective lengths. Written in terms of the notation used in Part 1 of the British Standard (Fig. 8.38), these equations are:

$$L_E = \pi L/(12 - 36(k_1 + k_2 - k_1 k_2)/(4 - k_1 k_2))^{0.5} \tag{8.9}$$

or

$$L_E = \pi L(1.05 - 0.05 k_1 k_2)/(12 - 36(k_1 + k_2 - k_1 k_2)/(4 - k_1 k_2))^{0.5} \tag{8.10}$$

The second equation gives a slightly closer answer to the graphs in the Code, although in normal design situations the errors caused by using the first equation are unlikely to be significant.

(b)   The moments obtained from the horizontal loading may be amplified by the factor $\lambda_{cr}/(\lambda_{cr} - 1)$, where λ_{cr} is the elastic critical load factor, and all the effective lengths are taken as unity.

The elastic critical load factor λ_{cr} is determined by applying the notional 0.5% horizontal loads at all floor levels and determining the sway index ϕ_s at any level.
The sway index ϕ_s is defined as

$$\phi_s = (\delta_U - \delta_L)/h \tag{8.11}$$

where δ_L and δ_U are the lower and upper horizontal deflections of a particular storey and h is the storey height.
The elastic critical load factor λ_{cr} is given by

$$\lambda_{cr} = 0.005/\phi_{s,max} \tag{8.12}$$

where $\phi_{s,max}$ is the maximum sway index.
Note, the elastic analysis required to determine the sway index will already have been needed to determine the classification of the frame as sway or non-sway.
The background to this method of calculating the elastic critical load factor is given in Horne.
An example of the elastic design of a sway frame will be found in the next chapter (Example 9.4).

(ii)   Plastic design

Either an incremental elasto-plastic analysis may be used or a simplified method can be adopted.
The frame should be checked under all possible combinations of loading including where appropriate the notional horizontal loading with the effective length factor for the columns being taken as unity in the plane of the frame.
Stability should then be checked by determining the plastic load factor λ_p, defined as the ratio by which the factored loading should be increased to cause collapse determined in accordance with (a) and (b) below:

(a) The mechanism assumed should be a sway mechanism with hinges in the beams and base of each column and with no other hinges in the columns.

(b) The lower lengths of the columns should be designed to remain elastic under the theoretical hinge assumed in (a).

(c) Under all combinations of unfactored loading including where appropriate the notional horizontal loading it should be possible by means of moment redistribution to produce sets of moments and forces throughout the frame that are in equilibrium with the applied loads and under which all members remain elastic.

A modified Rankine–Marchant approach is used to determine the relationship between the elastic critical load factor and the plastic load factor determined from the assumptions above (Horne, Wood).

For clad frames where no account is taken of the cladding:

$$\lambda_{cr} \geqslant 4.6$$

when $4.6 \leqslant \lambda_{cr} < 10$; $\lambda_p \geqslant 0.9\lambda_{cr}/(\lambda_{cr} - 1)$

when $\lambda_{cr} \geqslant 10$; $\lambda \geqslant 1.0$ (8.13)

Note that where the stanchions are encased in concrete, the increased second moment of area may be taken into account when calculating λ_{cr}.

For unclad frames or in clad frames where the stiffness of the cladding is taken into account:

$$\lambda_{cr} \geqslant 5.75$$

when $5.75 \leqslant \lambda_{cr} < 20$; $\lambda_p \geqslant 0.95\lambda_{cr}/(\lambda_{cr} - 1)$

when $\lambda_{cr} \geqslant 20$; $\lambda_p \geqslant 1.0$ (8.14)

8.11   Tutorial problem

8.11.1   Portal frame

A fabricator wishes to use a 356 × 171 × 67 UB for the rafter and a 457 × 191 × 82 UB (both Grade 50) for a portal frame to span 30 m, with a height to eaves of 6 m and a 10° pitch to the symmetric roof. The frames are at 7.5 m centres, and the rafters are haunched to twice their depth at the eaves and the haunch extends for 3 m from the face of the stanchion. The loadings are as follows: total dead load 0.45 kN/m² and imposed load 0.6 kN/m². The purlins are at 1.25 m centres (on plan) and are 230 mm deep. The sheeting rails are 200 mm deep. The feet of the stanchions may be assumed to act as pinned feet. Carry out a full check on the frame for strength, stability and serviceability.

Answers

(a)   Plastic checks:

UDL on frame at ULS $= 7.5 \times (1.6 \times 0.6 + 1.4 \times 0.45) = 11.93$ kN/m.
Vertical reaction $= 179$ kN.

$M_{ps} = 650$ kNm, $M_{pr} = 430$ kNm   (both sections are plastic).

Distance from hinge below haunch to base $= 6000 - 1.5 \times 358.6 = 5462$ mm, horizontal reaction $= 119$ kN.

Assuming critical point in rafter is second purlin point ($l = 12.5$, $r = 8.204$) then required $M_p = 329$ kNm, actual is 430.

Check first purlin point ($l = 13.75$, $r = 8.425$), $M = 333.8$ kNm (less than plastic capacity).

Maximum load carrying capacity based on total hinge formation at below eaves and second purlin point, $q = 12.86$ kN/m.

Check moment at end of haunch ($l = 3 + 0.460/2$, $r = 6.610$ m), $M = -270.7$ kNm. $0.87M_{pr} = 374.1$ kNm, therefore satisfactory.

The frame is thus satisfactory at ultimate load assuming that the full plastic moments can be generated.

(b) Stability checks:

1 Frame

(i) Equation (8.3)

$$275/p_{yr} = 0.775$$

$$\rho = 19.03$$

$$L_r/L = 1.105$$

$$\Omega = 1.56 \ (q_0 = 7.64 \text{ kN/m})$$

$$(L - L_h)/D = 83.1$$

Actual value $= (30000 - 3000)/358.6 = 75.3$, therefore satisfactory.

(ii) 0.5% load check (Equation (8.4))

$$P = 0.5 \times 11.93 \times 30/100 = 1.79 \text{ kN}$$

$$L_r/(h + L_r) = 0.717$$

$$1 + L_r I_c/(h I_r) = 5.83$$

$$\delta_h = 7.1 \text{ mm}$$

allowable value $h/1000 = 6$ mm.

Frame fails on this criterion (except that calculated deflection does not allow for haunch).

2 Stanchion

(i) Approximate approach

$$L_m = 1380 \text{ mm}$$

Moment at point of restraint is 486 kNm, check $F/A_g p_c + m M_x/M_b$

$F/A_g p_c$:
$F = 179$ kN; y–y axis, $L_E = 4082$ mm, $\lambda = 97$, $p_c = 179$ MPa; x–x axis, $L_E = 5462$, $= 29$, $p_c = 344$ MPa; $F/A_g p_c = 0.10$

$m M_x/M_b$:
$\beta = 0$, $m = 0.57$; $\lambda = 97$, $\lambda_{LT} = 77$, $p_b = 199$ MPa; $m M_x/M_b = 0.76$

$F/A_g p_c + m M_x/M_b = 0.86$, and is therefore satisfactory

(ii)   Exact check on whole stanchion $(F/P_c + M/M_b)$

F/P_c:

$\lambda = 5462/42.3 = 129;\ h_s = 460.2 - 16.0 = 444.2;\ a = 200 + 460.2/2 = 430.1;$
$y = 0.919;\ \lambda_{TC} = 119;\ p_c = 117$ MPa.

$F/P_c = 0.5$

M/M_b:

$\beta_t = 0;\ \ \ y = 0.92;\ \ \ m_t = 0.53;\ \ \ M = 0.53 \times 650$

$v_t = 0.931;\ n_t = 1.0;\ c = 1.0;\ \lambda_{TB} = 105;\ p_b = 129$ MPa; $M/M_b = 1.46$, thus indicating an additional restraint is required.

3   Rafter

Axial force at eaves 148.3 kN, at ridge 117.2 kN–use value at eaves throughout.

$f_c/130 = 0.134$

$L_m = 1598$, greater than purlin spacing.

4   Haunch

(i)   Bending resistance at haunch:
Moment at face of stanchion $= 678$ kNm $(= \bar{M},\ n = 1.0)$

$L_E = (1250 - 230)/\cos 10 = 1036$

Sectional properties at the haunch (in mm units);

$A = 11777,\ \ I_x = 0.946E9,\ \ I_y = 13.64E6,\ \ S = 3.04E6,\ \ r_y = 34.0,\ \ \gamma = 0.986,\ \ u = 0.920,$
$J = 0.622E6,\ x = 54.5$

$\lambda = 30.5,\ \lambda_{LT} = 28,\ M_b = 1079$ kNm, which is satisfactory.

(ii)   Tapered section

Divide haunch into 5 slices starting at $s = 0.23$ m and ending at $s = 3.23$ m.

Values of M/S (in MPa) starting at $s = 0.23$, are 221, 222, 230, 222 and 219. Values of F/A are 13, 13, 15, 16 and 17. Maximum value of $M/S + F/A$ is 245 MPa.

$u = 1.0,\ v_t = 0.814\ (a = 412,\ L_E = 3194$ mm, $h_s = 348.3$ mm), $c = 1.175\ (q = 3/3.52,\ R = 2)$, $n_t = 0.765$ (values of N/M from 1 to 5 are 0.623, 0.625, 0.649, 0.625 and 0.618 respectively), $\lambda_{TB} = 59,\ p_b = 254$ MPa, which is satisfactory.

5   End of haunch to point of contraflexure

point of contraflexure occurs at $s = 5.78$ m

purlin point beyond is at $s = 6.25$ m, therefore check stability from $s = 3.23$ to $s = 6.25$ m, i.e. at $s = 3.23,\ 3.985,\ 4.74,\ 5.495$ and 6.25 m.

Values of M/S are 220, 148, 82, 21 and zero. F/A is 17.

Values of N/M are 0.619, 0.416, 0.230, 0.060 and zero, giving $n_t = 0.497,\ \lambda_{TB} = 11$, $p_b = 355$ MPa and is therefore satisfactory.

(c) Deflection check

$q = 4.5 \text{ kN/m}$, $h = 6.0 \text{ m}$, $h_1 = 2.645 \text{ m}$, $\theta = 10$, $L_r = 15.228 \text{ m}$

$h_1/h = 0.441$

$hI_r/I_c L_r = 0.208$

$\delta_v = 0.189 \text{ m (span/160)}$

suggested limit is span/240 or 125 mm

$\delta_H = 0.033 \text{ m (height/182)}$

Suggested limit is $h/150$ or 40 mm

(d) Davies' stability check

$$m = 1.441, \quad R = 4.808, \quad N = 9.45$$
$$P_r = 134.4 \text{ kN}, \quad P_c = 179 \text{ kN}$$
$$\lambda_{cr} = 4.67, \quad \text{and} \quad \lambda_p = 1.15$$

Actual value of $\lambda_p = 12.86/11.93 = 1.08$ which is slightly below the required value of 1.15.

References

Allen P. H. and Lovejoy E. G. (1989). Designing for safety, *The Structural Engineer*, **67** (5), 83–7.

Alexander S. J. and Lawson R. M. (1981). *Movement design in buildings*, Technical Note 107, Construction Industry Research and Information Association.

British Constructional Steelwork Association (1989). *National structural steelwork specification for building construction* (1st Edition), BCSA.

British Steel Company – General Steels (undated). *Design concepts – single storey buildings*, BSC.

British Steel Company – General Steels (1985). *Design concepts – multi-storey buildings*, BSC.

BS 2573 Part 1. *Rules for the design of cranes – Part 1 – Specification for classification, stress calculations and design criteria for structures*, British Standards Institution, London.

BS 5531. *Safety in erecting structural frames*, British Standards Institution, London.

BS 5628 Part 3 *British standard code of practice for use of masonry – Part 3 – Materials and components, design and workmanship*, British Standards Institution, London.

BS 5977 Part 1. *Lintels – Part 1 – Method for assessment of load*, British Standards Institution, London.

BS 6399 Part 1. *Loading for buildings – Part 1 – Code of practice for dead and imposed loads*, British Standards Institution, London.

BS 6399 Part 3. *Loading for buildings – Part 3 – Code of practice for imposed roof loads*, British Standards Institution, London.

BS 8110 Part 1. *Structural use of concrete – Part 1 – Code of practice for design and construction*, British Standards Institution, London.

Coates R. C., Coutie M. G. and Kong F. K. (1988). *Structural analysis* (3rd edition), Van Nostrand.

CP 3 Ch V Part 2. *Code of basic data for the design of buildings – Chapter V Loading – Part 2 Wind loads*, British Standards Institution, London.

Cunningham R. (1990). Some aspects of semi-rigid connections in structural steelwork, *The Structural Engineer*, **68**, 85–92.

Curtin W. A. *et al.* (1988). *Structural masonry designers handbook* (2nd edition), Granada, London.

Davies J. M. (1990). Inplane stability in portal frames, *The Structural Engineer*, **68**, 141–7.

Davies J. M and Bryan E. R. (1982). *Manual of stressed skin diaphragm construction*, Granada, London.

Ghali A. and Neville A. M. (1989). *Structural analysis*, Chapman and Hall.

Hendry A. *et al.* (1981). *An introduction to loadbearing brickwork design*, Ellis Horwood.

Henry F. D. C. (Editor) (1986). *Design and construction of engineering foundations*, Chapman and Hall.

Horne M. R. (1975). An approximate method for calculating the elastic critical loads of multi-storey plane frames, *The Structural Engineer*, **53**, 242–8.

Horne M. R. and Morris L. J. (1981). *Plastic design of low rise frames*, Collins.

Horridge J. F. (1985). *The design of industrial buildings*, Civil Engineering Steel Supplement, 13–6.

Institution of Structural Engineers/Institution of Civil Engineers (1989). *Manual for the design of steelwork building structures*, Institution of Structural Engineers.

Irwin A. W. (1984). *Design of shear wall buildings*, Report 112, Construction Industry Research and Information Association.

Jones L. L. and Wood R. H. (1967). *Yield line analysis of slabs*, Chapman and Hall.

Marshall W. T. and Nelson H. M. (1977). *Structures*, Pitman.

Morris L. J. (1981). A commentary on portal frame design, *The Structural Engineer*, **59A** (12), 394–404.

Morris L. J. and Randall A. L. (1979). *Plastic design*, Constrado/Steel Construction Institute.

Moy S. S. J. (1981). *Plastic methods for steel and concrete structures*, Macmillan.

Neale B. G. (1977). *Plastic methods of structural analysis* (3rd edition), Chapman and Hall.

Williams F. W. and Sharp G. (1990). *Simple elastic critical load and effective length calculations for multi-storey rigid sway frames*, Proc ICE, **90**, 279–87.

Wood R. H. (1974). Effective lengths of columns in multi-storey buildings, *The Structural Engineer*; **52** (7), 235–44; **52** (8), 295–302; **52** (9), 341–6.

Wood R. H. (1974). *A new approach to column design*, Her Majesty's Stationery Office.

Woolcock S. T. and Kitipornachai S. (1986). *Deflection limits for portal frames*, Steel Construction, **20** (3), 2–10.

9

Trusses

9.1 Introduction

This chapter is concerned with the design of both triangulated and non-triangulated trusses. It also covers the design of purlins and sheeting rails.

Two of the most far-reaching developments in structural steelwork in recent years have been the availability of both rolled hollow sections and cold formed thin steel sections.

Rolled hollow sections are structurally very efficient as the material is as far away from the neutral axis as possible, allowing high loads to be carried both in flexure, as lateral torsional buckling seldom occurs, and axially as both radii of gyration are high and therefore slenderness ratios are low. They are also easy to maintain, as no internal surfaces need protection and the external surfaces are amenable to protection using spray rather than brush systems.

Cold formed steel sections are also efficient in the use of material, having a high strength to weight ratio and are also lightweight, thus easing problems of handling on site. The advances in cold formed sections have come in two areas; purlins and sheeting rails (covered in this chapter) and profile sheet steel decking used in composite construction (covered in the next chapter). The examples in this chapter will therefore concentrate on the design of trusses using rolled hollow sections and purlins and sheeting rails using cold formed sections.

As was pointed out in the previous chapter, large span single storey steel structures are normally either pitched roof portal frames, the design of which was undertaken in Example 8.1, or trussed non-parallel chord rafters, the design of which is covered in Example 9.3. To complete the range of truss systems the design of a Vierendeel girder is presented in Example 9.4. However, before covering the design of trusses, the design of purlins and sheeting rails is considered.

9.2 Purlins and sheeting rails

Common cross sections for cold formed sheeting rails and purlins are given in Fig. 9.1, where it is seen that there are essentially two basic types in common use–the Sigma and Zed shape–although some manufacturers have produced variants on these basic shapes to give slightly higher strengths at increased complexity of section.

Typical metal thicknesses are around 1.5 to 3 mm, and as the name implies the sections are made by cold forming thin steel sheets through roller systems to achieve the required

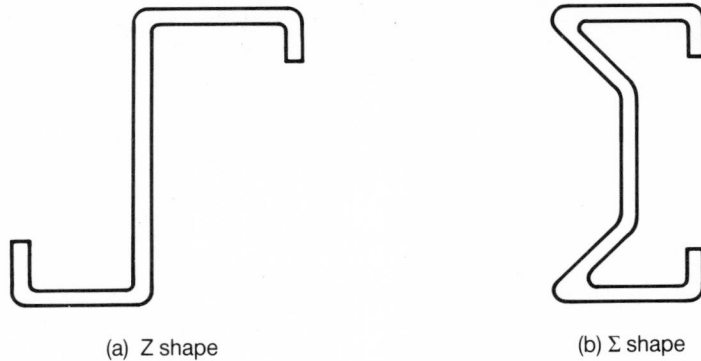

(a) Z shape   (b) Σ shape

Fig. 9.1   Common cold formed purlin cross sections.

section profiles. The design of such members is covered by Part 5 of BS 5950. BS 2994 gives section properties for the standard range of all available cold formed sections except the Sigma profile.

BS 5950 Part 5 allows three methods of design:

1   Full calculations involving consideration of buckling, which will be far more likely in thin sections, and of shear lag, where the stress will not be constant over the width of the section. The full calculation procedure is complex as the effective width used in determining the section properties is stress dependent and therefore the calculations are iterative. The background to the full calculation approach is given in Walker.
2   The use of manufacturers' certified test results.
3   Simplified rules.

It is proposed to consider the last of these here, as it is generally not worthwhile carrying out full design calculations. It should be noted that only the design of lipped Zed purlins may be carried out using these rules. If Sigma profile purlins are to be used reference should be made to specific manufacturers' design tables.

Manufacturers' safe load tables, based on results from tests where the purlin or sheeting rail is tested as part of a complete system including the cladding which will provide restraint to the flange of the purlin and increase its effective stiffness, will give as a result lower section sizes for a given load, or higher loads on a given section size than the approximate design rules in the Code.

It should be noted that the empirical rules for purlin design given in Section 4.12 of BS 5950 Part 1 are not to be used for purlins of cold formed sections.

BS 5950, Pt 5, Section 9 imposes certain constraints on the use of the simplified design rules for both purlins and sheeting rails. These are summarized in Sections 9.2.1 and 9.2.2.

9.2.1   Design of 'Z' purlins (cl 9.2, Pt 5)

(i)   Service loading should be used.
(ii) Loading should be derived from BS 6399, Pt 3 for imposed loading and from CP 3, Ch V for wind loading except that local wind load coefficients need not be considered and that a minimum imposed load of 0.6 kN/m^2 should be taken.

(iii) The cladding and fixings should be able to supply full torsional restraint to the purlins. The purlin should be designed to resist the component of load normal to the roof slope. The purlin may also be used as part of the wind bracing provided the axial stress so induced is not greater than 6 N/mm^2.

(iv) The rules only apply to purlins of up to 8 m span on roof slopes of less than 22.5°. For spans up to 5 m the purlins may be designed with a nominally simply supported connection. Over 5 m, the purlins must have fixings capable of supplying continuity. For multi-span conditions, the design coefficients may be used if the adjacent spans differ by less than 20%.

(v) For spans greater than 4.6 m, sag rods anchored to or supported from the ridge beam must be supplied such that the laterally unsupported length does not exceed 3.8 m.

The design rules impose geometrical constraints which are given in Fig. 9.2, and also give minimum elastic section moduli Z_{min} (in cm^3), for simply supported purlins,

$$Z_{min} = WL/1400$$

and for continuous or semi-rigidly jointed purlins,

$$Z_{min} = WL/1800$$

where W is the load (kN) and L the span (mm).

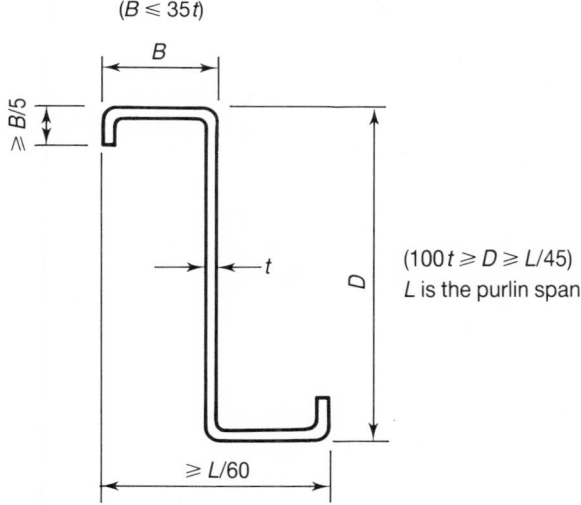

$(B \leqslant 35t)$

$\geqslant B/5$

$(100t \geqslant D \geqslant L/45)$
L is the purlin span

$\geqslant L/60$

Fig. 9.2   Geometric constraints for 'Z' purlins.

The maximum allowable wind uplift should be taken as not greater than 50% of the imposed dead and live load. Unless the imposed loading is high the geometrical constraints will usually govern the design.

EXAMPLE 9.1   Purlin design

Design a suitable lipped Zed purlin for the frame similar to that of Example 8.1, except that the frame spacing is to be taken as 7.5 m.

Vertical loading: Sheeting 0.18 kN/m^2
                                    Purlins   0.06
                                    Snow   0.60
                                    Total 0.84 kN/m^2

Span 7.5 m, purlin spacing (on plan) 1.25 m.

Vertical load on each purlin, W_v:

$$W_v = 0.84 \times 7.5 \times 1.25$$
$$= 7.88 \text{ kN}.$$

Rafter slope is 10°, so component normal to roof W is given by

$$W = W_v \cos 10$$
$$= 7.88 \times \cos 10$$
$$= 7.76 \text{ kN}.$$

Geometric constraints:

Total width $(2B) > L/60 = 7500/60 = 125$ mm
Depth $(D) > L/45 = 7500/45 = 167$ mm
Thickness $(t) > D/100 = 167/100 = 1.67$ mm

$$35 > B/t$$

$$B/5 > C$$

For a continuous purlin,

$$Z_{min} = WL/1800$$
$$= 7.76 \times 7500/1800$$
$$= 32.3 \text{ cm}^3$$

A 215 $(D) \times 75$ $(B) \times 20$ $(C) \times 2.5$ (t) purlin $(Z_x = 62.4)$ satisfies the conditions.

This particular purlin has a mass of 7.56 kg/m, which gives an equivalent UDL of 0.06 kN/m^2 which is the same as the assumed value.

Check wind load conditions:

The values for the wind pressure and the values for the pressure coefficients are those from Example 8.1, as the frame spacing does not affect these values.

Maximum upward wind $= (1.01 \times 0.315) \times 1.27 \times 7.5$
                                          $= 3.03 \text{ kN}.$

In the above calculation, 1.01 is the total pressure coefficient, 0.315 the dynamic wind pressure and 1.27 the purlin spacing measured along the roof.

The upward wind should not exceed 50% of the dead and imposed load,

$$0.5(\text{dead} + \text{imposed}) = 0.5 \times 7.76 = 3.88 \text{ kN}.$$

This condition is not therefore critical.

Note, it is usual in the end bay of a frame to fit a double thickness purlin (two normal purlins placed on each other) partly to allow for no continuity at one end of the bay and partly to reduce axial stresses since the purlin system will be used as part of the wind bracing system (this comment also applies to sheeting rails in the end bay of the frame only).

9.2.2 Design of sheeting rails (BS 5950 Pt 5 cl 9.3)

(i) Service loading must be used for design.

(ii) The rules only apply up to spans of 8 m. For spans up to 6.5 m the rails may be nominally supported, but for spans in excess of this the rails must be continuous or supplied with sleeves or splices capable of generating full moment capacity. The coefficients for the continuous case may be used where adjacent spans differ by less than 20%.

(iii) The self weight of the sheeting must be carried by an auxiliary support system (typical examples of which are illustrated in Fig. 9.3).

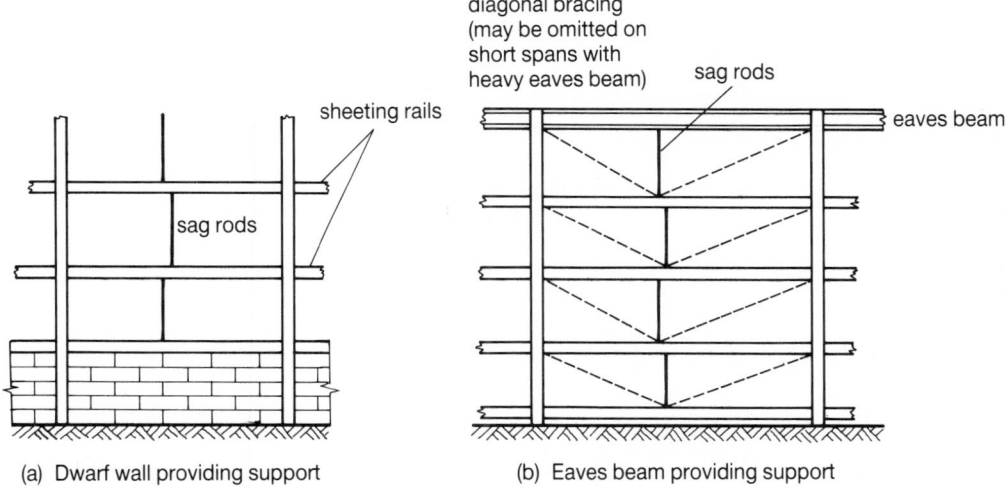

Fig. 9.3 Support systems to resist self weight of sheeting systems.

(iv) Vertical supports should be placed at mid-span for spans not exceeding 4.6 m and at third points for larger spans and should be designed to take an axial load force of 2 kN (tensile or compressive).

Dimensional constraints are imposed (see Fig. 9.2). Minimum elastic section moduli (in cm^3) are imposed using the following equation:

$$W_w L/(aZ_x) + W_d L/(bZ_y) \leqslant 1 \tag{9.1}$$

where W_d is the wind load causing tension on the purlin edge remote from the sheeting (kN), W_d the dead load (not less than 2 kN) (kN), L the span (mm), Z_x the elastic section modulus about the x axis (cm^3) and Z_y the elastic section modulus about the y axis.

The coefficients a and b take the following values:

Simply supported, single vertical support, 1750 and 6000

simply supported, two vertical supports, 1750, 16880

continuous, single support, 2250, 6000

continuous, two vertical supports, 2250, 16880 respectively.

The maximum wind pressure causing compression on the flange remote from the cladding is limited to 50% of that causing tension on the same flange.

EXAMPLE 9.2   Sheeting rail design

Design sheeting rail in lipped Zed cold formed sections for the side of the frame in Example 8.1, except that the frame spacing is to be 7.5 m, and the gable end for the frame designed in Example 8.2.

(a)   Side

Data are taken, where appropriate, from Example 8.1.

Span is 7.5, thus two intermediate supports are needed.

Sheeting rail spacing is 1.5 m.

Dynamic wind pressure is 0.315 kN/m^2.

Total wind coefficient producing tension on the inside is 0.89, and on the outside is -0.12. Since the wind coefficient producing compression on the inside face (-0.12) is numerically less than half that on the outside face (0.89), no special consideration need be made for wind reversal.

Geometry:

Total width $(2B) > L/60 = 7500/60 = 125$

Depth $(D) > L/45 = 7500/45 = 167$

Thickness $(t) > D/100 = 167/100 = 2$

$$35 > B/t$$

$$B/5 > C$$

Based on these constraints, try a $215 \times 75 \times 20 \times 5$ lipped Zed section ($Z_x = 62.4$ and $Z_y = 14.4$ cm^3).

Weight of sheeting W_d:

$W_d = 0.2 \times 7.5 \times 1.5 = 2.25$ kN > 2 kN minimum loading.

Wind load W_w:

$W_w = (0.89 \times 0.315) \times 7.5 \times 1.5 = 3.15$ kN.

Using the interaction equation $W_w = L/aZ_x + W_d L/bZ_y$ for the continuous case with two vertical supports ($a = 2250$ and $b = 16880$):

$$3.13 \times 7500/7500 \times 62.4 + 2.25 \times 7500/16880 \times 14.4 = 0.24.$$

This is less than unity and is therefore satisfactory.

The same comment, as made in Example 9.1 with reference to the doubling of the section in an end bay, also applies to sheeting rails.

(b)   End

Data are taken, where appropriate from Example 8.2.

Span is 6.25 m, so needing two vertical supports.

The spacing is 1.5 as for the side rails.

Total wind coefficient producing tension on the inside is 1.08, and that producing compression is -0.41. Again as for the rails on the side, there is no problem with wind reversal.

Geometry:

Total width $(2B) > L/60 = 6250/60 = 104$

Depth $(D) > L/45 = 6250/45 = 139$

Thickness $(t) > D/100 = 139/100 = 1.5$

$$35 > B/t$$

$$B/5 > C$$

Select a $180 \times 60 \times 20 \times 2$ $(Z_x = 34.9$ and $Z_y = 8.31$ cm$^3)$

Wind:

$$W_w = 1.08 \times 0.315 \times 6.250 \times 1.5 = 3.19 \text{ kN}$$

Dead load:

$$W_d = 0.2 \times 6.250 \times 1.5 = 1.88 \text{ kN} < 2.0 \text{ kN minimum load.}$$

Applying the interaction equation with the same coefficients as for the side sheeting rails,

$$3.19 \times 6250/2250 \times 34.9 + 2.0 \times 6250/16880 \times 8.31 = 0.34$$

This is less than unity and is therefore satisfactory.

9.3   Triangulated trusses

In general, it is permissible to analyse triangulated trusses on the assumption that the nodes act as pin joints, i.e. an analysis using simple statics may be adopted. However, effective lengths should be assessed allowing for the rigidity and stiffness of the nodal connections and presence of longitudinal ties (including purlins). Where the purlins do not align with the nodes of the truss, allowance must be made for the additional bending caused by this non-alignment.

Cl 4.10 (c), Pt 1   suggests that where exact positions of such load points is not known a local bending moment of   *WL/6*   should be adopted.

Even in fully triangulated trusses, secondary moments owing to joint fixity can arise, and it is advisable to check the magnitudes of these by undertaking an elastic analysis which allows for joint continuity of the truss under factored ultimate loads. The member capacities can then be rechecked. Cl 4.10, Pt 1 indicates that if the chord members have a slenderness ratio in the plane of the truss of more than 50 and that most of the web members 100, then secondary moments will be negligible. It is probable that the above figures were set for conventional roof trusses fabricated from angles and 'T' sections, and thus they may not be applicable to trusses fabricated from rolled hollow sections. It should be noted that the same analysis set up to calculate secondary moments can be used to assess deflections. This may be done by multiplying the deflections found in the case used for calculating secondary forces which uses factored imposed and dead load, by the ratio of unfactored imposed load to the factored dead and imposed load. This procedure can be adopted since the analysis is elastic and the principle of superposition can be applied.

An assumption of pin joints in calculating deflections will be unrealistic, as it will produce answers which are far too high. However for an initial calculation the assumption of pin joints may suffice as the answer produced will be conservative.

The analysis is more easily carried out if the service loading for each of the separate load cases (dead, imposed, wind upthrust and wind downthrust) are applied and the results from these separate analyses combined using the appropriate load factors to give the final output.

The design of members in a truss is carried out in a similar manner to column members under the effect of axial loading and, where necessary, bending moments. The background to this has been given in Chapter 6.

EXAMPLE 9.3   Triangulated truss design

Prepare a design of the members in Grade 50 Rolled Hollow Sections for the truss whose geometry is given in Fig. 9.4.

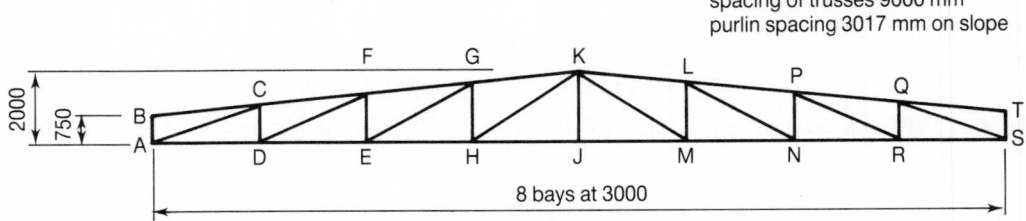

Fig. 9.4   Truss geometry.

Loading.

Dead loads: Sheeting 0.22 kN/m²
              Purlins 0.08
              Total 0.30 kN/m²

UDL for each truss $= 0.30 \times 9.0 = 2.70$ kN/m.
Assume self weight of truss $= 1.50$ kN/m.
Nodal load $= (2.70 + 1.50) \times 3.0 = 12.45$ kN.
Imposed load (BS 6399, Pt 3, cl 4.3.1)
Snow load $= 0.60$ kN/m²
Nodal load $= 0.60 \times 9.0 \times 3.0 = 16.2$ kN.
Wind loading (CP 3 Ch V Pt 2)
Note all Code references in this section are to the wind loading code.
Location of building–Birmingham
Basic wind speed $V = 42$ m/s (Fig. 1)

$\quad S_1 = 1.0$ (cl 5.4)
$\quad S_3 = 1.0$ (Design life 50 yrs, cl 5.6)
$\quad S_2$:

Structure is city centre (Category 4), length greater than 50 m (Class C), height above ground level 7.0 m, thus $S_2 = 0.53$ (Table 3).
Conservatively, set $S_4 = 1.0$

Calculation of characteristic wind speed (V_s) (cl 5.1)

$$V_s = S_1 \times S_2 \times S_3 \times S_4 \times V$$
$$= 1.0 \times 0.53 \times 1.0 \times 1.0 \times 42 = 22.3 \text{ m/s}$$

Calculation of dynamic pressure q (cl 6)

$$q = 0.613 V_s^2$$
$$= 0.613 \times 22.3^2$$
$$= 305 \text{ N/m}^2$$

Calculation of global C_{pe} values:
Walls (long sides only) (Table 7).
height to width ratio $h/W = 7/24 < 0.5$
length to width ratio L/W: assume $2 < L/W < 4$
Wind normal to long side ($\alpha = 0$): $C_{pe} = 0.7$ and -0.25
wind normal to end ($\alpha = 90°$): $C_{pe} = -0.5$ and -0.5
Roof (Table 8).
$h/w < 0.5$, roof slope is $6°$
for $\alpha = 0$, $C_{pe} = -1.0$ and -0.4
for $\alpha = 90°$, $C_{pe} = -0.9$ and -0.4

Calculation of C_{pi} (Appendix E).
   Since the proportion of any dominant openings in relation to the total wall area of any face is not known at this stage, C_{pi} should be taken as 75% of the value of the external coefficient C_{pe} outside the opening (i.e. from values relating to the wall).

Thus,

for $\alpha = 0$, $C_{pi} = 0.75 \times (-0.25) = -0.19$ (Case 1) or $0.75 \times (0.7) = 0.52$ (Case 2)
for $\alpha = 90°$, $C_{pi} = 0.75 \times (-0.50) = -0.38$

Total pressure coefficients ($C_{pe} - C_{pi}$)
$\alpha = 0$: Case 1: $C_{pe} - C_{pi} = -1.0 + 0.19 = -0.81$ and $-0.4 + 0.19 = -0.21$
$\qquad$ Case 2: $C_{pe} - C_{pi} = 1.0 - 0.52 = 0.48$ and $0.4 - 0.52 = -0.12$
$\alpha = 90°$: $C_{pe} - C_{pi} = -0.9 + 0.38 = -0.52$ and $-0.4 + 0.38 = -0.02$

Consider the uplift case first. Although the imbalance is greater in the $\alpha = 90°$ case than that in the $\alpha = 0$ case, the forces in the latter are higher and therefore should be used. The remaining case of the wind causing downthrust will also need considering.

Nodal forces due to wind:
The wind pressure is calculated from $q(C_{pe} - C_{pi})$

Upthrust ($\alpha = 0$, Case 1): $(-0.81 \times 0.304) \times (3/\cos 6) \times 9.0 = -7.06$ and $(-0.21 \times 0.304) \times (3/\cos 6) \times 9.0 = -1.73$ kN.

Downthrust $\quad (\alpha = 90°)$: $\quad (0.48 \times 0.304) \times (3/\cos 6) \times 9.0 = 3.96 \quad$ and $\quad (-0.12 \times 0.304) \times (3/\cos 6) \times 9.0 = -0.99$ kN.

The load cases to be considered are:

$$1.4g_k + 1.6q_k$$
$$1.0g_k + 1.4w_k \qquad \text{(Wind upthrust)}$$
$$1.2 \times (g_k + q_k + w_k) \quad \text{(Wind downthrust)}$$

However, in this particular design owing to the low magnitudes of both sets of wind forces, neither the upthrust case with its possible force reversals nor the downthrust case will be critical. They will not, therefore, be considered further.

The total factored nodal dead and imposed load is given by

$$1.4 \times 12.60 + 1.6 \times 16.2 = 44.5 \text{ kN}.$$

The results from the force analysis assuming pin joints are given in Table 9.1.

Table 9.1   Member forces

Member	End	Pinjoint analysis	Rigid analyis		
		Axial force kN	Axial kN	Shear kN	Moment kNm
AB	A	22.1	26.7	15.6	−7.12
	B	22.1	26.7	−15.6	−4.45
AD	A	−440.0	−429.0	−3.12	4.00
	D	−440.0	−429.0	3.12	4.90
DE	D	−581.7	−580.0	0.45	0
	E	−581.7	−580.0	−0.45	1.78
EH	E	−590.1	−593.0	0	−0.89
	H	−590.0	−593.0	0	0.89
HJ	H	−529.9	−534.0	−0.45	−0.89
	J	−529.9	−534.0	0.45	0
BC	B	0	16.0	3.11	4.45
	C	0	16.0	−3.11	4.00
CD	C	−64.6	−62.7	−9.79	−5.79
	D	−64.6	−62.7	9.79	−4.45
CF	C	442.7	441.0	0	−1.34
	F	442.7	441.0	0	0.89
FG	F	585.2	584.0	−0.45	−1.34
	G	585.2	584.0	0.45	0.45
GK	G	595.4	595.0	−0.45	−1.34
	K	595.4	595.0	0.45	−0.45
AC	A	466.4	438.7	1.78	3.12
	C	466.4	438.7	−1.78	3.12
CD	C	−64.6	−62.7	−9.8	5.79
	D	−64.6	−62.7	9.8	4.45
DF	D	155.8	155.8	0.45	−0.45
	F	155.8	155.8	−0.45	1.34
EF	E	−5.3	−5.8	0.89	−0.45
	F	−5.3	−5.8	−0.89	0.89
EG	E	11.1	13.2	0	−0.89
	G	11.1	13.4	0	0.45
HG	H	39.8	38.7	−0.89	0.89
	G	39.8	38.7	0.89	0.45
HK	H	−71.7	−69.0	0	−0.89
	K	−71.7	−69.0	0	0

Notes

1   Member KJ carries no forces.
2   Compression is positive and moments are positive clockwise.

Member design (assuming purely axial loads)

Top Chord (Members BC, CF, FG, GK)

Maximum force (GK) = −595.4 kN

Using a rectangular hollow section turned through 90° (Fig. 9.5).

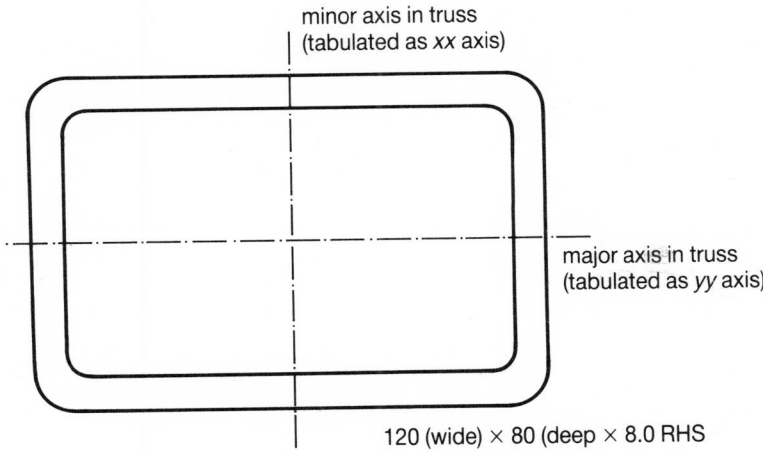

minor axis in truss
(tabulated as *xx* axis)

major axis in truss
(tabulated as *yy* axis)

120 (wide) × 80 (deep × 8.0 RHS

Fig. 9.5   Orientation of bottom chord truss member.

Try a 120 × 80 × 8.0 ($A = 29.1$ cm², $r_x = 42.9$ mm, $r_y = 30.9$ mm)

Out of plane buckling:

Effective length, L_{Ey} = purlin spacing

$$= 3.00/\cos 6$$
$$= 3.020 \text{ m}$$

$$\lambda_y = L_{Ey}/r_x$$
$$= 3020/42.9$$
$$= 70$$

In-plane buckling:

Effective length, $L_{Ex} = 0.85 \times$ (member length)

$$= 0.85 \times 3.02$$

$$\lambda_x = L_{Ex}/r_y$$
$$= 0.85 \times 3020/30.9$$
$$= 83$$

Table 25 specifies the use of Table 27(a) to calculate allowable axial stresses for both axes of a rolled section, so the largest value of λ (83) should be used.

From Table 27(a), with $\lambda = 83$, $p_c = 224$ N/mm^2

Section capacity $= 29.1E2 \times 224/1E3$
$$= 652 \text{ kN.}$$

This is in excess of the applied load of 595.4 and is therefore satisfactory.

Bottom chord (Members AD, DE, EH, HJ)

Maximum force (EH) $= 590.1$ kN.

For economic reasons, together with ease of fabrication of the truss, the bottom chord will also be a $120 \times 80 \times 8$ RHS.

Member capacity $= 29.1E2 \times 355/1E3$
$$= 1033 \text{ kN.}$$

This is obviously satisfactory.

End verticals (AB and TS)
For aesthetic reasons and ease of fabrication, these will also be $120 \times 80 \times 8$ RHS. Since the maximum force is -22.1 kN, there is no need to check the suitability of this member.

Internal members.

Maximum force (KH) $= 71.1$ kN
$$(AC) = -466.6 \text{ kN.}$$

The compression member AC will be the critical member.

Effective length, $L_E = 0.85L_{AC}$

$\qquad L_{AC} = (1.063^2 + 3.0^2)^{0.5}$
$\qquad\quad\ = 3.183$ m

Try a $100 \times 100 \times 5$ ($A = 18.9$ cm^2, $r = 38.7$ mm)

$\qquad \lambda = L_E/r$
$\qquad\quad = 0.85 \times 3183/38.7$
$\qquad\quad = 70.$

From Table 27(a), $p_c = 270$ N/mm^2

Member capacity $= 18.9E2 \times 270/1E3$
$$= 510.3 \text{ kN.}$$

This is greater than the applied force and is therefore satisfactory.

Note, British Steel in their publicity literature on structural rolled hollow sections indicate that it is possible to adopt an end fixity factor of 0.7 in calculating effective lengths. It has not been used here in order to allow some spare capacity for secondary moment effects.

Secondary Moments:
For the top and bottom chord members the in-plane slenderness ratio of 83 is greater than the suggested limit of 50, and should therefore be satisfactory, whereas the web members which have values ranging from 80 in member KH to 70 in member AC do not

satisfy the suggested minimum criterion of 100. It is unlikely, however, that secondary moments will be important. It is, however, proposed to carry out a check.

The forces obtained from an elastic analysis of the truss are also be be found in Table 9.1, where is will be noticed that the axial forces do not differ significantly from those calculated assuming pin joints, except member BC where the axial force found in the rigid analysis is entirely due to secondary moments. The moments and especially the shear forces are low.

Deflection:

A full analysis allowing for joint rigidity with the full dead and imposed load applied gave a central deflection of 36 mm.

An analysis assuming pin joints gives a value of 163 mm.

It is however possible to carry out an approximate analysis of the deflection. This is done by treating the truss as consisting only of its top and bottom chords and allowing for the slope by taking the mean of the squared depth.

$$\text{Mean squared depth, } (d^2)_{\text{mean}} = 0.5 \times (0.75^2 + 2.0^2)$$
$$= 2.28 \text{ m}^2$$

$$I_{\text{approx}} = A_f \times (d^2)_{\text{mean}}/2 \quad (\text{where } A_f \text{ is the flange area})$$
$$= 29.1\text{E-}4 \times 2.28/2$$
$$= 3.32\text{E-}3 \text{ m}^4$$

Nodal loading due to service imposed load = 16.2 kN.

Total imposed service load on the truss, $W = 16.2 \times 8 = 129.6$ kN.

$$E = 205 \text{ GN/m}^2 \text{ (cl 3.1.2)}$$

$$\delta = 5WL^3/384EI$$
$$= 5 \times 129.6 \times 24^3/384 \times 205\text{E6} \times 3.32\text{E-}3$$
$$= 0.034 \text{ m}$$

Maximum allowable deflection (Table 5) is span/200.

$$\text{Span}/200 = 24/200 = 0.120 \text{ m}.$$

The estimated deflection calculated using an approximate method is a reasonable approximation to the exact deflection calculated from the computer program, and would be perfectly adequate as a preliminary estimate.

Check on self weight of the truss (Table 9.2)

9.4   Non-triangulated trusses

The most common form of the non-triangulated truss is the Vierendeel girder in which the loading is carried by a combination of pure flexure and flexure due to the shear induced by the relative vertical deformation between the ends of the top and bottom chord members. This effect is also found in castellated beams (Section 5.8). A Vierendeel girder may have either parallel or non-parallel top and bottom chords.

Although traditionally the Vierendeel girder was fabricated from 'I' sections, its load carrying capacity is much enhanced when rolled hollow sections are used together with butt or fillet welded connections between the top and bottom chord members and the

Table 9.2   Self weight of truss

Member	Length	Total length (m)
$120 \times 80 \times 8$ members		
Top chord	$2 \times (12^2 + (2 - 0.75)^2)^{0.5}$	24.1
Bottom chord	24	24
End posts	2×0.75	1.5
Total		49.6

Total mass $= 49.6 \times 22.9 = 1136$ kg

	$100 \times 100 \times 50$ members	
KJ	1×2.0	2.0
KH	$2 \times (3^2 + 2^2)^{0.5}$	7.2
HG	2×1.688	3.4
GE	$2 \times (3^2 + 1.688^2)^{0.5}$	6.9
EF	2×1.375	2.8
FD	$2 \times (3^2 + 1.375^2)^{0.5}$	6.6
CD	2×1.063	2.1
CA	$2 \times (3^2 + 1.063^2)^{0.5}$	6.4
Total		37.4

Total mass $= 37.4 \times 14.8 = 554$ kg

Total mass of truss $= 1136 + 554$
$= 1690$ kg

Weight of truss $= 16.9$ kN, or weight/unit length $= 16.9/24 = 0.7$ kN/m.

The original calculations overestimate the self weight as 1.5 kN/m.

vertical panel members, since for normal panel sizes there will be no reduction in moment carrying capacity due to the effects of lateral torsional buckling. Also it is unlikely that any reduction will need to be made in the moment capacity for the effects of axial force or shear.

The Vierendeel girder must be classified as a sway frame by virtue of the manner in which the loads are carried, even though the method given in cl 5.1.3, Pt 1 is not applicable as it was written for conventional frames in which the imposed loading is applied to the beams. This therefore also means that the method proposed by Horne for calculating the elastic critical load factor and adopted in cl F.2.3, Pt 1 cannot be used. As the axial forces will be low it is probable that the elastic critical load factor will be high indicating that there will be little problem with this. However to be certain, an elastic analysis of the structure will be used and the member capacities checked using effective length factors for frames not braced against sidesway. The alternative option of magnifying the moments obtained from the horizontal loading cannot be used owing to the problems mentioned earlier of determining the elastic critical load factor. Redistribution will not be carried out, as the maximum redistribution allowed of 10% will not significantly affect the design. The elastic analysis may also be used to determine the deflection of the frame under service imposed loading. Note that although it is preferable to use a plane frame computer analysis package, it is possible to use Naylor's modified moment distribution method to determine the forces in a frame subject of pure sway (see Marshall and Nelson, Ghali and Neville, or Coutes *et al.*).

EXAMPLE 9.4   Design of a Vierendeel girder

Prepare a design in Grade 50 rolled hollow sections for the girder in Fig. 9.6.

girder spacing 8500 mm
all members are the same size
purlin spacing 3000 mm

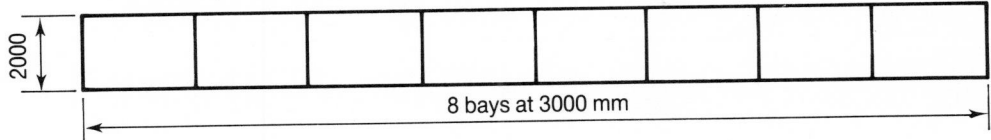

8 bays at 3000 mm

Fig. 9.6   Vierendeel girder data.

Loading.
Dead load: Sheeting 0.22 kN/m²
                     Purlins 0.1
                          Total 0.32 kN/m²
UDL loading for each truss $= 0.32 \times 8.5 = 2.72$ kN/m
Estimate self weight of truss $= 2.0$ kN/m
Nodal dead load $= (2.00 + 2.72) \times 3.0 = 14.16$ kN.
Imposed load (BS 6399 Pt 3 cl 4.3.1)
Snow load $= 0.60$ kN/m²
Nodal imposed load $= 0.60 \times 8.5 \times 3.0 = 15.3$ kN
Total factored load $= 1.4 \times 14.16 + 1.6 \times 15.3 = 44.3$ kN.

Note, in this example only dead and imposed load will be considered, whereas in practice the effects of wind ought also to be considered. It is however unlikely that wind loading will be critical.
    The bending moment, shear force and axial force diagrams for half the frame are given in Fig. 9.7.
    From the force diagrams, the critical members for designing are EE′, ED, E′D′ and DD′. Member A′B′ should also be checked as this carries the highest axial compressive load.
    Member DD′ carries the highest bending moment so design this first.

Member DD′
$F = 22.2$ kN, $F_v = 186.4$ kN, $M_x = 186.4$ kNm.
Since $M_y = 0$, all the checks are much simplified.
Try a $250 \times 150 \times 10$ rectangular hollow section.

Check for plastic action:
From Fig. 3 (BS 5950 Part 1)
for a RHS, $b = B - 3t$, and $d = D - 3t$, so

$$b/T = (B - 3t)/t = B/t - 3$$

Actual b/T ratio $= 150/10 - 3 = 12$
allowable b/T ratio

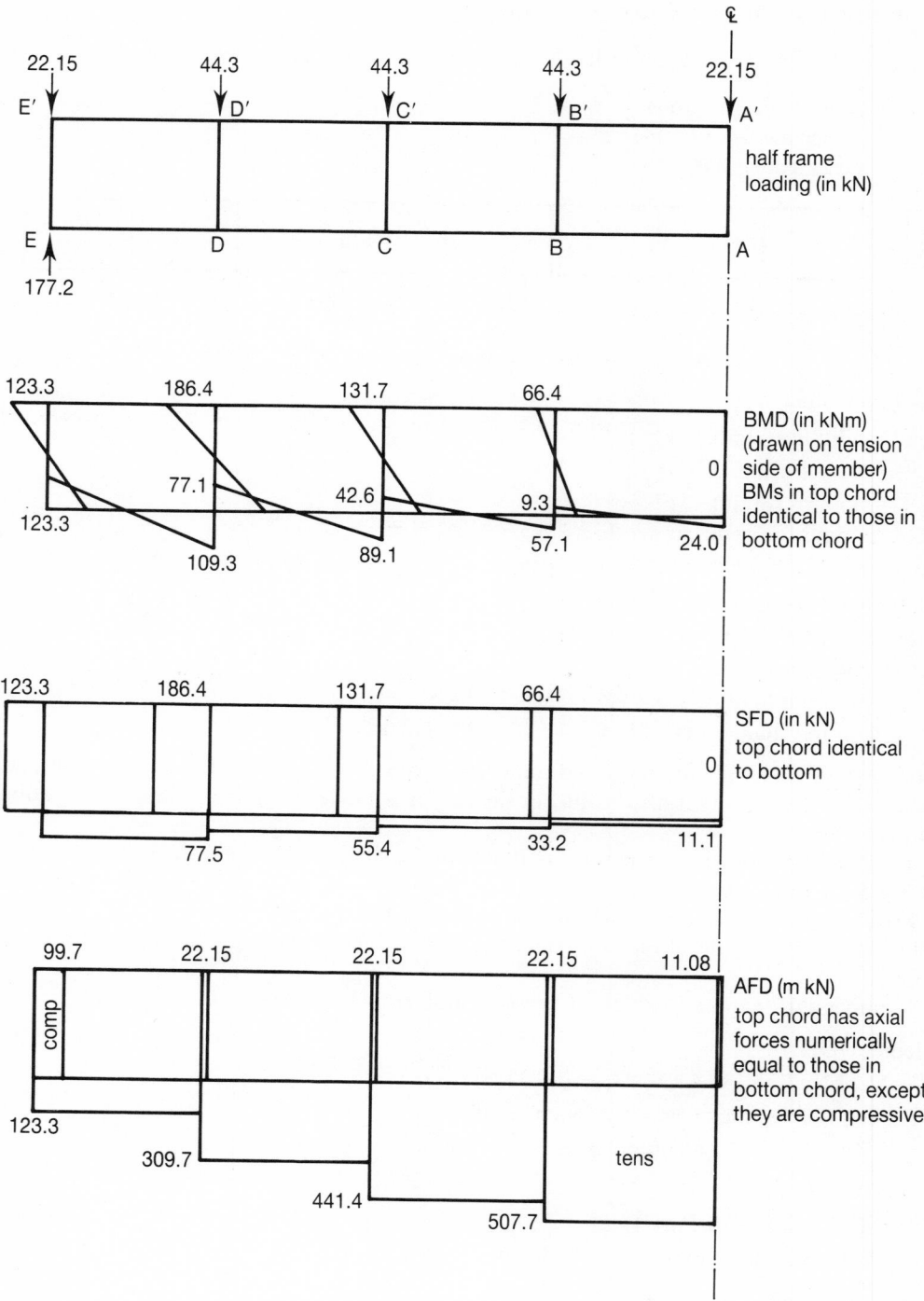

Fig. 9.7   Analysis results for girder.

For internal compression flanges for rolled sections (Table 7),

$$b/T = 26\varepsilon = 28 \times 0.88 = 22.9 \text{ as}$$
$$\varepsilon = (275/p_y)^{0.5} = (275/355)^{0.5} = 0.88$$
$$d/t = (D - 3t)/t = D/t - 3$$
$$= 250/10 - 3 = 22$$

Allowable d/t for web with NA at mid-depth is 79ε,

$$\varepsilon = 79 \times 0.88 = 69.5$$

It should be noted that there is a small axial force in the web which will displace the neutral axis from mid-depth, but this shift will be small and since the actual value of d/t is well below the allowable no further check will be made.

The member needs checking under cl 4.8.3, Pt 1 for both local capacity check and overall buckling check.

In view of the note in BS 5950, Pt 1 following cl 4.8.3.3 the simplified approach of cl 4.8.8.3.3.1 will be used as $M_y = 0$.

Local capacity check:
For plastic sections the following condition needs checking,

$$(M_x/M_{rx})^{z1} + (M_y/M_{ry})^{z2} \leqslant 1 \tag{6.13}$$

which for uniaxial bending reduces to

$$M_x \leqslant M_{rx}$$

where M_x is the bending moment in the x direction and M_{rx} the moment capacity of the section allowing for the effects of axial load.

For a $250 \times 150 \times 10$ RHS
$I_x = 6260 \text{ cm}^4$, $A = 75.5 \text{ cm}^2$, $S_x = 618 \text{ cm}^2$, $r_y = 6.07$ cm.
The axial stress p is given by

$$p = F/A = 22.2E3/75.5E2 = 2.94 \text{ MPa}.$$

This is very low and will cause no reduction in the plastic moment capacity, so M_{rx} can be taken as M_c.

$$M_c = S_x p_y$$
$$= 618 \times 355/1E3$$
$$= 219.4 \text{ kNm}$$

This exceeds the applied moment and is therefore satisfactory.

Overall buckling check (Equation (6.14)):
With $M_y = 0$ this reduces to

$$F/A_g p_c + mM_x/M_b \leqslant 1.0$$

Calculation of p_c:

$$L_E = 0.85 \times 2000 = 1700$$
$$\lambda = L_E/r_y$$
$$= 1700/60.7$$
$$= 28$$

From Table 27(a), $p_c = 344$ MPa

$$F/A_g p_c = 22.2E3/(75.5E2 \times 344)$$
$$= 0.01$$

Calculation of M_b:

For bending, $L_E = 2000$, and

$$\lambda = L_E/r_y = 2000/60.7 = 33$$

From Table 38 determine limiting values of λ

Actual value of $B/D = 250/150 = 1.67$

Use the value given for $D/B = 2$,

$$\lambda = 350 \times 275/p_y = 350 \times 275/355 = 271$$

Thus $M_b = M_c = 219.4$ kN

$$\beta = -1, \quad \text{so} \quad m = 0.43 \quad \text{(Equation (5.30))}$$

$$mM_x/M_b = 0.43 \times 186.4/219.4$$
$$= 0.37$$

Thus final value of the overall buckling check $= 0.01 + 0.37 = 0.38$ which is satisfactory.

Shear check (cl 4.2.3):

For a RHS,

$$P_v = 0.6p_y A(D/(D + B))$$
$$= 0.6 \times 355 \times 75.5E2(250/(150 + 250))/1E3$$
$$= 1006 \text{ kN}$$

This is very much in excess of the applied shear, and therefore no reduction need be made in the moment capacity due to shear.

Members Ed and E′D′:

$M_x = 123.3$ kNm, $F = 123.3$ kN (numerically), $F_v = 123.3$ kN.

The shear check is satisfactory, since the shear capacity is that found for member DD′.

(a)   Member ED

The axial force in tensile, and therefore only a capacity check need be carried out (cl 4.8.2).
   For plastic sections the following relationship should be satisfied:

$$(M_x/M_{rx})^{z1} + (M_y/M_{ry})^{z2} \leqslant 1 \tag{6.13}$$

which for uniaxial bending reduces to

$$M_x \leqslant M_{rx}$$

where M_x is the bending moment in the x direction and M_{rx} the moment capacity of the section allowing for the effects of axial load.

$$p = F/A = 123.3E3/75.5E2$$
$$= 16.3 \text{ MPa}$$

The reduced section modulus, S_{rx} is given by (SCI Guide Part 1, Member Properties and

Safe Load Tables)

$$S_{rx} = S_x - (An)^2/8t$$

with $n = p/p_y \leqslant 2t(D - 2t)/A$

Limiting value of n:

$$2t(D - 2t)/A = 2 \times 12(250 - 2 \times 10)/75.5E2$$
$$= 0.61$$

actual value of n:

$$n = p/p_y = 16.3/355$$
$$= 0.046$$

Working in cm units,

$$S_{rx} = 618 - (75.5 \times 0.046)^2/(8 \times 1.0)$$
$$= 616 \text{ cm}^3$$

$$M_{rx} = S_{rx}p_y$$
$$= 616 \times 355/1E3$$
$$= 218.7 \text{ kNm}.$$

It can be seen that in this case the reduction in moment capacity is quite small, and could have been ignored.

The applied moment exceeds the moment capacity.

(b) E'D'

The axial force is compressive.

The local capacity check is the same as that required if the axial force were tensile, and this check is therefore satisfactory.

Overall buckling check:

Calculation of p_c.

The effective length factor of E'D' must be obtained using Appendix E of the Code (Fig. 24).

To obtain the effective length factor the stiffnesses of the members framing into the column and the column itself need calculating.

In Fig. 9.8 a comparison is made between the situation in the actual frame and the situation in the Code.

Since all the second moments of area of all the members are the same, then for the purpose of calculating relative stiffnesses they can be set equal to unity.

Using the Code notation,

$$K_{TL} = K_{BL} = 1/2 = 0.5$$

$$K_c = K_L = 1/3 = 0.333$$

$$K_{TR} = K_{BR} = K_U = 0$$

$$k_1 = (K_c + K_U)/(K_c + K_U + K_{TL} + K_{TR})$$
$$= 0.333/(0.333 + 0.5)$$
$$= 0.40$$

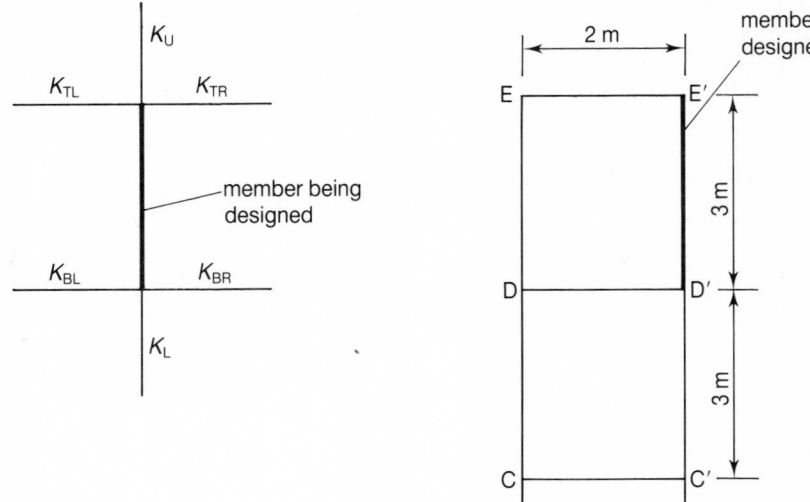

(a) BS 5950 subframe (Fig 8.27)   (b) Actual case

Fig. 9.8   Determination of member stiffness for buckling of member E'D'.

$$k_2 = (K_c + K_L)/(K_c + K_L + K_{BL} + K_{BR})$$
$$= (0.333 + 0.333)/(0.333 + 0.333 + 0.500)$$
$$= 0.57$$

Using these values of k_1 and k_2 a value of 1.46 is obtained from Fig. 24 for the effective length factor. As a comparison use the equations developed by Williams and Sharp (1990):

$$L_E = L/(12 - 36(k_1 + k_2 - k_1 k_2)/(4 - k_1 k_2))^{0.5} \tag{8.9}$$

Substituting in values of k_1 and k_2 gives

$$L_E = 1.42L$$
$$L_E = L(1.05 - 0.05k_1 k_2)/(12 - 36(k_1 + k_2 - k_1 k_2)/(4 - k_1 k_2))^{0.5} \tag{8.10}$$

or

$$L_E = 1.47L$$

Both these values are in close agreement with the value of 1.46 obtained from Fig. 24 in the Code.

Using the code value of 1.46,

$$\lambda = L_E/r_y$$
$$= 1.46 \times 3000/60.7$$
$$= 72$$

From Table 27(a), $p_c = 264$ MPa, so

$$F/A_g p_c = 123.3E3/(75.5E2 \times 264)$$
$$= 0.06$$

$m = 0.43$ and $M_b = 219.4$ kNm as before, so

$$mM_x/M_b = 0.43 \times 123.3/219.4$$
$$= 0.24$$

Final value of the overall buckling check is $0.06 + 0.24 = 0.30$, which is satisfactory. Member EE′ will also be found to be satisfactory.

Member A′B′

Overall capacity check:

$$p = F/A = 507.7E3/75.5E2$$
$$= 67.2 \text{ MPa}$$

$$S_{rx} = S_x - (An)^2/(8t)$$

$$n = p/p_y = 67.2/355$$
$$= 0.19$$

this is less than the critical value of 0.61

$$S_{rx} = 618 - (75.5 \times 0.19)^2/(8 \times 1.0)$$
$$= 592.3 \text{ cm}^3$$

$$M_{rx} = 592.3 \times 355/1E3$$
$$= 210.3 \text{ kNm}$$

This is greater than the applied moment of 24.0 kNm.

Local capacity check:
Fig. 9.9 indicates the comparison between the Code and the actual situation for calculating the relative end stiffnesses.

$$K_{TL} = K_{BL} = 0.5$$

$$K_c = K_U = K_L = 0.333$$

$$K_{TR} = K_{BR} = 0$$

$$k_1 = (K_c + K_U)/(K_c + K_U + K_{TL} + K_{TR})$$
$$= 2 \times 0.333/(2 \times 0.333 + 0.5)$$
$$= 0.57$$

$$k_2 = (K_c + K_L)/(K_c + K_L + K_{BL} + K_{BR})$$
$$= 2 \times 0.333/(2 \times 0.333 + 0.500)$$
$$= 0.57$$

From Fig. 24, the effective length factor is 1.6.

Note Equations (8.9) and (8.10) give values of 1.57 and 1.62 respectively.

$$\lambda = 1.6 \times 3000/60.7$$
$$= 79$$

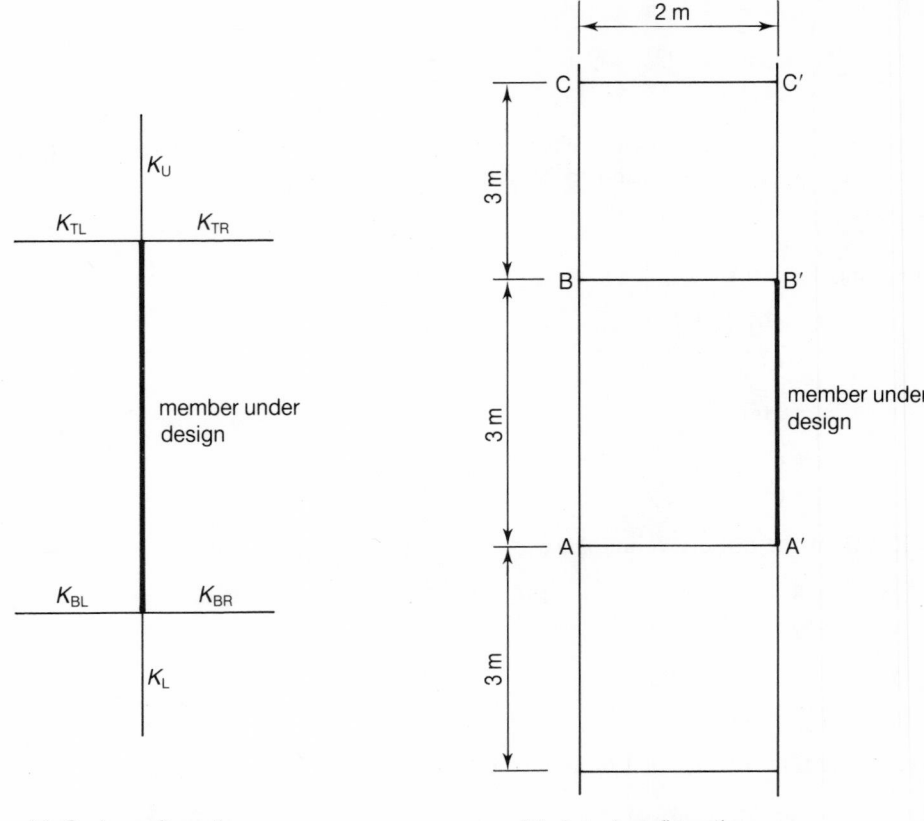

(a) Code configuration   (b) Actual configuration

Fig. 9.9   Determination of member stiffness for buckling of member A′B′.

From Table 27(a), $p_c = 238$ MPa

$$F/A_g p_c = 507.7\text{E}3/(75.5\text{E}2 \times 238/1\text{E}3)$$
$$= 0.28$$

$$m = 0.43$$

$$mM_x/M_b = 0.43 \times 24.0/219.4$$
$$= 0.05$$

thus a final value for the interaction equation is 0.33, which is satisfactory.

Web bearing check (cl 4.5.3, Pt 1).
   This will be needed at E, where the support reaction is applied.

Bearing capacity is $(b_1 + n_2)t p_{yw}$

Calculation of n_2 for both webs

$$n_2 = 2 \times 2.5t = 5 \times 10 = 50 \text{ mm}$$

Beam component $= n_2 t p_{yw}$
$$= 50 \times 10 \times 355/1E3$$
$$= 178 \text{ kN}$$

This is only just greater than the applied reaction of 177 kN.

In practice, there would be a greater reserve of strength, however, as the bottom flange would be stiffened by additional plates to enable the truss to be bolted down to the supports (either steel stanchions or pad stones in masonry construction). A typical detail for this is given in Fig. 9.10.

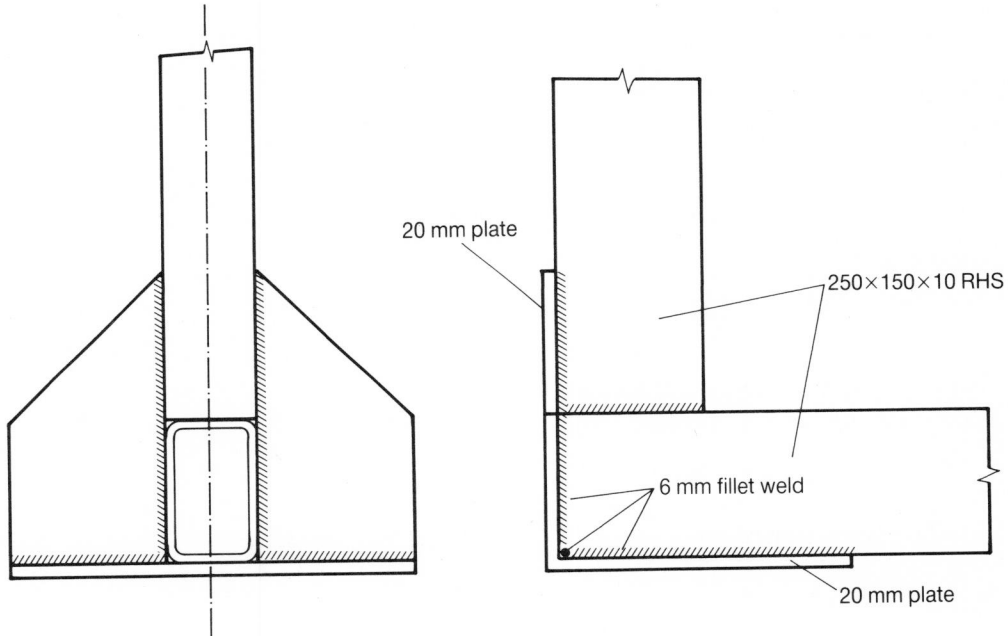

Fig. 9.10 Typical end detail.

Web buckling check at E

The SCI Guide indicates that for RHS sections the slenderness ratio λ to be used is given by

$$\lambda = 1.5\sqrt{3(D - 2t)}/t$$

This differs from the value given in BS 5950 as it is suggested that the web is restrained against rotation less in a RHS than in an 'I' or 'H' section

$$\lambda = 1.5\sqrt{3(250 - 2 \times 10)}/10$$
$$= 60$$

From Table 27(c), $p_c = 247 \text{ MPa}$

The web buckling resistance P_w is given by

$$P_w = (b_1 + n_1)t p_c$$

Calculating the beam component first,

$n_1 = D/2$,   so

beam component $= (125/2) \times (2 \times 10) \times 247/1E3$
$= 309$ kN.

This exceeds the reaction of 177 kN therefore no additional bearing is needed to handle web buckling.

Deflection check.
This was carried out using a plane frame analysis package and gave an answer of 25.0 mm. The allowable deflection is span/200 (Table 5)

span/200 $= 24/200 = 0.120$ m

Self weight check.
Top and bottom chords

$2 \times 24 \times 59.3 = 2846$ kg

Vertical members:

$9 \times 2 \times 59.3 = 1067$ kg.

Total mass $= 2846 + 1067 = 3913$ kg.
Mass/unit length $= 3913/24 = 163$ kg/m.
Weight/unit length $= 1.63$ kN/m.
The assumed weight was slightly overestimated at 2.0 kN/m.

9.5   Tutorial problem

9.5.1   Monopitch roof truss

Prepare a design in Grade 50 rolled hollow sections for the truss detailed in Fig. 9.11.
   The effect of wind loading is not to be considered, and do not check deflection. It is also sufficient to consider the truss as pinpointed.

Answer

Unfactored nodal loads; dead 18.75 kN, imposed load 22.50 kN.

Truss analysis for unit nodal load is given in Fig. 9.12.

Design forces:
Internal members: CD 62.3 kN (tens), BC 139.4 kN (comp)
External members: JH 311.3 kN (tens), 324.9 kN (comp) + bending
Minimum size internal member (using 0.85 for effective length factor) $100 \times 100 \times 5$ ($L_E = 0.85 \times 6.73$, $r = 38.7$ mm, $p_c = 85$ MPa, $P_c = 161$ kN).
External member: Minimum size $200 \times 100 \times 5$
Calculate max BM assuming member to be simply supported at the end (this is conservative). Purlin load 15.7 kN, max BM 39.3 kNm.

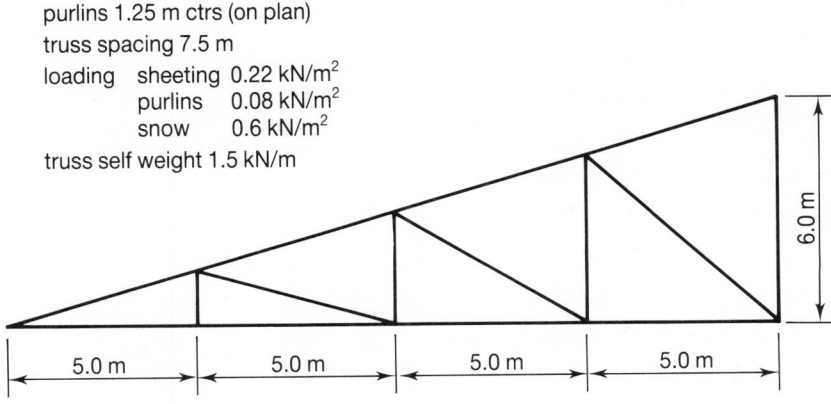

purlins 1.25 m ctrs (on plan)
truss spacing 7.5 m
loading   sheeting 0.22 kN/m²
          purlins   0.08 kN/m²
          snow   0.6 kN/m²
truss self weight 1.5 kN/m

Fig. 9.11.

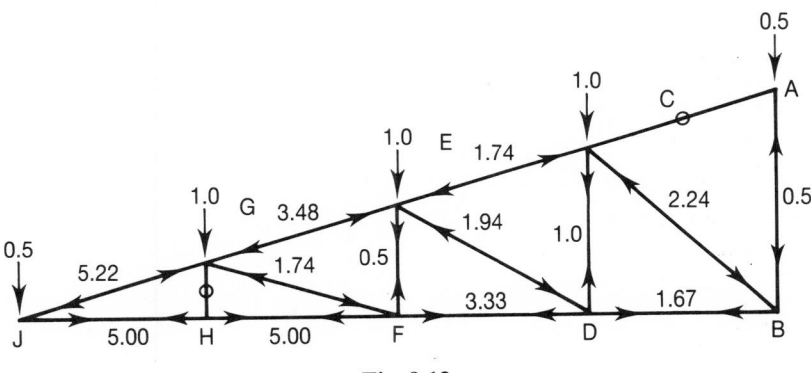

Fig. 9.12.

Moment capacity (not corrected for axial load) 64.3 kNm.

Reduced moment capacity, $p = 112$ MPa, $n = 0.32$ (limiting value of $n = 0.66$), reduced section modulus $= 186 - 21 = 165$, $M = 58.6$ kNm

Local capacity check is clearly satisfied.

Buckling check:

Axial term:

nodal length of member $= 5.22$ m, purlin spacing along member $= 1.305$ m

Slenderness ratios: x–x axis: $0.85 \times 5220/72.3 = 61$: y–y axis: $1305/42 = 31$

$$F/A_g p_c = 324.9\text{E}3/(28.9\text{E}2 \times 295) = 0.38$$

Moment term:

under first purlin, $M = 29.4$ kNm, under middle purlin, $M = 39.3$ kNm, $\beta = 0.75$, $m = 0.87$, $M_b = 64.3$ kNm

$$mM_x/M_b = 0.87 \times 39.3/64.3 = 0.53$$

Final value $= 0.38 + 0.53 = 0.91$.

References

BS 2994. *Cold rolled sections*, British Standards Institution, London.

BS 5950 Part 5. *Structural use of steelwork in building – Part 5 – Code of practice for design of cold formed sections*, British Standards Institution, London.

BS 6399 Part 1. *Loading for buildings – Part 1 – Code of practice for dead and imposed loads*, British Standards Institution, London.

Coates R. C., Coutie M. G. and Kong F. K. (1988). *Structural analysis* (3rd edition), Van Nostrand.

CP 3 Ch V Part 2. *Code of basic data for the design of buildings*, Chapter V *Loading*, Part 2 *Wind loads*, British Standards Institution, London.

Ghali A. and Neville A. M. (1989). *Structural analysis*, Chapman and Hall.

Horne M. R. (1975). An approximate method for calculating the elastic critical loads of multi-storey plane frames, *The Structural Engineer*, **53** (6), 242.

Marshall W. T. and Nelson H. M. (1977). *Structures*, Pitman.

Walker A. C. (Editor) (1975). *Design and analysis of cold formed sections*, Intertext.

Williams F. W. and Sharp G. (1990). *Simple elastic critical loads and effective length calculations for multi-storey rigid sway frames*, Proc ICE, **90**, 279–87.

10

Composite Construction

10.1 Introduction

Although the combination of a concrete slab connected by shear connectors to the supporting steel beam has been a viable proposition for some thirty years, it is only within the last decade that composite construction has become widespread.

Originally composite construction was achieved by the casting of an in situ normal-weight concrete slab onto the top flange of a steel beam to which shear connectors in the form of studs had been shop welded before erection of the steelwork (Fig. 10.1(a)). This system had the disadvantage of needing formwork for casting the soffit of the slab. This formwork, if permanent, was generally only capable of supporting the weight of the wet concrete over relatively short spans or, if temporary, although allowing the main support beams in the steel frame to be at larger centres, needed extensive propping from the floors below causing inherent time delays. This type of construction was often combined with the use of concrete encasement to provide the structure with adequate fire resistance (see Chapter 11).

It was following a series of developments initiated in the United States (Walker), that the idea of the use of cold formed steel trough decking could be used both as permanent shuttering for the concrete slab and tension reinforcement for the composite slab spanning between the main beams. Composite action between the steel beam and the concrete slab is achieved by the use of site welding the shear studs after the placing of the steel profile decking onto the steel beams (Fig. 10.1(b)).

This system of construction has a series of advantages:

1   The use of propping systems is rarely needed as it is generally more economic to reduce the spacing of intermediate support beams. This means that lower floors are relatively clear and that other building trades gain faster access.
2   The construction depth is less due both to the availability of composite action and to the general use in constructions of this type of lightweight concrete for the floor slabs.
3   The use of lightweight concrete itself has advantages:
    a   better fire performance (Section 11.6)
    b   reduced load on the foundations, thereby giving less problems with foundation design,
    c   it is better adapted owing to its inherent high workability to placement by pumping rather than by conventional techniques such as a skip.

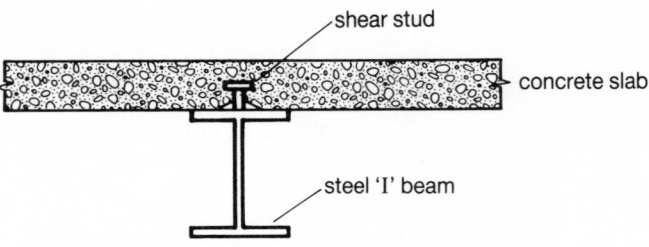

(a) Conventional composite construction

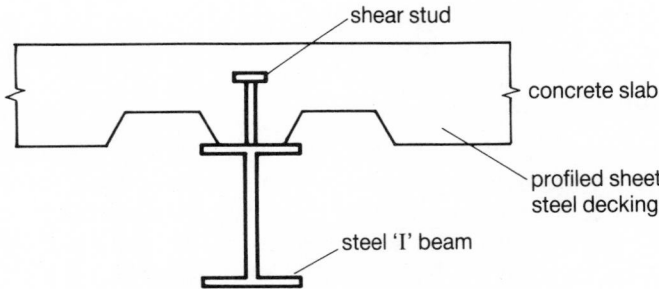

(b) Composite construction with profiled sheet steel decking

Fig. 10.1   Composite beam construction.

4   Plant requirements, most notably cranage, are often reduced since the profiled sheet steel decking is relatively light and is generally delivered in bundles and pumping is used for concrete placement.
5   An in situ slab can be given a reasonable finish by power floating, whereas precast concrete floor units will need screeding to take out slight variations in the slab surface levels and to provide a good finish.

It can therefore be appreciated that modern composite construction can allow a structure to be made weathertight to enable finishing trades earlier access, and gain earlier completion and occupancy of the finished structure. This, coupled with recent trends in relative cheapness of steelwork compared to a total in situ concrete design, has made composite construction a viable solution for multi-storey construction (Gray and Walker).

Concrete encasement is very much a hybrid as the concrete case, in design, is deemed to contribute nothing directly toward the flexural capacity of the member but may be used to assist the carrying of axial load. The original use of concrete encasement as mentioned earlier was to insulate the steel to reduce the temperature rise in a fire and thus ensure adequate fire resistance. This method has two inherent drawbacks:

1   It is very heavy, thus increasing by substantial amounts the loads on foundations, and indeed the loads to be carried by the members themselves.
2   It is very expensive, both in terms of time, in that formwork is needed and cannot be stripped until the concrete has cured, and in terms of basic material costs.

Concrete encasement has now been generally superseded by the use of vermiculite plaster spray, plaster boarding or intumescent paints (see Section 11.1) for periods of fire resistance up to around two hours. For longer periods of fire resistance (up to four hours) it may still be necessary to encase steelwork in concrete, although modern site practice is to cast the concrete off site and leave only in situ casting at the connections.

10.2   Composite slabs

10.2.1   Introduction

A composite slab comprises the profile sheet steel decking which acts both as permanent formwork and as tension reinforcement for the sagging moments in the slab and an in situ concrete slab cast on the decking.

There are essentially two patterns of profiled decking, an open trapezoidal section (Fig. 10.2(a)), and a re-entrant trapezoidal section (Fig. 10.2(b)).

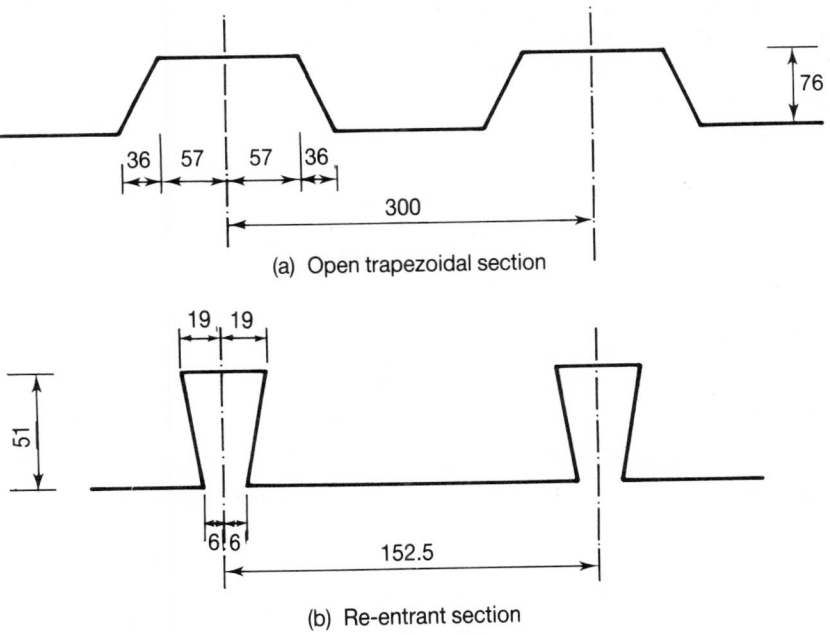

(a) Open trapezoidal section

(b) Re-entrant section

Fig. 10.2   Typical dimensions of sheet steel decking.

The latter has a slight advantage in that proprietary wedge fixings placed in the re-entrant slot may be used to hang suspended ceilings and light services from the soffit of the slab. For further details of the deck profiles available and for an overview of the design process reference should be made to Lawson (1983).

The design criteria fall into two categories; design of the decking acting as formwork, and design of the composite deck slab.

Where profile steel decking is only required to act as shuttering and is not being used for composite action only the first category need be considered.

The design of profile decking and composite slabs is covered by BS 5950, Part 4.

10.2.2   Design of profile decking as permanent formwork

The design criteria for trough steel decking acting as permanent formwork are covered by Section 5 of BS 5950, Part 4.

The loading that should be considered is

(a) The weight of the wet concrete, and
(b) the effect of construction loading to allow for plant, men, and for localized piling of concrete during placement, etc. Also should the erected decking be used for storage, or for storage on the slab before composite action can be achieved, this must be taken account of in the design.

The construction (and storage) loading should be taken as a minimum of 1.5 kN/m^2 (cl 13.2 and 13.3, Pt 4). If the spans are less than   3 m,   an alternative loading of a line load of intensity   2 kN/m   parallel to the supports should be considered. The load factor to be applied to such loading is 1.4 (Table 1).

The profile steel decking has a design strength p_y of   0.93 times the characteristic strength or the specified yield strength,   and a Young's modulus of   210 GPa.

Although cl 16, Pt 4 gives detailed information on the calculation of section properties to determine the effects of bending, shear, web buckling and deflection, such calculations are complex to carry out (they are of a similar nature to those for the design of cold formed sections for, say purlins) as the section properties are dependent on the stress state within the section, and are thus iterative. It should also be noted that the moment carrying capacity is limited to that just causing yield, i.e based on the elastic section modulus (cl 16.4, Pt 4). It is more normal therefore to have recourse to test data and to a manufacturer's safe load tables based on such test data and calculation.

The criteria that need checking are:

10.2.2.1   *Deflection due to the wet concrete and decking self-weight (cl 17, Pt 4)*

It is not necessary to check deflection under the effect of construction loading as it is deemed to be very short term. The allowable deflection is to be taken as the lesser of   span/180   or   overall slab depth/10.   If the effect of ponding is taken into account in the design of the slab and supporting structure, the deflection need only be limited to span/130   (Fig. 10.3).

10.2.2.2   *Compression in the bottom flange*

This check is only needed when the sheeting is designed as continuous over internal supports.

Compression will always be more critical than tension as the compression flange width is reduced to an effective width to allow for shear lag.

10.2.2.3   *Compression in the top flange*

This will need checking under maximum sagging moment.

10.2.2.4   *Web shear and buckling*

The former is critical at the support, and the latter, which is a measure of the ability of the sheeting to transfer forces in bending, is critical at points of high moment.

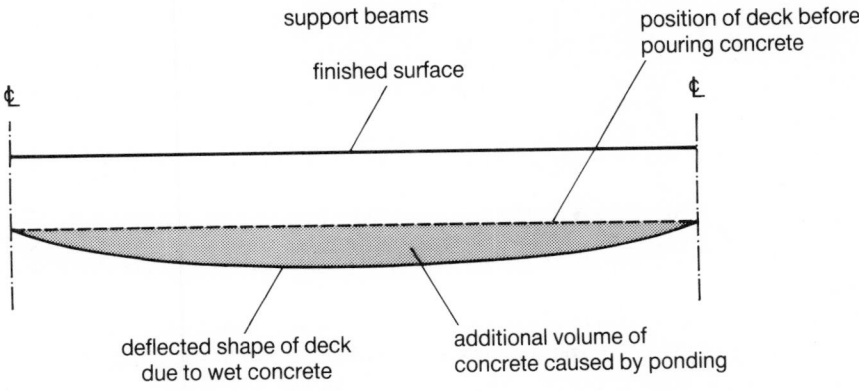

Fig. 10.3   Effect of deflection of sheet steel decking.

10.2.2.5· *Combined bending and shear*

At points where high bending stresses and high shear stresses co-exist the section needs to be checked under combined forces. The check is in the form of an interaction equation,

$$\text{if } W_f/P_w \leqslant 0.25, \ M/M_{cp} \leqslant 1.0$$

$$\text{if } W_f/P_w > 0.25,$$

$$M/M_{cp} + 0.64 W_f/P_w \leqslant 1.16 \tag{10.1}$$

10.2.3   Design of the composite slab

There are three basic modes of failure of a composite slab:

1   Flexural failure at midspan.
It is simpler to design composite slabs as simply supported between the support beams and not to allow for continuity even though there will be anti-crack reinforcement in the top face of the slab. The minimum mesh reinforcement is specified as not less than 0.1% of the gross concrete area (cl 25, Pt 4). The cover to such reinforcement is generally 25 mm, although a smaller cover may be taken in accordance with Table 3.4 of BS 8110 Part 1 for normal-weight concrete and Table 5.1 of BS 8110 Part 2 for lightweight concrete.
    Should the composite slab be designed as continuous then flexure will also need checking at the support.

2   Shear bond failure between the decking and the concrete.
Most profile decking is manufactured with indents or small shear keys in order to generate the bond between the concrete slab and the decking to give strain compatibility at the interface and hence provide the composite action needed. The parameters needed to calculate the shear bond behaviour are in general found from tests and are detailed by the manufacturer for a particular type of decking.

3   Flexural shear failure.
This type of failure is analogous to that in reinforced concrete and is essentially a measure of the inability to carry the derived shear force. It is generally only critical for short span deep slabs carrying very high loading.

In addition it is necessary to check the deflection under imposed load and also determine the fire performance of the composite slab.

Consideration can now be given to the detailed calculations required for each of the design criteria.

10.2.3.1   *Flexural design*
(a)   Sagging (cl 20, Pt 4)

The flexural capacity of the section should be determined using a rectangular stress block for the concrete with an average strength of $0.4f_{cu}$, the steel decking considered as acting at its centroid of area with a strength p_y, and any additional reinforcement a strength of $0.87f_y$. The depth of the neutral axis should be limited to $0.5d_s$, where d_s is the effective depth of the slab, unless there is compression reinforcement in which there is no limit on the neutral axis depth.

(b)   Hogging

This should be treated as for reinforced concrete with no allowance for the effect of the decking.

10.2.3.2   *Shear bond (cl 21, Pt 4)*

The shear bond capacity V_s is given by,

$$V_s = B_s d_s (m_r A_p / B_s L_v + k_r (f_{cu})^{0.5})/1.25 \qquad (10.2)$$

where B_s is the width of the slab, L_v the shear span, m_r and k_r are empirical factors obtained from tests, and A_p the cross-sectional area of the profiled sheet, with a partial safety factor of 1.25.

The shear span L_v is taken as

$L/4$ for a UDL   (where L is the effective span), and

the distance from the support to the nearest load for a symmetric two point load system.

For other cases reference should be made to cl 21.2, Pt 4.

10.2.3.3   *Flexural shear capacity*

The shear capacity V_v is given by

$$V_v = b d_s v_c \qquad (10.3)$$

where b is defined as the width of the base of the trough for open profile decking and the width between the tops of slots for re-entrant decking (Fig. 10.4), and v_c the shear strength of the concrete, calculated with $b = B_s$ and $d = d_s$, from Table 3.9 of BS 8110 (modified as appropriate for lightweight concrete (BS 8110 Part 2 Section 5)).

Punching shear will also need checking where concentrated loads act. This check is performed according to cl 22.2, Pt 4.

10.2.3.4   *Deflection (cl 23, Pt 4)*

The deflection should be limited to

(a)   the lesser of span/350 or 20 mm   under imposed load, and

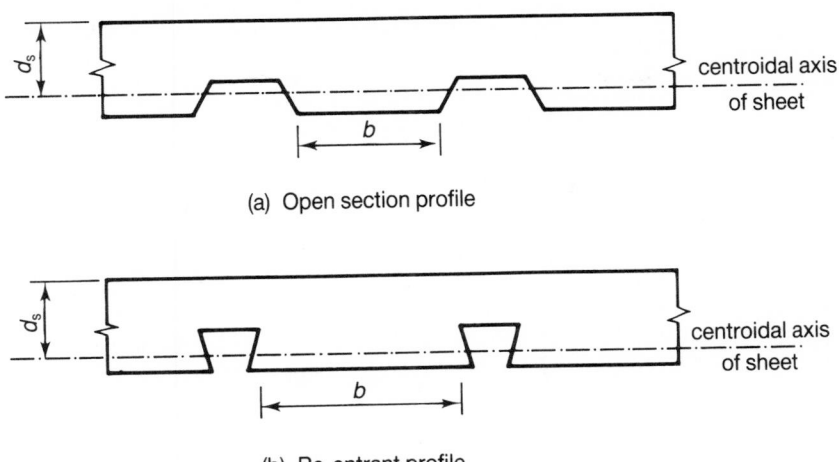

(a) Open section profile

(b) Re-entrant profile

Fig. 10.4 Definition of dimensions for calculation of shear capacity.

(b) span/200 for the deflection calculated for the total load less that due to deflection from the self weight of the slab plus that due to removal of the props if any be used.

The second moment of area should be calculated as the average of the cracked and uncracked second moment of areas of the concrete slab and decking. A modular ratio of 15 should be adopted for normal-weight concrete and 20 for lightweight concrete.

10.2.3.5 *Fire performance*

This is either dealt with by reference to manufacturers' data based on tests or by performing calculations (Fire Safety Engineering Approach). This topic is covered in Section 11.6.

EXAMPLE 10.1 Composite slab design

A floor system for an office block is to be designed using composite construction. The beam layout is given in Fig. 10.5.

The profile decking to be used is Richard Lees Super Holorib, 0.9 mm gauge. This is a re-entrant profile decking which is more convenient for fixing suspended ceilings etc, as mentioned earlier in the chapter. The slab is to be a 105 mm lightweight concrete slab with a characteristic strength of 30 MPa.

Note, the slab depth is generally determined by the insulation requirement for fire resistance (see Section 11.6).

Since the maximum sheet length is 12.5, and the width of the building is 12 m advantage will be taken of being able to design the sheeting and deck as continuous, with 2 spans of 2.5 m either side of a central span of 2.0 m.

Clause 11.3 states that patterned loading must be considered with either all spans fully loaded or with two adjacent spans fully loaded and the remainder under dead load only. The resultant bending moment and shear force diagrams for the three possible load cases are given in Fig. 10.6(a) to (c).

The additional load case of a line load of 2 kN/m run also needs to be considered as the spans are less than 3 m (cl 13.2). The bending moment diagrams for this case are plotted in Fig. 10.7(a) to (c).

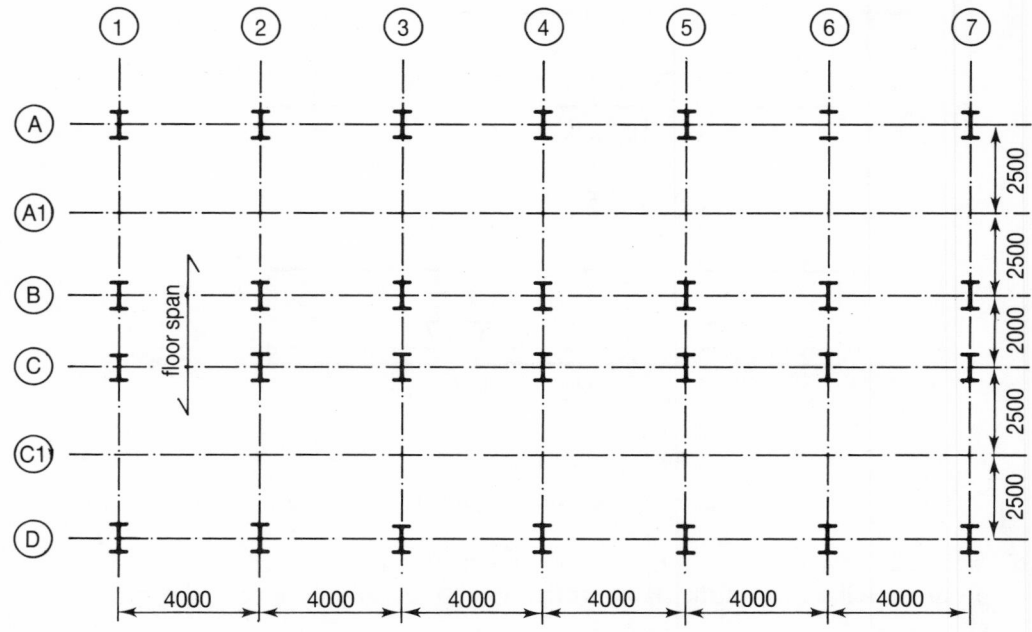

Fig. 10.5   Grid layout for composite deck design.

(a)   Design for the decking as shuttering:

Flexural design.

Construction loading

For the line load case;

Maximum sagging moment $= 1.0$ kNm/m, and

maximum hogging moment $= 0.51$ kNm/m

For the UDL case;

$$
\begin{aligned}
\text{maximum sagging moment} &= 0.438 \times 1.5 \\
&= 0.66 \text{ kNm/m}
\end{aligned}
$$

$$
\begin{aligned}
\text{maximum hogging moment} &= 0.72 \times 1.5 \\
&= 1.08 \text{ kNm/m}
\end{aligned}
$$

Thus design moments due to the construction loading are 1.0 kNm/m (sagging) and 1.08 kNm/m (hogging).

Dead loading

Take specific weight of concrete as 20 kN/m³

UDL due to concrete $= 20 \times 0.105 = 2.1$ kN/m²

Sagging moment $= 0.421 \times 2.1 = 0.88$ kNm/m

hogging moment $= 0.72 \times 2.1 = 1.51$ kNm/m

Total sagging moment $= 1.0 + 0.88 = 1.88$ kNm/m

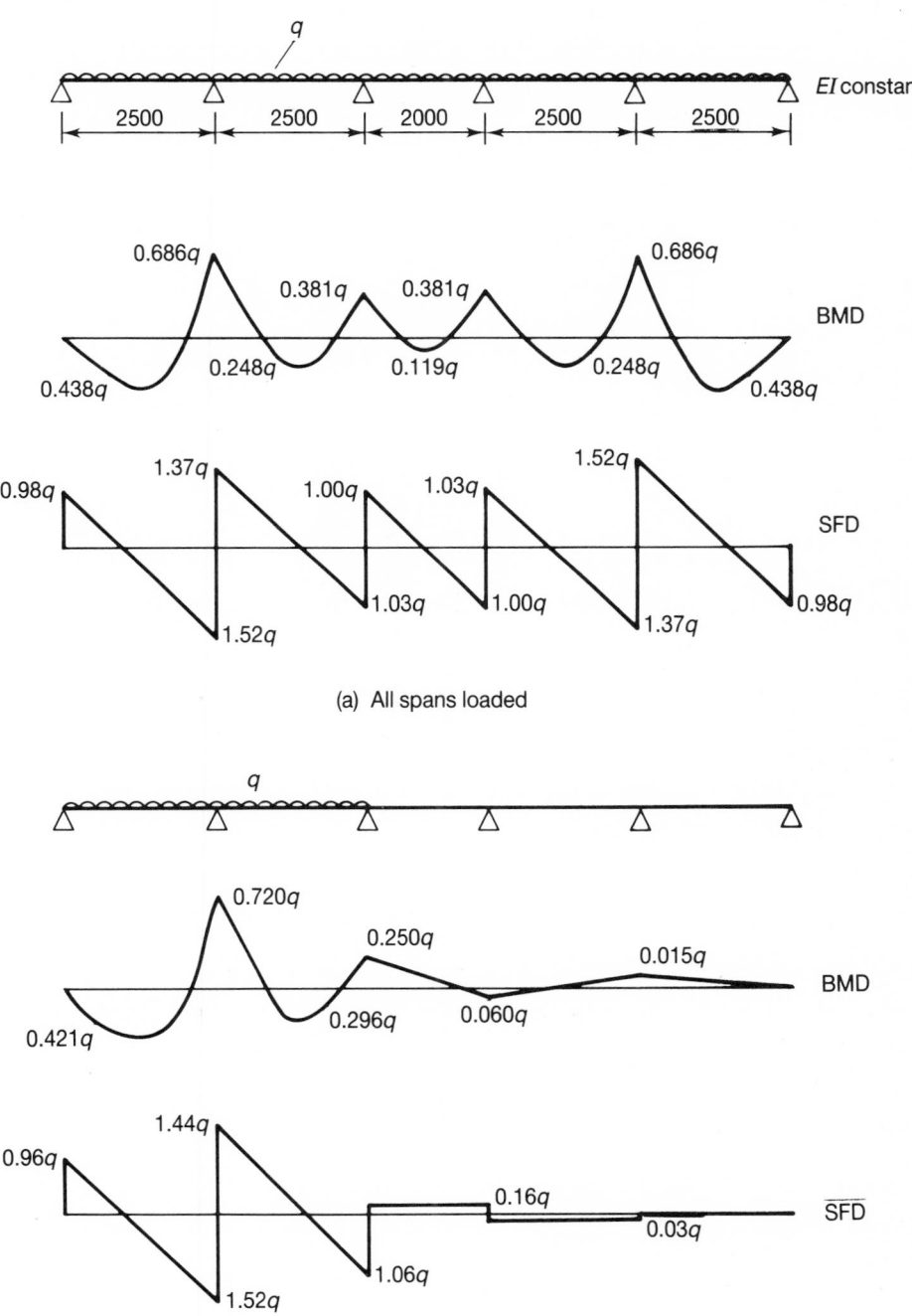

Fig. 10.6 Bending moment and shear force diagrams for the uniformly distributed load case.

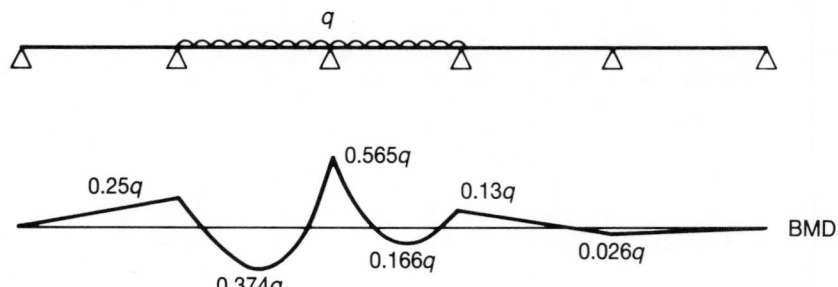

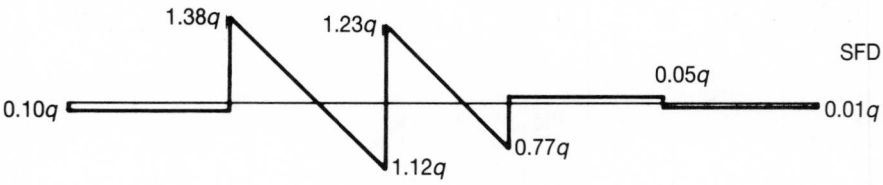

(c) Second and third spans loaded

Fig. 10.6   (c).

From manufacturer's catalogue, service load moment capacity $= 4.51$ kNm/m

Total hogging moment, $M = 1.08 + 1.51 = 2.59$ kNm/m
service load moment capacity, $M_{cp} = 4.31$ kNm/m

Shear design (Maximum reaction).

From line load case,
maximum reaction (unfactored) $= 2.0 + 2.1 \times (1.44 + 1.54)$
$$= 8.26 \text{ kN/m}$$

From UDL

maximum reaction (unfactored) $= (2.1 + 1.5) \times (1.44 + 1.54)$
$$= 10.73 \text{ kN/m}$$

Maximum factored reaction, $W_f = 1.4 \times 10.73$
$$= 15.02 \text{ kN/m}$$

Maximum allowable, $P_w = 45.43$ kN/m

Combined bending and reaction check:

Calculate W_f/P_w

$W_f/P_w = 15.02/45.43$
$$= 0.33$$

Since W_f/P_w is greater than 0.25, Equation (10.1) must be satisfied,

$$M/M_{cp} + 0.64 W_f/P_w \leqslant 1.16 \tag{10.1}$$

$$M/M_{cp} = 2.59/4.13 = 0.63$$

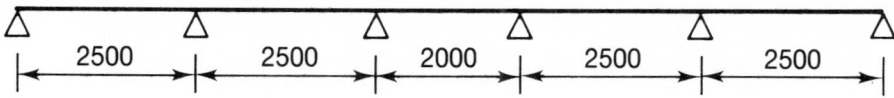

point load 2 kN per metre width of slab

|← 2500 →|← 2500 →|← 2000 →|← 2500 →|← 2500 →|

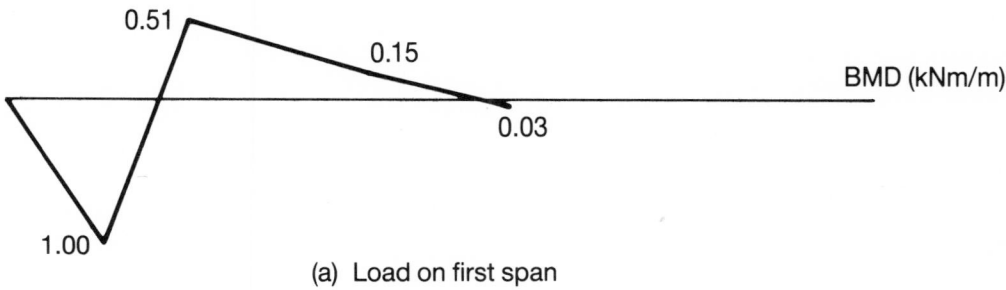

0.51

0.15

BMD (kNm/m)

0.03

1.00

(a) Load on first span

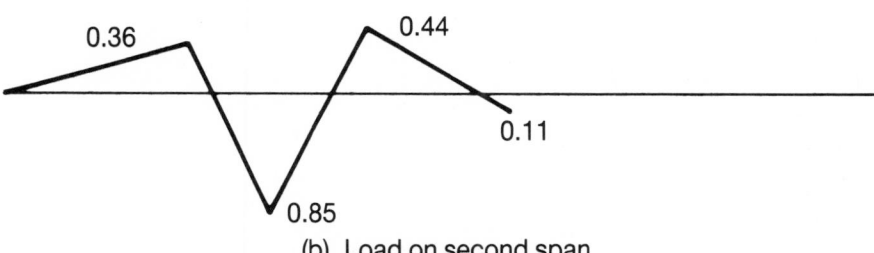

0.36

0.44

0.11

0.85

(b) Load on second span

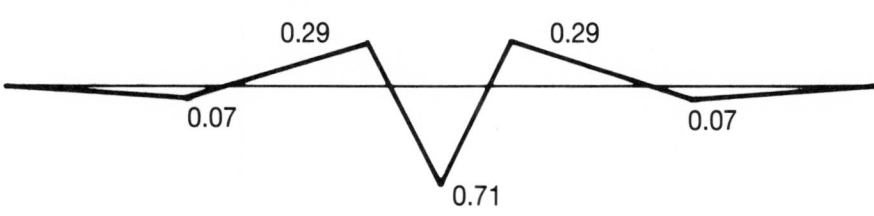

0.29

0.29

0.07

0.07

0.71

(c) Load on third span

Fig. 10.7   Bending moments due to line load.

so substituting values into Equation (10.1)

$$0.63 + 0.64 \times 0.33 = 0.84 < 1.16$$

and is therefore satisfactory

Deflection check:

Conservatively, assume all spans are simply supported for the calculation of deflection.

$E = 210$ GPa (cl 14)

load due to wet concrete, $W_w = 2.1$ kN/m (per metre width of sheet)

$I = 640.4E3$ mm^4 (per metre run of sheet)

$L = 2.5$ m

$$\delta = (5/384)(W_w L^4 / EI)$$
$$= (5 \times 2.1 \times 2.5^4 \times 1E6)/(384 \times 210 \times 640.4E3)$$
$$= 7.9E - 3 \text{ m}$$

Maximum allowable is lesser of

span$/180 = 2.5/180 = 0.014$ m

or

$D_s/10 = 0.105/10 = 0.011$ m

Deflection is satisfactory.

(b)   Design as a composite slab.

Loading (kN/m^2)

Self weight of slab   2.1

Finishes   2.5

Imposed   4.0

Factored dead load $= 1.4 \times (2.1 + 2.5)$
$$= 6.44$$

Factored imposed load $= 1.6 \times 4.0$
$$= 6.4 \text{ kN/m}^2$$

The Code also implies that patterned loading should be used for the design of the composite slab, and does not mention whether any redistribution is allowed following an elastic analysis. Thus use the elastic bending moment diagrams (Fig. 10.7) derived from the construction cases.

Maximum moments:

Sagging;

The largest sagging moment occurs with all spans fully loaded,

maximum moment $= 0.438 \times (6.44 + 6.4)$
$$= 5.63 \text{ kNm/m}$$

Hogging;

This occurs when two spans are fully loaded and the remainder have unfactored dead load only.

This can be simulated using the principle of superposition by using the all spans loaded case for 1.0 times dead load, and the two adjacent spans loaded taking $(1.4 - 1.0)$ times dead load plus 1.6 times the imposed load.

Maximum moment $= 0.686 \times 4.6 + ((1.4 - 1.0) \times 4.6 + 1.6 \times 4.0) \times 0.72$
$$= 9.09 \text{ kNm/m}$$

Maximum shear.

With only a very slight loss in accuracy, this can be taken to occur when all spans are fully loaded.

$$\text{Maximum shear} = 1.52 \times (6.4 + 6.44)$$
$$= 19.52 \text{ kN/m}$$

Flexural design

Sagging (Fig. 10.8)

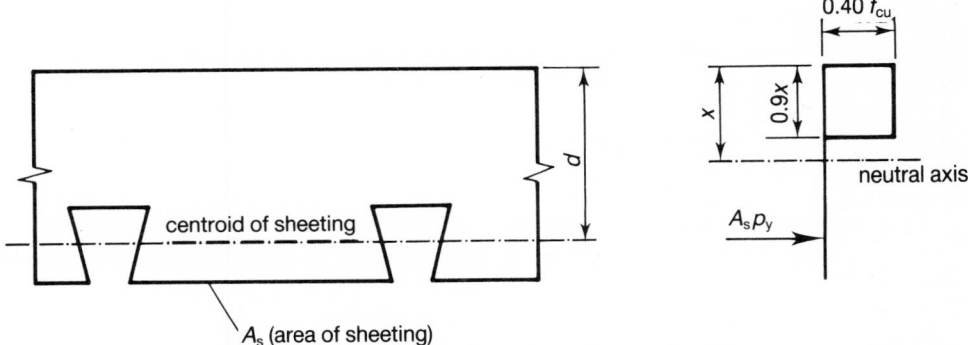

Fig. 10.8   Calculation of sagging moment capacity.

$$0.4f_{cu}b_s 0.9x = p_y A_s$$

$$x = 260.4 \times 1550/(0.4 \times 30 \times 1000 \times 0.9)$$
$$= 37.4 \text{ mm}$$

$$z = d - 0.45x$$
$$= 88.4 - 0.45 \times 37.4$$
$$= 71.6 \text{ mm}$$

$$M_u = A_s p_y z$$
$$= 1550 \times 260.4 \times 71.6/1E6$$
$$= 28.9 \text{ kNm/m}$$

This is higher than the applied moment and is therefore satisfactory.

Hogging

The average overall depth assuming the trapezoidal voids to be replaced by a void of constant depth is given by,

$$D = 105 - (38 + 12) \times (51/2) = 96 \text{ mm}$$

With 25 mm cover and a bar diameter of 10 mm, the effective depth d is given by,

$$d = 96 - 25 - 5 = 66 \text{ mm}$$

Using the IStructE/ICE Concrete design manual,

$$M_{uc} = K' f_{cu} bd^2$$

$$K' = 0.156$$

$$M_{uc} = 0.156 \times 30 \times 1000 \times 66^2/1E6$$
$$= 20.4 \text{ kNm/m}$$

This is higher than the applied moment, so calculate

$$K = M/(bd^2 f_{cu})$$
$$= 9.09E6/(1000 \times 66^2 \times 30)$$
$$= 0.070$$

From Table 24 (IStructE/ICE Manual), $z = 0.91d$

$$A_s = M/(0.87 f_y z)$$
$$= 9.09E6/(0.87 \times 460 \times 0.91 \times 66)$$
$$= 378 \text{ mm}^2/\text{m}$$

Fix B385 mesh with an area of 385 mm²/m.

This is higher than the minimum requirement of $0.1\% A_c = 0.1 \times 105 \times 1000/100$ $= 105 \text{ mm}^2/\text{m}$.

The final amount of reinforcement fixed will also be dependent on the transverse reinforcement required to stop shear failure at the beam–slab interface (Example 10.2) and on the reinforcement required for the fire performance of the slab (Example 11.5).

Flexural Shear Capacity.

The shear capacity V_v is given by

$$V_v = (b_s d)(0.8 \times 0.79(400/d)^{0.25}(100A_p/(B_s d))^{0.33}(f_{cu}/25)^{0.33}/1.25)$$

$$b_s = 1000 - 3 \times (1000/152.5)$$
$$= 750.8 \text{ mm}$$

$$V_v = (750.8 \times 88.4)(0.8 \times 0.79 \times (400/88.4)^{0.25}(100 \times 1550/(1000 \times 88.4))^{0.33}$$
$$\times (30/25)^{0.33}/1.25)/1E3$$
$$= 62.7 \text{ kN/m}$$

This is in excess of the applied shear of 19.52 kN/m.

Shear bond

The shear bond capacity V_s is given by

$$V_s = B_s d_s (m_r A_p/B_s L_v + k_r(f_{cu})^{0.5})/1.25 \qquad (10.2)$$

From manufacturer's catalogue, $k_r = -0.0037$ and $m_r = 217.70$
for a UDL, $L_v = L/4 = 625$ mm

$$V_s = 1000 \times 105 \times ((217.71 \times 1550/1000 \times 625) - 0.0037 \times 30^{0.5})/1.25$$
$$= 43600 \text{ N/m}$$
$$= 43.6 \text{ kN/m}$$

This is in excess of the applied shear and is therefore satisfactory.

Deflection.

To calculate the second moments of area required for deflection calculations, the second moment of area of the profiled sheeting about its own centroidal axis has been ignored as it is negligible compared with the remaining terms in the calculation.

The equations used for calculating the required second moments of area are taken from Martin *et al.* (Chapter 4, Equations (4.5) to (4.9)).

Uncracked, gross second moment of area.

$$I_G/bd^3 = (x/d)^3/3 + (h/d - x/d)^3/3 + \rho\alpha_e(1 - x/d)^2$$

with

$$x/d = ((h/d)^2/2 + \rho\alpha_e)/((h/d) + \rho\alpha_e)$$

$$d = 88.4$$

$$h = 105$$

$$h/d = 105/88.4 = 1.188$$

$$\rho\alpha_e = (1550/(88.4 \times 1000)) \times 20 = 0.35$$

$$x/d = (1.188^2/2 + 0.35)/(1.188 + 0.35)$$
$$= 0.686$$

$$I_G/bd^3 = 0.686^3/3 + (1.188 - 0.686)^3/3 + 0.35 \times (1 - 0.686)^2$$
$$= 0.184$$

$$I_G = 0.184 \times 1000 \times 88.4^3$$
$$= 0.127E9 \text{ mm}^4$$

Cracked second moment of area:

$$I_T = (x/d)^3/3 + \rho\alpha_e(1 - x/d)^2$$

with,

$$x/d = -\rho\alpha_e + ((\rho\alpha_e)^2 + 2\rho\alpha_e)^{0.5}$$
$$= -0.35 + (0.35^2 + 2 \times 0.35)^{0.5}$$
$$= 0.557$$

$$I_T/bd^3 = 0.557^3/3 + 0.35 \times (1 - 0.557)^2$$
$$= 0.126$$

$$I_T = 0.126 \times 1000 \times 88.4^3$$
$$= 0.087E9 \text{ mm}^4$$

$$I_{NA} = (I_G + I_T)/2$$
$$= (0.127E9 + 0.087E9)/2$$
$$= 0.107E9 \text{ mm}^4$$

Note, this second moment of area is in 'concrete' units, and thus the E value used to calculate deflections must be E_s/α_e.

As in the deflection of the sheeting, it will be assumed that the spans are simply supported.

Service load deflection:

$$q = 4 \text{ kN/m (per metre run of deck)}$$

$$I_{NA} = 0.107\text{E9 mm}^4 \text{ (per metre run of deck)}$$

$$E = 210/20 = 10.5 \text{ GPa}$$

$$L = 2.5 \text{ m}$$

$$\delta = (5/384)(4 \times 2.5^4 \times 1\text{E6})/(10.5 \times 0.107\text{E9})$$
$$= 1.8\text{E} - 3 \text{ m}$$

$$\text{allowable deflection} = \text{span}/350$$
$$= 2.5/350$$
$$= 7.1\text{E} - 3 \text{ m}$$

Total deflection check:

Since no account has been taken of the duration of loading in calculating the second moments of area, the total deflection minus the deflection due to the slab self weight reduces to a calculation of the deflection due to service imposed load plus finishes. There is no deflection in this example due to propping.

Imposed load plus finishes = $4.0 + 2.5 = 6.5$ kN/m (per m run of slab)

Thus the deflection is given by,

$$\delta = (6.5/4.0) \times 1.8$$
$$= 3.0 \text{ mm}$$

$$\text{allowable deflection} = \text{span}/250$$
$$= 2500/250$$
$$= 10 \text{ mm}$$

Both deflection checks are satisfactory.

This completes the design of the slab, except for the calculations required to assess its fire performance. These will be found as Example 11.4.

10.3   Composite beams

10.3.1   Introduction

As was pointed out in the general introduction to this chapter, composite action between the slab and beam is generally achieved by the use of stud shear connectors. Alternative forms of shear connectors such as hoops or channel sections are not commonly used in building construction and thus will not be considered further. Also only simply supported composite beams with profiled steel decking giving composite slab action will be considered. This restriction concerning the use of profiled decking has little effect on the basic design approach except to remove the need for certain checks on tranverse shear. Currently in UK practice most frames are designed using simple construction and continuous beams with regions of hogging moments are little used. Thus the restriction to simply supported beams only is reasonable, although it will be necessary to consider cantilevers. Reference should be made to Johnson for details on the design of continuous members. The relevant Part of BS 5950 is Part 3 Design in Complete Construction, Section 3.1 Code of Practice for Design of simple and continuous beams (referred to in this text as Pt 3.1). Section 3.2 dealing with columns and frames is still to be published.

For full composite action there should be no slip at the interface between the beam and the composite slab. However any shear connector system has a finite stiffness and thus any composite beam has a slight degree of slip, but it was found that this small degree of slip had little effect on either the beam deflection or its ultimate load behaviour (Yam and Chapman). This led to the concept of designing the shear connectors in composite beams to act at 80% of their design strength. A drawback with full interaction (which assumes the connectors are at 80% design strength) is that there may be excessive requirements for transverse reinforcement to resist the transverse shear effects and that the shear connectors may be very closely spaced. In order to overcome these drawbacks, the concept of partial interaction was developed in which the number of shear connectors is reduced, and hence also the transverse reinforcement requirements (see Section 10.3.3.4). The effect of partial interaction is to increase deflections and to possibly increase the size of the steel beam required since the magnitude of the force that can be transmitted across the concrete–steel interface is reduced and therefore the ultimate moment of resistance is also reduced (Johnson and May).

10.3.2 Section classification

The limits for composite construction are more onerous than those for plain steelwork since the ductility of a composite beam is closer to that of reinforced concrete than plain steelwork. It is necessary to make sure that there will be sufficient ductility to ensure that plastic hinges can occur and that the resultant redistribution of moments can take place as those hinges occur. Clearly this is more important for continuous construction than for simple construction. Some further discussion on the background to this is given in Johnson.

It is therefore recommended that only sections which comply with the criteria for plastic classification of flanges (Class 1) and Class 1 webs should be used.

Where the compression flange of the beam is restrained by effective attachment to a concrete slab by shear connectors it may be assumed to be 'plastic' (cl 4.5.2, Pt 3.1).

The web slenderness ratio must be determined with allowance made for the position of the neutral axis. In areas of hogging moment the reinforcement is to be included in calculating the neutral axis position. For webs to be considered 'plastic', the d/t ratio should not exceed $64\varepsilon/(1 + r)$, where $\varepsilon = (275/p_y)^{0.5}$ and $r = (y_c - y_t)/d < 1.0$, where y_c is the depth of the web in compression, y_t the depth of the web in tension and d the clear depth of the web. Note, for a neutral axis at mid-depth of the web $r = 0$ and the limiting d/t value is 64ε.

10.3.3 Design criteria

Besides the customary checks for bending, vertical (or flexural) shear and deflection, there are additional checks to be made for transverse shear and the shear connector design. It will be necessary to distinguish between cases where the profile sheeting spans at right angles to the support beam and where the profile decking runs parallel to the support beams.

10.3.3.1 *Flexural design*

A distinction must be made between propped and unpropped construction.

In propped construction, the beam is propped during the pouring of the concrete and thus the load from the self weight of the slab does not act until the removal of the props and, moreover, it will act on the composite section. This also has the effect of reducing the overall deflection.

In unpropped construction the self weight of the wet concrete acts on the steel beam section alone, and it is therefore essential that the steel beam remains elastic during the construction stage. Since the self weight of the slab is taken on the steel beam alone high deflections due to dead load may result, and it may be advisable to precamber the beam to negate these deflections. It is usually only necessary to precamber an amount equal to about 2/3 to 3/4 of the calculated deflection as no account will have been taken in the calculations of the stiffening effect of either the profile decking or the connections. During the construction stage an allowance of 0.5 kN/m^2 (cl 2.2.3, Pt 3.1) should be added to the self weight of the beam as a construction load. This is less than the construction load applied during decking design (Section 10.2.2) as it is recognized that local effects will be less severe on the beam. A beam with profile decking spanning perpendicular to the beam may be considered as being fully laterally restrained during the construction process, whereas a beam with the decking running parallel must be checked for lateral torsional buckling (Lawson and Nethercot) (cl 2.3.2, Pt 3.1).

At ultimate conditions, i.e. when the full imposed loading is applied to the slab, the behaviour is the same whether propped or unpropped construction has been used. Owing to the stress distribution across the flange of a composite beam, it is not possible to utilize the whole width of the flange of the slab to resist compression. This is analogous to the behaviour of 'T' beams in reinforced concrete where an effective width which is assumed to be under constant stress is defined. For composite construction, the effective breadth B_e is to be taken as the sum of the effective breadths b_e of the portions either side of the beam centre-line (cl 4.6, Pt 3.1).

For a slab spanning perpendicular to the beam:
for a simply supported beam:

$$b_e = L/8 < b$$

for a cantilever:

$$b_e = 1.5L/8 < b$$

For a slab spanning parallel to the beam:
Simply supported beam:

$$b_e = L/8 < 0.8b$$

for a cantilever:

$$b_e = 1.5L/8 < 0.8b$$

where L is the effective span of the beam or cantilever, and b is the actual breadth of the half slab, taken as half the centre-line distance between the beams, except that at a free edge the distance b is measured from the beam centre-line to the free edge. Note, these values appear higher than those for reinforced concrete 'T' beams until it is remembered that the values of span/10 are for the outstand of the flange measured from the face of the stem of the 'T'.

To calculate the flexural capacity at ultimate limit state of a composite section the following assumptions are made (cl 4.4.1 and 2, Pt 3.1).

1 For a slab with ribs running perpendicular to the beam only the concrete above the ribs should be considered. For slabs with ribs running parallel to the beam the full cross section of concrete should be taken.
2 Concrete in compression should be taken as being uniformly stressed to $0.45f_{cu}$, whereas concrete in tension should be ignored.
3 Any contribution by the profiled steel decking is ignored. The reinforcement is assumed to be acting at its design strength of $0.87f_y$ and the steel section at its design strength p_y but not greater than 355 MPa (cl 3.1, Pt 3.1). This latter limit is due to there being little experimental data on the behaviour of composite beams with Grade 55 steel (Johnson). Any welded mesh reinforcement or reinforcement of diameter less than 10 mm is ignored in the effective section.

Flexural Capacity.

Sagging (or positive moments)

All symbols are defined in Fig. 10.9(a).

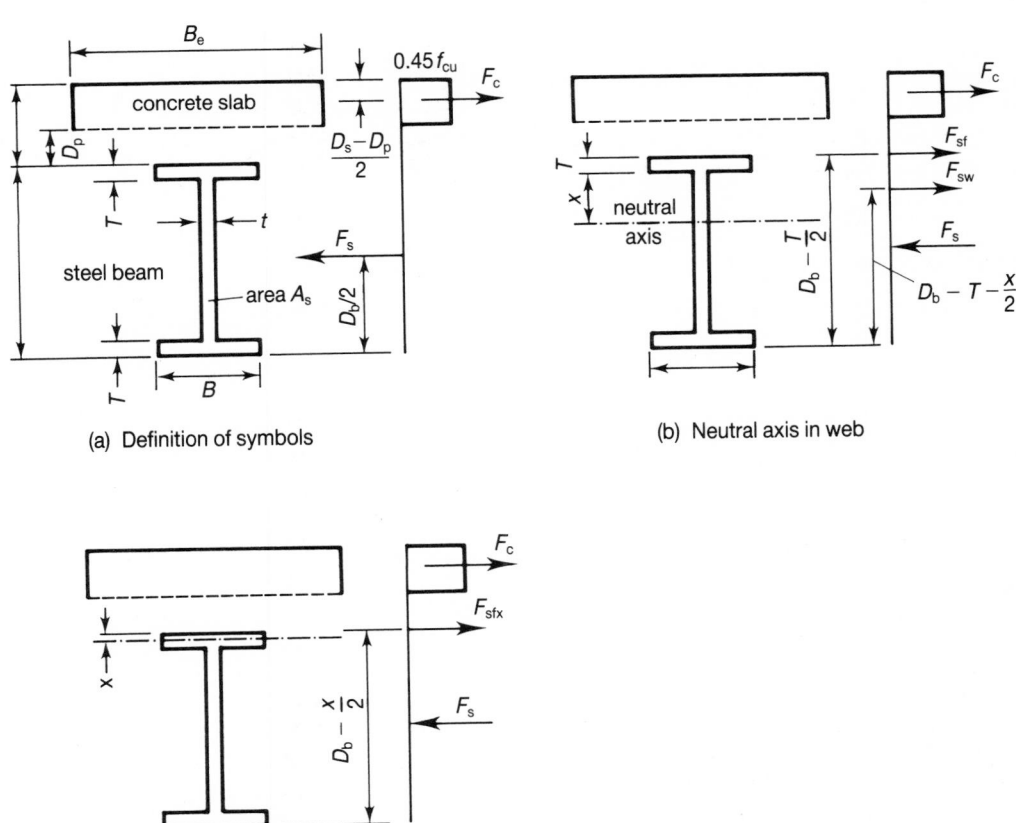

(a) Definition of symbols

(b) Neutral axis in web

(c) Neutral axis in top flange

Fig. 10.9 Calculation of plastic moment capacity for a composite beam in the sagging region.

The force in the concrete F_c is given by

$$F_c = 0.45 f_{cu} B_e (D_s - D_p) \tag{10.4}$$

The force in the steel beam F_s is given by

$$F_s = A_s p_y \tag{10.5}$$

In general the plastic neutral axis will lie in the steel beam, i.e. $F_s > F_c$.
The case where the neutral axis lies in the slab will not be considered further.

With the neutral axis in the beam, either of two conditions will exist, i.e. the neutral axis may be in the top flange or the web.
The difference between capacities of the steel beam and the concrete flange F_{sc} is given by

$$F_{sc} = F_s - F_c \tag{10.6}$$

The capacity of the beam flange ignoring fillets F_{sf} is given by

$$F_{sf} = BT p_y \tag{10.7}$$

If $F_{sc} > 2F_{sf}$, the neutral axis is in the web, or
if $F_{sc} < 2F_{sf}$, the neutral axis is in the flange.

Case 1–neutral axis in the web (Fig. 10.9(b)).
Depth of web x in compression is given by

$$x = (F_{sc} - 2F_{sf})/2t p_y \tag{10.8}$$

The compression force in the web F_{sw} is given by

$$F_{sw} = x t p_y \tag{10.9}$$

The ultimate moment of resistance of the section M_c is given by,

$$M_c = F_c(D_b + (D_p + D_s)/2) + 2F_{sf}(D_b - T/2) + 2F_{sw}(D_b - T - x/2) - F_s D_b/2 \tag{10.10}$$

Case 2–neutral axis in the flange (Fig. 10.9(c))
Depth of flange in compression x is given by

$$x = F_{sc}/2B p_y \tag{10.11}$$

The force F_{sfx} in the flange is given by

$$F_{sfx} = x B p_y \tag{10.12}$$

The ultimate moment of resistance of the section M_c is given by,

$$M_c = F_c(D_b + (D_p + D_s)/2) + 2F_{sfx}(D_b - x/2) - F_s D_b/2 \tag{10.13}$$

Hogging (negative moment)
All symbols are defined in Fig. 10.10(a).

The force in the reinforcement F_r is given by

$$F_r = 0.87 f_y A_{st} \tag{10.14}$$

The force in the steel F_s is given by

$$F_s = A_s p_y \tag{10.5}$$

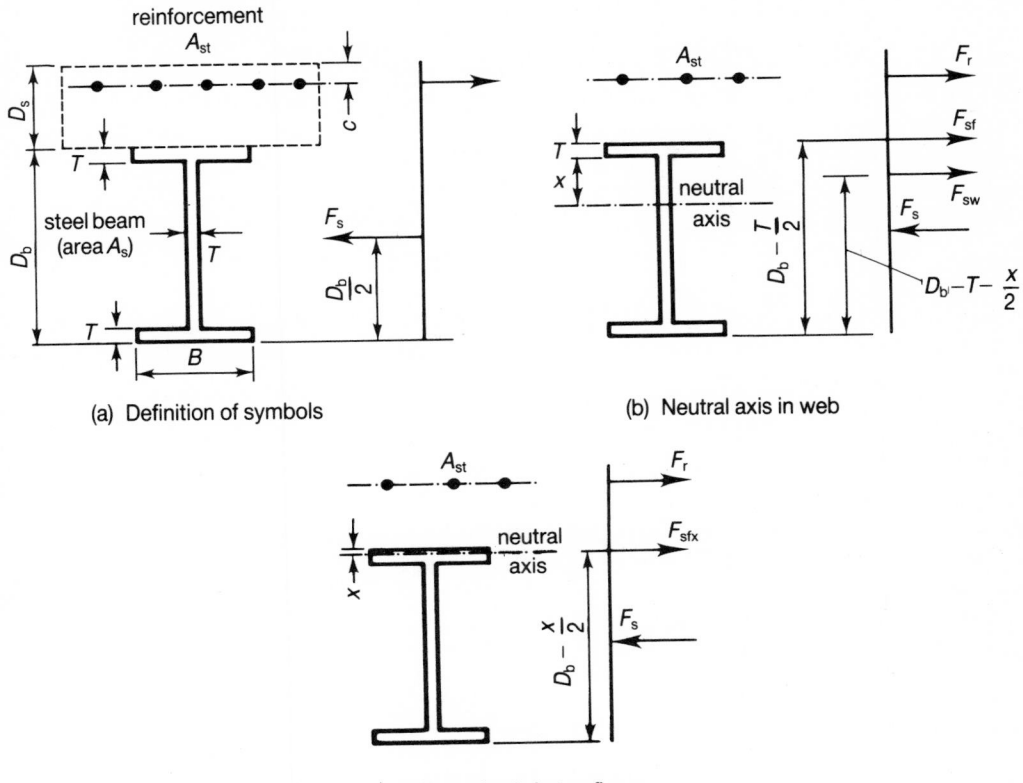

(a) Definition of symbols

(b) Neutral axis in web

(c) Neutral axis in top flange

Fig. 10.10 Calculation of the plastic moment capacity for a composite beam in the hogging region.

The force in a flange F_{sf} is given by

$$F_{sf} = BTp_y \tag{10.7}$$

If $F_s > F_r + 2F_{sf}$ the neutral axis is in the web, or
if $F_s < F_r + 2F_{sf}$ the neutral axis is in the flange

Case 1–Neutral axis in web (Fig. 10.10(b))
The depth of web x in tension is given by

$$x = (F_s - F_r - 2F_{sf})/2tp_y \tag{10.15}$$

The capacity of the web in tension F_{sw} is given by

$$F_{sw} = xtp_y \tag{10.9}$$

The ultimate moment capacity M_c of the composite section is given by

$$M_c = F_r(D_b + D_s - c) + F_{sf}(D_b - T/2) + F_{sw}(D_b - T - x/2) - F_s D_b/2 \tag{10.16}$$

Case 2–Neutral axis in flange (Fig. 10.10(c))
Depth of flange x in tension is given by

$$x = (F_s - F_r)/2Bp_y \tag{10.17}$$

Force in the flange in tension F_{sfx} is given by

$$F_{sfx} = xBp_y \tag{10.12}$$

The ultimate loading moment capacity M_c of the composite section is given by

$$M_c = F_r(D_b + D_s - c) + F_{sfx}(D_b - x/2) - F_s D_b/2 \tag{10.18}$$

Note that when the applied shear exceeds 0.5 times the shear capacity of the section, the moment capacity should be reduced (cl 5.3.4, Pt 3.1).

10.3.3.2 *Flexural shear*

This is taken on the web of the beam and is calculated exactly the same as for normal rolled sections used non-compositely.

10.3.3.3 *Design of shear connectors*

The design of shear connectors will initially be covered for solid slabs and then the relevant modifications presented to deal with ribbed slabs.

The basic strength of stud shear connectors is determined by the use of pull-out tests. The basic test set up is illustrated in Fig. 10.11 (BS 5400 Part 5). It is to be noted that bending of the shear connector is avoided by testing in double shear.

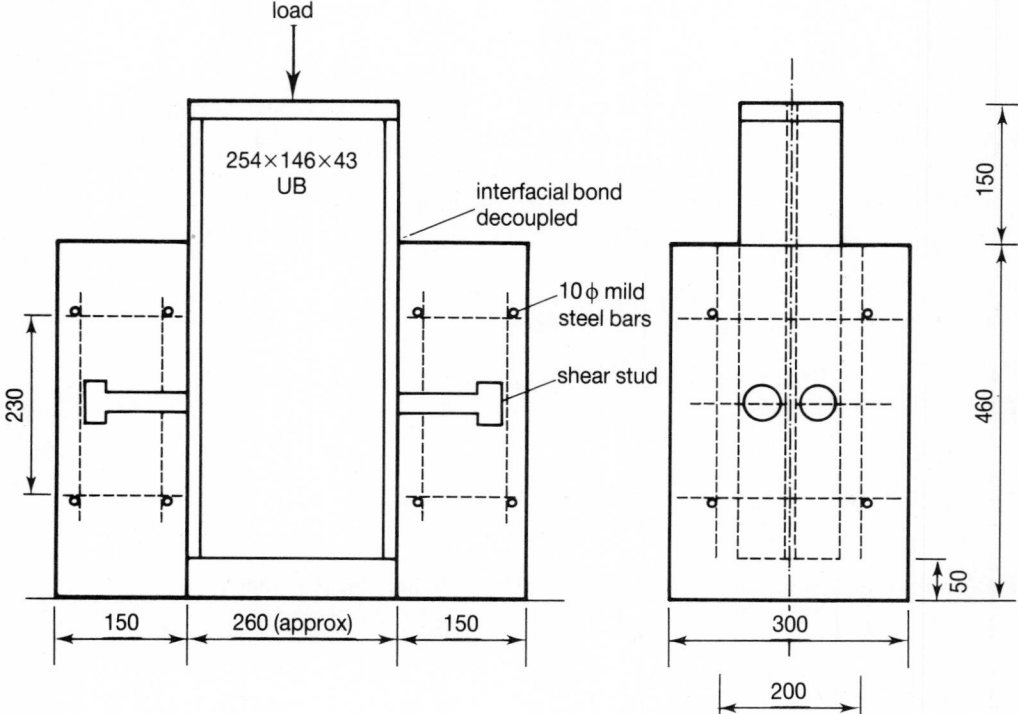

Fig. 10.11 Typical test set up for determining stud connector strength.

The results of such tests indicate that the strength P_d of the shear connection is proportional to the square root of the concrete strength f_{cu} and the square of the stud diameter ϕ, i.e.

$$P_d = k\phi^2(f_{cu}E_c)^{0.5}/1.25 \tag{10.19a}$$

where E_c is the short term elastic modulus of concrete, and k is an empirical constant and has a value of 0.32 for $h/d \geqslant 4.2$, and 0.25 for $h/d = 3.0$ (intermediate values may be interpolated), h is the height of the stud, and the 1.25 factor is a partial safety factor (Olgaard *et al.*).

However a further analysis by Oehlers and Johnson (1987) of the test data from Olgaard *et al.*, together with additional tests, gives the strength of a shear connector P_p as

$$P_p = 4.1f_u A(E_c/E_s)^{0.4}(f_{cu}/f_u)^{0.35} \tag{10.19b}$$

where $f_u A$ is the tensile capacity of the shear stud. This equation (10.19(b)) is the basis of the characteristic resistances for stud connectors Q_k for normal-weight concrete, given in Table 5. For lightweight concrete these values should be multiplied by 0.9 (cl 5.4.6). This is in recognition that lightweight concrete will crush more easily than normal-weight concrete.

The design shear capacities are given by:

in areas of sagging (or positive) moments

$$Q_p = 0.8Q_k \tag{10.20a}$$

in areas of hogging (or negative) moments

$$Q_n = 0.6Q \tag{10.20b}$$

The 0.8 factor in Equation (10.20(a)) follows work by Yam and Chapman (referred to earlier) which indicated that full interaction could be obtained with the shear connectors at 80% capacity. The factor of 0.6 in Equation (10.20(b)) indicates that shear connectors are less efficient in areas of hogging moments as the concrete is in tension and is likely to be cracked (Johnson *et al.*).

Provision of shear connectors.

Where the loading is sensibly uniform the spacing of shear connectors may be taken as constant over the span of the beam.

In areas of sagging moments, the number of connectors N_p is given by

$$N_p = F_p/Q_p \tag{10.21}$$

where the force F_p is given by, for full interaction, the lesser of F_s or F_c as given by Equations (10.5) and (10.4).

For partial interaction the number of shear connectors may be reduced from that calculated by Equation (10.21), such that the number of shear connectors actually provided, N_a, satisfies the following relationship,

$$N_a/N_p \geqslant (L-6)/10 \geqslant 0.4 \tag{10.22}$$

where L is the span (cl 5.5.2, Pt 3.1).

Partial interaction may only be used where the beam is of uniform section, the shear connection is in the form of stud connectors and the span is less than 16 m. The moment capacity of the section should be calculated with the concrete force F_c replaced by $N_a Q_p$, the force transmitted by the shear connectors.

In areas of hogging (or negative moments) the number of connectors N_n is given by

$$N_n = F_n/Q_n \tag{10.23}$$

where $F_n = 0.87 f_y A_{st}$.
Note, partial interaction cannot be used in areas of hogging moments.

It is important to note that the number of shear connectors N_p or N_n, calculated from Equations (10.21) and (10.23) is for the region EITHER side of the point of maximum moment.

Detailing Requirements:

These fall into three categories–dimensional details, additional checks, and additional requirements for trough decking.

1   Dimensional details (cl 5.4.8, Pt 3.1)

Maximum spacing: the lesser of 600 or $4D_s$

Minimum spacing: $6d$ along the beam and $4d$ across the beam, where d is the stud diameter.

Maximum stud diameter: Unless placed directly over the web, $d < 2.5T$

Mimimum edge distance: 20 mm.

2   Additional checks (cl 5.4.5.2, Pt 3.1)

These are necessary where there are heavy point loads or exceptionally large concrete flanges.

A heavy point load is defined as one whose free moment calculated on the assumption of a simply supported span exceeds 10% of the moment capacity of the composite section. The additional checks need to be carried out between such loads and the adjacent support.

A large concrete flange is defined as one where the plastic moment capacity of the composite section exceeds 2.5 times that of the steel section alone. In this case the additional check should be carried out midway between the points of maximum moment and the supports.

The check takes the form of a calculation of the number of shear connectors N_i between the point of check and the adjacent support.

In regions of positive moment:

$$N_i = N_p(M - M_s)/(M_c - M_s) + N_n \geqslant N_n \tag{10.24}$$

in regions of negative moment:

$$N_i = N_n(M_c - M)/(M_c - M_s) \leqslant N_m \tag{10.25}$$

where M_c is the moment capacity of the composite section calculated for negative or positive action as appropriate, M is the moment at the intermediate point and M_s the moment capacity of the steel section.

3   Additional checks for ribbed composite slabs (cl 5.4.7, Pt 3.1)

The following section only applies if

(a)   the overall depth of the profile steel decking is not greater than 80 mm or less than 35 mm
(b)   the mean width of the troughs is not less than 50 mm
(c)   nominal diameter of the studs is not greater than 19 mm
(d)   the height of the studs is at least 35 mm greater than the overall depth of the profiled sheeting.

The basic strength of the shear studs should be multiplied by a reduction factor k which is given by:

for ribs perpendicular to the beam

$$k = (0.85/N^{0.5})(b_r/D_p)((h/D_p) - 1) \leqslant 1.0 \tag{10.26}$$

for ribs parallel to the beam

if $b_r/D_p \geqslant 1.5$

$$k = 1.0 \tag{10.27a}$$

if $b_r/D_p < 1.5$

$$k = 0.6(b_r/D_p)((h/D_p) - 1) \tag{10.27b}$$

where b_r is the breadth of the rib, D_p the overall depth of the profile sheet, h the overall depth of the stud (taken for the purpose of calculating k as not greater than $2D_p$ or $D_p + 75$), and N the number of studs per rib on the decking (taken for the purpose of calculating k as not greater than 3).

The breadth of the concrete rib b_r is taken as the mean width of the trough for open profile sections or the minimum width of the trough for re-entrant sections provided that in either case the stud is placed centrally in the rib. Where the stud is eccentric, b_r is limited to $2e$, where e is the least distance from the stud to the edge of the trough, but not less than 25 mm.

This reduction factor is necessary to compensate for the loss of concrete adjacent to the base of the stud. Since the maximum bearing stress is adjacent to the base of the stud a loss of strength will result. The reduction factor k was derived empirically (Grant *et al.*). It is suggested by Mottram and Johnson (1990) that Equation (10.26) should be replaced by an equation proposed by Lawson,

$$k = (0.75r/N^{0.5})(h/(h + D_p))$$

where r is the lesser of b_r/D_p or 2.0 when $e \geqslant b_r/2$
or
the least of b_r/D_p, $(e/D_p) + 1$, and 2.0 when $e < b_r/2$

10.3.3.4   *Transverse shear (cl 5.6, Pt 3.1)*

The transverse shear arises from the transfer of longitudinal shear across the interface between the steel beam and the concrete slab. It is therefore necessary to check that the concrete slab can resist this shear, which is partly resisted by dowel action of the reinforcement, partly by aggregate interlock, and for composite slabs with profile sheet steel decking, the steel decking itself. Since the concrete in the zone above the beam will be cracked in a direction transverse to the beam by the action of the loading on the slab, it is therefore necessary to provide an estimate of the amount of shear that may safely be carried by this cracked zone. Where the slab spanning between the beams has been designed as simply supported, these cracks are likely to be large, even though some nominal reinforcement is required (see Section 10.2.3), and thus will have very little carrying capacity. For continuous slabs, the capacity will be increased.

It was shown by Mattock and Hawkins that the maximum shear v_u (in MPa) that could be carried across a pre-existing crack was given by,

$$v_u = 1.4 + 0.8pf_y \qquad (10.28)$$

where p is the ratio of reinforcement to concrete area.

Applying material partial safety factors of 1.5 to the first term, which represents the contribution of the concrete, and replacing the 0.9 parameter which results by $0.03f_{cu}$, since Equation (10.28) was derived for a 30 MPa concrete, and a factor of 1.15 to the second term due to the reinforcement gives,

$$v_{u'} = 0.03f_{cu} + 0.7pf_y \qquad (10.29)$$

Multiplying both sides of Equation (10.28) by the concrete area A_{cv} gives

$$v_r = 0.7A_{sv}f_y + 0.03A_{cv}f_{cu} \qquad (10.30)$$

In addition it has been recognized that the profile steel sheeting will also contribute to the shear strength, giving an additional term v_p and the concrete term is modified by a factor η to allow for the use of lightweight concrete. Thus Equation (10.30) becomes

$$v_r = 0.7A_{sv}f_y + 0.03\eta A_{cv}f_{cu} + v_p \leqslant 0.8\eta A_{cv}(f_{cu})^{0.5} + v_p \qquad (10.31)$$

where f_{cu} is the cube strength of the concrete (taken as not greater than 40 MPa), f_y the reinforcement strength, A_{sv} the cross-sectional area of reinforcement crossing the shear surface, A_{cv} the mean cross-sectional area of concrete, and η is taken as 1.0 for normal-weight concrete and 0.8 for lightweight concrete. The upper limit is taken as the sum of the ultimate shear on the concrete calculated in accordance with BS 8110 and the contribution of the decking (v_p).

The critical planes that need checking for transverse shear are illustrated in Fig. 10.12.

The contribution of the profile sheet steel decking (cl 5.6.4, Pt 3.1), v_p, is calculated as follows:

For ribs running perpendicular to the beam and the sheeting continuous across the beam,

$$v_p = t_p p_{yp} \qquad (10.32)$$

where t_p and p_{yp} are the thickness and design strength of the sheet, respectively.

For discontinuous sheeting running in either direction,

$$v_p = (N/s)(ndt_p p_{yp}) \leqslant t_p p_{yp} \qquad (10.33)$$

where s is the longitudinal spacing of the shear connectors, d the diameter of the shear connectors, N the number of connectors in a group, and n takes the value 4 unless a higher value can be justified from tests.

For ribbed slabs the concrete area A_{cv} should be calculated taking account of the ribs. Where the ribs run perpendicular to the span of the beam the concrete between the ribs is included in the calculation of A_{cv}.

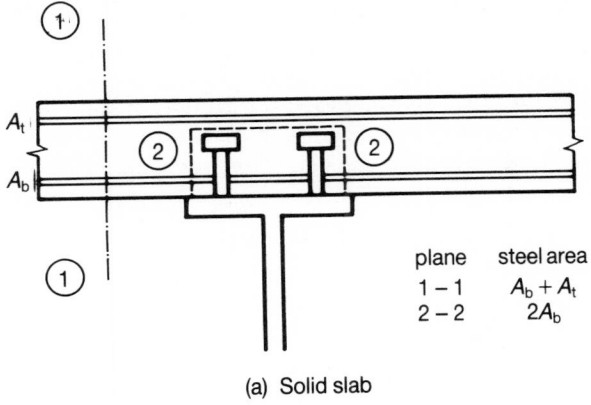

plane	steel area
1 – 1	$A_b + A_t$
2 – 2	$2A_b$

(a) Solid slab

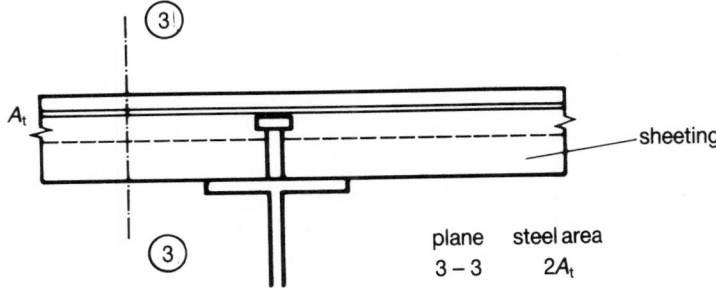

plane	steel area
3 – 3	$2A_t$

(b) Composite slab (sheeting perpendicular to span)

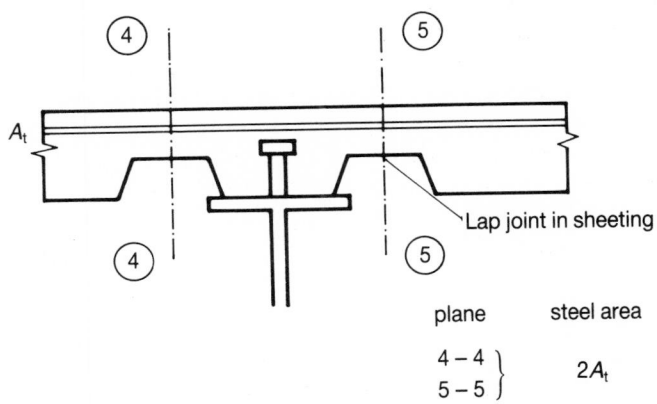

plane	steel area
4 – 4 ⎫ 5 – 5 ⎭	$2A_t$

(c) Composite slab (sheeting parallel to span)

Fig. 10.12 Critical transverse shear planes.

The transverse shear to be resisted v is given by

$$v = NQ/s \qquad (10.34)$$

where Q is taken as Q_p or Q_n (Section 10.3.3.3) as appropriate, s is the spacing of the shear connectors, and N the number of shear connectors in a group.

It is the total transverse shear which is calculated in Equation (10.34), and it may be reduced by

(a)   dividing the value obtained by the number of shear planes on which the transverse shear acts (Fig. 10.13), and
(b)   in areas of positive moments only, multiplying the value of v acting on the shear plane being checked by $(1 -$ (the ratio of the area contained withing the shear plane or planes being checked, to the area of the concrete contained within the effective width of the slab)).

10.3.3.5   *Serviceability checks*

(a)   Deflection (cl 6.1, Pt 3.1)

Deflection calculations should be based on the gross uncracked second moment of area of the composite section except for dead load in unpropped construction which should be taken on the steel section alone, and for cantilevers where the second moment of area is based on the steel beam and the reinforcement only.

The modular ratios to be used are given in Table 10.1 (after Table 1 of BS 5950, Pt 3). The fourth column is an effective modular ratio given by Equation (10.35)

$$\alpha_e = \alpha_s + \rho_L(\alpha_L - \alpha_s) \qquad (10.35)$$

where α_s is the short term modular ratio, α_L the long term ratio, and ρ_L the proportion of loading to be taken as long term.

For the purpose of calculation, one third of floor loading should be taken as long term (cl 4.1). This value has been used for the final column of Table 10.1. Storage loads and permanent loads should be taken as long term, and roof loads and wind loads as short term.

Table 10.1   Modular ratios

Type of concrete	Short term	Long term	Effective (1/3 long term)
Normal-weight	6	18	10
Lightweight	10	25	15

The gross second moment of area I_c can be calculated as follows for simply supported beams:

All symbols are defined in Fig. 10.13

$$I_c = \alpha_e I_s + \alpha_e A_s(D_b/2 - x)^2 + B_e(d_s)^3/12 + B_e d_s(D_b + d + d_s/2)^2 \qquad (10.36)$$

where

$$x = (\alpha_e A_s D_b/2 + B_e d_s(D_b + d + d_s))/(\alpha_e A_s + B_e d_s) \qquad (10.37)$$

The deflection limit should be taken as span/200.

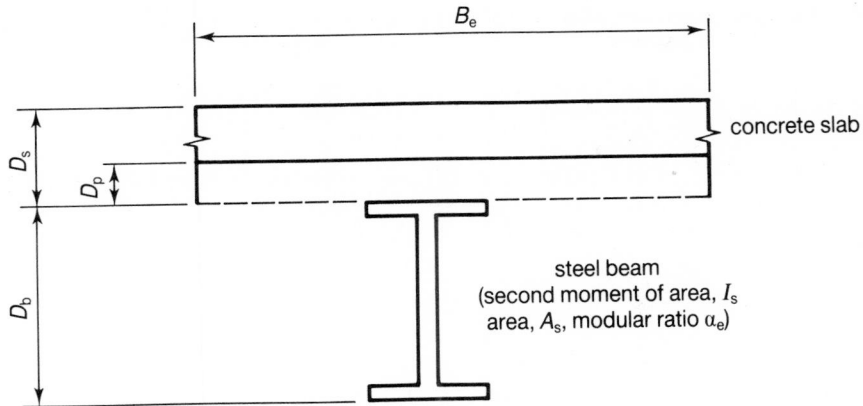

Fig. 10.13 Second moment of area calculation for a composite beam.

For partial interaction of shear connection, the deflection δ is calculated as follows:

for propped construction,

$$\delta = \delta_c + 0.5(1 - N_a/N_p)(\delta_s - \delta_c) \tag{10.38}$$

for unpropped construction,

$$\delta = \delta_c + 0.3(1 - N_a/N_p)(\delta_s - \delta_c) \tag{10.39}$$

where N_a and N_p are the actual and design shear connectors for positive moments, and δ_c and δ_s are the deflections calculated on the composite section and the bare steel beam respectively.

(b) Service stress limits
The service stress limits to avoid irreversible deformations are p_y for the steel, unless lateral torsional buckling can occur when it should be taken as p_b, and $0.5f_{cu}$ for the concrete calculated on the elastic section modulus under service loading.

(c) Vibration
It is necessary to limit the frequency of vibration since if the frequency is too low discomfort will be felt by the occupants. It is generally accepted that if the frequency is greater than around 4 to 5 Hz, excitation will not be noticeable (Lawson).
The frequency of vibration f_q may be calculated from

$$f_q = (\pi/2)(gE_cI_c/(W_dL^3))^{0.5} \tag{10.40}$$

where L is the span (in m), I_c the second moment of area in concrete units (in m^4), E_c the elastic modulus for concrete (in GPa), W_d the total service dead load of the slab and finishes (in kN). Note that Johnson indicates that the second moment of area should be calculated on the whole section taking the actual and not the effective slab width. This will have the effect of giving a slightly higher value of frequency. Use of the effective width will give a slightly conservative result.

In order to illustrate the design of composite beams, two examples will be undertaken. The first is a beam supporting the sheeting, with the ribs perpendicular to the span, and the beam itself spanning on to the main beams (Beam 4CD, in Fig. 10.5). The second is one of the long span beams (span 12 m, beam 3AD (Fig. 10.5)) which carries the column loads from an upper floor.

EXAMPLE 10.2   Composite beam design (decking perpendicular to span)

Design a composite beam in Grade 50 steel to carry the profile steel decking designed in Example 10.1. As in Example 10.1 the concrete is lightweight with a characteristic cube strength of 30 MPa. The span of the beam is 4 m, and supports the two 2.5 m spans of the sheeting.

Loading

All coefficients are taken from Fig. 10.6

Maximum reaction on the support beam, $R = (1.54 + 1.44)q$
$$= 2.98q$$

Loading due to wet concrete 2.1 kN/m²
Construction loading   0.5 kN/m²
Finishes   2.5 kN/m²
Imposed loading   4.0 kN/m²

Note, the dry and wet densities of the concrete will be assumed to be the same, although in practice the dry density will be slightly less by around 10%. This difference will also allow for the effect of the self weight of the decking and the beam.

Load combinations:

Service limit state:

Dead load: $2.98 \times 4 \times 2.1 = 25.0$ kN

Dead load plus finishes: $2.98 \times 4 \times (2.1 + 2.5) = 54.8$ kN

Dead load plus construction load: $2.98 \times 4 \times (2.1 + 0.5) = 31.0$ kN

Imposed load $= 2.98 \times 4 \times 4 = 47.7$ kN

Ultimate limit state:

Dead plus construction load $= 1.4 \times 31 = 43.4$ kN

Dead plus finishes plus imposed load $= 1.4 \times 54.8 + 1.6 \times 47.7 = 153.0$ kN

The beam will be designed as unpropped, thus the dead load and construction load is taken on the steel section alone and the total load on the composite section.

Maximum moment at ULS $= 153 \times 4/8$
$$= 76.5 \text{ kNm}$$

Try a 203 × 133 × 25 UB

The effective width b_e is given by

$$b_e = L/8 = 4000/8 = 500$$

Actual half spacing $2500/2 = 1250$ mm, so use 500 mm

Total effective width, $B_e = 2b_e = 2 \times 500 = 1000$ mm

The construction stage check will be performed later, and the ultimate load check will be performed first.

Ultimate moment of resistance:
All relevant dimensions are given in Fig. 10.14.

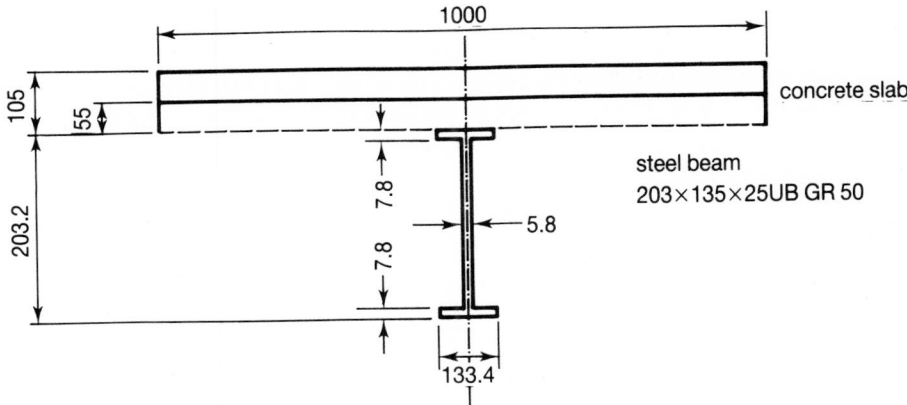

Fig. 10.14 Design dimensions for Example 10.2.

Using the theory in Section 10.3.3.1,

$$F_c = 0.45 f_{cu} B_e (D_s - D_p) \tag{10.4}$$
$$= 0.45 \times 30 \times 1000 \times (105 - 55)/1E6$$
$$= 0.675 \text{ MN}$$

$$F_s = A_s p_y \tag{10.5}$$
$$= 32.3E2 \times 355/1E6$$
$$= 1.147 \text{ MN}$$

$$F_{sc} = F_s - F_c \tag{10.6}$$
$$= 1.147 - 0.675$$
$$= 0.472 \text{ MN}$$

$$F_{sf} = BT p_y \tag{10.7}$$
$$= 133.4 \times 7.8 \times 355/1E6$$
$$= 0.369 \text{ MN}$$

Since $F_{sc} < 2F_{sf}$, the plastic neutral axis is in the flange.

$$x = F_{sc}/2B p_y \tag{10.11}$$
$$= 0.472E6/(2 \times 133.4 \times 355)$$
$$= 4.98 \text{ mm}$$

$$F_{sfx} = xB p_y \tag{10.12}$$
$$= 4.98 \times 133.4 \times 355/1E6$$
$$= 0.236 \text{ MN}$$

$$M_c = F_c(D_b + (D_p + D_s)/2) + 2F_{sfx}(D_b - x/2) - F_s D_b/2 \tag{10.13}$$
$$= 0.675(203.2 + (105 + 55)/2) + 2 \times 0.236(203.2 - 4.98/2) - 1.147 \times 203.2/2$$
$$= 169.4 \text{ kNm}$$

This exceeds the applied moment, therefore it will be possible to use partial interaction.
Check the b/T and d/t ratios for the beam:
During construction BS 5950, Pt 1 applies.
Construction;

$$\varepsilon = (275/355)^{0.5} = (275/355)^{0.5} = 0.88$$

Actual $b/T = 7.38$
Allowable $= 8.5\varepsilon = 8.5 \times 0.88 = 7.5$
Actual $d/t = 52.7$
Allowable $d/t = 79\varepsilon = 79 \times 0.88 = 70$
Both these are satisfactory for a 'plastic' section.

Composite action;
the b/t ratio does not need checking as the flange is restrained by through deck welding
limiting $d/t = 64\varepsilon/(1 + r)$
$d = 187.6$ mm, $y_c = 0$, as the whole web is in tension, so $r = -1.0$, and so there is no
restriction on the web slenderness.

Flexural shear:

$$P_v = 0.6p_y D_t$$
$$= 0.6 \times 355 \times 203.2 \times 5.8/1\text{E}3$$
$$= 251 \text{ kN}$$

Applied shear $= 153/2 = 76.5$ kN
Shear Connectors.

Use 19 mm diameter by 100 long connectors. This size of stud satisfies the detailing
requirements for ribbed slabs.
From Table 5 (Pt 3.1), the characteristic strength of the shear stud Q_k is 100 kN.
This must be multiplied by 0.9 as lightweight concrete is in use to give a strength of
90 kN.
For regions of positive moment, the capacity of connectors is given by Equation
(10.20(a)),

$$Q_p = 0.8Q_k$$
$$= 0.8 \times 90$$
$$= 72 \text{ kN}$$

Calculation of reduction factor k which is given by Equation (10.26)

$$k = (0.85/N^{0.5})(b_r/D_p)((h/D_p) - 1) \tag{10.26}$$

Assume 1 stud per group, i.e. $N = 1$, $D_p = 55$ mm, $h = 100$ mm, $b_r = 152.5 - 38 = 114.5$ mm

$$k = (0.85/1^{0.5})(114.5/55)((100/55) - 1)$$
$$= 1.45$$

Maximum value of k is unity, so there is no reduction due to the profile.

The number of studs per half beam is given by Equation (10.21),

$$N_p = F_p/Q_p \tag{10.21}$$

F_p is taken as the lesser of F_s or F_c,
i.e. the lesser of 1.147 MN and 0.675 MN

$$N_p = 675/72$$
$$= 9.4 \text{ studs}$$

This gives a spacing s of

$$s = (L/2)/(N_p - 1)$$
$$= 2000/8.4$$
$$= 238 \text{ mm}$$

In practice this would probably mean placing one stud per trough and reducing the connector force below the design value.

In view of the fact that the moment capacity of the section is much higher than the imposed moment, employ partial interaction and set the centres of studs at 305 mm, i.e. every second trough.

This spacing of 305 satisfies the maximum spacing criteria of 600 mm or $4D_s$

No. of studs $= 1 + 2000/305 = 7.6$.

This is less than 60% reduction over full interaction and is therefore satisfactory.

$$F_p = N_p Q_p = 7.6 \times 72/1E3 = 0.547 \text{ MN}.$$

The moment capacity of the section with $F_c = 0.547$ MN is 158.4 kNm

Transverse Reinforcement:
The force to be resisted is given by Equation (10.34),

$$v = NQ/s \tag{10.34}$$

$$N = 1.0, \quad s = 305, \quad Q = Q_p = 72 \text{ kN}$$

$$v = 1 \times 72E3/305$$
$$= 236 \text{ N/mm}$$

The shear plane needing to be checked is 1–1 in Fig. 10.15.

The number of shear planes is 2.

The area of concrete contained within the shear plane and the centre-line of the beam is $50 \times 66.7 \text{ mm}^2$

Area contained between the effective half width of the slab $= 500 \times 50 \text{ mm}^2$

Thus the design transverse shear is

$$v = 236 \times (1 - 50 \times 66.7/(500 \times 50))/2$$
$$= 102 \text{ N/mm}$$

The maximum shear that may be carried is given by

$$v_{r,max} = 0.8\eta A_{cv}(f_{cu})^{0.5} + v_p \tag{10.31}$$

First calculate the maximum shear that may be carried by the concrete

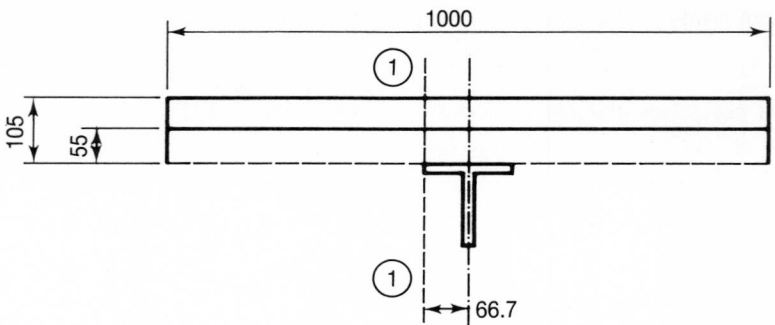

Fig. 10.15   Critical shear plane.

A_{cv} is calculated as the total area of the concrete in the slab, which is given by the total area minus the area occupied by the voids,

$$A_{cv} = 1000 \times 105 - (1000/152.5)(51(38 + 12)/2)$$
$$= 96640 \text{ mm}^2$$

or expressing the concrete area per mm run of the beam,

$$A_{cv} = 96.64 \text{ mm}^2/\text{mm}$$

$$\eta = 0.8, \quad f_{cu} = 30 \text{ MPa}$$

$$v_{r,max} = 0.8 \times 0.8 \times 96.64 \times 30^{0.5}$$
$$= 339 \text{ N/mm}.$$

This is greater than the applied shear v.

Calculation of v_r (Equation (10.31))

First calculate the component due to profile decking (v_p):

For sheeting being continuous and perpendicular to the beam, v_p is given by Equation (10.32),

$$v_p = t_p p_{yp} \tag{10.32}$$

$$t_p = 0.9 \text{ mm} \quad \text{and} \quad p_{yp} = 260.4 \text{ MPa}, \quad \text{giving}$$

$$v_p = 0.9 \times 260.4$$
$$= 234 \text{ N/mm}$$

This is greater than the design transverse shear force, so no contribution is required by either the concrete or the reinforcement, even though reinforcement is provided by virtue of the slab having been designed as continuous.

Deflection

As the dead load is taken on the steel beam alone, which may be pre-cambered to take account of this, the section need only be checked under imposed load.

Second moment of area I_c (Equation (10.36)

For lightweight concrete with office loading α_e is taken as 15.

The elastic neutral axis x is given by Equation (10.37),

$$x = (\alpha_e A_s D_b/2 + B_e d_s(D_b + d + d_s))/(\alpha_e A_s + B_e d_s) \qquad (10.37)$$
$$= (15 \times 32.3\text{E}2 \times 203.2/2 + 1000 \times 50(203.2 + 55 + 50/2))/$$
$$(15 \times 32.3\text{E}2 + 1000 \times 50)$$
$$= 193.8 \text{ mm}$$

$$I_c = \alpha_e I_s + \alpha_e A_s(D_b/2 - x)^2 + B_e(d_s)^3/12 + B_e d_s(D_b + d + d_s/2 - x)^2 \qquad (10.36)$$
$$I_c = 15 \times 2356\text{E}4 + 15 \times 32.3\text{E}2 \times (203.2/2 - 198.8)^2 + 1000 \times 50^3/12$$
$$+ 1000 \times 50 \times (203.2 + 55 + 50/2 - 193.8)^2$$
$$= 1.18\text{E}9 \text{ mm}^4$$

$$E_c = E_s/\alpha_e = 205/15 = 13.67 \text{ GPa}$$

$$\delta_c = (5/384)(47.7 \times 4^3)/(13.67\text{E}6 \times 1.18\text{E} - 3)$$
$$= 2.5\text{E} - 3 \text{ m}$$

$$\delta_s = \delta_c(I_c/\alpha_e)/I_s = 2.5(1.18\text{E}9/15)/2356\text{E}4 = 8.4 \text{ mm}$$

$$\delta = \delta_c + 0.3(1 - N_a/N_p)(\delta_s - \delta_c) \qquad (10.38)$$
$$= 2.5 + 0.3(1 - 7.6/9.4)(8.4 - 2.5)$$
$$= 2.8 \text{ mm}$$

Allowable deflection = span/200 = 0.020 m.

Serviceability stress check:

Dead load:

dead load = 25 kN

$$M = WL/8 = 25 \times 4/8 = 12.5 \text{ kNm}$$
$$f_d = M/Z = 12.5\text{E}3/231.9 = 54 \text{ MPa}$$

Dead load plus construction load:

$$W = 31 \text{ kN}, \quad M = 15.5 \text{ kNm}, \quad \text{and} \quad f_{dc} = 67 \text{ MPa}$$

These stresses are below the design strength, and are therefore satisfactory.

Composite stage:

$$Z_{conc} = I_c/(D_b + d + d_s - x)$$
$$= 1.18\text{E}9/(203.2 + 55 + 50 - 193.8)$$
$$= 10.31\text{E}6 \text{ mm}^3$$

$$Z_s = (I_c/\alpha_e)/x$$
$$= (1.18\text{E}9/15)/193.8$$
$$= 0.406\text{E}6 \text{ mm}^3$$

Service load due to finishes and imposed load = 75.4 kN

Service imposed moment,

$$M_{imp} = 37.7 \text{ kNm}.$$

Stress in the concrete, f_c is given by

$$f_c = M_{imp}/Z_{conc}$$
$$= 37.7E6/10.31E6$$
$$= 3.7 \text{ MPa}$$

limiting stress $= 0.5 f_{cu} = 15$ MPa
This is satisfactory.

Stress in the steel f_{si} is given by

$$f_{si} = M_{imp}/Z_s$$
$$= 37.7E6/0.406E6$$
$$= 93 \text{ MPa}$$

Total stress in the steel f_s is given by the sum of the stresses on the steel beam under dead load and on the composite beam under the imposed load and finishes.

$$f_s = f_d + f_{si}$$
$$= 54 + 93$$
$$= 147 \text{ MPa}$$

This is below the design strength for Grade 50 steel and is therefore satisfactory.

Vibration check:

This may be checked using Equation (10.40)

$$f_q = (\pi/2)(gE_c I_c/(W_d L^3))^{0.5} \tag{10.40}$$

W_d is taken as the full dead load plus finishes.

$$W_d = 2.98 \times (2.1 + 2.5) \times 4 = 54.8 \text{ kN}$$

$$f_q = (\pi/2)(9.81 \times 1.18E9 \times 13.67E-6/54.8 \times 4^3)^{0.5}$$
$$= 10.6 \text{ Hz}$$

This is satisfactory.

EXAMPLE 10.3   Composite beam design (decking parallel to span)

Prepare a design in Grade 50 steel with the composite deck as designed in Example 10.1 for beam Mark 3AD (Fig. 10.5).

Due to the long span the beam will be designed as propped so as to eliminate high dead load deflections. Thus the total loading will be taken on the composite section. The resultant loading at ULS due to dead load, finishes and imposed load together with the resultant bending moment and shear force diagrams are given in Fig. 10.16.

Effective width, b_e

$$b_e = L/8 = 12000/8 = 1500 \text{ mm}$$

Total effective width, $B_e = 2b_e = 2 \times 1500 = 3000$ mm

This is less than the beam spacing of 4 m.

Since the ribs of the decking run parallel to the beam, the area of the concrete that can be taken is the full area minus the area of the re-entrant ribs. This area loss is small and may be ignored.

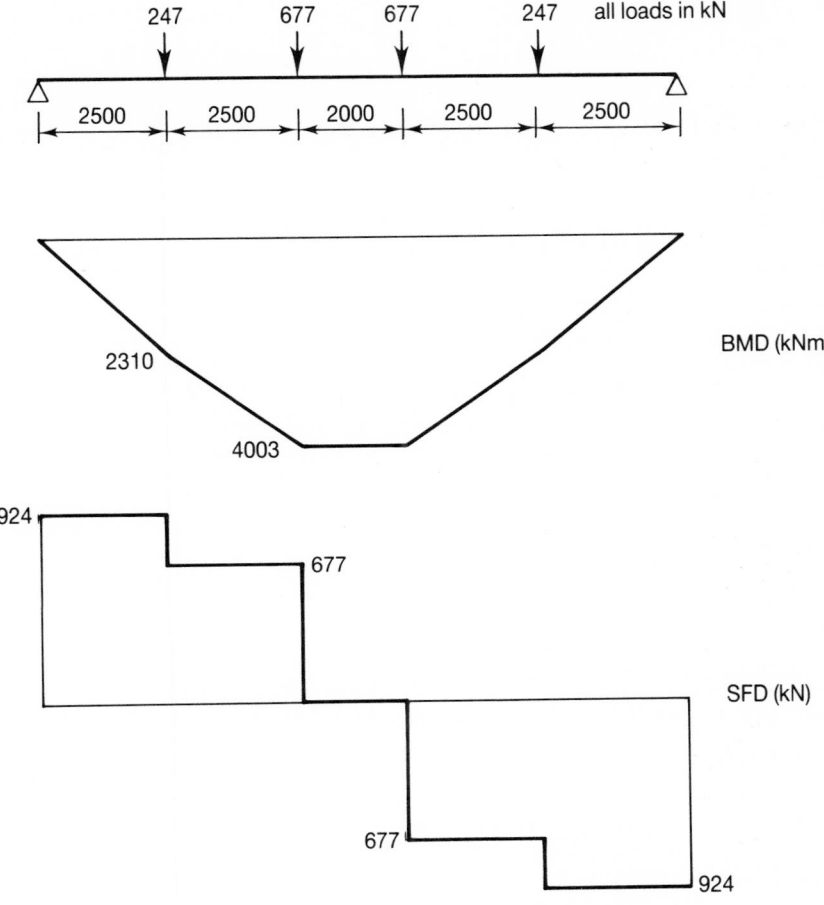

Fig. 10.16 Loading for Example 10.3.

Try a $914 \times 305 \times 201$ UB. Since $T = 20.2$ mm, $p_y = 345$ MPa, (Table 6, Part 1)

Calculation of moment capacity:

Using the theory in Section 10.3.3.1 with $D_p = 0$, as the concrete occupies the whole of the slab depth.

$$F_c = 0.45 f_{cu} B_e D_s$$
$$= 0.45 \times 30 \times 3000 \times 105/1E6$$
$$= 4.253 \text{ MN}$$

$$F_s = A_s p_y \tag{10.5}$$
$$= 256E2 \times 345/1E6$$
$$= 8.832 \text{ MN}$$

$$F_{sf} = BT p_y \tag{10.7}$$
$$= 303.4 \times 20.2 \times 345/1E6$$
$$= 2.114 \text{ MN}$$

$$F_{sc} = F_s - F_c \tag{10.6}$$
$$= 8.832 - 4.253$$
$$= 4.579 \text{ MN}$$

This is greater than twice the flange capacity, so the plastic neutral axis lies in the web.

$$x = (F_{sc} - 2F_{sf})/2tp_y \tag{10.8}$$
$$= (4.579 - 2 \times 2.114) \times 1E6/(2 \times 15.2 \times 345)$$
$$= 33.5 \text{ mm}$$

$$F_{sw} = xtp_y \tag{10.9}$$
$$= 33.5 \times 15.2 \times 345/1E6$$
$$= 0.176 \text{ MN}$$

$$M_c = F_c(D_b + (D_p + D_s)/2) + 2F_{sf}(D_b - T/2) + 2F_{sw}(D_b - T - x/2) - F_s D_b/2$$
$$= 4.253 \times (903 + 105/2) + 2 \times 2.114 \times (903 - 20.2/2) \tag{10.10}$$
$$+ 2 \times 0.176 \times (903 - 20.2 - 33.5/2) - 8.832 \times 903/2$$
$$= 4156 \text{ kNm}$$

This exceeds the applied moment.

Moment capacity of the bare steel section M_s:

$$M_s = S_x p_y$$
$$= 8360 \times 345/1E3$$
$$= 2884 \text{ kNm}$$

Check the ratio M_c/M_s as the flange is large (cl 5.4.5.4, Pt 3.1)

$$M_c/M_s = 4156/2884$$
$$= 1.48$$

This is less than the critical value of 2.5

Flexural Shear:

$$P_v = 0.6p_y Dt$$
$$= 0.6 \times 345 \times 903 \times 15.2/1E3$$
$$= 2841 \text{ kN}$$

The maximum applied shear of 924 kN is well below 50% of the shear capacity, so there is no reduction in the calculated moment capacity.

Design of shear connectors:

As in Example 10.3, use 19 diameter by 100 mm long shear studs. There is no reduction in capacity owing to the deck as b_r/D_p exceeds 1.5 and thus $k = 1.0$ (Equation (10.27a)).

$$Q_p = 72 \text{ kN}$$
$$N_p = F_p/Q_p \tag{10.21}$$
$$F_p = F_c = 4.253 \text{ MN}$$
$$N_p = 4253/72 = 59$$
$$s = 6000/(59 - 1) = 103 \text{ mm}$$

There is spare moment capacity, so partial interaction may be used, also owing to the width of the flanges of the beam the studs may be grouped in pairs.

If the number of studs is reduced to 48 and set in pairs at 260 centres, the ratio N_a/N_p is 0.8.

The maximum value allowed of N_a/N_p is given by Equation (10.22),

$$N_a/N_p = (L - 6)/10 = (12 - 6)/10 = 0.6.$$

With $N_a/N_p = 0.8$, $M_c = 4030$ kNm (with $x = 114.6$).

Additional checks need to be made owing to the existence of the point loads:

(a)   at the 306 kN load

Free bending moment M_0 due to this load

$$\begin{aligned} M_0 &= Wab/L \\ &= 306 \times 2.5 \times 9.5/12 \\ &= 606 \text{ kNm.} \end{aligned}$$

This exceeds 10% of the maximum moment in the beam, and the number of connectors between the load and the support must therefore be checked using Equation (10.24), with $N_n = 0$,

$$N_i = N_p(M - M_s)/(M_c - M_s) \tag{10.24}$$

$M = 2310$ kNm (From Fig. 10.16), but $M - M_s$ is negative and thus the check is inoperative.

(b)   the 677 kN load

$$\begin{aligned} N_i/N_p &= (4003 - 2884)/(4156 - 2884) \\ &= 0.88 \end{aligned}$$

thus 88 percent of the shear connecter requirement calculated on a full interaction basis is required in the end 5 m of the beam, i.e. place 52 studs in pairs at 200 mm centres.

For the central 2 m, use partial interaction of 80% and place 14 studs in pairs at 330 mm centres.

Check d/t ratio:

$$d = 824.5, \quad y_c = 114.6, \quad y_t = 824.5 - 114.6 = 709.9 \text{ mm}$$

$$\begin{aligned} r &= (y_c - y_t)/d \\ &= (114.6 - 709.9)/824.5 = -0.72 \end{aligned}$$

$$d/t = 824.5/15.2 = 54.2$$

Check the limit of $64\varepsilon/(1 + r)$:

$\varepsilon = (275/345)^{0.5} = 0.89$, thus $64\varepsilon/(1 + r) = 64 \times 0.89/(1 - 0.72) = 203$

Thus the web d/t ratio is satisfactory.

Transverse Shear:

The highest value of the transverse shear to be resisted is given by considering the first 5 m of beam.

$$v = NQ/s \tag{10.34}$$

$$Q = Q_p = 72 \text{ kN}$$

$$N = 2, \quad s = 200 \text{ mm}$$

$$\begin{aligned} v &= 2 \times 72/200 \\ &= 720 \text{ N/mm} \end{aligned}$$

This acts on two shear planes, and the critical region for checking is at the edge of the beam, where a rib is aligned with the beam centre-line, giving a resultant design shear of

$$v = 720 \times (1 - (303.4/2)/1500)/2$$
$$= 324 \text{ N/mm}$$

Design of transverse reinforcement (Equation (10.31)):

The transverse reinforcement may be designed by first calculating the shear that may be resisted by the concrete and the sheeting and then taking the balance on the reinforcement.

Capacity of sheeting (v_p):

Since the sheeting is discontinuous and runs parallel to the beam, Equation (10.33) must be used.

Calculate the factor $(N/s)nd$

$$N = 2, \quad s = 200 \text{ mm}, \quad n = 4, \quad d = 19 \text{ mm}$$

$$(N/s)nd = (2/200) \times 4 \times 19$$
$$= 0.76$$

This is less than unity, thus the maximum limit in Equation (10.33) does not apply, so

$$v_p = 0.76 \times 0.9 \times 260.4$$
$$= 178 \text{ N/mm}$$

(b)   due to the concrete $(v_{r,c})$:

Consider the concrete only above the re-entrant rib.

This gives $A_{cv} = 54 \text{ mm}^2/\text{mm}$

$$v_{r,c} = 0.03\eta \times A_{cv}f_{cu}$$
$$= 0.03 \times 0.8 \times 54 \times 30$$
$$= 39 \text{ N/mm}$$

Sum of contributions from the decking and the concrete is 217 N/mm, leaving $324 - 217 = 107 \text{ N/mm}$ to be resisted by the reinforcement $(v_{r,s})$.

$$v_{r,s} = 0.7A_{sv}f_y$$

$$f_y = 460 \text{ MPa}, \quad \text{so}$$

$$A_{sv} = 107/(0.7 \times 460)$$
$$= 0.332 \text{ mm}^2/\text{mm}, \quad \text{or} \quad 332 \text{ mm}^2/\text{m}.$$

This could be accomplished by fixing T8-150 (355) or using B385 mesh (385 mm^2/m)

Deflection check:

Second moment of area:

This calculation follows the same pattern as that in Example 10.2, except $d = 0$, so the results will only be summarized.

$$x = 678.6 \text{ mm} \quad \text{and} \quad I_c = 9.31\text{E}10 \text{ mm}^4$$

Since there is a degree of partial interaction, the final deflection is given by Equation 10.38, and the deflection must be calculated on the composite section and the steel beam alone.

The service imposed loads are given in Fig. 10.17(a).

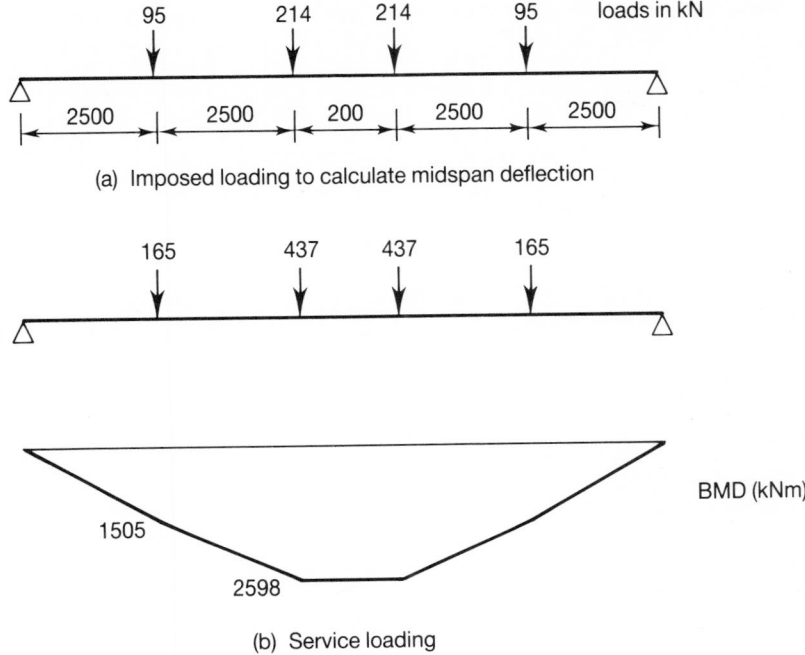

(a) Imposed loading to calculate midspan deflection

(b) Service loading

Fig. 10.17 Loadings for deflection and service stress check.

The total mid-span deflection may be calculated by superposition using for each load the following deflection formula:

$$\delta = WL^3/48EI)(3(a/L) - 4(a/L)^3) \quad \text{with} \quad a < L/2$$

To ease the calculation, determine $W(3(a/L) - 4(a/L)^3)$ for each load and sum the results

(a) the 95 kN load

$W = 95, \quad a = 2.5 \text{ m}, \ L = 12 \text{ m}, \quad$ so

$W(3(a/L) - 4(a/L)^3) = 54$

(b) the 214 kN load

$W = 214, \quad a = 5.0 \text{ m}, \quad \text{and} \quad L = 12.0 \text{ m}, \quad$ so

$W(3(a/L) - 4(a/L)^3) = 202$

The effect may now be summed over the four loads which are symmetric, i.e.
$2 \times (54 + 202) = 512$

$$\delta_c = \sum (WL^3/48EI_c)(3(a/L) - 4(a/L)^3)$$

From above $\Sigma W(3(a/L) - 4(a/L)^3) = 512$, thus

$\delta_c = 512 \times 12^3 \times 1E6/(48 \times 13.67 \times 9.31E10)$

$\quad = 0.0145 \text{ m}$

$\delta_s = \delta_c(I_c/\alpha_e)/I_s$

The α_e is needed in the above expression since the composite second moment of area has been calculated in 'concrete' units.

$$\delta_s = 0.0145 \times (9.31E10/15)/326000E4$$
$$= 0.0276 \text{ m}$$

$$\delta = \delta_c + 0.5(1 - N_a/N_p)(\delta_s - \delta_c) \tag{10.38}$$
$$= 0.0145 + 0.5 \times (1 - 0.8)(0.0276 - 0.0145)$$
$$= 0.0158 \text{ m}$$

Allowable deflection = span/200 = 12/200 = 0.060 m
Thus the deflection is satisfactory.

Service stress check:
The service loading is given in Fig. 10.18(b).

$$Z_{conc} = I_c/(D_b + d + d_s - x)$$
$$= 9.31E10/(903 + 105 - 678.6)$$
$$= 283E6 \text{ mm}^3$$

$$M_s = 2598 \text{ kNm}$$

$$f_{conc} = M_s/Z_{conc}$$
$$= 2598E6/283E6$$
$$= 9.2 \text{ MPa}$$

Limiting stress = $0.5f_{cu} = 0.5 \times 30 = 15$ MPa, and thus is satisfactory.

$$Z_s = (I_c/\alpha_e)/x$$
$$= 9.15E6 \text{ mm}^3$$

$$f_s = M_s/Z_s$$
$$= 2598E6/9.15E6$$
$$= 284 \text{ MPa}$$

This is less than the design strength allowing for the effects of lateral torsional buckling of 345 MPa, and is satisfactory.

This now completes the design checks.

10.4   Tutorial problems

10.4.1   Composite deck

Part of the floor layout of a structure is given in Fig. 10.18.

The imposed loading is 4 kN/m^2, the finishes etc 2.5 kN/m^2, construction load 1.5 kN/m^2. The concrete is lightweight with a strength of 30 MPa and a specific weight of 20 kN/m^3 and has a thickness of 105 mm. The decking which is Lees Super 'Holorib' decking (depth 51 mm, gauge 1.2 mm, self weight 17.69 kg/m^2, area 2124 mm^2/m, second moment of area 863.5E3 mm^4/m, height to NA 17.28 mm, serviceability moment capacity 6.44 kNm/m (sagging) and 5.77 kNm/m (hogging), allowable reaction 76.87 kN/m) is continuous over two 3 m spans. Check the suitability of decking to act as a composite decking and design any additional reinforcement as appropriate.

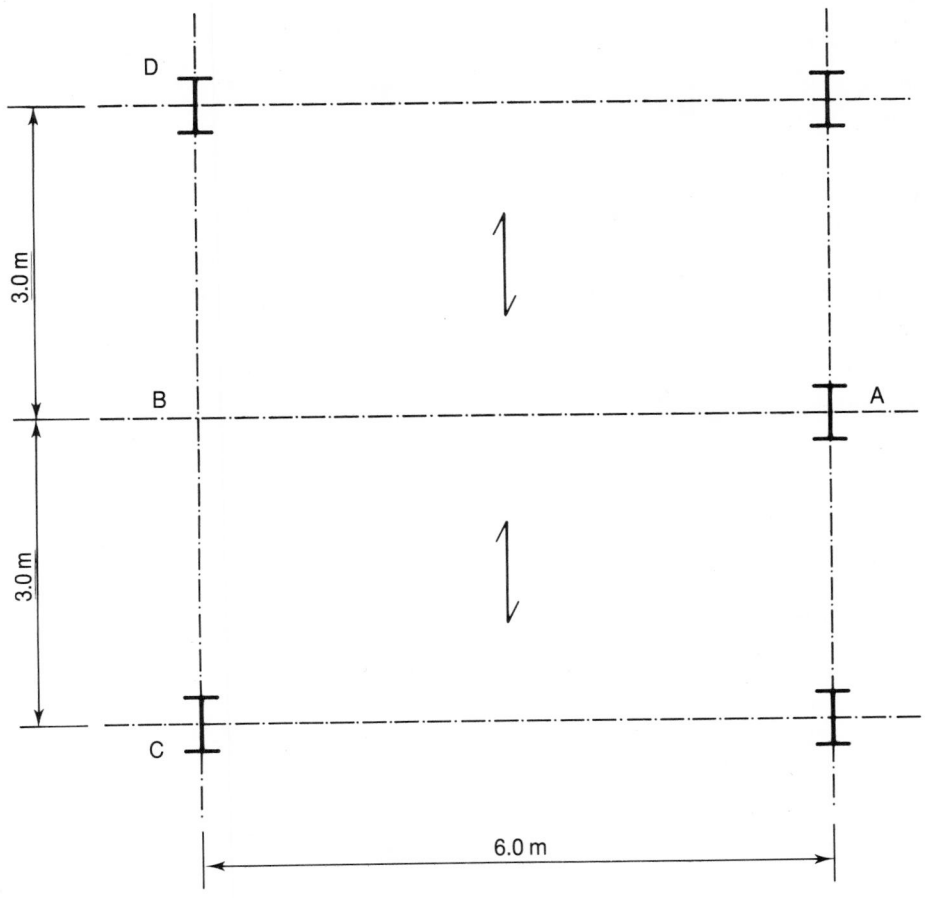

D

A

B

C

3.0 m

3.0 m

6.0 m

Fig. 10.18.

Answer

Patterned loading must be considered, the resultant bending moment and shear force diagrams are given in Fig. 10.19.

Critical cases for construction loading: hogging–both spans loaded, sagging–one span loaded.

Dead load $= 0.105 \times 20 = 2.1$ kN/m^2

Unfactored design moments due to construction load 1.69 kNm/m (hogging), 1.29 kNm/m (sagging); due to deadload (including deck self weight), 2.55 (hogging), 1.44 (sagging).

Although the maximum sagging moments from each load case do not occur at the same point it will be conservative to add the maximum values. Both the imposed moments do not exceed the capacity of the decking.

Unfactored reaction $= 3 \times (10/8)(1.4 + 2.27) = 14.1$ kN/m; factored reaction $= 19.74$.

This is below the allowable, but Equation (10.1) needs checking; moment ratio 0.66, shear ratio 0.26, final value $0.83 < 1.16$.

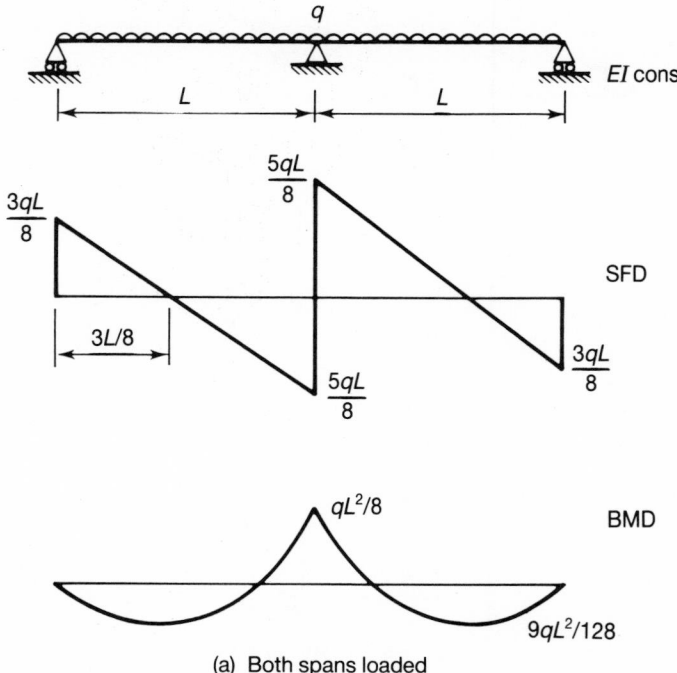

(a) Both spans loaded

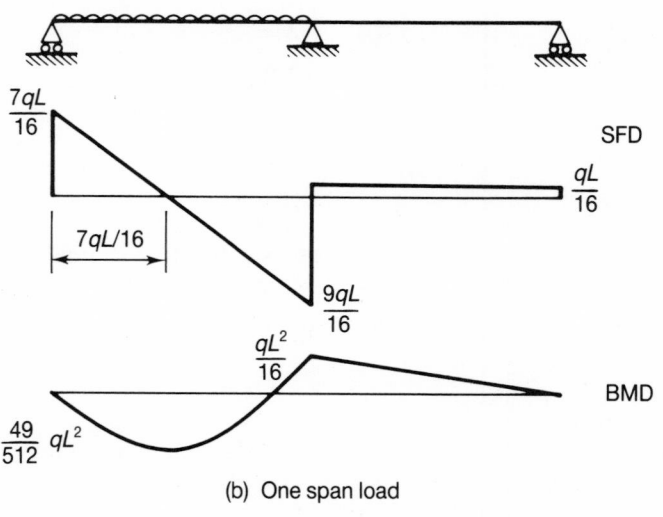

(b) One span load

Fig. 10.19.

Deflection (assuming span is simply supported):

$I = 863.5E3$ mm^4, $E = 210$ GPa, $W = 2.1$ kN/m (wet concrete), $L = 3$, giving a deflection of 0.012 m

Allowable deflection: span/180 (0.017), $D_s/10$ (0.011).

The second of the two criteria is not satisfied by a small margin. If the decking is checked as continuous over the support the deflection is reduced to 5 mm.

Composite deck checks:

Factored dead loading: $1.4(2.1 + 2.5) = 6.44$ kN/m^2, factored imposed load $1.6 \times 4 = 6.40$

Maximum support moment (both spans loaded);

$$(3 \times 3/8)(6.44 + 6.4) = 14.45 \text{ kNm/m}$$

Maximum sagging moment 8 dead load both spans, imposed load one span:

$$(3 \times 3)((49/152) \times 6.4 + (9/128) \times 6.44) = 9.59 \text{ kNm/m}$$

Shear (both spans fully loaded)

$$2 \times (5/8) \times 3 \times (6.4 + 6.44) = 48.2 \text{ kN/m}$$

Moment capacity:

Sagging: $d = 87.7$ mm, $x = 51.2$ mm, $z = 67.9$ mm, $M_u = 37.6$ kNm/m.
Satisfactory.

Hogging:

$d = 66$ (25 cover), $0.156bd^2 f_{cu} = 20.4 < 14.45$, $K = 0.11$, $z = 0.86 \times 66 = 56.8$, $A_s = 636$ mm^2/m (T10-200 or B785 mesh)

Shear:

Flexural shear:

$$b_s = 750.8 \text{ mm}, \quad V = 69.5 \text{ kN/m}$$

Shear bond:

$$L_v = 750, \quad V_s = 50.1 \text{ kN/m}$$

Deflection:

$d = 87.72$ mm; Gross, $x/d = 0.714$, $I_G/bd^3 = 0.188$, $I_G = 1.269E8$; Cracked, $x/d = 0.612$, $I_T/bd^3 = 0.149$, $I_T = 1.005E8$; $I_{NA} = 1.137E8$ mm^4

$$E_c = 210/20, \quad L = 3 \text{ m}$$

Assuming simply supported:

Imposed load only ($q = 4$), deflection $= 3.5$ mm, span/350 $= 8.6$
total load ($q = 4 + 2.5$), deflection $= 5.7$ mm, span/250 $= 12$ mm

10.4.2 Composite beam (ribs perpendicular to span)

It is proposed to use a Grade 50 254 × 146 × 31 UB for the beam Mark AB in Fig. 10.18. Carry out all the checks required on the beam. The reactions of the composite decking and its loading may be taken from Fig. 10.19. 100 × 19 mm diameter shear studs are to be used.

Answer

Effective width of the flange is taken as $L/8$ either side.

Loading   (unfactored):   dead, $2.1 \times (6 \times (10/8) \times 3) = 2.1 \times 22.5 = 47.25$ kN;   finishes, 2.5×22.5; construction, 0.5×22.5; imposed, 4.0×22.5.

Loading (factored): dead plus construction, $1.4(47.25 + 11.25) = 82.7$ kN; dead plus finishes plus imposed, $1.4(47.25 + 56.25) + 1.6 \times 90 = 288.9$ kN

Applied moment: $288.9 \times 6/8 = 216.7$ kNm

Effective width $= 2 \times L/8 = 1500$ mm

Calculation of M_c (Fig. 10.20).

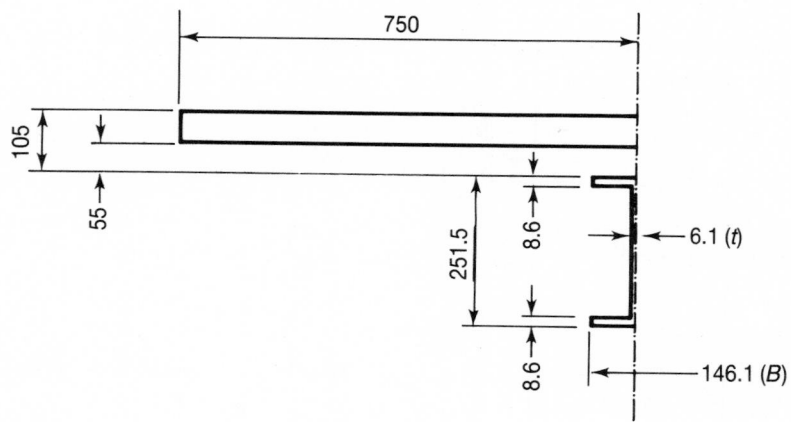

Fig. 10.20.

$F_c = 0.45 \times 30 \times (105 - 55) \times 1500 = 1.013$ MN, $F_s = 355 \times 40E2 = 1.420$ MN, $F_{sc} = 0.407$ MN

NA is in the flange, $x = 3.92$ mm, $M_c = 359.3$ kNm

This is in excess of the applied moment and so partial interaction may be used.

Capacity of shear studs: 72 kN, $k = 1.0$, number for half beam $= 14$ at 230 mm spacing. Increase spacing to 305 mm, i.e. a stud every other trough. Ratio of studs provided to studs required on full interaction < 0.4.

No. of studs on whole beam on partial interaction $= 21$, reduced $F_c = 10.5 \times 72 = 756$ kN so re-checking beam: NA is still in flange, $x = 6.40$ mm, $M_c = 236.9$ kNm, which is satisfactory.

Both the d/t ratio and flexural shear capacity are satisfactory.

Transverse shear:

$$v = 1.0 \times 72/305 = 0.236 \text{ kN/mm}$$

Area of concrete plane $= 50 \times 146.1/2$, area of half width of slab $= 750 \times 50$, design shear/shear plane $= 0.236(1 - 3653/37500)/2 = 0.107$ kN/mm, $v_{r,max} = 0.8 \times 0.8 \times 96.64 \times 30^{0.5} = 0.339$ kN/mm, $v_p = 1.2 \times 260.4 = 312 > 107$, therefore no contribution required from reinforcement or concrete.

Deflection:

$\alpha_e = 15$, $x = 240.1$, $I_c = 2.093E9$ mm^4, $E_c = 205/15$, $N_a/N_p = 0.75$, $\delta_c = 8.85$ mm,
$\delta_s = 27.8$ mm, $\delta = 11.2$ mm, span/200 = 30 mm

Service stress checks:

W due to dead plus finishes plus imposed = 193.5 kN, $M = 145.1$ kNm; $Z_c = 18.0E6$ mm^3,
$f_c = 8.1$ MPa (limit 0.5×30), $Z_s = 0.581E6$, $f_s = 250$ MPa (limit = 355)

Vibration check:

$$W_D = 103.5 \text{ kN}, \quad f_q = 5.6 \text{ Hz}$$

10.4.3   Composite beam (decking parallel to span)

For beam CD in Fig. 10.18(a) Grade 50 305 $\times$ 165 $\times$ 54 UB is to be used acting compositely
with the slab. Composite decking is on both sides of the centre-line of beam CD, i.e. the
loading may be taken as the total loading from Beam AB.

Answer

Total effective width is 1500 mm (span/8 each side), and the whole depth of the concrete
may be taken, i.e. $D_p = 0$.

Moment capacity:

$F_c = 2.126$ MN, $F_s = 2.428$ MN, NA is in the flange with $x = 2.55$ mm and $M_c = 489$ kNm
Applied moment = $229 \times 6/4 = 343$ kNm, i.e. partial interaction may be used.
Number of studs/half beam = $2.126E3/72 = 29.5$, i.e. $s = 105$ mm
Increase spacing to 200 mm, $N = 16$ (ratio is greater than 0.5), $F_c = 16 \times 72 = 1.152$ MN,
NA in flange with $x = 10.77$ mm and $M_c = 431$ kNm.
d/t ratio and flexural shear are satisfactory.

Transverse shear:

$v = 1 \times 72/200 = 360$ kN/m; design shear = $360 \times (1 - 83.4/750)/2 = 160$ kN/m
Capacity of sheeting: $(1/200) \times 4 \times 19 \times 1.2 \times 260.4 = 119$ kN/m
Contribution of concrete (at rib) = $0.03 \times 0.8 \times 30 \times (1 \times 50) = 36$ kN/m
Contribution from reinforcement = $160 - 119 - 36 = 5$ kN/m. This will need only nominal
reinforcement.

Deflection:

$\alpha_e = 15$, $x = 281.4$ mm, $I_c = 4.588E9$ mm^4, $E_c = 210/15$, $\delta_c = 6.5$ mm ($W = 90$ kN, central
point load), $\delta_s = 17$ mm, $\delta = 8$ mm (span/200 = 30 mm)
Service stress check is satisfactory.

References

BS 5400, Part 5. *Steel, concrete and composite bridges – Part 5 – Code of practice for design
   of composite bridges*, British Standards Institution, London.
BS 5950, Part 3. *Structural use of steelwork in buildings – Part 3 – Design in Composite
   Construction – Section 3.1 – Code of practice for design of simple and continuous
   composite beams*, British Standards Institution, London.

BS 5950, Part 4. *Structural use of steelwork in buildings – Part 4 – Code of practice for design of floors with profiled steel sheeting*, British Standards Institution, London.

BS 8110, Part 1. *Structural use of concrete – Part 1 – Code of practice for design and construction*, British Standards Institution, London.

Grant J. A., Fisher J. W. and Slutter R. G. (1977). Composite beams with formed metal decks, *Engineering Journal American Institute of Steel Construction*, **14** (1), 24–42.

Gray B. A. and Walker H. B. (1985). *Steel framed multi-storey buildings – the economics of construction in the UK*, Constrado.

Gray B. A. *et al.* (1983). *Steel framed multi-storey buildings – design recommendations for composite floors and beams using steel decks*, Constrado.

Institution of Structural Engineers and Institution of Civil Engineers (1985). *Manual for the design of reinforced concrete building structures*, Institute of Structural Engineers.

Johnson R. P. (1982). *Composite structures of steel and concrete*, Volume 1 *Beams, columns, frames and application in buildings*, Constrado Monographs/Granada.

Johnson R. P. and May I. M. (1975). Partial interaction design of composite beams, *The Structural Engineer*, **53** (8), 305–11.

Johnson R. P., Greenwood R. D. and Van Dalen K. (1969). Shear stud connectors in hogging moment regions of composite beams, *The Structural Engineer*, **47** (9), 345–50.

Lawson R. M. (1983). *Composite beams and slabs with profiled steel sheeting*, Report 99, Construction Industry Research and Information Association.

Lawson R. M. and Nethercot D. A. (1985). Lateral stability of I beams restrained by profile sheeting, *The Structural Engineer*, **63B** (1), 1–7 and 13.

Martin L. H., Croxton P. C. L. and Purkiss J. A. (1989). *Concrete design to BS 8110*, Edward Arnold.

Mattock A. H. and Hawkins N. M. (1972). Shear transfer in reinforced concrete – recent research, *Journal Prestressed Concrete Institute*, **17** (2), 55–75.

Mottram J. T. and Johnson R. P. (1990). Push tests on studs welded through profiled steel sheeting, *The Structural Engineer*, **68**, 187–93.

Oehlers D. J. and Johnson R. P. (1987). The strength of stud shear connections in composite beams, *The Structural Engineer*, **65B** (2), 44–8.

Olgaard J. E., Slutter R. G. and Fisher J. W. (1971). Shear strengths of stud connectors in normal and light-weight concrete, *Engineering Journal, American Institute of Steel Construction*, **8** (2), 55–64.

Walker H. B. *et al.* (1981). *Design and construction methods for multi-storey office buildings in North America*, Constrado.

Yam L. P. C. and Chapman J. C. (1968). *The inelastic behaviour of simply supported composite beams of steel and concrete*, Proc. ICE, **41**, 651–83.

11

Fire Safety Engineering Design of Steel Structures

11.1 Introduction

All structures unless specifically exempted by the Building Regulations or other statutory authority need to be designed to contain the effect of fire and to remain stable during a fire. The period for which this resistance is required is laid down by the relevant regulatory authority and currently is in the range 30 to 240 minutes. The regulations are only concerned with the ability to evacuate the structure in a fire, to avoid the spread of the fire to adjacent property and to allow the fire to be fought in safety. The regulations are not concerned with effect of the fire on the contents of the structure, this is a matter for the insurers of the property and its owner.

Traditionally fire protection to steel structures was provided with concrete encasement. This method however is uneconomical partly due to the extra weight imposed on the structure and foundations, and partly due to the long construction period owing to the need for the erection of formwork and the curing of the concrete. The development of alternative methods such as plaster or gypsum spray for beams, plasterboard encasement for columns and the use of intumescent paints led to the need to assess these methods. A further development was to allow the use of bare, or unprotected, steelwork where the temperatures developed in a fire are insufficient to cause the steelwork to attain temperatures to cause collapse.

It is usual to design the fire protection for a steel structure after the main structural calculations have been carried out, although it should be noted that it may be possible to avoid the need for fire protection if methods of construction such as shelf-angle floors are used (Section 11.10) or that the basic steel member is overdesigned to give what is in effect sacrificial protection (Section 11.6.3). It should be noted that in a multi-storey office block typical costs of fire protection are around 15–20% of the total costs, and the steel frame around 10–15%.

The traditional approach to the assessment of steelwork performance in a fire was to use the standard fire test. The fire test is expensive, will only give limited data since the result is dependent on the section size and the thickness of the protection applied to the section, and as the test is likely to have been instigated by a manufacturer of the protection system being evaluated there will be confidentiality of the result. It thus became necessary to develop alternative methods of assessing the fire performance of steel structures. Much early work in this field was undertaken in Europe, and led to the concept of Fire Safety Engineering.

Prior to the 1985 Building Regulations in the UK, the only approach to assessing fire performance was the use of data derived from standard tests and contained within the Building Regulations. With the development of limit state design under ambient conditions, it became an obvious move to extend the concept to the limit state design of steel structures under fire conditions. This development has now been encapsulated in the current (1985) Building Regulations which quite simply state 'the building shall be so constructed that, in the event of fire, its stability will be maintained for a reasonable period' (cl B3–(1)). Clauses B3–(2) and B4–(1) deal respectively with the use of compartmentation to control fire spread internally and the spread of fire from one building to an adjacent structure.

The period over which the structure is required to remain stable is laid down in Approved Document B to the Building Regulations and is a function of the structure size (or the size of the fire compartment) and the use of the structure. The Building Regulations indicate that the responsibility rests with the Engineer to ensure adequate stability to the structure, although it is expected, but not necessary, that the Engineer will use the documents indicated as 'Approved'. Currently these are BS 5950, Part 8 and the BRE Guidelines.

BS 5950, Part 8 effectively allows three methods to satisfy the fire performance of steel structures:

1   Use of tabulated data for proprietary protection materials derived from tests to BS 476, Part 8 prior to 1986, or, from 1986, BS 476, Parts 20–22.
2   Calculation methods based on the exposure of the steel element to a natural fire.
3   Calculation methods based on exposure to a pseudo-fire based on the furnace curve in BS 476.

The usual method used is the third, except where it is possible to calculate the effect of a natural fire with a high degree of accuracy, together with the knowledge that there cannot be a change of usage of the structure within its design life (Section 11.11).

Before going on to consider such methods, it will be advantageous to discuss the behaviour of a natural fire, to assess the standard fire test and the behaviour of steel structural elements therein.

11.2   Compartment temperature–time response in a natural fire

From measurements taken in controlled compartment fires, it was noted that the temperature–time response showed three phases; a growth period, the fully developed fire, and a decay period (Fig. 11.1).

The growth period, although possibly of a long duration, will only give a very low rise in the compartment gas temperature compared with the substantial rise encountered in the fully developed fire. For this reason the growth period is usually ignored in the calculation of the compartment temperature–time response, and the fire is considered to start at flashover. It should also be noted that flashover is generally considered to be an instantaneous event. This is not so, however, since a substantial temperature rise can occur, albeit in a very short period.

Pettersson *et al.* demonstrated that by considering a heat balance equation (represented diagramatically in Fig. 11.2), it is possible to generate compartment temperature–time responses.

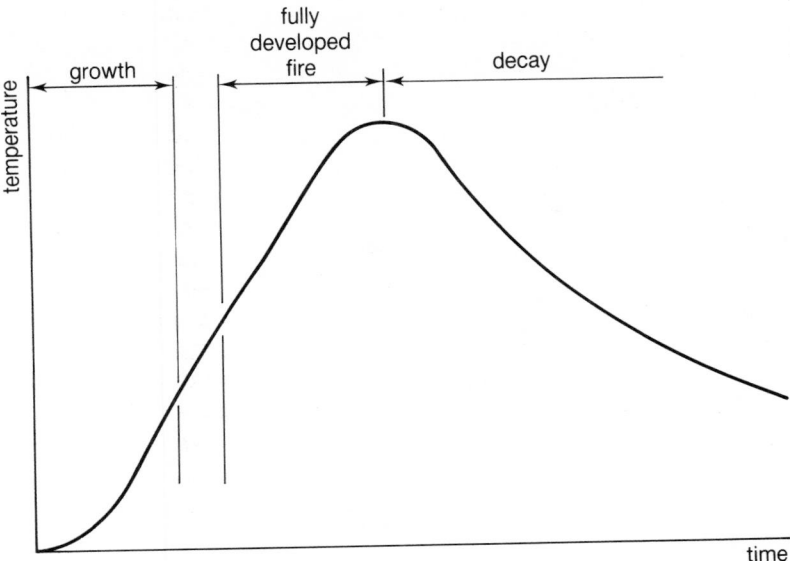

Fig. 11.1   Phases in a developed fire.

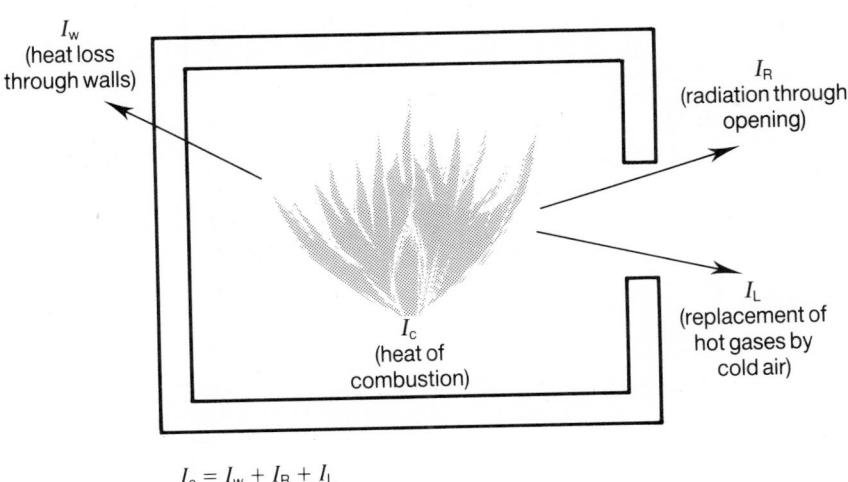

$$I_c = I_w + I_R + I_L$$

Fig. 11.2   Compartment heat balance.

The main assumptions made in this simulation were that:

(i)   the fire load was uniformly spread over the compartment floor;

(ii)   all the fire load (comprising wood cribs) ignited together; and

(iii)   the values of heat loss through the compartment walls and roof could be calculated using average values for the thermal properties of normal building construction. The heat loss through the walls and roof is characterized by the thermal inertia $(\rho c k)^{0.5}$ where ρ is the density, c the specific heat, and k the thermal conductivity. It is generally adequate to take average values over the fire compartment. The results given as a series of curves in Fig. 11.3 are for a standard fire compartment corresponding to brickwork or normal-weight concrete. It should be noted that the heat loss through the walls is sensibly independent of the wall thickness. For a compartment with a different thermal inertia both the fire load and the ventilation factor need to be multiplied by a correction factor k_f values of which are given in Pettersson *et al.* Typically, $0.5 < k_f < 3$, where the lower values are for poor insulators and the higher values are for good insulators such as lightweight concrete.

It had been noted prior to the work by Pettersson *et al.*, that the major parameters governing the compartment temperature–time response are:

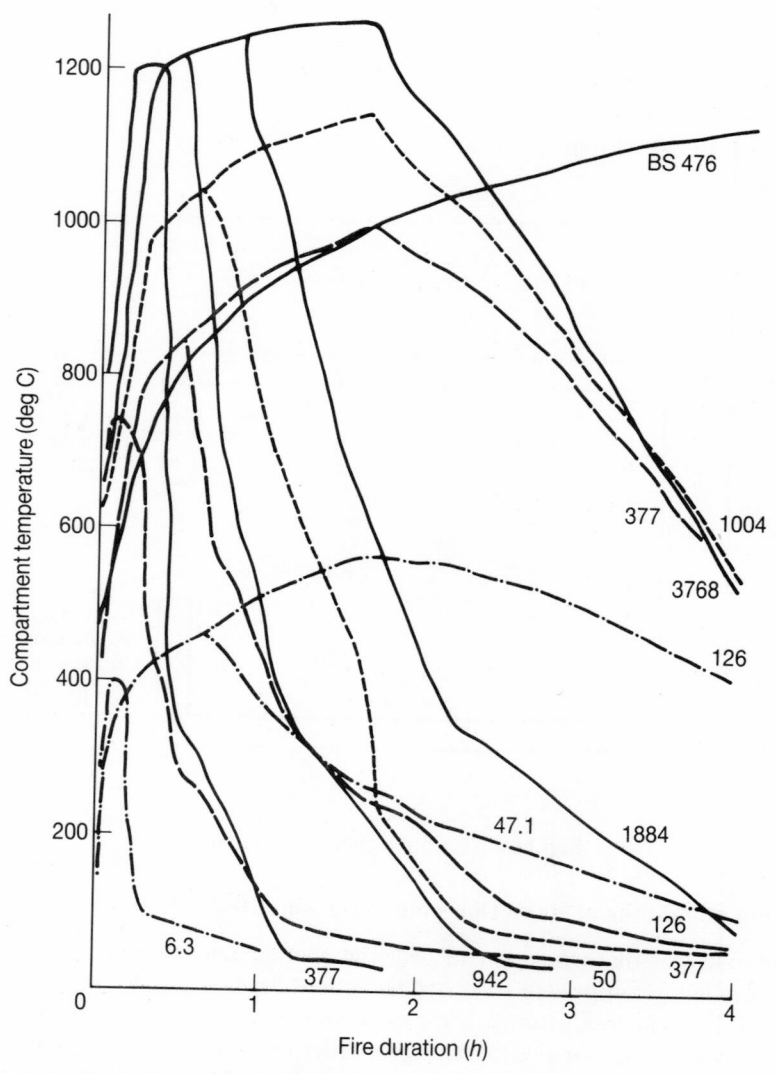

Fig. 11.3 Typical compartment temperature–time response curves for varying fire loads and ventilation. Opening factors ($\sqrt{m}$): ——, 0.3; – – – –, 0.08; —— ——, 0.04; —·—·—, 0.01. The fire load is indicated thus, 1004 MJ/m².

(i)   the fire load/unit floor area (q_f) expressed in terms of wood equivalent, and
(ii) the ventilation factor ($A_w\sqrt{H_w}$), where A_w is the window area and H_w the window height.

A natural fire may either be ventilation controlled, i.e. the availability of air controls the rate of burning, or fuel controlled, i.e. the availability of the fuel controls burning rate.

The maximum temperature $T_{f,max}$ reached in a natural fire can easily be calculated (Law, Kirby (1986)):

$$T_{f,max} = 6000(1 - \exp(-0.1\eta))\eta^{-0.5} \tag{11.1}$$

where

$$\eta = (A_t - A_w)/(A_w\sqrt{H_w})$$

and A_t is the total area of the containing compartment.

Where the fire load is low the temperature $T_{f,max}$ predicted by Equation (11.1) may not be reached and must be modified to give a value T_f given by

$$T_f = T_{f,max}(1 - \exp(-0.05\psi)) \tag{11.2}$$

where

$$\psi = q_f A_f/(A_w(A_t - A_w))^{0.5}$$

and A_f is the floor area.

The duration t_c of a natural compartment fire can be estimated (Law), and is given by,

$$t_c = q_f A_f/(0.1 A_w\sqrt{H_w}) \tag{11.3}$$

It should be noted that Equation (11.3) will tend to underestimate t_c for fires with low fire load and high ventilation and then will be conservative if used to assess the fire performance of a structural element.

Equation (11.3) was derived by taking the rate of burning R as

$$R = 0.1 A_w\sqrt{H_w} \tag{11.4}$$

A better estimate of the rate of burning is given by (Law and O'Brien),

$$R = 0.18 A_w\sqrt{H_w}(W/D)^{0.5}(1 - \exp(-0.36\eta)) \tag{11.5}$$

where W/D is the width to depth ratio of the compartment and η is defined as in Equation (11.1).

The duration of the fire t_c is then given by,

$$t_c = q_f A_f/(0.18 A_w\sqrt{H_w}(W/D)^{0.5}(1 - \exp(-0.36\eta))) \tag{11.6}$$

The fire load in a structural compartment must be expressed in terms of wood equivalence (in kg/m^2) or as a fire load density (in MJ/m^2). The fire load may either be obtained as an estimate for the total compartment or as a mass weighted average for individual components within the compartment. For further data the reader is referred to the Design Guide on Structural Safety, or the Workshop on Structural Safety (January 1983). Typical values for fire load (from the Workshop report) are given in Table 11.1.

To attempt to assess experimentally the behaviour of structures, or more properly, structural elements, under natural fires would be very difficult if not near impossible owing to the large variation in the compartment temperature–time responses obtained in natural fires. It thus became necessary to establish a form of standard test procedure.

Table 11.1   Typical values of fire load

Type of compartment	Average MJ/m^2	Standard deviation MJ/m^2
Dwellings	150.0	24.7
Offices	124.0	31.4
Schools	80.4	23.4
Hospitals	116.0	36.0
Hotels	67.0	19.3

11.3   Standard fire test

The standard fire test, now covered by BS 476, Parts 20 to 22, was developed over a large period of time. For the history of the UK standard fire test reference should be made to Malhotra.

Essentially the element under test is heated under load in a test furnace with the furnace temperature–time curve given by

$$T_f - T_0 = 345 \log (8t + 1) \tag{11.7}$$

where T_f is the furnace temperature, T_0 is the ambient temperature and t the time (in minutes). Equation (11.7) is plotted in Fig. 11.3.

The end point of the test occurs when any of the limit states listed below is attained:

(i)   The element can no longer carry the applied loading or reaches a limiting deflection or rate of deflection.
(ii)   The temperature on the unexposed face of the element under test reaches a critical value.
(iii)   The element ceases to act as a flame barrier or loses its integrity.

The load to be applied during the test is defined as that which would be needed to be applied to produce the same stress in the member as would be imposed under service conditions. This definition needs slight modification to the characteristic loading multiplied by the relevant load factors as limit state methods are now in use for ambient design.

There is a series of drawbacks to the standard test:

(i)   Replication of results between different furnaces is difficult since although the standard imposes the same temperature–time regime in the furnace, different furnaces will have different characteristics for the heat flux to the members under test, i.e. the temperature rise in similar specimens will be different. It should be noted that the temperature–time response in BS 476 is the same as that in the international standard ISO 834.
(ii)   The furnace used for the test may well impose restrictions on the specimen size (restricting possible values of member slenderness) or restrictions in loading (either problems with modelling the effects of end restraint to the member or the type of loading such as columns under axial load only).
(iii)   The standard temperature–time curve is not representative of the compartment response of natural fires (see Fig. 11.3), since it is possible to have fire loads and ventilation factors which give temperatures that are below those in the standard test. Equally the converse is true in that some natural fires exceed the standard curve.

In spite of the above comments the fire test remains a valuable method of comparing the performance of differing methods of construction, but may not be a good indicator of the performance of those methods of construction in a real fire. However the standard fire test has provided some very interesting results on the behaviour of steel elements, some of which will now be described.

11.4 The behaviour of steel elements in the standard fire test

A comprehensive series of furnace tests, on both protected and unprotected beam and column sections, was instigated partly to assess the validity of the concept of a 'Critical Temperature' of 550°C, which had arisen from observations in earlier tests whereby failure appeared to occur when the steel temperature reached 550°C, partly to assess the effect of stress level, and partly to assess the effect of shielding due to blockwork or floor slabs. The idea of a critical temperature of around 550°C also arose since stress–strain (or strength) tests on steel at elevated temperatures using the classical technique whereby the specimens were heated to a specified temperature, allowed to stabilize at that temperature and then strained to failure, indicated that at around 550°C the strength dropped to 0.6 of the ambient temperature strength. This drop is equivalent to the erosion of the factor of safety of 1.7 applied to design strengths in the then current steel design code (BS 449). Further information on the high temperature mechanical properties of steel is given in Section 11.9.3.1.

The results of these tests to determine the behaviour of steel elements in the standard furnace test are taken from Robinson and Latham, Robinson and Walker, and Kirby (1986).

11.4.1 Tests on unprotected members

It has long been realized, from a consideration of the mechanism of heating (see Section 11.6.2.1), that a large number of tests is not needed since sections with the same ratio of heated perimeter (H_p) to cross-sectional area (A) attain the same temperatures at a given time.

From such results it is possible to derive a series of heating curves for unprotected steel sections tested under full service load. Some typical results are given in Fig. 11.4(a) for the lower flanges of steel beams and Fig. 11.4(b) for the average taken over the web and both flanges for steel columns.

These results form the basis of Tables 6 and 7 in BS 5950, Part 8, with the H_p/A ratio replaced by the flange thickness.

It was also shown that the effect of halving the applied stress on beams was to increase the temperature of the lower flange from around 650 to around 750°C, and that for intermediate values the variation was linear.

A further comprehensive source for data on the fire performance of unprotected steelwork is given in Wainman and Kirby.

11.4.2 Tests on partially protected beams

A series of tests was carried out on beams with varying proportions of the web shielded from the fire and with differing stress levels. The reason for the increase in the temperature of the lower flange and the increase in fire performance is due to the fact that the more the web is shielded, the lower the temperature of the top flange will be and therefore the higher flexural capacity of the member before the total section yields. The results of these tests are given in Fig. 11.5.

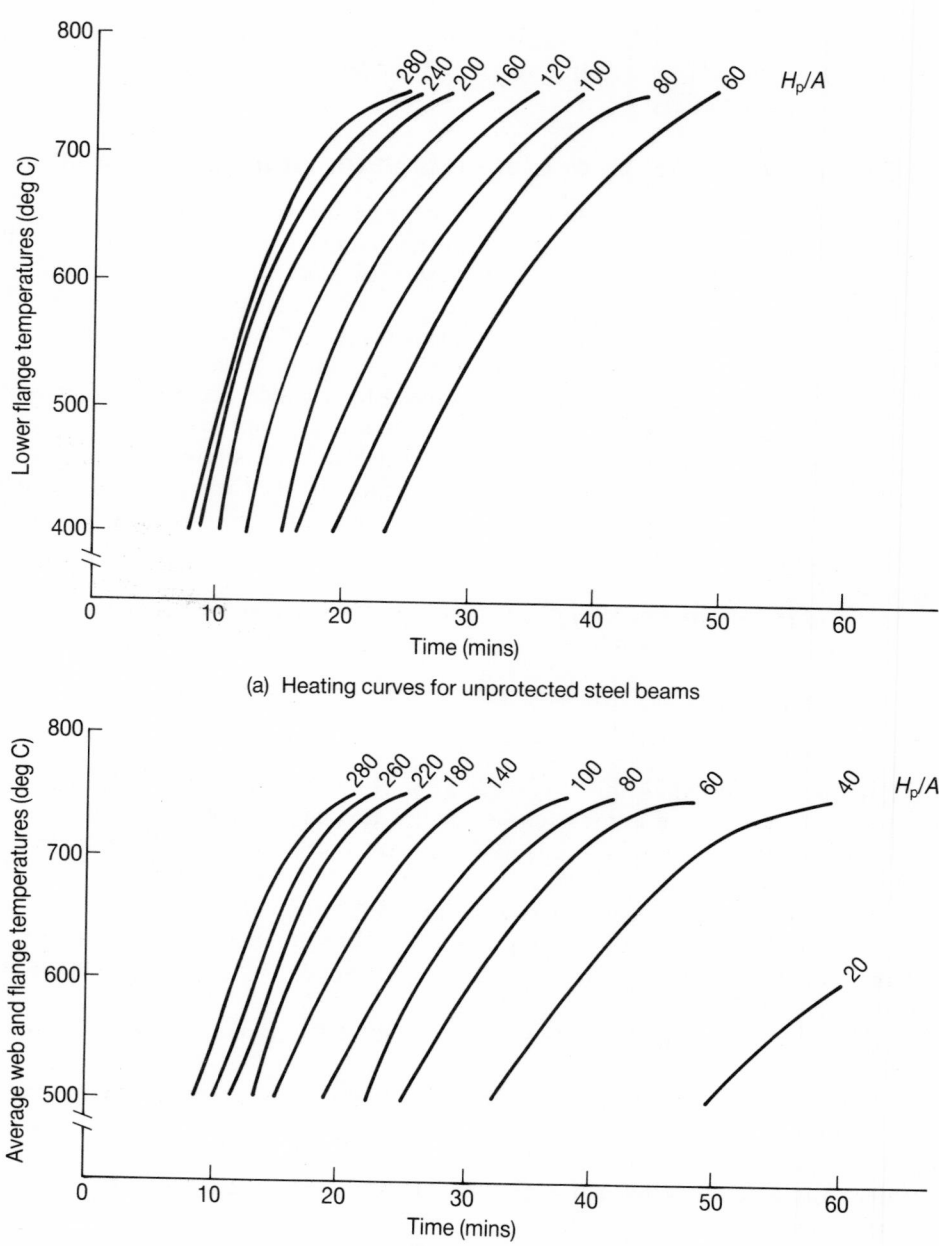

(a) Heating curves for unprotected steel beams

(b) Heating curves for unprotected steel columns

Fig. 11.4 Heating curves for unprotected steelwork to BS 476.

11.4.3 Tests on partially protected columns

A series of tests on Grade 43A 203 × 203 × 52 columns was carried out with conditions ranging from no protection to the column, the column infilled with free-standing block-work, and the column partially and totally built into a blockwork wall. In the first two

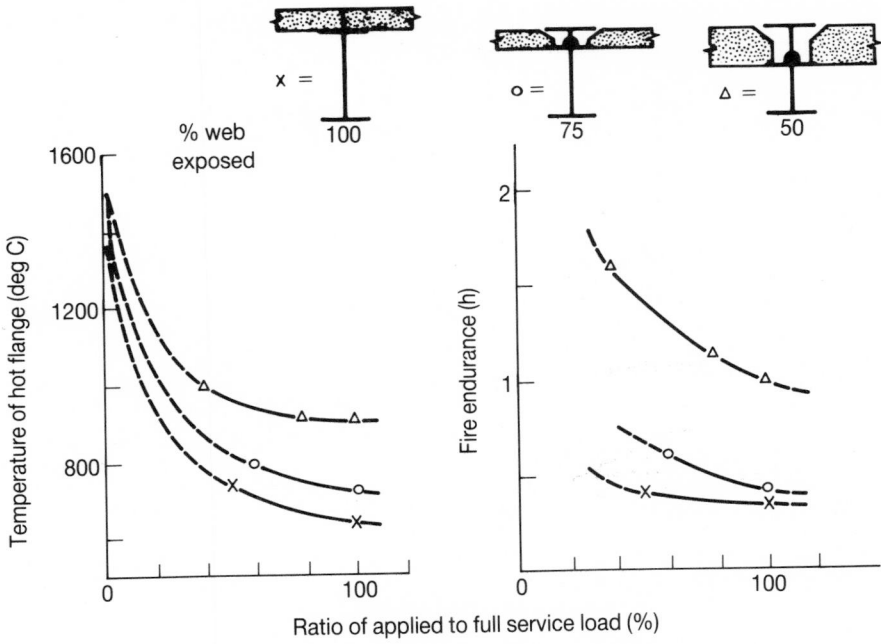

Fig. 11.5   Fire endurance of beams with partially exposed webs.

cases (no protection and free standing blockwork) the columns were heated on four sides, whereas with the column built into the blockwork wall, heating was on one side. In all cases the columns were axially loaded, with no load taken on the blockwork. The results are presented in Fig. 11.6, where it is seen that a reduction in the loading and an increase in the amount of shielding both increase the fire performance.

As mentioned earlier BS 5950, Part 8 allows three approaches to the assessment of fire performance. It is convenient to discuss the method involving the use of tabulated data first before continuing by examining the principles behind the calculation approach, whether exposure to a natural or a BS 476 fire is considered.

11.5   Tabular approach to the fire performance of steel structures

There are essentially two sources of data for this approach; the BRE Guidelines and specific data relating to a manufacturer's product.

11.5.1   BRE guidelines

This is a document produced, and updated at reasonable intervals, by the Fire Research Station. It provides tabular data on the performance of generic fire protection systems and gives values of thickness of material required on a particular element to give the desired fire resistance period. Such data are derived from the standard fire test which may be carried out on an unloaded specimen and the end point of the test taken when the steel member reaches 550°C. Thus no account is taken of the beneficial effect of allowing a higher flange temperature if the member is loaded to less than its design strength (or

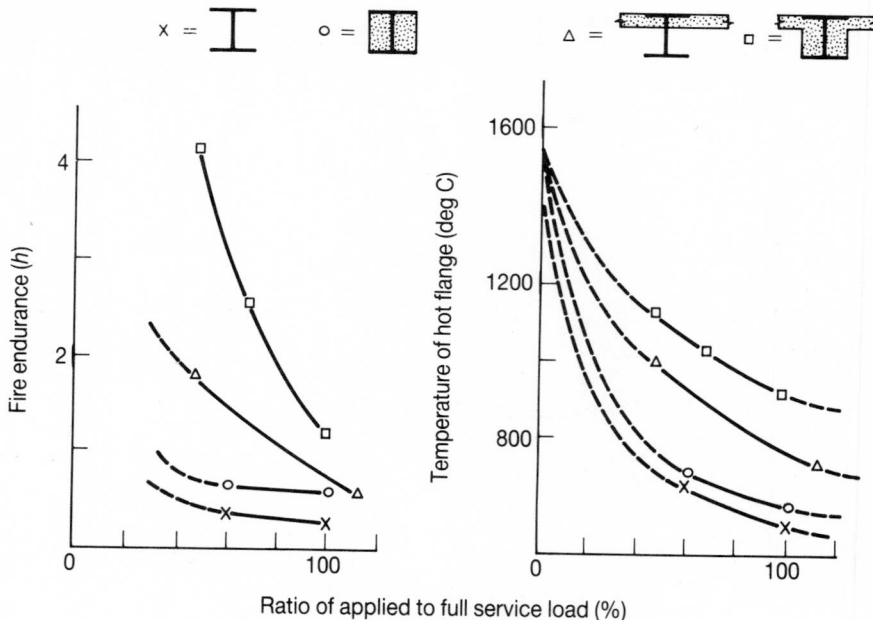

Fig. 11.6 Effect of load intensity and degree of encasement on the fire endurance of stanchions.

service loading). There may also be a loss in economy since the result from the fire test is expressed in terms of fixed durations (0.5, 1, 1.5, 2, 3 and 4 hrs), with a result intermediate to these periods being rounded down.

11.5.2 Manufacturers' data

These may either be obtained from the manufacturer for a specific product, of from the compendium issued by the Association of Structural Fire Protection Contractors and Manufacturers Ltd, Steel Construction Institute and Fire Test Study Group (ASFPCM/ SCI/FTSG). This latter document also contains data on H_p/A values for unprotected and protected members and further has the convenience of allowing the designer to make a more reasoned choice of protection system as all the data are available together. It should be noted that these data are based on a steel temperature of 550°C being attained.

11.6 Calculation approach

11.6.1 Introduction

The design procedure, once the decision whether to consider exposure of the steel element to either a natural fire or the BS 476 pseudo-fire has been made, essentially is to calculate the temperature rise within the structural element and to assess whether the maximum temperature reached will cause collapse. The procedure can be further simplified if the structure has been designed as 'simple' (i.e. all the connections offer little restraint to rotation), as the structural elements–beams and columns–can be considered separately. It should be noted that most simple connections will develop some moment capacity, thus allowing some moment to be transferred from the beam to the column (Lawson (1990)). The situation for frames which are designed as 'rigid' is much more complex, and indeed Part 8 does not cover this type of frame, except for single storey portal frames. For completeness some information on the behaviour of rigid jointed frames is presented in Section 11.9.2.

Since only 'simple' construction is being considered, certain simplifications can be made in the calculation procedure, these are listed below:

(i) The steel element is assumed to be at a uniform mean temperature over its cross-section and throughout its length.

This is not correct as either tests or finite element analyses will show that temperature gradients exist where either the flange is in contact with a concrete slab, or the web of a column is partially protected. Although the protection of a flange or web is allowed for in the calculation of the H_p/A ratio, the above assumption will over-estimate mean temperatures, giving reduced, therefore conservative, fire performance.

(ii) The sections are checked for reduction in strength only or, more correctly, the load ratio is checked.

This approach is possible since the strength reduction parameters used are taken from anisothermal creep tests (i.e. tests in which specimens loaded at constant load are heated to a given strain level, and the resultant strain temperature plot obtained) (see Section 11.9.3.1).

(iii) The thermal properties of both the insulation and the steel are assumed constant with respect to temperature.

This again is incorrect for some insulation materials and for steel, but will produce a conservative result, since the design values used are those measured at steel temperatures of around 550°C, which corresponds very roughly to the limiting temperature reached in the steel in a fire.

11.6.2 Calculation of temperatures

This may be done in one of two ways; either by considering the heat flow equation, or by fitting curves to experimental data. The former approach was advocated in the ECCS Draft Recommendations of 1983, and must in any case by used where the exposure conditions are natural, the latter is used in the 1985 ECCS Design Guide, which forms the basis of BS 5950, Part 8.

11.6.2.1 *Direct calculation*

If the heat flow to steel element is considered, then rate of heat flow $\dot{Q}$ is given by

$$\dot{Q} = KF(T_t - T_s) \tag{11.8}$$

where K is the coefficient of total heat transfer, F the surface area of the element exposed to heat, T_t the ambient gas temperature at time t, and T_s the steel temperature at time t.

The coefficient of heat transfer has three components:

(i) the coefficient of heat transfer due to convection α_c

This mode is not very important and a constant value of 25 W/m²/deg C may be taken.

(ii) the coefficient of heat transfer due to radiation α_r

This is given by the Stefan-Boltzman law. It is usual to take a resultant emissivity, ε_r, of 0.5

$$\alpha_r = (5.77E - 8\alpha_r/(T_t - T_s))((T_t + 273)^4 - (T_s + 273)^4) \qquad (11.9)$$

(iii) the heat transfer through the insulation (d_i/λ_i), where d_i is the thickness of the insulation, and λ_i is an effective thermal conductivity of the insulation (in W/m/deg C).

The total coefficient of heat transfer K is given by

$$K = 1/(1/(\alpha_c + \alpha_r) + d_i/\lambda_i) \qquad (11.10)$$

The rate of heat flow, $\dot{Q}$, may also be written as

$$\dot{Q} = c_s \rho_s V \Delta T_s / \Delta t \qquad (11.11)$$

where c_s is the specific heat of steel, ρ_s the density of steel and V the volume per unit length.

Equating the values of $\dot{Q}$ in Equations (11.8) and (11.11) and replacing F/V by H_p/A gives,

$$\Delta T_s = (K/c_s \rho_s)(H_p/A)(T_t - T_s)\Delta t \qquad (11.12)$$

Equation (11.12) forms the basis for calculating temperatures in either uninsulated or insulated members, and is in both cases capable of being simplified.

For unprotected members, Equation (11.12) reduces to

$$\Delta T_s = (\alpha/c_s \rho_s)(H_p/A)(T_t - T_s)\Delta t \qquad (11.13)$$

where

$$\alpha = \alpha_c + \alpha_r$$

For insulated members the term d_i/λ_i is much larger than the terms due to convection and radiation and thus K reduces to

$$K = \lambda_i/d_i \quad \text{and Equation (11.2) to}$$

$$\Delta T_s = ((\lambda_i/d_i)/c_s \rho_s)(H_p/A)(T_t - T_s)\Delta t \qquad (11.14)$$

For members where the insulation has a substantial heat capacity, Equation (11.12) is modified to

$$\Delta T_s = ((\lambda_i/d_i)/c_s \rho_s)(H_p/A)(T_t - T_s)\Delta t/(1 + \xi) - \Delta T_t/(1 + 1/\xi) \qquad (11.15)$$

where

$$\xi = c_i \lambda_i d_i H_p/(2c_s \rho_s A) \qquad (11.16)$$

This additional modification is only needed when $\xi > 0.25$

Equations (11.12) to (11.15) are all suitable for use in hand calculations, or spreadsheet, which are illustrated in Examples 11.1 and 11.2.

If too large a value of Δt is used in these calculations, then inaccuracies will occur, and it is therefore recommended (ECCS Recommendations) that Δt be limited such that

$$\Delta t < 25000/(H_p/A) \tag{11.17}$$

11.6.2.2 *Empirical approach*

For members heated in a furnace, the ECCS Design Guide indicates that the time t_f to reach a steel temperature T_s is given by

$$t_f = 40(T_s - 140)((d_i/\lambda_i)(A/H_p + d_i\rho_i/\rho_s))^{0.77} \tag{11.18a}$$

The background to this equation and discussion of its accuracy is given in Melinek and Thomas, and Melinek.

The ECCS recommendations allow for the effect of moisture in the insulation, which will increase the time to achieve a given temperature (since heat is required to turn the moisture into steam and thus reduce the heat input to the steel), by introducing a time delay term t_v which is defined by

$$t_v = 0.2c\rho_i d_i^2/\lambda_i \tag{11.18b}$$

where c is the percentage moisture content by weight in the insulation.

Thus the total time to reach a given temperature is given by the sum of t_f from Equation (11.18a) and t_v from Equation (11.18b). This approach is cumbersome for the purposes of design as an explicit relationship for d_i cannot be obtained. Melinek (1989) has shown that the same result may be obtained by using Equation (11.18a) with a moisture dependent density of the insulation, defined in Equation (11.21).

Equation (11.18a) forms the basis of the method in Appendix D of BS 5950, Part 8.

In all the examples generic values of the various thermal parameters will be used for insulation rather than values relating to specific products. In addition values will be needed for steel.

The values to be used are:

For steel:

Density $\rho_s = 7850 \text{ kg/m}^3$

Specific heat $c_s = 520 \text{ J/kg/deg C}$ (cl 2.1, Pt 8)

Thermal conductivity $a_s = 37.5 \text{ W/m/deg C}$ (cl 2.1, Pt 8).

For insulation:

The values required are given in Malhotra and are presented in Table 11.2.

11.6.2.3 *Calculation of H_p/A values*

The cross-sectional area A is always taken as the area of the steel member. The heated perimeter H_p will depend on the type of insulation, e.g. sprayed insulation or intumescent paint, which is applied to the profile of the section or board insulation which boxes the section, and also on whether the member is heated on three or four sides. The calculation of the heated perimeter (H_p) to cross-sectional area (A) ratio may either be performed from first principles using the data in Fig. 11.7 or by reference to the ASFPCM/SCI/FTSG Manual, to Wainman and Kirby, or to Table 3 of Part 8.

Table 11.2 Thermal properties of insulation materials

Material	Density kg/m³	Specific heat J/kg/deg C	Thermal conductivity W/m/deg C	Moisture content % by weight
Sprayed mineral fibre	250–350	1050	0.10	1.0
Vermiculite slabs	300	1200	0.15	7.0
Vermiculite/gypsum slabs	800	1200	0.15	15.0
Gypsum plaster	800	1700	0.20	20.0
Mineral fibre sheets	500	1500	0.25	2.0
Aerated concrete	600	1200	0.30	2.5
Lightweight concrete	1600	1200	0.80	2.5
Dense concrete	2200	1200	1.5	1.5

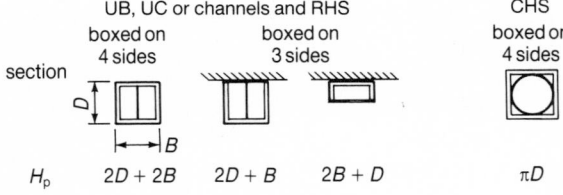

(a) Values of H_p for unprotected steelwork or steelwork protected by sprayed insulation or intumescent paint

(b) Values of H_p for boxed sections

Notes

1 A is always taken as the cross-sectional area of the steel member in calculating H_p/A.
2 The units of H_p/A are m^{-1} and the value is conveniently rounded to the nearest $5m^{-1}$.
3 Fillets and external radii have been ignored in calculating H_p.
4 Values for other configurations than those shown here may be calculated from analogous principles.

Fig. 11.7 Typical H_p/A values.

EXAMPLE 11.1 Temperature calculations on unprotected steel

Calculate the temperature rise on a 203 × 203 × 46 UC heated on 4 sides after exposure for 15 minutes to the BS 476 standard temperature–time curve.

For a bare 203 × 203 × 46 UC, $H_p/A = 202$ /m
The maximum value of Δt to be used is given by,

$$\Delta t = 25000/(H_p/A) \tag{11.17}$$
$$= 25000/202$$
$$= 123.8$$

Use $\Delta t = 120$ s

The governing equation for this case is Equation (11.13),

$$\Delta T_s = (\alpha/c_s\rho_s)/(H_p/A)(T_t - T_s)\Delta t \tag{11.13}$$

with

$$\alpha = 25 + (0.577\text{E} - 7 \times 0.5/(T_t - T_s))((T_t + 273)^4 - (T_s + 273)^4)$$

Substituting numerical values for c_s, ρ_s, H_p/A and Δt gives

$$\begin{aligned}\Delta T_s &= (\alpha/(520 \times 7850)) \times 202 \times (T_t - T_s) \times 120\\ &= (\alpha/168.4)(T_t - T_s)\end{aligned}$$

At $t = 0$, both T_0 and T_s are 20°C

T_t is given by,

$$T_t = T_0 + 345 \log (8t + 1) \tag{11.7}$$

The values of T_s are calculated at $t = 0, 2, 4, 6, 8 \ldots$ mins and the values of T_t, α, $(T_t - T_s)$, and ΔT_s at $t = 1, 3, 5, 7 \ldots$ mins.

The calculations are best carried out in tabular form and are presented in Table 11.3.

Table 11.3   Temperature profile of an unprotected section

t min	T_t deg C	α W/m^2/deg C	$(T_t - T_s)$ deg C	ΔT_s deg C	T_s deg C
0					20.0
	349.2	37.5	329.2	73.3	
2					93.3
	502.3	49.2	409.0	119.5	
4					212.8
	576.4	61.9	363.6	133.7	
6					346.5
	626.8	77.3	280.3	128.7	
8					475.2
	662.8	94.7	187.6	105.5	
10					580.7
	692.5	112.2	111.8	74.5	
12					655.2
	717.3	127.0	62.1	46.8	
14					702.0
	738.6	138.1	36.6	30.0	
16					732.0

Taking the mean of the values for 14 and 16 minutes, the steel has reached a temperature of 717°C.

A check on this result is given by an empirical equation derived from both British and Continental tests (Twilt and Witteveen),

$$t = 0.54(T_s - 50)(H_p/A)^{-0.6} \tag{11.19}$$

where t is the time for the steel to reach a temperature T_s.

Equation (11.19) can be rewritten as,

$$T_s = 50 + t(H_p/A)^{0.6}/0.54 \tag{11.20}$$

Thus calculating the steel temperature at 15 minutes,

$$T_s = 50 + 15 \times 202^{0.6}/0.54$$
$$= 721.3°C$$

This compares well with the calculated value of 717°C using the heat transfer equation.

EXAMPLE 11.2   Calculation of temperature response for an insulated section

Determine the temperature history for a 203 × 203 × 46 UC with fire protection formed by 30 mm thick mineral fibre boarding encasing the column on all four sides (Fig. 11.8).

The density used in the calculation should be modified to an effective density, ρ_i', allowing for the moisture, c (in % by weight), and is given by

$$\rho_i' = \rho_i(1 + c) \tag{11.21}$$

$$\rho_i' = 500(1 + 0.03 \times 2)$$
$$= 530 \text{ kg/m}^3$$

Determine whether the insulation has a substantial heat capacity using Equation (11.16), with the density of insulation modified to allow for moisture content,

$$\xi = c_i \rho_i' d_i H_p/(2c_s \rho_s A)$$

$$H_p/A = 140/\text{m}, \quad \text{so}$$

$$\xi = ((1500 \times 530 \times 0.03)/(2 \times 7850 \times 520)) \times 140$$
$$= 0.41$$

This value is greater than the limiting value of 0.25 and thus the thermal capacity of the section must be considered.

Using Equation (11.15)

$$\Delta T_s = ((\lambda_i/d_i)/(c_s \rho_s))(H_p/A)(T_t - T_s)\Delta t/(1 + \xi) - \Delta T_t/(1 + 1/\xi) \tag{11.15}$$

Calculate the allowable value of Δt from Equation (11.17),

$$\Delta t = 25000/140$$
$$= 178.6 \text{ s}$$

Use a time step of 3 mins.

Substituting values for all the parameters, with $\lambda_i = 0.25$ W/m/deg C, gives

$$\Delta T_s = ((0.25/0.030)/(7850 \times 520)) \times 140 \times (T_t - T_s) \times 180/(1 + 0.41)$$
$$\quad - \Delta T_t/(1 + 1/0.41)$$
$$= 0.0365(T_t - T_s) - 0.291\Delta T_t$$

As in Example 11.1 the reference temperature is taken as 20°C and the values of T_t are calculated from Equation (11.7).

This calculation is again best performed in a table. The results are presented in Table 11.4.

Note that in the very early stages ΔT_s is found to be negative, as the heat flux is being absorbed by the insulation. Negative values have been set equal to zero and indicated thus '*'.

These results, together with those calculated on the assumption that the insulation has negligible thermal capacity, are plotted in Fig. 11.8.

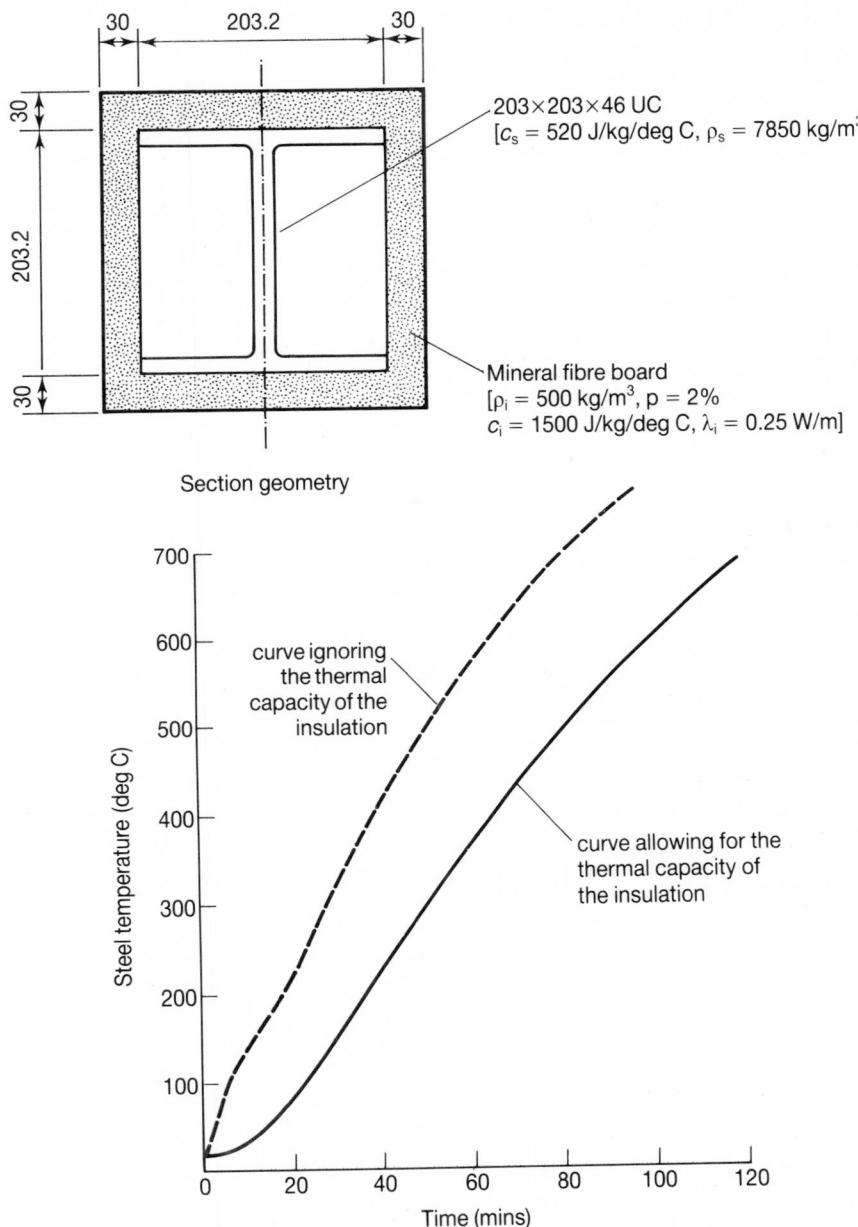

30 | 203.2 | 30

203 × 203 × 46 UC
[c_s = 520 J/kg/deg C, ρ_s = 7850 kg/m³]

Mineral fibre board
[ρ_i = 500 kg/m³, p = 2%
c_i = 1500 J/kg/deg C, λ_i = 0.25 W/m]

Section geometry

curve ignoring
the thermal
capacity of the
insulation

curve allowing for the
thermal capacity of
the insulation

Steel temperature (deg C)

Time (mins)

Fig. 11.8 Temperature–time response.

11.6.3 Calculation of structural response

Initially only members designed in simple construction will be considered as the method is essentially the same irrespective of the type of member. Separate consideration will need to be given to composite slabs designed with profile sheeting, shelf angle floors, waterfilled structures and frames.

Table 11.4   Temperature profile for a protected steel section

t min	T_t deg C	ΔT_t deg C	$(T_t - T_s)$ deg C	ΔT_s deg C	T_s deg C
0					20
	404.3	384.3	384.3	*	
3					20
	561.0	156.3	541.0	*	
6					20
	635.9	74.9	615.9	0.7	
9					20.7
	685.6	49.7	664.9	9.8	
12					30.5
	722.9	37.3	692.4	14.4	
15					44.9
	752.7	29.8	707.8	17.2	
18					62.1
	777.6	24.9	715.5	18.9	
21					81.0
	798.9	21.3	717.9	20.0	
24					101.0
	817.6	18.7	716.6	20.7	
27					121.7
	834.1	16.5	712.4	21.2	
30					142.9

This calculation is then continued and will be skipped to the final lines in the table

t min	T_t deg C	ΔT_t deg C	$(T_t - T_s)$ deg C	ΔT_s deg C	T_s deg C
90					532.3
	1008.5	5.0	476.2	15.9	
93					548.2
	1013.3	4.8	464.8	15.6	
96					563.8
	1018.0	4.7	454.2	15.2	
99					579.0
	1022.5	4.5	443.5	14.9	
102					593.9
	1026.9	4.4	433.0	14.5	
105					608.4
	1031.2	4.3	422.8	14.2	
108					622.6
	1035.3	4.1	412.7	13.9	
111					636.5
	1039.4	4.1	402.9	13.5	
114					650.0
	1043.3	3.9	393.3	13.2	
117					663.2
	1047.2	3.9	384.0	12.9	
120					676.1

Part 8 adopts two methods of calculating the response of a structural element in a fire. The first method is only applicable to beams and is to calculate the moment capacity of the beam using a temperature profile calculated using finite difference or finite element techniques, or obtained from fire tests and the strength reduction factors in Fig. 11.19 (cl 4.4.4, Pt 8). The method is illustrated in Example 11.7.

The second method is to calculate the load ratio R of the effect of the loading specified to be applied during a fire, determine the maximum steel temperature that can be allowed with this value of R and hence determine the thickness of insulation to limit the temperature rise in the element to the maximum allowed.

It is possible, using this method to design a section to have no protection or a reduced level of protection by keeping the value of the load ratio R low, i.e. by overdesigning the section at ambient conditions. It is thus vital that when a steel structure is either checked for change of usage involving increased loading, extension or other alterations that the fire protection requirements are also checked, since in the original design the designer may have kept the load ratio low by deliberate overdesign. If so, it must be realized that an increase in the load ratio will mean an increase in the required level of fire protection assuming that the required fire resistance period remains the same.

11.6.3.1 *Calculation of the loading to be applied for fire performance assessment.*

The limit state to be considered is that of accidental loading which gives rise to a series of reduced load factors compared with those at the normal ultimate limit state. Table 2 (Pt 8) is summarized below:

Dead load: 1.00

Imposed loads:

(a) Permanent loads: these comprise those allowed for in design, e.g. plant, fixed partitions and storage loads; 1.0
(b) Non-permanent loads: the load factor is taken as 0.8 except for loading on escape stairs and lobbies when the value is 1.0.

Note: loading due to snow may be disregarded

Wind loads: 0.33

Wind loads need only be considered in buildings in which the height to eaves exceeds 8 m, and should only be considered to be carried by primary members of the framework.

Material partial factors of safety should be taken as 1.00 for steel (including reinforcement) and 1.3 for concrete.

The load factors to be applied in Part 8 give a ratio of around 1.7 between the ultimate load at ambient conditions and the ultimate load in the fire limit state. This is higher than the equivalent ratio for the fire design of reinforced concrete of around 1.5 and raises an anomaly when composite slabs are considered (Section 11.7). It is almost certain that the load factors have been set to give a ratio of 1.7 since all the tests described in Section 11.4 were performed on specimens designed to BS 449 which operated under a factor of safety of 1.7, and the results from these tests form the basis of the interrelation between load ratio and maximum temperature used in the Code.

The analysis of the structure to determine the loading on elements in the structure for the purpose of assessing their fire resistance should be performed using the same assumptions and methods that would be used under ambient conditions except for the changed load factors and the fact that snow loading is not considered.

11.6.3.2 *Calculation of the load ratio R*

(a) Beams (cl 4.4.4.2, Pt 8)

The load ratio R for beams is defined as the lesser of

$$R = M_f/M_c \quad \text{or} \quad R = mM_f/M_b \tag{11.22}$$

where M_f is the moment on the beam due to the loading applied in a fire, and M_c is the moment capacity of the section at ambient conditions, m is defined by Equation (5.30) and M_b is the moment capacity allowing for the effects of lateral torsional buckling.

This method may only be used for rolled sections, or compound sections fabricated from rolled sections and composite beams provided that there is no reduction in moment capacity due to shear.

(b)   Columns in simple construction (cl 4.4.2.3(a), Pt 8)

$$R = F_f/A_g p_c + M_{fx}/M_b + M_{fy}/p_y Z_y \tag{11.23}$$

where the member capacities (the denominators of Equation (11.23)) are calculated according to the relevant clauses in Part 1, and the forces applied to the column are calculated from the loads applied during the fire.

(c)   Tension members (cl 4.4.2.4, Pt 8)

$$R = F_f/A_s p_y + M_{fx}/M_{cx} + M_{fy}/M_{cy} \tag{11.24}$$

where the capacities are calculated using the relevant clauses of Part 1, and the forces under the effect of the fire loading.

11.6.3.3   *Determination of limiting temperatures*

As noted earlier a series of tests led to the conclusion that the limiting temperature at collapse was a function of the applied stress and also the temperature gradient on the specimen at failure, thus beams which are heated on three sides and columns which are heated on four need to be differentiated. It will also be necessary to limit the strain in the member in order to ensure integrity of the insulating medium at these elevated temperatures.

The limiting temperatures are given in Table 5 of Part 8. It should be noted that where the maximum temperatures for unprotected elements given in Tables 6 and 7, adjusted in accordance with Table 8 if necessary, are below the limiting temperatures in Table 5 the element needs no protection. It should be noted that Tables 6 and 7 only apply to members heated by a fire equivalent to the BS 476 curve.

11.6.3.4   *Determination of insulation thickness (Appendix D, Pt 8)*

It will be recalled that in Section 11.6.2.2, Equation (11.18) was presented which gave the time to failure for an insulated member heated in a furnace test,

$$t_f = 40(T_s - 140)((d_i/\lambda_i)(A/H_p + d_i\rho_i/\rho_s))^{0.77} \tag{11.18}$$

Changing the notation slightly and replacing d_i by t, T_s by T_{lim} (from Table 5, Pt 8), λ_i by k_i and the dry density of the insulation, ρ_i, by the moisture dependent density, ρ_i' defined by Equation (11.21), and defining t_f as the required fire resistance period to give

$$t_f = 40(T_{lim} - 140)((t/k_i(A/H_p + t\rho_i'/\rho_s))^{0.77} \tag{11.25}$$

or

$$t(A/H_p + t\rho_i'/\rho_s) = k_i(t_f/(40(T_{lim} - 140)))^{1.3} \tag{11.26}$$

Define an insulation factor I_f which is given by,

$$I_f = (t_f/(40(T_{lim} - 140)))^{1.3} \tag{11.27}$$

Equation (11.26) is the basis of Table 16 (Pt 8), except that all values in Table 16 have been multiplied by 1E6 and rounded to the nearest 5 units. This of course necessitates the use of dividing by 1E6 in subsequent calculations. This multiplication by 1E6 will be ignored during the derivation of the theory.

Equation (11.26) can be written as a quadratic equation in t

$$(\rho_i'/\rho_s)t^2 + (A/H_p)t - k_i I_f = 0 \tag{11.28}$$

which may be solved to give, after a little manipulation,

$$t = ((1 + 4(A/H_p)^2 I_f k_i \rho_i'/\rho_s)^{0.5} - 1)/(2(\rho_i'/\rho_s)(H_p/A)) \tag{11.29}$$

Define

$$\mu = k_i(\rho_i'/\rho_s)I_f(H_p/A)^2 \tag{11.30}$$

and substitute Equation (11.30) into Equation (11.29) to give

$$t = k_i I_f(H_p/A)((1 + 4\mu)^{0.5} - 1)/(2\mu) \tag{11.31}$$

The protection material density factor, F_w is given by

$$F_w = ((1 + 4\mu)^{0.5} - 1)/(2\mu) \tag{11.32}$$

Values of F_w for various values of μ are given in Table 17 (Pt 8).

The insulation thickness t is given by

$$t = k_i I_f F_w(H_p/A) \tag{11.33}$$

Note that the derivation of Equation (11.33) and its progenitor, Equation (11.18), makes the assumption that the specific heat of insulation materials is twice that of steel. This is a reasonable assumption if Table 11.2 is consulted, and further, the result is not very sensitive to the ratio of specific heats.

It should be further noted that Equation (11.33) is not dimensionless and the result for the insulation thickness in in metres.

Before performing two examples on fire protection design, it is necessary to note a few special cases where extra provisions are required.

11.6.3.5 *Special cases*

(a) Castellated beams (cl 4.3.3.3, Pt 8)

The value of H_p/A should be calculated on the parent section from which the castellated beam is fabricated, and the thickness of insulation determined from Equation (11.33) should be increased by 20%.

(b) Hollow sections (cl 4.3.3.4, Pt 8)

Calculations for the required thickness for either board systems or spray systems follow that for 'I' or 'H' sections with no modification.

Where the protection system requirements are derived from tests on 'I' or 'H' sections with the same H_p/A ratio, no extrapolation of results is allowed for intumescent paint systems, and the thickness of other insulation systems must be modified,

$$\text{for} \quad H_p/A < 250 \text{ /m}, \quad t_{RHS} = t(1 + 0.0001 H_p/A) \tag{11.34a}$$

$$\text{for} \quad H_p/A \geqslant 250, \quad t_{RHS} = 1.25t \tag{11.34b}$$

where t_{RHS} is the thickness to be applied to the hollow section and t the thickness on the equivalent 'I' or 'H' section.

(c)   Tapered sections (cl 4.2.4, Pt 8)

The value of H_p/A used in the calculation must be the maximum value along the length of the member.

Two examples, one a beam and the other a column, are presented. The fire protection calculation for the beam following the calculation of the load ratio R is covered in detail. The second example is covered in less detail. Both elements are deemed to be exposed to the BS 476 standard furnace curve.

EXAMPLE 11.3   Fire protection design for beams

Design suitable fire protection using a gypsum plaster spray to give a fire resistance of 1.5 hours for the composite beam designed in Example 10.2.

To recapitulate design data:
Dead loading 4.6 kN/m²
Imposed load 4.0 kN/m²
Span 4.0 m
$M_c = 148.6$ kNm
From Example 10.2, the line load reaction due to the UDL is given by $4.0 \times 2.98 \times q$
thus using the load factors from Section 11.6.3.1,

$$\begin{aligned} W_f &= 2.98 \times 4(4.6 \times 1.0 + 4.0 \times 0.8) \\ &= 93.0 \text{ kN} \end{aligned}$$

$$\begin{aligned} M_f &= W_f L/8 \\ &= 93.0 \times 4/8 \\ &= 46.5 \text{ kNm} \end{aligned}$$

$$\begin{aligned} R &= M_f/M_c \\ &= 46.5/148.6 \\ &= 0.31 \end{aligned}$$

From Table 5 (Pt 8), assuming the insulation is capable of resisting a strain of 2.0% at failure, $T_{lim} = 720°C$.
   From Equation (11.27) (or Table 16), for a fire resistance period of 1.5 hours,

$$\begin{aligned} I_f &= (t_f/(40(T_{lim} - 140)))^{1.3} \\ &= (90/(40(720 - 140)))^{1.3} \\ &= 7.33E - 4 \end{aligned} \tag{11.27}$$

Table 16 after multiplication by 1E6 gives 738 using linear interpolation.

The H_p/A ratio for a $203 \times 133 \times 25$ UB heated on three sides with profile insulation is 240 /m (from the ASFPCM/SCI/FTSG Manual or Fig. 11.7)

For gypsum plaster,

$$\rho_i = 800 \text{ kg/m}^3, \quad c = 20\%, \quad \text{and } k_i = 0.20 \text{ W/m/deg C (Table 11.2)}$$

Calculate the moisture dependent density, ρ_i', from Equation (11.21),

$$\rho_i' = \rho_i(1 + 0.03c) \tag{11.21}$$
$$= 800(1 + 0.03 \times 20)$$
$$= 1280 \text{ kg/m}^3$$

Calculate μ, using the calculated value of I_f, from Equation (11.30),

$$\mu = k_i(\rho_i'/p_s)I_f(H_p/A)^2 \tag{11.30}$$
$$= 0.20(1280/7850) \times 7.33\text{E} - 4 \times (240)^2$$
$$= 1.377$$

Calculate F_w using Equation (11.32) giving,

$$F_w = ((1 + 4\mu)^{0.5} - 1)/(2\mu) \tag{11.32}$$
$$= ((1 + 4 \times 1.377)^{0.5} - 1)/(2 \times 1.377)$$
$$= 0.563$$

Calculate the required thickness of insulation t using Equation (11.33),

$$t = k_i I_f F_w(H_p/A) \tag{11.33}$$
$$= 0.2 \times 7.33\text{E} - 4 \times 0.563 \times 240$$
$$= 0.020 \text{ m}$$

EXAMPLE 11.4 *Fire protection design for columns*

Design the fire protection to give 90 mins fire resistance for a column using mineral fibre boarding.

The unfactored dead and imposed loading, ignoring snow load, is given in Fig. 11.9. Calculation of applied loading during the fire:

(a) Axial load (F_f)

Floor	Dead load	Imposed load
Roof	78	0
Second Floor	134	70
Total Loads	212	70

$$F_f = 212 + 0.8 \times 70$$
$$= 268 \text{ kN}$$

(b) Moments about y–y axis (M_{fy})

$M_{fy} = 0$ for both floors

(c) Moments about x–x axis (M_{fx})

These are calculated exactly the same way as for ambient conditions by assuming the reaction acts at 100 mm from the face of the column flange.

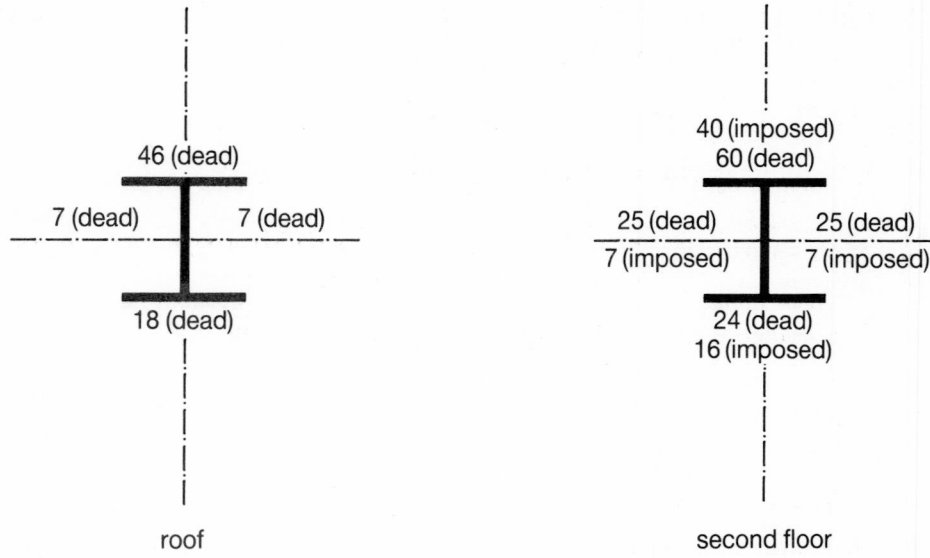

Fig. 11.9   Beam reactions (in kN) on stanchion.

Roof:

$$M_{fx} = ((46 - 18)(100 + 203.2/2)/1E3$$
$$= 5.64 \text{ kNm}$$

Second floor:

$$M_{fx} = ((60 - 24) + 0.8(40 - 16))(11 + 203.2/2)/1E3$$
$$= 11.13 \text{ kNm}$$

From the ambient condition design,

$$A_g p_c = 58.8E2 \times 180/1E3$$
$$= 1019 \text{ kN}$$

$$M_b = S_x p_b = 497 \times 289/1E3$$
$$= 138.4 \text{ kNm}$$

Conservatively assume no distribution of M_{fx} between the upper and lower stories, and thus apply the total moment to the stanchion between first and second floor.

Calculate the load ratio R using Equation (11.23)

$$R = F_f/A_g p_c + M_{fx}/M_b + M_{fy}/p_y Z_y$$
$$= 268/1019 + 11.13/138.4 + 0$$
$$= 0.34$$

From Table 5 (Pt 8), with an axial slenderness of 51.1, $T_{lim} = 639$

Using Equation (11.27)

$$I_f = (90/(40(639 - 140)))^{1.3}$$
$$= 8.92E - 4$$

For mineral fibre boarding (Table 11.2):

$$\rho_i = 500 \text{ kg/m}^3, \quad c = 2\%, \quad k_i = 0.25 \text{ W/m/deg C}$$

From Equation (11.21) calculate ρ_i'

$$\rho_i' = 500(1 + 0.03 \times 2)$$
$$= 530 \text{ kg/m}^3$$

From the ASFPCM/SCI/FTSG Manual (or Fig. 11.7), $H_p/A = 140$ /m

Calculate μ from Equation (11.30)

$$\mu = 0.25 \times (530/7850) \times 8.92\text{E} - 4 \times 140^2$$
$$= 0.295$$

Calculate F_w from Equation (11.32)

$$F_w = ((1 + 4 \times 0.295)^{0.5} - 1)/(2 \times 0.295)$$
$$= 0.808$$

Calculate t from Equation (11.33)

$$t = 0.25 \times 8.92\text{E} - 4 \times 0.808 \times 140$$
$$= 0.025 \text{ m}$$

Thus a board thickness of 25 mm would suffice.

If the nominal connection moment M_{fx} were to be divided between the upper and lower storeys, then

$$R = 0.30, \quad T_{\lim} = 655°\text{C}, \quad I_f = 8.56\text{E} - 4, \quad \mu = 0.283, \quad F_w = 0.813, \quad t = 0.024$$

This produces very little difference in the result, and as would be expected as the limiting temperature is slightly reduced, the insulation thickness is equally slightly reduced. The reason for only a very small change is that in this example the moment term contributes little to the value of R.

For certain types of member such as composite slabs, shelf angle floors, external steelwork, water filled structures and rigid frames the methods described above are not generally applicable, and recourse must be made to more sophisticated methods which generally involve a direct analysis of the structural element during a fire.

11.7   Composite slabs

11.7.1   Introduction

Most composite slabs with steel decking have no protection applied to the steel profile decking and the steel decking suffers a large temperature rise and loses strength rapidly and thus cannot be used to carry any applied loading. Therefore the slab will behave like a normal reinforced concrete slab without profile decking. This means that if the required fire resistance period is longer than 30 mins, additional tension reinforcement may be needed to carry the sagging moment, and, since there will be a redistribution of moments away from mid-span (as the loss in moment capacity will be more rapid here than at the support) to the support, it may be necessary to detail reinforcement additional to the minimum specified if the slab is designed as simply supported or to check the capacity at the support if the slab has been designed as continuous. For a continuous slab the critical span will be the end span since continuity will only exist over one support.

It has been shown from tests (Lawson (1985)) that for fire resistance periods up to 90 mins no additional tension reinforcement may be necessary. Subsequent to the tests reported by Lawson and additional tests, reported in Cooke *et al.*, CIRIA produced a simplified design guide (Special Publication 42).

There are two design criteria to be considered for the slab. These are:

1   The slab should be sufficiently thick to limit the temperature rise on the unexposed surface to an average value of 140°C. This requirement is directly specified by BS 476.
2   The slab should be able to carry the applied load without collapse, or excessive deflection, throughout the duration of the fire.

Note that throughout this section, exposure to fire means exposure to the BS 476 furnace temperature–time curve.

11.7.2   Insulation requirement (cl 4.9.2.2, Pt 8)

This is covered by specifying a minimum thickness of concrete above the top of the profile for open profile decking and total slab thickness for re-entrant profile decking, with a stipulation that the minimum depth should be 50 mm and that the width of the opening at the base of the re-entrant section of the profile should not exceed 10% of the spacing of the re-entrant sections. The required data are given in Table 11.5 which is a conflation of Tables 13 and 14 of Part 8.

Table 11.5   Insulation requirements for profile decking

Fire resistance (hours)	Minimum concrete dimension (mm)			
	Trapezoidal decks		Re-entrant decks	
	NW	LW	NW	LW
0.5	60	50	90	90
1.0	70	60	90	90
1.5	80	70	110	105
2.0	100	80	125	115
3.0	130	110	150	135
4.0	150	130	170	150

Note
NW = Normal-weight concrete and LW = lightweight concrete.

It will be noted that in recognition of its better performance as an insulator the thicknesses specified for lightweight concrete are lower. This together with the reduced specific weight means that dead loads on the structure will be lower.

11.7.3   Structural capacity (cl 4.9.2, Pt 8)

This clause either allows the use of the CIRIA Special Publication 42 or direct calculation. Only the latter will be considered here. For the background to the Fire Safety Engineering approach for reinforced concrete slabs reference should be made to Chapter 5, Section 5.9.3 of Martin *et al.*, or to Newman (1989).

Cooke *et al.* suggested that in a fire the flexural capacity of a composite deck with no additional sagging reinforcement could be calculated using the strength of the decking reduced to 5% of its ambient strength. This enhancement has not been used in Example 11.5, and so the results quoted therein are conservative.

The temperatures in the slab should be determined using either Fig. 11.10(a) and (b) (which are reproduced from Fig. 14 in the IStructE/Concrete Society Design Guide or Table 12 in Part 8, which is a tabular version of the IStructE/Concrete Society data). Note the cover to the bar must be taken as the shortest distance measured normal from the fire-exposed steel profile decking. The curves for loss of strength, in both the steel reinforcement and the concrete, are those in the IStructE/Concrete Society Design Guide (given in this text as Fig. 11.11) also to be found in BS 8110, Part 2.

The load factors to be applied are those specified by BS 8110, Part 2, i.e. 1.05 on dead loading and 1.00 on imposed loading. Note Newman (1989) suggests values of unity for both dead and imposed.

At first glance it would appear that these load factors are at variance with those used for the fire limit state on steelwork and that different loading is applied to the slab and its supporting beam system. Whilst this apparent anomaly is regrettable in terms of producing a totally unified design approach irrespective of materials, it arose because the calculated performance of structural steelwork in a fire was calibrated against tests on members designed to BS 449 (an elastic design code) with a safety factor on stresses of 1.7, and that the calculated performance of concrete structures was calibrated against a limit state code (CP 110), whose load factors were taken over into BS 8110. In both cases the partial safety factors on the loads and the material design strengths were set to give a correspondence between test and calculation, albeit on a different basis.

The design procedure for composite slabs is best illustrated by an example.

EXAMPLE 11.5   Fire safety engineering design of a composite slab

Check the suitability of the slab designed in Example 10.1 for 90 mins fire resistance.

For convenience the design parameters are repeated:

Span 2.5 m
Imposed load 4.0 kN/m^2
Dead load (including finishes, etc) 4.6 kN/m^2
Overall thickness of deck 105 mm
Concrete: Lightweight Grade 30.
Decking: Super Holorib, 0.9 mm thick
Reinforcement at the support: B283 mesh giving 283 mm^2/m with a cover of 25 mm

Insulation check:
from Table 11.5, minimum thickness for 90 min fire resistance with lightweight concrete is 105 mm.
Actual thickness 105 mm.

Structural check:
The loading q_f during the limit state of fire is given by,

$$q_f = 1.05 \times 4.6 + 1.0 \times 4.0$$
$$= 8.83 \text{ kN/m}^2$$

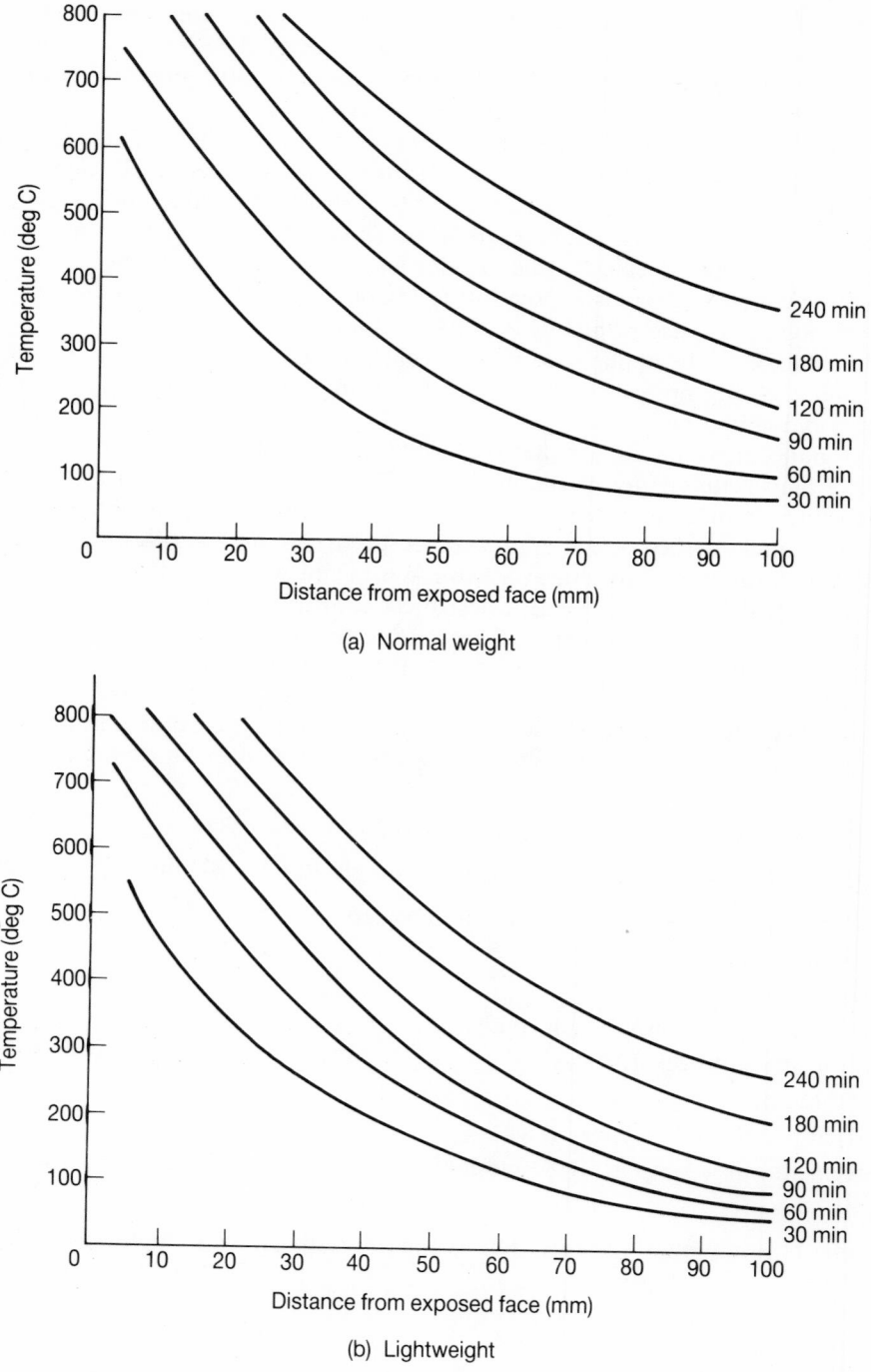

(a) Normal weight

(b) Lightweight

Fig. 11.10   Temperature profiles for normal and lightweight concrete.

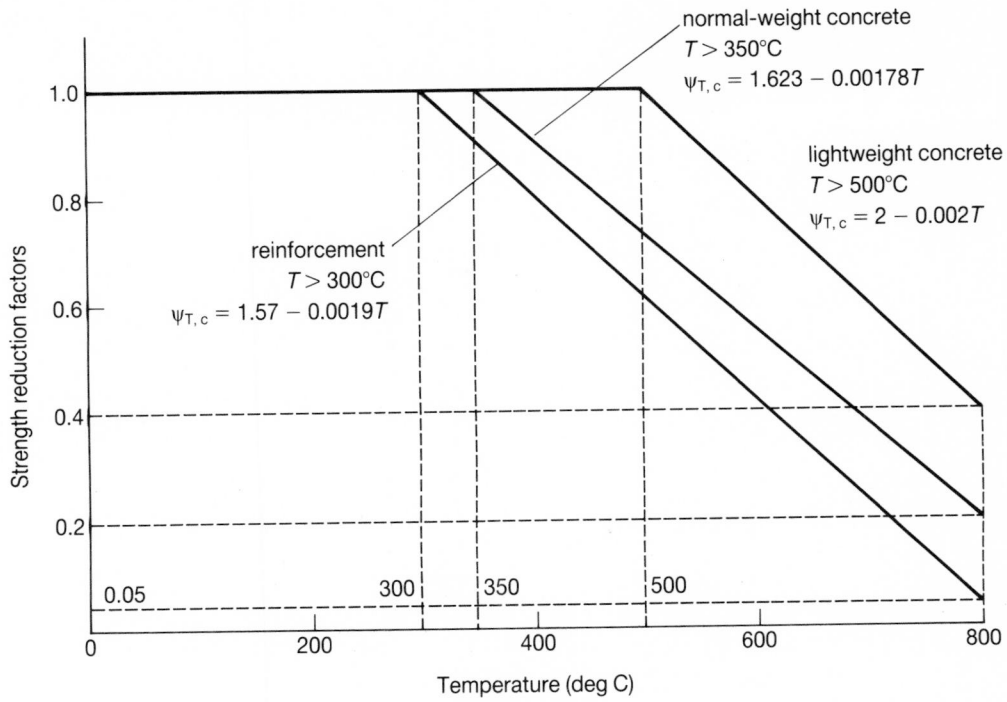

Fig. 11.11   Design strength reduction factors for concrete ($\psi_{T,c}$) and reinforcement ($\psi_{T,s}$).

The free bending moment M_f is given by

$$M_f = q_f L^2 / 8$$
$$= 8.83 \times 2.5^2 / 8$$
$$= 6.9 \text{ kNm/m}$$

Check the moment capacity at the support.

This will need an iterative calculation since the concrete is temperature-affected and the depth of the neutral axis is not known in advance. The symbols are defined in Fig. 11.12.

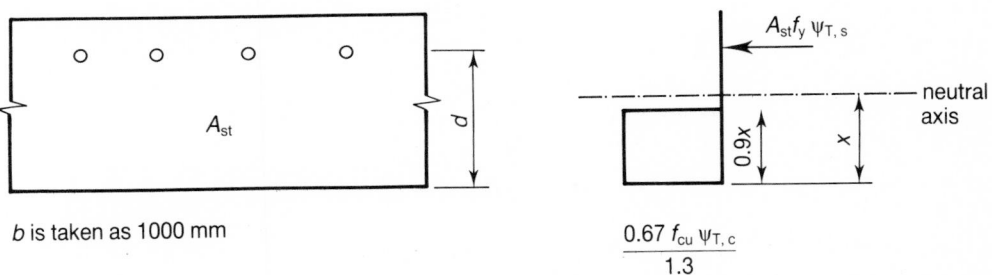

Fig. 11.12   Calculation of support (hogging) moment capacity.

Assume $x = 22$ mm,

centroid is at $0.45x = 0.45 \times 22 = 10$ mm

From Fig. 11.10, the concrete at the centre of the compression block is at $710°C$ and the reinforcement temperature is less than $300°C$, and therefore suffers no strength loss.

From Fig. 11.11 the strength loss factor for the concrete, $\psi_{T,c} = 0.58$

Equating forces:

$$0.67b(0.9x)\psi_{T,cfcu}/1.3 = A_{st}f_y\psi_{T,s} \qquad (11.34)$$

Substituting in values gives,

$$0.67 \times 1000 \times 0.9x \times 30 \times 0.58/1.3 = 385 \times 460$$

or $x = 21.9$ mm

This result is sufficiently accurate so take $x = 22$ mm

$$z = d - 0.45x < 0.95d$$
$$= 66 - 0.45 \times 22 = 56.1 \text{ mm}$$

$$0.95d = 0.95 \times 66 = 62.7$$

$$M_H = A_{st}f_y\psi_{T,sz}$$
$$= 385 \times 460 \times 1.0 \times 56.1/1E6$$
$$= 9.94 \text{ kNm/m}$$

The free bending moment M_f is less than this and therefore for an internal span no additional reinforcement is necessary.

This will not be true for the end span, since the bending moment diagram at collapse is as Fig. 11.13.

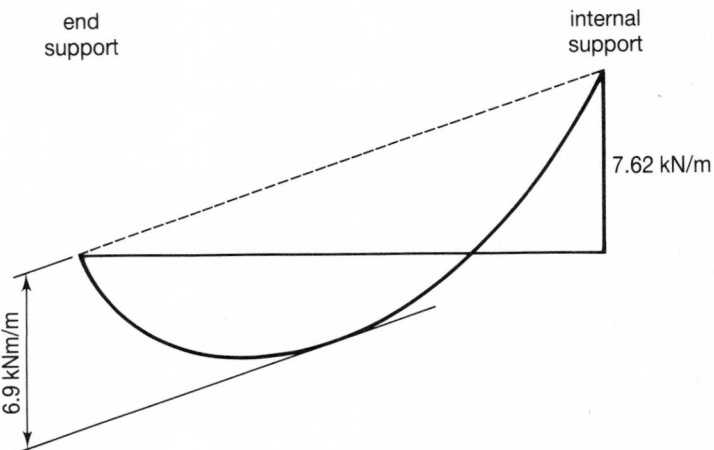

Fig. 11.13   BMD for end span.

This means that the hogging reinforcement will be insufficient to resist the effects of the fire, and that additional reinforcement will be needed in the bottom of the deck level with the top of the re-entrant profile (Fig. 11.14).

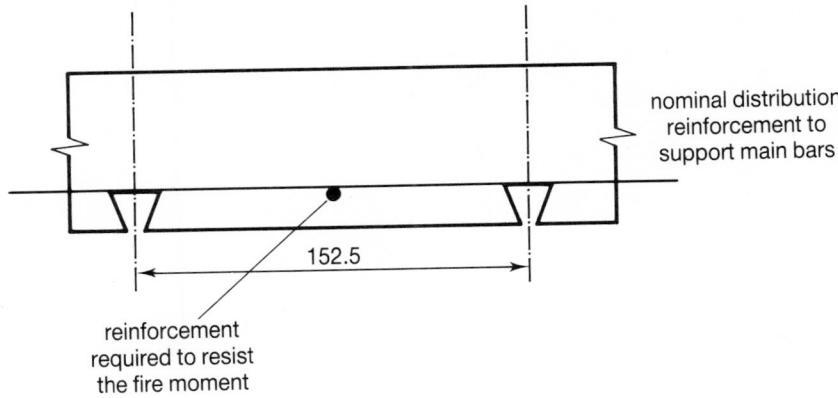

Fig. 11.14 Additional reinforcement to deck.

An initial estimate of the required moment capacity M_s may be made by assuming that the plastic hinge occurs at mid-span and is given by,

$$M_s = M_f - M_H$$
$$= 6.90 - 9.94/2$$
$$= 1.93 \text{ kNm/m}$$

The cover to each bar measured normal to the surface of the profile is 50 mm, thus the steel temperature, T_s from Fig. 11.10(b) is 220°C giving no strength loss. Determine the area of reinforcement using an effective depth of 55 mm and a lever arm z of 0.95d or 47.5 mm (given that the low effective depth the neutral axis depth will be small).

$$A_{st} = M_s/(zf_y)$$
$$= 1.93\text{E}6/(47.5 \times 460)$$
$$= 88.3 \text{ mm}^2/\text{m}$$

Assuming there is one bar per trough, the spacing s will be 152.5 mm and the area of each bar A is given by

$$A = A_{st}/(1000/s)$$
$$= 88.3/(1000/152.5)$$
$$= 13.5 \text{ mm}^2$$

giving a bar diameter of 6 mm ($A = 28.3$)

For 6 mm bars at 152.5 mm centres, $M_u = 4.05$ kNm/m

A more accurate check can be made on the end span:

From Newman (1989) or Cooke *et al.*, with changed notation,

$$M_s + 0.5M_H(1 - M_H/(8M_f)) \geqslant M_s \tag{11.35}$$

Substituting in values gives,

$$M_s + 0.5M_H(1 - M_H/(8M_f)) = 4.05 + 0.5 \times 9.94(1 - 9.94/(8 \times 6.9))$$
$$= 8.13 \text{ kNm/m}$$

This exceeds the free moment and is therefore satisfactory.

11.8   External steelwork (cl 4.8 and 4.9, Pt 8)

For architectural reasons the frame to the structure may need to be external to the structure and the cladding line and the windows are therefore within the shell of the building. This means that were a fire to occur in such a structure the external frame would be exposed to the flames and the resultant heat of the fire emerging through the window openings. Clearly conventional methods of fire protection would be inappropriate in such a situation and it is therefore necessary to develop an alternative approach, in which either the frame can be situated in a zone such that the temperature of the steelwork is limited or that it may be kept cool during a fire.

The first approach was formulated by Law and O'Brien in which calculations are performed to assess the maximum temperature reached by the steelwork frame for a particular fire severity and window size, and gives design tables for the spacing of columns and beams away from the openings so as to limit the temperature of the steel members to below 550°C.

The second approach led to the concept of a water-filled structure, which is only available if the external frame is fabricated from rolled hollow sections in which the internal void is continuous throughout the frame. In this the frame is permanently full of water (with additives to prevent freezing), kept topped up by a system of gravity fed water tanks. The water will heat up in the area of a fire, and start a thermally fed circulation cycle in which colder water will replace the hotter water and conduct the heat away, thus keeping the steel cool. Further information on this is given by Bond.

11.9   Frames

11.9.1   Introduction

It is not possible to use simple methods of assessment of the fire performance of frames except for a very limited number of cases. The main reasons for this are that simple methods are incapable of handling any redistribution of forces within the frame during the fire. Also problems will be caused by buckling or instability of members due either to increased forces due to redistribution, or to loss of stiffness caused by the material degrading at elevated temperatures.

Since the furnace test is expensive to carry out for individual members it is even more prohibitive to carry out tests on frameworks, even assuming the test facilities to be available. This means that for frameworks much more reliance must be placed on computer simulation, which needs a comprehensive set of data on both the thermal and mechanical properties of steel, and in the case of composite structures, concrete. However, before considering the background to computer simulation, some experimental results on both full size and model frameworks are presented.

11.9.2   Frame tests

11.9.2.1   *Small scale tests*

Tests were carried out on model pinned feet portal frames of 400 mm span and 100 mm height fabricated from various sized members to give different slenderness ratios (Witteveen *et al.*). Frames were tested with both no sway and with sidesway allowed, with the column members and the rafter being heated uniformly. The loading conditions used are illustrated in Fig. 11.15. The values of the loads K and H were sufficient to generate moments equal to fifty percent of the collapse moment at the eaves. Fig. 11.15 also indicates the collapse mechanisms that were noted in the tests. The interesting case is that of the no-sway frames in which four hinges formed, those in the stanchions formed first due to buckling and then those at the eaves, giving collapse.

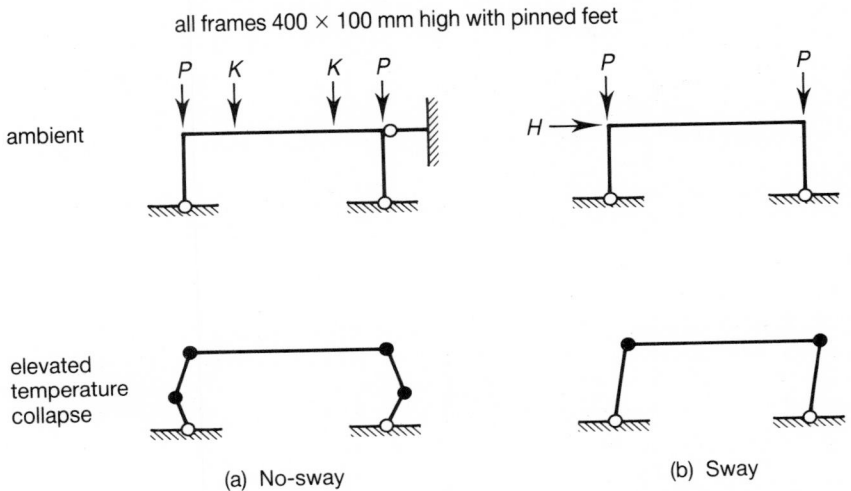

all frames 400 × 100 mm high with pinned feet

ambient

elevated
temperature
collapse

(a) No-sway   (b) Sway

Fig. 11.15 Loading patterns and frame geometry for model tests.

The experimental results were compared with those predicted by an incremental elasto-plastic analysis to determine the temperatures at which collapse would occur. The results are presented in Table 11.6, where it is seen that increasing the axial load on the stanchion significantly decreases the collapse temperature, and increasing the slenderness ratios decreases the collapse temperature, but that for the sway cases the effect of increasing the rate of heating did not significantly affect the results.

11.9.2.2 Medium scale tests

A series of tests on frames with spans of 1200 and heights of 1170 mm were carried out using a steel of strength 400 MPa (Rubert and Schaumann). The ratio of test load to failure load varied between 0.38 and 0.79, with the frame slenderness (defined as the square root of the ratio of the frame collapse load to the elastic buckling load) between 0.33 and 1.0. The bare steel frames were heated uniformly at a constant rate until failure ensued. The calculated failure temperatures were derived from an elasto-plastic analysis using stress–strain data derived from tests on an isostatic beam system. These stress–strain data therefore include the effect of creep. The correlation between calculated collapse temperatures and experimental results is very close as is seen in Fig. 11.16.

11.9.2.3 Full size tests

For reasons mentioned above, very few tests have been carried out on full size frames, although one significant test reported by Cooke and Latham was on an unprotected frame, although the column webs were infilled with blockwork. The portal frame was fabricated

Table 11.6   Model frame test results

Frame type	Effective slenderness ratio	P/P_{20}	$T_{crit,anal}$ °C	$T_{crit,test}$ °C	Heating rate deg C/min
No-sway	44	0.4	525	500	10
		0.6	380	350	10
	28	0.4	590	515	10
		0.6	530	425	10
Sway	120	0.4	310	325	10
			345	325	10
		0.6	260	275	10
			250	275	10
	76	0.2	535	505	10
			390	400	10
		0.6	270	285	10
			260	285	10
	76	0.2	535	505	5
			530	505	10
		0.2	535	505	50

Note
(1) $T_{crit,anal}$ is the calculated failure temperature and $T_{crit,test}$ is the experimental failure temperature.
(2) P/P_{20} is the ratio of the load applied at the top of the stanchion to that causing failure at ambient conditions.

from 203 × 203 × 52 UC 3.53 m long for the stanchions and a 406 × 178 × 54 UB 4.55 m long for the rafter. All steelwork was Grade 43A. The frame had pinned feet and flexible end plate connections between the rafter and stanchions. The frame was loaded by four loads of 39.6 kN applied at fifth points on the rafter and loads of 552 kN applied to the top of each stanchion.

The rafter alone, with no end restraint would have a fire resistance of 20 minutes, and the column with blocked in webs of 36 mins in a standard fire test. The test had to be terminated after 22 mins as the loading could no longer be applied owing to excessive deflections occurring, with plastic hinges having occurred some 600 mm from the ends of the beam. The maximum temperature of the rafter was 775°C and the stanchion 600°C. The effect of the natural fire was deemed to be equivalent to 32.5 minutes in a standard fire test. This indicates that a framework can achieve a longer fire resistance than that for the individual elements.

11.9.3   Calculation of frame behaviour in a fire

It is not proposed to give full details of such methods but rather to produce an overview. In order to carry out any computer simulation it is necessary to have complete data on materials at elevated temperatures. An excellent compendium of data is provided by two RILEM reports (Anderberg, Schneider), one on concrete and one on steel, but the steel report needs to be carefully used as the differing chemical compositions of various steels will affect results.

11.9.3.1   *Mechanical behaviour of steel at elevated temperature*

1   Thermal expansion:

This may be taken as constant at 14 microstrain/deg C. A value of 12 is given in cl 3.1.2, Pt 1, which only applies at temperatures close to ambient conditions.

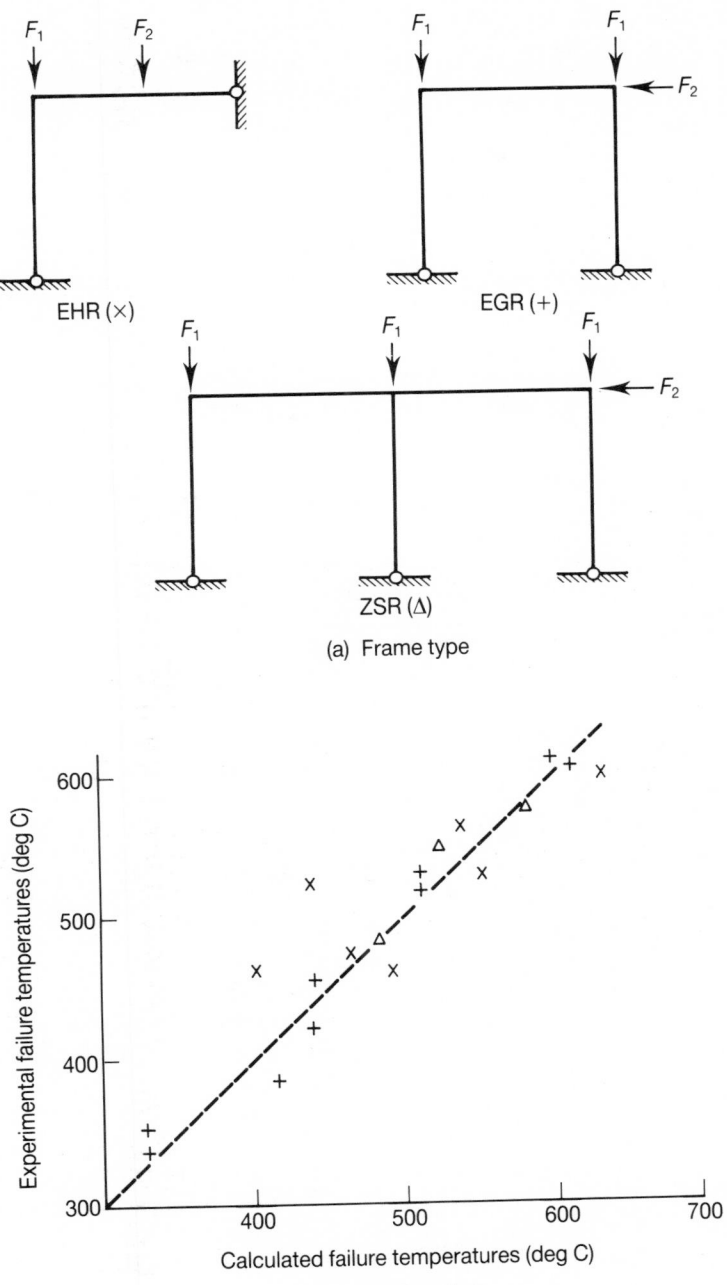

(a) Frame type

(b) Results

Fig. 11.16 Frame details and results for large scale frame tests.

2   Stress–strain data

Traditionaly these have always been obtained from tests in which the specimens are heated to the given test temperature, allowed to stabilize at that temperature and then tested either under constant strain rate or constant stress rate. Data for Grade 43A steel are presented in Fig. 11.17 (after Cooke).

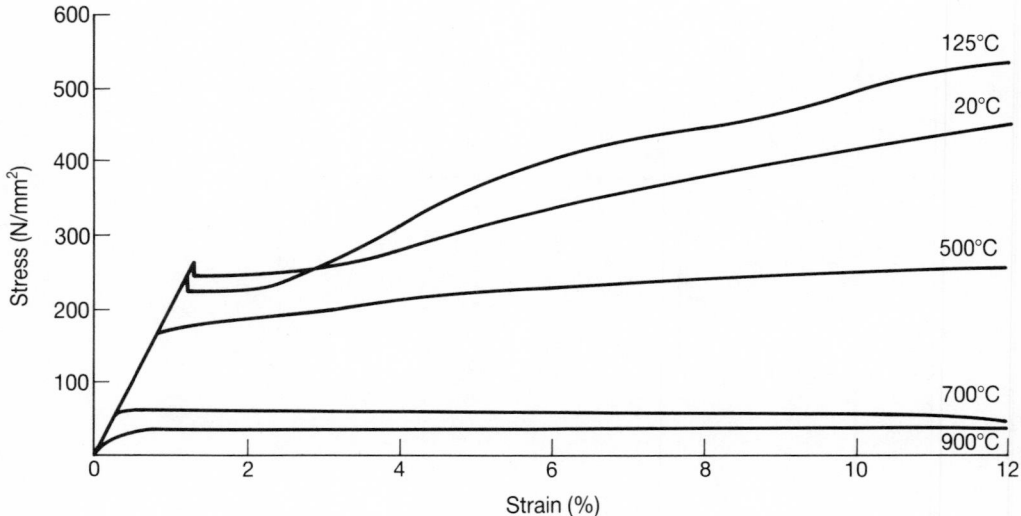

Fig. 11.17   Stress–strain data for Grade 43A steel at constant temperature.

From these results it will be observed that the yield plateau that occurs at room temperature (or only marginally increased temperatures) does not exist at much higher temperatures. The problem with data of this type is that there will be certain amounts of creep strains within the results which are not easy to calculate.

Traditional creep tests are not easy to carry out as little creep occurs below approximately 500°C and it is very difficult given the paucity of data to predict the amount of creep. It is easier to allow for creep by carrying out anisothermal creep tests in which the load (or stress) is kept constant, the specimen heated and the resultant strains measured after deduction of the initial elastic strains on loading.

Typical data (from Kirby and Preston) for Grade 43A and 50B are given in Fig. 11.18 (a) and (b).

From such isothermal data it is possible to construct traditional stress–strain curves at constant temperature by taking values of strain from the intersection between the line of constant temperature and the plot for a given stress level. These curves may be non-dimensionalized on the stress axis by dividing by the steel strength. Non-dimensionalized stress–strain data obtained from anisothermal creep data are presented in Kirby and Preston. From such stress–strain curves data for the variation of proof stress taken at strains of 1.0, 2.0, and 5.0% strains with temperature may be prepared. The reasons why the 0.2% proof stress is not used and a range of proof stresses are required is that the strains measured in fire tests at failure are much higher than those that would be acceptable in ambient conditions.

The proof stress data from Table 1 of Part 8 are plotted in Fig. 11.19.

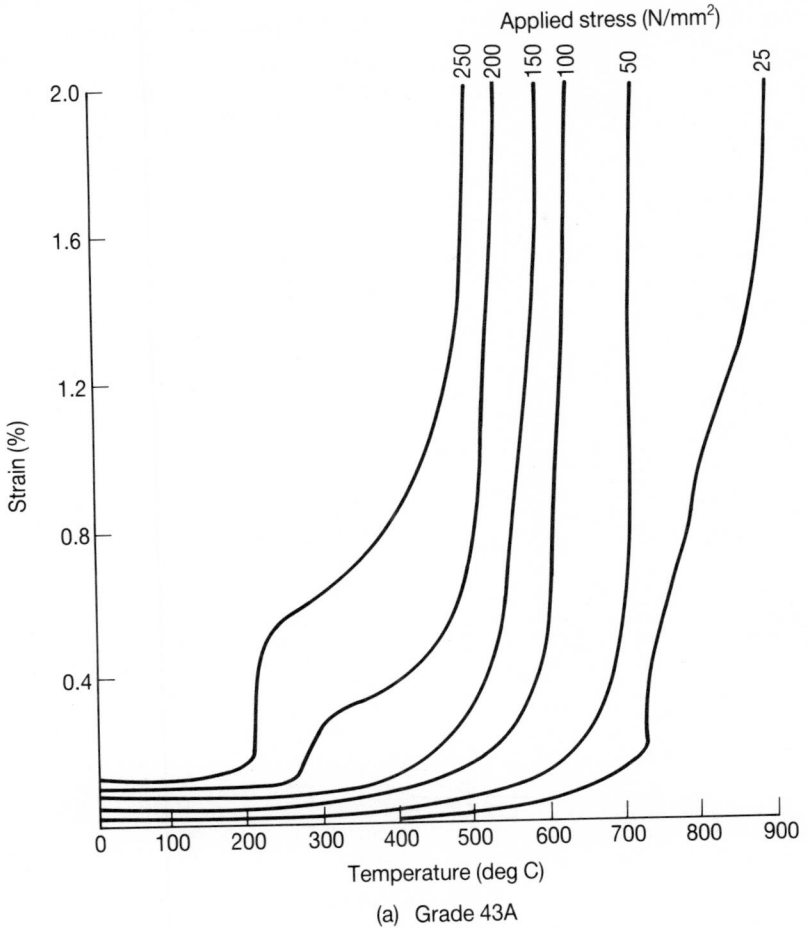

Fig. 11.18   Anisothermal creep data.

11.9.3.2   *Computer simulation of frame behaviour*

There are two phases to the computer simulation of the behaviour of steelwork in fire conditions. The first phase is the calculation of temperatures within the members of the frame, and the second is the calculation of structural response.

1   Calculation of temperature response

The method used is a lumped mass finite element analysis using temperature dependent thermal properties and the requisite boundary conditions. Examples of such programs are FIRES-T2 (Becker *et al.*), FIRES-T3 (Iding *et al.*) or TASEF-2 (Wickstrom). The output from these programs is then used as data for the structural analysis program.

2   Calculation of structural response

One of the first programs to be developed was FASBUS II, the theory behind which is given by Jeanes. There are some drawbacks with the current version, most notably the fact that the column support system is not fire-affected and is assumed to be fire-unaffected.

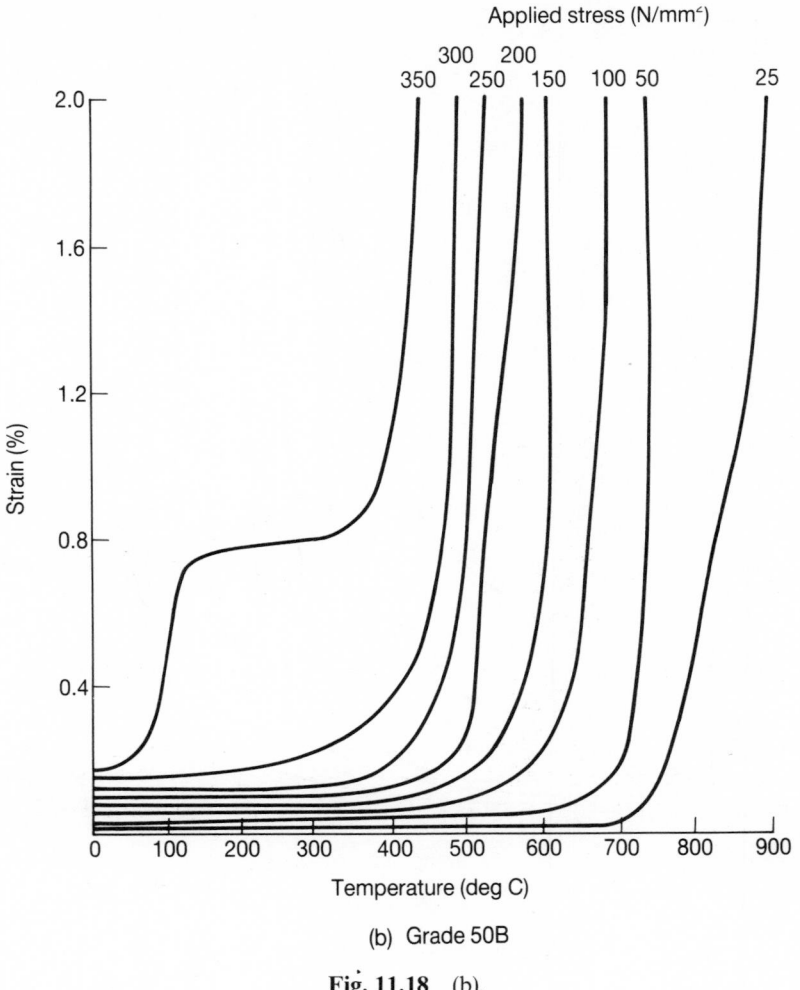

(b) Grade 50B

Fig. 11.18   (b).

FASBUS II has been shown to give excellent correlation on both single composite beams (Bresler and Iding) and on multi-bay composite slab structures (Jeanes).

For single storey frames the use of computer simulation is rarely necessary, and where calculations are needed they may be done much more simply.

11.9.4   Single storey frames

Under normal circumstances the rafters of a single storey portal frame are completely unprotected and thus lose their strength very quickly and are thus unable to support any dead load still acting from the roofing material. The problem is that when the rafters collapse into the structure the stanchions will also deflect. Ideally they should collapse inwards taking any cladding whether lightweight or masonry into the structure. For portal frames away from a boundary the direction of collapse is less important except that collapse outwards could endanger the fire services attempting to fight the fire. Collapse

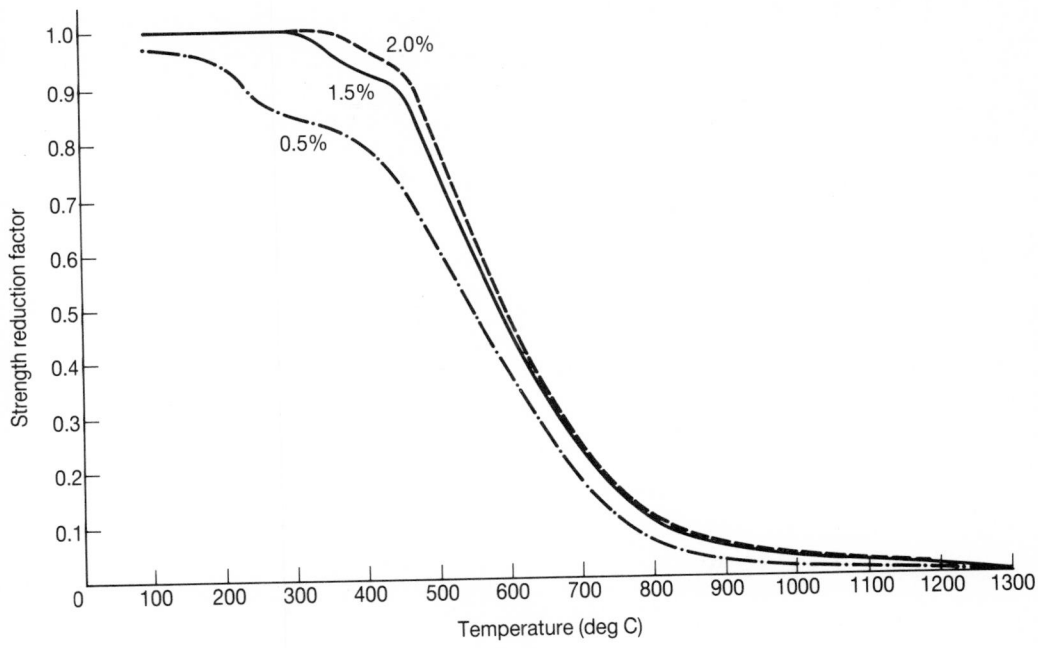

Fig. 11.19   Strength reduction factors for different levels (in %).

inwards can be initiated by only allowing a hinge to occur in the stanchion base plate on the face inside the structure and relying on some degree of fixity between the base plate and the stanchion to carry a moment at the base of the stanchion (Fig. 11.20).

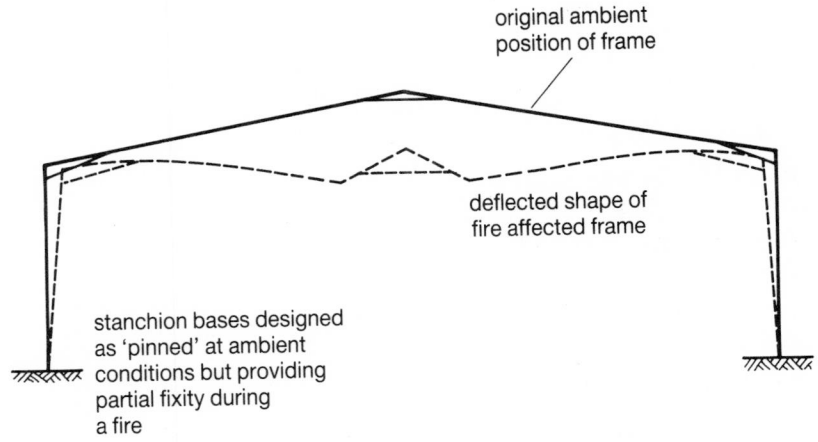

Fig. 11.20   Portal frame collapse during a fire.

For portal frames adjacent to a boundary such as a highway it is essential that the frame is stable during a fire and that if it should collapse it collapses inwards. A guide to the calculations involved is presented in Newman (1990). The calculations are summarized below.

11.9.4.1 *Portal frames in boundary conditions (Appendix F, Pt 8)*

When collapse occurs in a fire hinges will tend to occur in the rafter at either side of the apex and at the end of the haunch. With these hinges it is then possible to determine the magnitude of the moment needed at the base of the stanchion, the horizontal and the vertical reaction. The half frame at collapse is shown in Fig. 11.21.

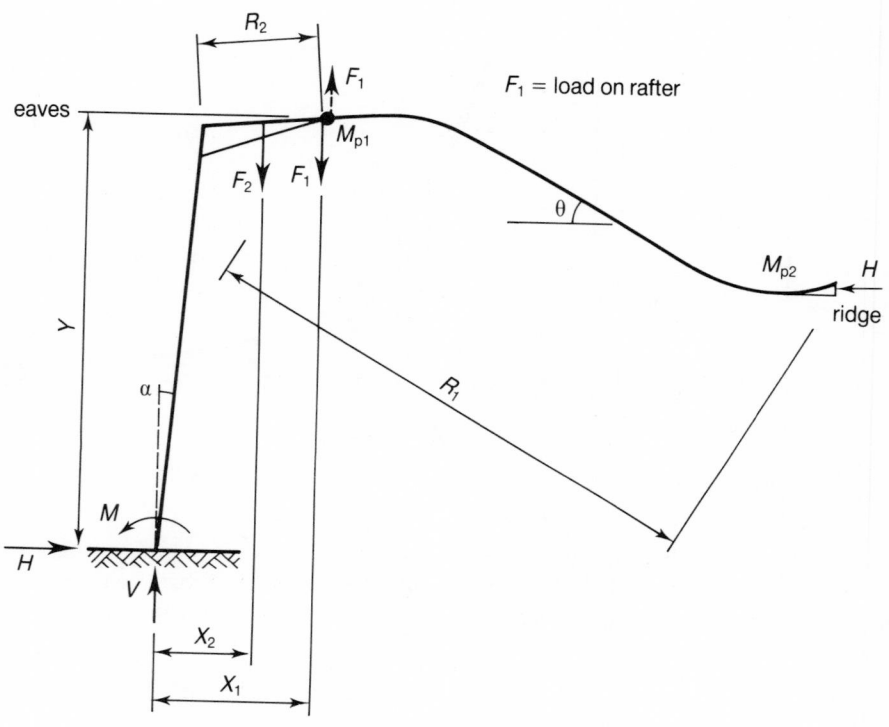

Fig. 11.21 Forces at collapse.

The geometry of the frame needs to be calculated in terms of the angle of rotation of the stanchion and the angle of slope of the collapsed rafter.

The vertical reaction V is given by

$$V = F_1 + F_2 \tag{11.36}$$

the horizontal reaction H by

$$H = (F_1 R_1 \cos \theta - 2(M_{p1} + M_{p2}))/2R_1 \sin \theta \tag{11.37}$$

the moment M at the base of the stanchion,

$$M = HY + F_1 X_1 + F_2 X_2 + M_{p1} \tag{11.38}$$

To use these equations, certain assumptions need to be made:

(a) In the fire situation the rafter will expand in length. It is suggested that a strain of 2% will be reasonable to take account of this.

(b)   The plastic moment capacity of the rafter will be very much reduced from ambient conditions. The suggested moment capacity in a fire is 0.065 of that at ambient conditions.

(c)   The stanchion will lean inwards and it is necessary to estimate this. A rotation of $1°$ is generally assumed.

With these assumptions Equations (11.36) to (11.38) may be simplified to,

$$V = 0.5W_f SL \qquad (11.39)$$

$$H = K(W_f SGA - CM_{pr}/G) \geqslant M_{pc}/(10Y) \qquad (11.40)$$

$$M = K(W_f SGY(A + B/Y) - M_{pr}(CY/G - 0.065)) \geqslant M_{pc}/10 \qquad (11.41)$$

where W_f is the load applied to the structure during the fire, S is the frame spacing, L the frame span, M_{pr} the plastic moment capacity of the rafter, M_{pc} the plastic moment capacity of the stanchion, G the distance between the ends of the haunches, $B = (L^2 - G^2)/8G$, A and C are factors dependent on the pitch of the rafter and are given in Fig. 11.22, K a multiplication factor (greater than unity for multibay frames, values of which are given in Table 2 of Newman (1990) and Table 21 (Pt 8), and unity for single bay frames).

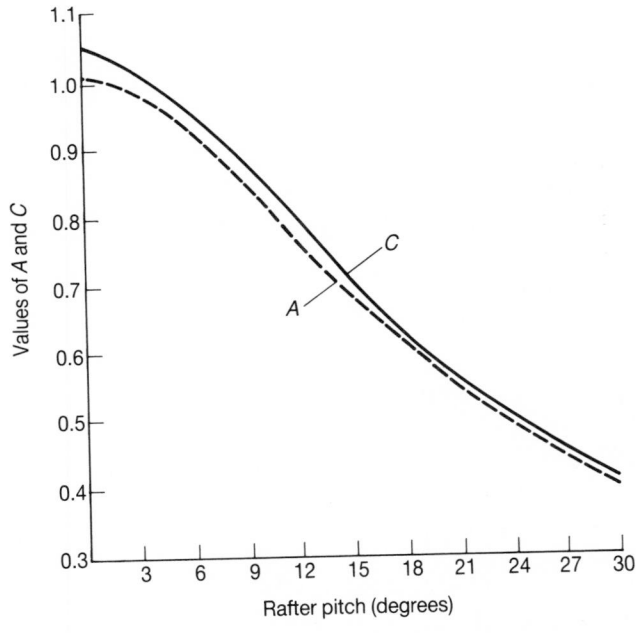

Fig. 11.22   Coefficients A and C as a function of rafter pitch.

The applied loading, W_f, is taken as the dead load on the roof minus any combustible materials. Guidance for the values of percentage loading remaining for various roofing systems is given in Table 3 of Newman (1990) and Table 22 (Pt 8).

Additional considerations:

(a)   Wind loading need only be considered on frames whose height to eaves exceeds 8 m, and a load factor of 0.58 is to be applied.

(b)   Longitudinal stability

This should be satisfied by the following:

Four holding down bolts should be provided, spaced symmetrically at a spacing of not less than 70% of the flange width, or if cast into a concrete base, the base must be capable of resisting a moment in the longitudinal direction equal to that resisted by the four bolts.

A protected wall with a height not less than 75% of the height to the eaves, or where the percentage is less, the provision in the unprotected area of horizontal steel members capable of resisting $0.025 \times$ (height of unprotected area to height of eaves) V, where the summation is over all the frames.

The holding down bolts may be designed to act at the lesser of their yield strength or their ultimate strength divided by 1.2. The base plate and columns are to be designed to load factors of 1.2. The foundations should be designed to take the ultimate design pressure, and no limitation should be placed on the ratio of the maximum to minimum bearing pressures.

EXAMPLE 11.6   *Portal frame check in boundary conditions*

Check the behaviour of the portal frame designed in Example 8.1 when exposed to fire in a boundary situation.

Resumé of design data:

Span $L = 25$ m

Spacing $S = 9.0$ m

Pitch $= 10°$

Height to end of haunch $Y = 5.48$ m

Distances between ends of haunch $G = 19.54$ m

$$M_{pr} = 317.7 \text{ kNm}$$

$$M_{pc} = 521.9 \text{ kNm}$$

Fire load W_f assuming all the cladding remains $= 0.31$ kN/m²

$K = 1.0$ (Single span)

$A = 0.82, \quad C = 0.93$ for a pitch of $10°$ (Fig. 11.22)

$B = (L^2 - G^2)/8G$
$\quad = (25^2 - 19.54^2)/(8 \times 19.54)$
$\quad = 1.56$

$M = K(W_f SGY(A + B/Y) - M_{pr}(CY/G - 0.065))$ (11.41)
$\quad = 1.0(0.31 \times 9 \times 19.54 \times 5.48(0.82 + 1.56/5.48) - 317.7$
$\quad \quad \times (0.93 \times 5.48/19.54 - 0.065))$
$\quad = 267.8$ kNm

Lower limit:

$M_{pc}/10 = 521.9/10 = 52.2$ kNm

Applying a load factor of 1.2 to the calculated value of M;

$$1.2M = 321.4 \text{ kNm}$$

This is less than the moment capacity and is therefore satisfactory.

$$V = 0.5W_f SL \quad\quad\quad (11.39)$$
$$V = 0.5 \times 0.31 \times 9 \times 25$$
$$\quad = 34.9 \text{ kN}$$

$$H = K(W_f SGA - CM_{pr}/G) \quad\quad\quad (11.40)$$
$$\quad = 1.0(0.31 \times 9 \times 19.54 \times 0.82 - 0.93 \times 317.7/19.54)$$
$$\quad = 29.6 \text{ kN}$$

Lower limit:

$$M_{pc}/(10Y) = 521.9/(10 \times 5.48) = 9.52 \text{ kN}$$

Note the foundation must be designed for the vertical load calculated above together with the weight of any masonry cladding and the moment due to the sum of the moment at the base of the stanchion and the effect of the horizontal reaction applied at the top of the foundation.

11.10 Shelf angle floors

11.10.1 Introduction

Shelf angle floors were developed to reduce the construction depth of floors using precast concrete units and steel beams. Traditionally in this type of flooring system the precast units are supported on the top flange of the beam. However in shelf angle flooring systems the precast units are supported on angles welded or bolted to the web of the beam (Fig. 11.23).

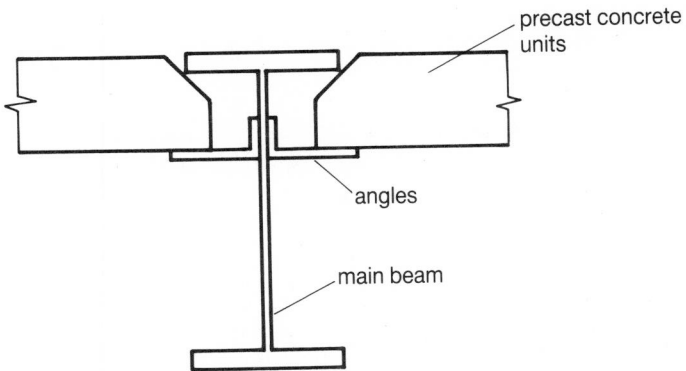

Fig. 11.23 Shelf angle flooring.

This system has the advantage that when the underside of the floor is exposed to a fire a large proportion of the web is shielded and thus temperatures in the upper part of the beam are much reduced (Section 11.4.2) and therefore it may be possible to achieve sufficient fire performance without the need for any fire protection. It is possible to assess the performance of such flooring systems by direct calculation.

11.10.2 Calculation of fire performance

This is covered by Appendix E of Part 8. It should be noted that calculations may only be carried out if the following conditions are met:

(a) The precast units should be cast using normal-weight concrete and the end 75 mm of the unit should be solid.
(b) The void between the precast slab and the beam should be filled with grout.
(c) The units should have at least 75 mm bearing on the angles.
(d) The angles should be Grade 50 and at least $125 \times 75 \times 12$ with the short leg positioned upwards (Fig. 11.23) and long leg supporting the units.
(e) The connections at the end of the beam should be within the depth of the floor slab or they should be protected to the same standard as the supporting member.
(f) The supporting angles should be designed such that the transverse bending capacity M_c at the fire limit state exceeds that due to the loads applied from the precast floor slab, where M_c is given by

$$M_c = 1.2 k_R p_y Z$$

where k_R is the temperature dependent strength reduction factor at a strain level of 1.5% (see Fig. 11.19).

(g) The connection between the angle and the web of the beam whether welded or bolted should be designed to resist the applied vertical shear and the longitudinal shear co-existing. Any weld on the underside of the angle should be ignored.

The reason for these restrictions is the limited nature of the available test data (Wainman and Kirby).

The method used is to determine the temperature profile at the required fire resistance period in the section and then to calculate the moment capacity based on temperature reduced strengths corresponding to a 1.5% strain level and compare it with that induced by the applied loading. Note that the angles are included in this calculation.

11.10.2.1 *Calculation of temperature response (cl E3, Pt 8)*

The section is divided into a series of zones of constant width. For this all fillets should be ignored and all sections treated as being rectangles.

The zones to be used are defined in Fig. 11.24(a).

The temperatures reached in each section are given in Table 18 (Pt 8). Within the depth of the slab it is first necessary to define the point at which the temperature reaches 300°C, as at temperatures below 300°C the steel suffers no strength loss.

The distance x_{300} of the 300°C isotherm above the top of the angle is given by

$$x_{300} = (\theta_R - 300)/G \tag{11.42}$$

where G takes values of 2.3 for a fire resistance of 0.5 hr, 3.8 for 1.0 hr and 4.3 for 1.5 hr.

If the 300°C isotherm falls within the short leg of the angle, then the angle is split into two zones, the lower with a temperature greater than 300°C and the upper with a temperature of 300°C (Fig. 11.24(b)), if the 300°C isotherm falls within the web above the

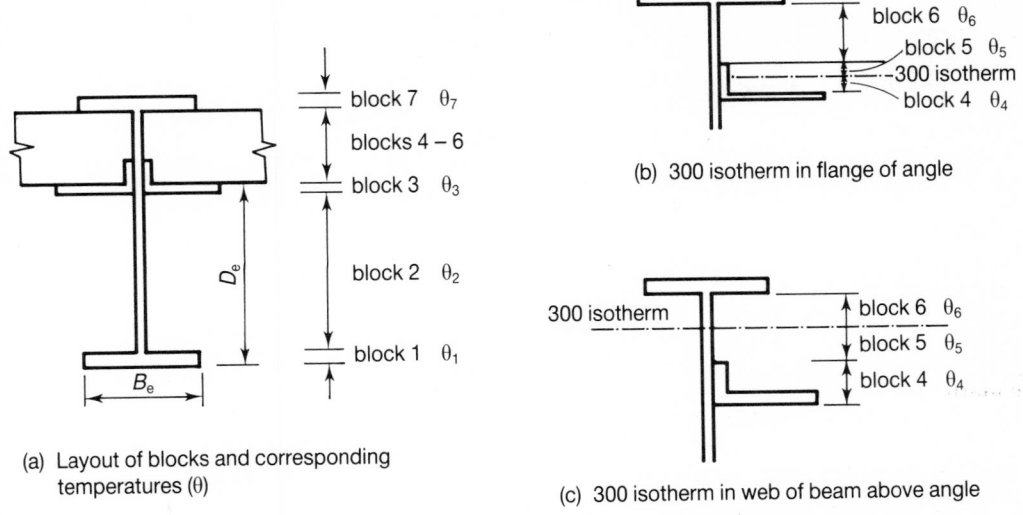

(a) Layout of blocks and corresponding
      temperatures (θ)

(b) 300 isotherm in flange of angle

(c) 300 isotherm in web of beam above angle

Fig. 11.24   Calculation of temperature zones in a shelf angle floor.

angle, the web is then divided into two zones, one at a temperature greater than 300°C and the other at a temperature equal to 300°C (Fig. 11.24(c)).

In any zone within the depth of the slab having a temperature greater than 300°C, the temperature θ_x should be calculated at the mid-height of the zone and this value is given by

$$\theta_x = \theta_R - Gx \tag{11.43}$$

where x is measured above the top surface of the long leg of the shelf angle.

11.10.2.2   *Calculation of moment capacity*

Using the temperature profile calculated above, the strength reduction factors corresponding to a strain level of 1.5% can be determined from Fig. 11.19, and hence the actual strengths can be calculated. From these strengths and the section dimensions the position of the plastic neutral axis and the resultant moment capacity can be determined and compared with the moment induced by the applied loading.

The method is best illustrated by an example.

EXAMPLE 11.7   Fire resistance of shelf angle floors

Determine the capacity of a shelf angle floor fabricated from a 406 × 178 × 54 UB (Grade 43A) with 125 × 75 × 12 Grade 50 angles welded with the upper face of the long leg of the angle 282 mm above the soffit of the UB.

Determine θ_D:
From Table 7 (Pt 8), $\theta_D = 767$°C for a fire resistance period of 0.5 hr and $T = 10.9$ mm.

Calculation of aspect ratio (D_e/B_e):

Using nominal dimensions, $D_e = 282$ mm, $B_e = 178$ mm

$$D_e/B_e = 282/178 = 1.58$$

with this value of D_e/B_e select the last line of Table 18 (Pt 8)

$$\theta_1 = \theta_D = 767$$

$$\theta_2 = \theta_D = 746$$

$$\theta_3 = 550$$

$$\theta_R = 425.$$

Calculation of position of 300°C isotherm (x_{300})

$$x_{300} = (\theta_R - 300)/G \tag{11.42}$$

$$G = 2.3, \quad \text{so}$$

$$x_{300} = (425 - 300)/2.3$$
$$= 54.3 \text{ mm}$$

This falls within the vertical leg of the angle, and thus the upper portion of the angle and the web within the depth of the deck and the top flange are all at a temperature of less than 300 and therefore their strengths are unaffected.

Calculation of the temperature at mid depth of the vertical leg of the angle (θ_4):

$$\theta_x = \theta_R - Gx \tag{11.43}$$

$$x = 54.3/2 = 27.1 \text{ mm}$$

$$\theta_4 = 425 - 2.3 \times 27.1$$
$$= 363$$

The resultant temperatures together with the relevant section dimensions are given in Fig. 11.25. Note that symmetry has been used and only the portion of the beam to the right of the centre line has been included.

The strength reduction factors are determined from Fig. 11.19 and are also indicated in Fig. 11.25.

Determination of the position of the plastic neutral axis:

For each portion of the beam and shelf angles the force in each is calculated as $bdk_R p_y$, where b and d are the width and depth of the portions with the portions being split where there are different design strengths, k_R the strength reduction factor and p_y the design strength. These calculations are found in the second and third columns of Table 11.7.

The total force in the section is given by the summation of the third column (ΣP_i) and the tensile or compressive force in the total section as half this value.

$$\Sigma P_i = 1042.78 \text{ kN}$$

$$0.5 \Sigma P_i = 521.39 \text{ kN}$$

The plastic neutral axis occurs in the vertical leg of the angle at a distance x above the

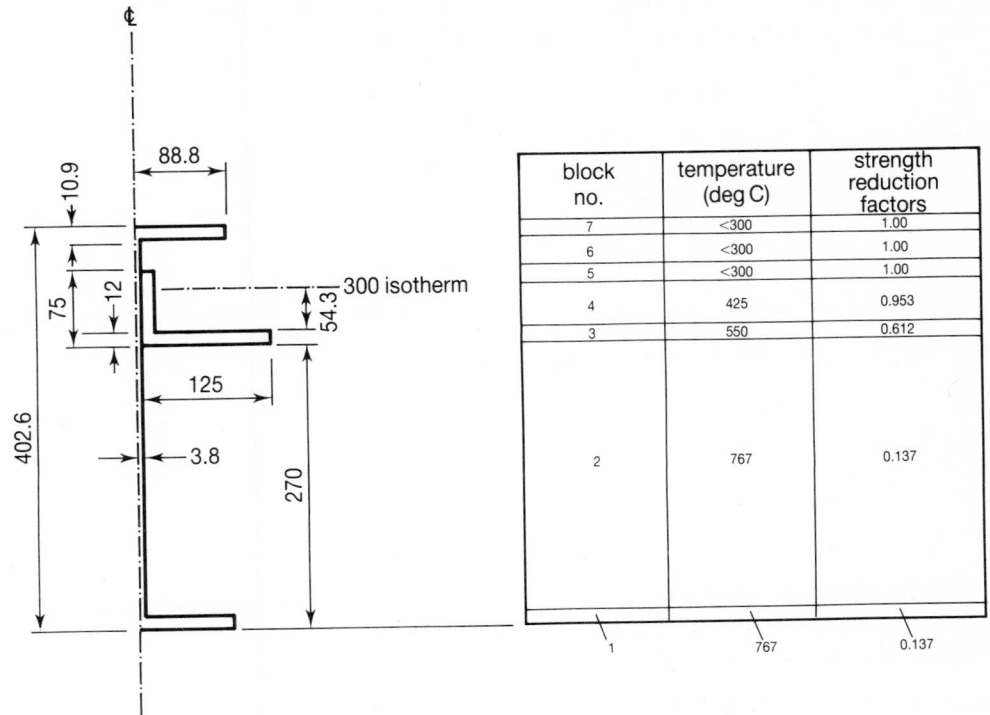

Fig. 11.25   Data for shelf angle floor calculations.

top of the flange given by

$$x = (521.39 - 36.47 - 37.09 - 325.89 - 7.67) \times 1E3/(12 \times 355 + 3.8 \times 275)$$
$$= 21.54 \text{ mm}$$

This means that part of section No. 4 is in tension and part in compression, and the calculations of P_i for each section are given in column 4 with an indication of T or C as appropriate.

The lever arms for the individual sections may be calculated and are to be found in column 5, and the values of $l_a P_i$ found in column 6 (with moments due to tensile forces indicated as negative)

$$\sum P_i l_a = 64.60 \text{ kNm}$$

Moment capacity M_c:

$$M_c = 2 \sum P_i l_a$$
$$= 2 \times 64.60$$
$$= 129.2 \text{ kNm}$$

This particular beam was the subject of a fire test to BS 476 and is reported in Data Sheet No. 35 in Wainman and Kirby, and lasted 29 mins with a UDL of 271.4 kN on a span of 4.5 m, or an applied moment of 152.7 kNm.

Part of the difference between the calculated and test result is that in the actual test the lower flange reached a temperature of only 733°C and the lower part of the web 715°C compared with values of 767°C assumed in the calculations. It should also be noted that, in general, calculation methods tend to be calibrated to give a lower bound to test results.

Table 11.7　Calculations for the fire resistance of a shelf angle floor

No.	Calculation	P_i kN	P_i kN	l_a mm	$P_i l_a$ kNm
1	88.8 × 10.9 × 0.137 × 275	36.47		5.45	−0.20
2	3.8 × 259.1 × 0.137 × 275	37.09		104.45	−3.87
3a	125 × 12 × 0.612 × 355	325.89		276.0	−89.95
3b	3.8 × 12 × 0.612 × 275	7.67		276.0	−2.12
4a	12 × 54.3 × 0.953 × 355	220.45	87.45T	292.77	−25.60
			133.00C	319.924	42.55
4b	3.8 × 54.3 × 0.953 × 275	54.08	21.45T	292.77	−6.28
			32.63C	319.92	10.44
5a	12 × 8.7 × 1.0 × 355	37.06		340.7	12.63
5b	3.8 × 8.7 × 1.0 × 275	9.09		340.7	3.10
6	3.8 × 46.7 × 1.0 × 275	48.8		368.4	17.98
7	88.8 × 10.9 × 1.0 × 275	266.18		397.2	105.73
	Summation	Σ 1042.78			Σ 64.60

11.11　Exposure to natural fires

This is an option available in Part 8 for determining the fire performance of steel structures. However it is not of universal applicability which is why the text has concentrated on exposure to the 'pseudo-fire' delineated by the BS 476 furnace curve.

For design using exposure to a natural fire it is absolutely essential that the fire load and the ventilation can be calculated with reasonable certainty. Given the layout of any structure the ventilation is unlikely to change unless the structure is modified during its history, but the position with respect to the fire load is less certain as a change of use during the lifetime of the structure could substantially change the fire load. Hence the possible maximum temperatures which the structure had been designed to resist during a fire could increase substantially. Thus design using exposure to natural fires should be restricted to structures which are specialist, thereby ensuring that the structure usage should not change during its lifetime. Typical examples where this procedure has been used include sports stadia, airport terminals, bus stations and shopping malls surrounding atria (Kirby (1986), Kirby (undated)).

Where calculations are carried out using exposure to natural fires it is often possible to show that temperatures reached in a fire will be sufficiently low for the steelwork to be left unprotected (Latham, Kirby and Thomson).

11.12　Tutorial problems

11.12.1　Fire protection for a column

A two storey column is required to have a fire resistance of 90 min provided by vermiculite/gypsum slabs. The loading and other design data are presented in Fig. 11.26. Determine the thickness of slabs required if the column is heated on four sides.

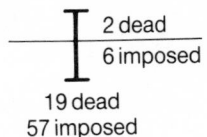

loading (kN)
applied at face of
web or flange

2 dead
6 imposed

19 dead
57 imposed

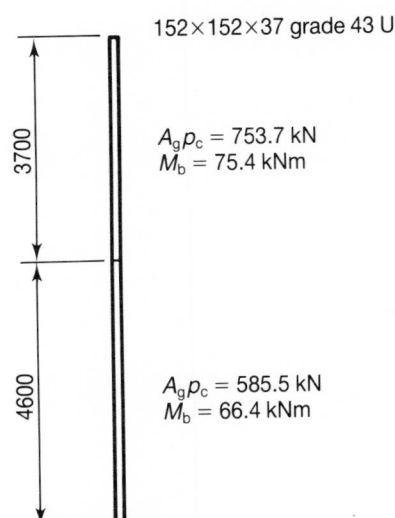

152×152×37 grade 43 UC

3700

$A_g p_c$ = 753.7 kN
M_b = 75.4 kNm

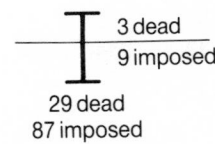

3 dead
9 imposed

29 dead
87 imposed

4600

$A_g p_c$ = 585.5 kN
M_b = 66.4 kNm

Fig. 11.26.

Answer

Factored fire loading (kN):

Top storey: 64.6 x axis, 6.8 y axis

Lower storey: 98.6 x axis, 10.2 y axis

Calculation of load ratio, R:

Top storey:

$$A_g p_c = 753.7 \text{ kN}, \quad M_b = 75.4 \text{ kNm}, \quad Z_y p_y = 25.2 \text{ kNm}$$

Applied moments:

x axis: $64.6(100 + 161.8/2) = 11.7$ kNm;   y axis: $6.8(100 + 8.1/2) = 0.71$ kNm:

$$N = 71.4 \text{ kN}$$

$$R = 0.28$$

Lower storey:

$$A_g p_c = 585.5 \text{ kN}, \quad M_b = 66.4 \text{ kNm}$$

Applied moments:

x   axis: $0.5 \times 98.6(100 + 161.8/2) = 8.92$ kNm;   y   axis: $0.5 \times 10.2(100 + 8.1/2)$
$= 0.53$ kNm: $N = 180.2$ kN

$$R = 0.46$$

$\rho = 800 \text{ kg/m}^3$, $\rho' = 800(1 + 0.03 \times 15) = 1160$, $k = 0.15$ W/m/deg C

$T_{\text{lim}} = 563$ (with $R = 0.46$): $I_f = 1.106\text{E} - 3$, $H_p/A = 135$ /m

$\mu = 0.447$: $F_w = 0.749$: $t = 0.017$ m

11.12.2   Composite decking

The composite deck designed for Tutorial Problem 10.4.1 is required to have a fire resistance of 90 mins, carry out the checks required.

Answer

The applied loading has been calculated using the partial safety factors from BS 8110.

$$q_f = 1.05(2.1 + 2.5) + 1.0 \times 4 = 7.78 \text{ kN/m}^2$$

Free bending moment $= 7.78 \times 3 \times 3/8 = 8.57 \text{ kNm/m}$

Slab depth:

the slab depth is sufficient to give a fire resistance of 90 mins.

Moment capacity:

Support (Fig. 11.27).

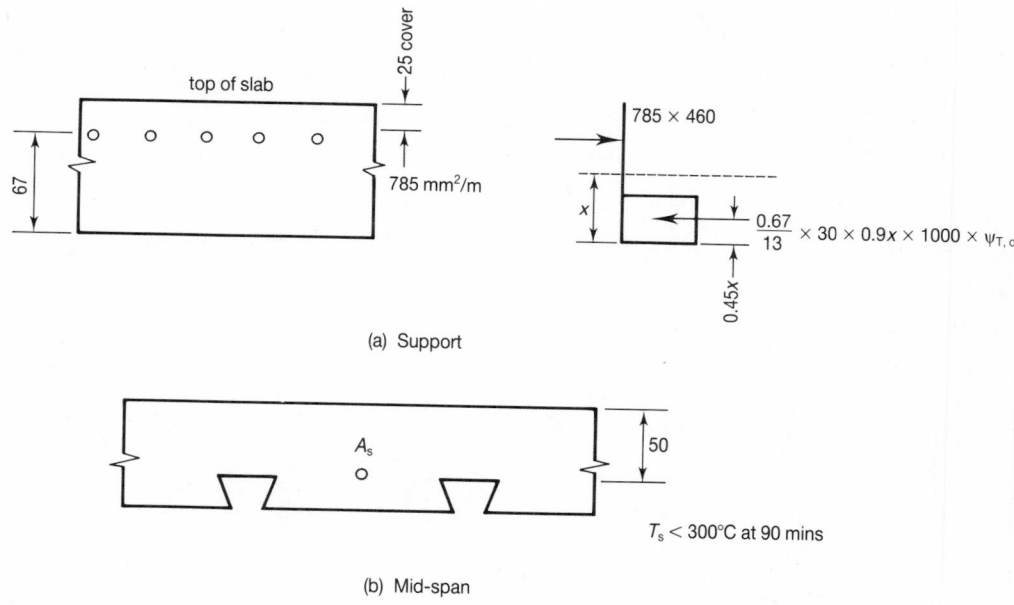

(a) Support

(b) Mid-span

Fig. 11.27.

The calculation procedure is iterative, and gives

$$x = 36 \text{ mm} \ (T_c = 630, \quad \psi_{T,c} = 0.74, \quad M_u = 18.3 \text{ kNm/m (with B 785 mesh)}$$

Mid-span:

the maximum sagging moment of 1.99 kNm occurs at 0.72 m from the outside support.

With one bar per rib centrally placed level with the top of the trapezoidal void, the steel temperature is below 300°C. Since depth of concrete is small, i.e. NA depth will be small as the bar size is also small, $z = 0.95d$ (with $d = 50$ mm (Fig. 11.27))

$$A_s = 1.99\text{E}6/(47.5 \times 460) = 91.1 \text{ mm}^2$$

bar spacing is 152.5 mm, bar area = 13.9, bar diam = T6 (28.3) with T6 bars spacing could be increased to 305 mm, i.e. a bar in alternate ribs, giving 92.8 mm^2/m.

11.12.3 Composite beam

The composite beam designed in Tutorial Problem 10.4.4 is to be given a 90 min fire resistance using sprayed mineral fibre to the beam profile. Determine the thickness of the protection required.

Answer

Beam loading:

Imposed 90 kN; dead 60 kN

Applied moment 198 kNm

$M_c = 431$ kNm (using the moment capacity with partial interaction, as this will give a conservative value of R)

$$R = 0.46$$

$$T_{lim} = 656, \quad I_f = 8.54\text{E} - 4$$

$$H_p/A = 160 \text{ /m (profile section heated on three sides)}$$

$$\rho = 300 \text{ kg/m}^3; \quad \rho' = 300(1 + 0.03 \times 1) = 309; \quad k = 0.1 \text{ W/m/deg C}$$

$$\mu = 0.086; \quad F_w = 0.926; \quad t = 0.0127 \text{ m}, \quad \text{i.e. specify a minimum of 13 mm.}$$

References

Anderberg Y. (Editor) (1983). RILEM Report, *Properties of materials at high temperatures – steel*, Lund Institute of Technology.

ASFPCM/SCI/FSTG (1988). *Fire protection for structural steel in buildings*, Association of Structural Fire Protection Contractors and Manufacturers.

Becker J., Bizri H. and Bresler B. (1974). *FIRES-T2 – a computer program for the fire response of structures – thermal*, Report UCBFRG 74–1, University of California, Berkeley.

Bond G. V. L. (1975). *Fire and steel construction – water cooled hollow columns*, Steel Construction Institute.

Bresler B. and Iding R. H. (1985). *Effect of fire exposure and fire proofing requirements of structural steel frame assemblies*, ASTM STP 882 (*Fire safety science and engineering*, Harmathy T. (Ed)), ASTM, 206–22.

BS 449. *The use of structural steel in buildings* (Part 2), British Standards Institution, London.

BS 476 Part 8. *Fire tests on building materials and structures – Part 8 – test methods and criteria for the fire resistance of elements of building construction*, British Standards Institution, London.

BS 476 Part 20. *Fire tests on building materials and structures – Part 20 – methods for the determination of the fire resistance of elements of construction (General principles)*, British Standards Institution, London.

BS 476 Part 21. *Fire tests on building materials and structures – Part 21 – methods for the determination of the fire resistance of load bearing elements of construction*, British Standards Institution, London.

BS 5950 Part 8. *Structural use of steelwork in buildings – Part 8 – Code of practice for fire resistance design*, British Standards Institution, London.

BS 8110 Part 2. *Structural use of concrete – Part 2 – Code of practice for special circumstances*, British Standards Institution, London.

CIB W14 Workshop (1983). A conceptual approach towards a probability based design guide on structural fire safety (CIB W14 Workshop on structural fire safety), *Fire Safety Journal*, **6**, 1–79.

CIRIA (1986). *Fire resistance of composite slabs with steel decking: data sheet, Special Publication 42*, Construction Industry Research and Information Association.

Cooke G. M. E. (1988). An introduction to the mechanical properties of structural steel at elevated temperatures, *Fire Safety Journal*, **13**, 45–54.

Cooke G. M. E. and Latham D. J. (1987). The inherent fire resistance of a loaded steel framework, *Steel Construction Today*, **1**, 49–58.

Cooke, G. M. E., Lawson R. M. and Newman G. M. (1988). Fire resistance of composite deck slabs, *The Structural Engineer*, **66** (16), 253–61 and 267.

ECCS (1983). *European recommendations for the fire safety of steel structures – calculation of fire resistance of load bearing elements and structural assemblies exposed to a standard fire*, Elsevier Applied Science.

ECCS (1985). *Design manual on the European recommendations for the fire safety of steel structures*, ECCS Technical Committee 3, Brussels.

Iding R. J., Bresler B. and Nizamuddin Z. (1977). *FIRES–T3, a computer program for the fires response of structures – thermal*, Report No. UCB FRG 77–15, University of California, Berkeley.

Institution of Structural Engineers and Concrete Society (1978). *Design and detailing of concrete structures for fire resistance*, Institution of Structural Engineers.

ISO 834 (1985). *Fire resistance tests – elements of building construction*, ISO.

Jeanes D. C. (1985a). *Computer modelling the fire endurance of floor systems in steel framed buildings*, ASTM STP 882 (*Fire safety science and engineering*, Harmathy T. (Ed)), ASTM, 223–38.

Jeanes D. C. (1985b). Application of the computer in modelling fire endurance of structural steel floor systems, *Fire Safety Journal*, **9**, 119–35.

Jeanes D. C. (1986). *Developing design concepts for structural fire endurance using computer models, Design of structures against fire*, Anchor R. D. *et al.* (Editors), Elsevier Applied Science, 127–53.

Kirby B. R. (1986). Recent developments in structural fire engineering design, *Fire Safety Journal*, **11**, 141–79.

Kirby B. R. (undated). *Fire engineering in sports stands*, British Steel.

Kirby B. R. and Preston R. R. (1988). High temperature properties of hot rolled structural steels for use in fire engineering design studies, *Fire Safety Journal*, **13**, 27–37.

Latham D. J., Kirby B. R. and Thomson G. (1987). The temperatures attained by unprotected structural steelwork in experimental natural fires, *Fire Safety Journal*, **12**, 139–52.

Law M. (1983). A basis for the design of fire protection for building structures, *The Structural Engineer*, **61A** (1), 25–33.

Law M. and O'Brien T. (1981). *Fire and steel construction – Fire safety of bare external steelwork*, Steel Construction Institute.

Lawson R. M. (1985). *Fire resistance of ribbed concrete floors*, Report 107, Construction Industry Research and Information Association.

Lawson R. M. (1990). Behaviour of steel beam to column connections in a fire, *Structural Engineer*, **68**, 263–71.

Malhotra H. L. (1982). *Design of fire resisting structures*, Surrey University Press.

Martin L. H., Croxton P. C. L. and Purkiss J. A. (1989). *Concrete design to BS 8110*, Edward Arnold.

Melinek S. J. (1989). Prediction of the fire resistance of insulated steel, *Fire Safety Journal*, **14**, 127–34.

Melinek S. J. and Thomas P. H. (1987). Heat flow to insulated steel, *Fire Safety Journal*, **12**, 1–8.

Morris W. A., Read R. E. H. and Cooke G. M. E. (1988). *Guidelines for the construction of fire-resisting structural elements*, Building Research Establishment.

Newman G. M. (1989). *The fire resistance of composite floors with steel decking*, Steel Construction Institute.

Newman G. M. (1990). *Fire and steel construction – the behaviour of steel portal frames in boundary conditions* (2nd Edition), Steel Construction Institute.

Pettersson O., Magnusson S. E. and Thor J. (1976). *Fire engineering design of steel structures*, Publication No. 50, Swedish Institute of Steel Construction.

Robinson J. T. and Latham D. J. (1986). *Fire resistant steel beam design – the future challenge, Design of structures against fire*, Anchor R. D. *et al.* (Ed), Elsevier Applied Science, 225–36.

Robinson J. T. and Walker H. B. (1987). *Fire safe structural design*, Construction and Building Materials, **1** (1), 40–50.

Rubert A. and Schaumann P. (1986). Structural steel and plane frame assemblies under fire action, *Fire Safety Journal*, **10**, 173–84.

Schneider U. (Editor) (1981). RILEM Report: *Properties of materials at high temperatures– concrete*, Department of Civil Engineering, Gesamthochschule, Kassel.

Thomas P. H. (Co-ordinator) (1986). Workshop CIB W14, *Design guide – structural fire safety*, Fire Safety Journal, **10**, 75–154.

Twilt L. and Witteveen J. (1986). *Calculation methods for fire engineering design of steel and composite structures, Design of structures against fire*, Anchor R. D. *et al.* (Editors), Elsevier Applied Science, 155–76.

Wainman D. E. and Kirby B. R. (1988). *Compendium of UK standard fire test data – unprotected structural steel – 1*, Report No. RS/RSC/S10328/1/87/B, British Steel Company.

Wickstrom U. (1979). *TASEF-2, A computer program for temperature analysis of structures exposed to fire*, Lund Institute of Technology.

Witteveen J., Twilt L. and Bijlaard F. S. K. (1977). *The stability of braced and unbraced frames at elevated temperatures*, Second International Colloquium – The stability of steel structures, Liege, 47–55.

Appendix

Appendix A1 Design strengths for fillet welds

Leg length(s) (mm)	Design strength per unit length (P_w) in kN/mm				
	grade 43	grade 50		grade 55	
	E43&E51 $p_w = 215$	E43 $p_w = 215$	E51 $p_w = 255$	E51 $p_w = 255$	E51* $p_w = 275$ N/mm^2
4	0.602	0.602	0.714	0.714	0.770
5	0.753	0.753	0.893	0.893	0.963
6	0.903	0.903	1.071	1.071	1.155
8	1.204	1.204	1.428	1.428	1.540
10	1.505	1.505	1.785	1.785	1.925
12	1.806	1.806	2.142	2.142	2.310
15	2.257	2.257	2.667	2.667	2.887
18	2.709	2.709	3.213	3.213	3.465
20	3.010	3.010	3.750	3.750	3.850
22	3.311	3.311	3.927	3.927	4.235
25	3.763	3.763	4.463	4.463	4.813

Notes
(1) $P_w = 0.7sp_w$.
(2) *Only applies to electrodes having a minimum tensile strength of 550 N/mm^2 and a minimum yield strength of 450 N/mm^2.
(3) E denotes electrode complying with BS639.

Appendix A2(a) Design strengths for ordinary bolts (grade 4.6)

Bolt size d (mm)	Tension strength $p_t = 195$ P_t (kN)	Single shear strength $p_s = 160$ P_s (kN)	Thickness of ply for bolt bearing $p_{bs} = 460$ N/mm^2 $t_{bb} = t_e$ (mm)
(M12)	16.4	13.5	2.4
M16	30.6	25.1	3.4
M20	47.8	39.2	4.3
(M22)	59.1	48.5	4.8
M24	68.8	56.5	5.1
(M27)	89.5	73.4	5.9
M30	109.4	89.8	6.5
(M33)	135.3	111.0	7.3
M36	159.3	130.7	7.9

Notes
(1) $P_s = A_t p_s$.
(2) $P_t = A_t p_t$.
(3) $t_{bs} = P_s/(dp_{bb})$.
(4) $t_e = 2P_s/(ep_{bb}) = 2P_s/(2dp_{bb}) = P_s/(dp_{bs})$.
(5) Sizes in brackets are not preferred.

Appendix A2(b)   Dimensions of ordinary bolts (grade 4.6) to BS 4190

Bolt size	Thread pitch	Width across flats (mm)	corners (mm)	Nom. depth of head (mm)	Nom. thickness of nut (mm)	Reduced area (mm^2)
M12	1.75	19	21.9	8	10	84.3
M16	2	24	27.7	10	13	157
M20	2.5	30	34.6	13	16	245
M22	2.5	32	36.9	14	18	303
M24	3	36	41.6	15	19	353
M27	3	41	47.3	17	22	459
M30	3.5	46	53.1	19	24	561
M33	3.5	50	57.7	22	26	694
M36	4	55	63.5	23	29	817

Appendix A2(c)   Dimensions of black washers to BS 4320

Bolt size	Form E outside dia. (mm)	thickness (mm)	Form F outside dia. (mm)	thickness (mm)
12	24	2.5	28	2.5
16	30	3	34	3
20	37	3	39	3
22	39	3	44	3
24	44	3	50	3
27	50	4	56	4
30	56	4	60	4
33	60	5	66	5
36	66	5	72	6

Appendix A3   Design strengths for ordinary bolts (grade 8.8)

Bolt size d (mm)	Tension strength $p_t = 450$ P_t (kN)	Single shear $p_s = 375$ P_s (kN)	Thickness of ply for plate bearing grade 43 $p_{bs} = 460$ $t_{bs} = t_e$ (mm)	grade 50 $p_{bs} = 550$ $t_{bs} = t_e$ (mm)	grade 55 $p_{bs} = 650$ $t_{bs} = t_e$ (mm)
M12	37.9	31.6	5.7	4.8	4.0
M16	70.7	58.9	8.0	6.7	5.7
M20	110.3	91.9	9.9	8.4	7.1
(M22)	136.4	113.6	11.2	9.4	7.9
M24	158.9	132.4	12.0	10.0	8.5
(M27)	206.6	172.1	13.9	11.6	9.8
M30	252.5	210.4	15.2	12.8	10.8
(M33)	312.3	260.3	17.1	14.3	12.1
M36	367.7	306.4	18.5	15.5	13.1

Notes
(1) $P_s = A_s p_s$.
(2) $P_t = A_s p_t$.
(3) $t_{bs} = P_s/(d p_{bs})$.
(4) $t_e = 3P_s/(e p_{bs}) = 3P_s/(3 d p_{bs}) = P_s/(d p_{bs})$.
(5) Sizes in brackets are not preferred.

Appendix A4(a)   Design strengths for parallel shank friction grip (PSFG) bolts

Bolt size d (mm)	Min. shank tension P_0 (kN)	Slip resistance P_{sL} (kN)	Tension capacity P_t (kN)	Thickness of ply for plate bearing		
				grade 43 $p_{bg} = 825$ $t_{bg} = t_e$ (mm)	grade 50 $p_{bg} = 1065$ $t_{bg} = t_e$ (mm)	grade 55 $p_{bg} = 1210$ $t_{bg} = t_e$ (mm)
M12	49.4	24.5	44.5	2.5	1.9	1.7
M16	92.1	45.6	82.9	3.5	2.7	2.4
M20	144	71.3	129.6	4.3	3.4	2.9
(M22)	177	87.6	159.3	4.8	3.7	3.3
M24	207	102.5	186.3	5.2	4.0	3.5
(M27)	234	115.8	210.6	5.2	4.0	3.5
M30	286	141.6	257.4	5.7	4.4	3.9
M36	418	206.9	376.2	7.0	5.4	4.7

Notes
(1) $P_{sL} = 1.1\mu P_0$ where $\mu = 0.45$.
(2) $P_t = 0.9 P_0$.
(3) $t_{bg} = P_{sL}/(d p_{bg})$.
(4) $t_e = 3 P_{sL}/(e p_{bg}) = 3 P_s/(3 d p_{bs}) = P_s/(d p_{bs})$.
(5) Sizes in brackets are not preferred.

Appendix A4(b)   Dimensions of parallel shank friction grip (PSFG) bolts to BS 4395

Bolt size	Thread pitch	Width across		Max. depth of head (mm)	Max. thick. of nut (mm)	Reduced area (mm^2)
		flats (mm)	corners (mm)			
M12	1.75	22	25.4	10.45	11.55	84.3
M16	2	27	31.2	10.45	15.55	157
M20	2.5	32	36.9	13.9	18.55	245
M22	2.5	36	41.6	14.9	19.65	303
M24	3	41	47.3	15.9	22.65	353
M27	3	46	53.1	17.9	24.65	459
M30	3.5	50	57.7	20.05	26.65	561
M36	4	60	69.3	24.05	31.8	817

Appendix A4(c)   Dimensions of washers for parallel shank and waisted shank friction grip bolts to BS 4395 Pt 1

Bolt size	Round flat	
	outside dia. (mm)	thickness (mm)
12	30	2.8
16	37	3.4
20	44	3.7
22	50	4.2
24	56	4.2
27	60	4.2
30	66	4.2
33	75	4.6
36	85	4.6

Appendix A5(a)   Design strengths for waisted shank friction grip (WSFG) bolts to BS 4395

Bolt size d (mm)	Min. shank tension P_0 (kN)	Slip resistance P_{sL} (kN)	Tension capacity P_t (kN)
M16	95.4	47.2	85.7
M20	150.5	74.5	135.5
M22	188.6	93.4	169.7
M24	216.5	107.2	194.9
M27	286.3	141.7	257.7
M30	347.6	172.0	312.8
M33	436.1	215.9	392.5

Notes
(1) $P_{sL} = 1.1\mu P_0$ where $\mu = 0.45$.
(2) $P_t = 0.9 P_0$.
(3) No bearing values.

Appendix A5(b)   Dimensions of waisted shank friction grip (WSFG) bolts to BS 4395

Bolt size	Thread pitch	Width across		Max. depth of head (mm)	Max. thickness of nut (mm)	Reduced area (mm^2)	Waisted shank (mm^2)
		flats (mm)	corners (mm)				
M16	2	27	31	10.45	15.55	157	123–113
M20	2.5	32	36.9	13.90	18.55	245	194–181
M22	2.5	36	41.6	14.4	19.65	303	243–230
M24	3	41	47.3	15.90	22.65	353	279–264
M27	3	46	53.1	17.9	24.65	459	369–351
M30	3.5	50	57.7	20.05	26.65	561	448–428
M33	3.5	55	63.5	22.05	29.65	694	562–541

Appendix A6(a)   Summary of edge distance and spacing rules for bolts (cl 6.2, Pt 1)

Situation	Distance equal to or less than
Min. edge and end distances for a rolled, machine cut or planed edge	1.25 × (dia. of hole)
Min. edge and end distances for a sheared or hand flame cut edge and end	1.4 × (dia. of hole)
Maximum edge distance (non-corrosive)	11 × (thickness of thinner element)
Maximum edge distance (corrosive)	40 mm + 4 × (thickness of thinner element)
Maximum spacing (non-corrosive)	14 × (thickness of thinner element)
Maximum edge distance (corrosive)	16 × (thickness of thinner element) or 200 mm
Minimum spacing	2.5 × (nominal bolt diameter)

Appendix A6(b)   Spacing of holes in standard sections

Beams, columns, joists and tees

Nominal flange widths (mm)	Spacings in millimetres				Recommended dia. of bolt (mm)	Actual b_{min} (mm)
	S_1	S_2	S_3	S_4		
419 to 368	140	140	75	290	24	362
330 to 305	140	120	60	240	24	312
330 to 305	140	120	60	240	20	300
292 to 203	140	–	–	–	24	212
190 to 165	90	–	–	–	24	162
152	90	–	–	–	20	150
146 to 114	70	–	–	–	20	130
102	54	–	–	–	12	98
89	50	–	–	–	–	–
76	40	–	–	–	–	–
64	34	–	–	–	–	–
51	30	–	–	–	–	–

Angles

Nominal leg length	Spacing of holes						Maximum diameter of bolt		
	S_1	S_2	S_3	S_4	S_5	S_6	S_1	S_2 and S_3	S_4, S_5 and S_6
(mm)	(mm)	(mm)	(mm)	(mm)	(mm)	(mm)	(mm)	(mm)	(mm)
200	–	75	75	55	55	55	–	30	20
150	–	55	55	–	–	–	–	20	–
125	–	45	50	–	–	–	–	20	–
120	–	45	50	–	–	–	–	16	–
100	55	–	–	–	–	–	24	–	–
90	50	–	–	–	–	–	24	–	–
80	45	–	–	–	–	–	20	–	–
75	45	–	–	–	–	–	20	–	–
70	40	–	–	–	–	–	20	–	–
65	35	–	–	–	–	–	20	–	–
60	35	–	–	–	–	–	16	–	–
50	28	–	–	–	–	–	12	–	–
45	25	–	–	–	–	–	–	–	–
40	23	–	–	–	–	–	–	–	–
30	20	–	–	–	–	–	–	–	–
25	15	–	–	–	–	–	–	–	–

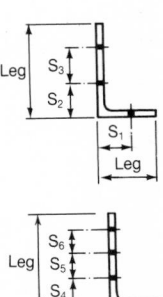

Channels

Nominal flange width (mm)	S_1 (mm)	Recommended dia. of bolt (mm)
102	55	24
89	55	20
76	45	20
64	35	16
51	30	10
38	22	–

Index